Ex libris

Supercritical Fluid Technology

Process Technology Proceedings

Vol. 1 Multi-Phase Flow and Heat Transfer III (Proceedings of the Third Multi-Phase Flow and Heat Transfer Symposium–Workshop, Miami Beach, Florida, U.S.A., April 18–20, 1983), edited by T.N. Veziroğlu and A.E. Bergles (Parts A and B)

Vol. 2 Industrial Crystallization 84 (Proceedings of the 9th Symposium on Industrial Crystallization, The Hague, The Netherlands, September 25–28, 1984), edited by S.J. Jančić and E.J. de Jong

Vol. 3 Supercritical Fluid Technology, edited by J.M.L. Penninger, M. Radosz, M.A. McHugh and V.J. Krukonis

Process Technology Proceedings, 3

Supercritical Fluid Technology

Edited by

J.M.L. Penninger

AKZO Zout Chemie Nederland B.V., P.O. Box 25, 7550 GC Hengelo, The Netherlands

M. Radosz

Exxon Research and Engineering Company, Clinton Township, Annandale, NJ 08801, U.S.A.

M.A. McHugh

Department of Chemical Engineering, The Johns Hopkins University, Baltimore, MD 21218, U.S.A.

V.J. Krukonis

Phasex Corporation, No. 1 Mill, 85 Factory Street, Nashua, NH 03060, U.S.A.

ELSEVIER– Amsterdam – Oxford – New York – Tokyo 1985

ELSEVIER SCIENCE PUBLISHERS B.V.
Sara Burgerhartstraat 25
P.O. Box 211, 1000 AE Amsterdam, The Netherlands

Distributors for the United States and Canada:

ELSEVIER SCIENCE PUBLISHING COMPANY INC.
52, Vanderbilt Avenue
New York, NY 10017, U.S.A.

Library of Congress Cataloging-in-Publication Data
Main entry under title:

Supercritical fluid technology.

(Process technology proceedings ; 3)
Bibliography: p.
1. Separation (Technology) 2. High pressure (Technology) I. Penninger, J. M. L. (Johan M. L.), 1942- . II. Series.
TP156.S45S87 1985 660.2'96 85-20625
ISBN 0-444-42552-7 (U.S.)

ISBN 0-444-42552-7 (Vol. 3)
ISBN 0-444-42382-6 (Series)

Printed in The Netherlands

PREFACE

This book reflects the most recent progress in the area of supercritical fluid technology. The amount of research in this area has increased substantially, as indicated by the large number of publications which have appeared within the last decade. On a commercial scale, the supercritical fluid technology has been applied to energy- and food-related separation processes. These applications demonstrate a lower energy consumption and a better product purity, when compared with conventional separations. Both energy- and food-related separations are covered in this book. Also covered are extraction of organic compounds from water, polymer separations, drug separations, and chemical reactions in the presence of supercritical fluids.

Thermodynamics, particularly high-pressure phase equilibria, is the single most important discipline of chemical engineering which is used to understand and develop new supercritical fluid technologies. Phase equilibrium data are needed to evaluate and design new processes. In most cases, such data have to be measured experimentally. However, the need for experimental effort can be reduced when thermodynamic models are available for correlating and predicting phase equilibrium data.

Thermodynamic fundamentals and modeling of supercritical fluids are the subject of the first part of this book. The second part covers experimental data and techniques. However, most of the papers in both parts deal with both modeling and experiment, and the only difference between the two parts is that of emphasis. It is worth noting that in addition to the phase equilibrium studies mass transfer studies are also covered in part two.

The third part of the book discusses process development issues and energy-related applications and the fourth part discusses special applications of supercritical fluids. While these two parts suggest many new opportunities for supercritical fluids, they also indicate that more data and further improvements are needed in such areas as solids handling at high pressures, metal corrosion, gasket material compatibility, and heat and mass transfer characteristics.

Johan M.L. Penninger
Maciej Radosz
Mark A. McHugh
Val J. Krukonis

ACKNOWLEDGEMENTS

This book is a joint effort of many people. We are particularly indebted to the authors for their contributions. Also, many colleagues reviewed various sections of the book and offered helpful advice; we are grateful to: C. Cozewith, P. Debenedetti, J. De Swaan Arons, D. Dimitrelis, C.A. Eckert, R.D. Gray, J.L. Heidman, K.P. Johnston, R. Kander, J.M.H. Levelt-Sengers, D.P. Maloney, A.K. McClellan, T. Panagiotopoulos, M.E. Paulaitis, R.C. Reid and C. Tsonopoulos. Finally, it is a pleasure to thank Bori Franko, chairman of the High Pressure Committee, and Toby Eubank, chairman of the Thermodynamics and Transport Properties Committee, both of the American Institute of Chemical Engineers, for their support in giving us the opportunity to organize the symposium at the 1984 AIChE Meeting which led to this book.

CONTENTS

PART I

Thermodynamic Modeling of Supercritical Fluids

Supercritical Fluid Technology, edited by J.M.L. Penninger, M. Radosz, M.A. McHugh and V.J. Krukonis
1985 Elsevier Science Publishers B.V., Amsterdam — Printed in The Netherlands

FUNDAMENTAL THERMODYNAMICS OF SUPERCRITICAL EXTRACTION

K. E. Starling, M.A. Khan, and S. Watanasiri
School of Chemical Engineering and Materials Science, University of Oklahoma, Norman, OK 73019

ABSTRACT

Supercritical extraction phenomena in vapor-liquid and liquid-liquid systems are discussed. The effects of operating conditions and solvent to solute ratio on product separation and solvent recovery are studied for the model vapor-liquid system, solvent (methane) plus solute (n-butane plus n-decane). A similar study for the model liquid-liquid system solvent (methane) plus solute (n-hexane) leads to the general conclusions that the best operating conditions for supercritical extraction processes are as near as possible to the mixture critical point (not the pure solvent critical point). Also, it is concluded that within process constraints, high solvent to solute ratios are preferred in supercritical extraction processes.

INTRODUCTION

Supercritical extraction has been used increasingly in recent years for specialized separation processes. These processes include separation of drugs from plants, caffeine from coffee, oils from vegetable seeds, impurities from labile materials, and chemical feedstocks from coal and petroleum residua. The popularity of supercritical extraction processes stems principally from the enhanced solubility characteristics of a solvent near its critical point and the ease with which the solvent can be recovered for recycle.

In a supercritical extraction process a solvent is contacted with a solute at conditions near a critical point of the solvent plus solute mixture. The mixture may exhibit multiphase behavior involving vapor, liquid, and solid phases, depending on the mixture composition and temperature and pressure conditions.

Although multiphase equilibria have long been observed experimentally (e.g., ref.1) the effects of industrial operations near mixture critical points were not clear in early applications. For example, multiphase behavior has been used for deasphalting and refining lubricants (ref.2) and fractionation of crude oils (ref.3). Vink, et al. (ref.4) observed that when propane and other solvents were added to crude oils three phase vapor-liquid-liquid systems sometimes resulted and the precipitation of asphaltic materials was feasible.

In the case of two phase vapor-liquid systems, the fact that retrograde condensation and vaporization phenomena occur probably has been known for over a century (ref.5). The retrograde phenomenon is quite important in petroleum reservoir fluid phase behavior (ref.6) and therefore extensive laboratory studies have been performed (ref.7, ref.8). However, only recently have the advantages of retrograde behavior been exploited extensively in industrial processing (ref.9).

The objective of the study presented here is to explain certain effects in supercritical extraction using examples for two and three component mixtures such that there is confidence that the phenomena referred to actually occur and phase splits and compositions at conditions considered are quantitatively accurate. The computations are performed using equation of state methods and have been compared with experimental data for the mixtures studied to assure accuracy of the predictions. Because interest here is in system behavior and not computation methods, discussion of the equation of state methodology is omitted; the equation of state may be considered merely to be a data interpolation tool in this application. The study is limited to vapor-liquid and liquid-liquid systems. Vapor-liquid-liquid systems were excluded because of the paucity of reliable experimental critical region data. Systems containing solids were excluded because only fluid phase effects are considered important.

The first major section of the paper deals with vapor-liquid systems. In contrast to the work of Brulé and Corbett (ref.9), which considered aromatic solvents plus heavy temperature-sensitive asphaltic materials, here relatively light, well defined systems for which more detailed experimental phase behavior data are available are studied. The methane + n-butane + n-decane ternary system was chosen for study because of the extensive data available (ref.8). This is an interesting system because it simulates petroleum fluid behavior. For high methane concentrations, the system exhibits retrograde behavior at pressures and temperatures similar to high pressure condensate gas reservoir fluids. Furthermore, it contains a light component, an intermediate component and a heavy component, analagous to the solvent and products from industrial critical solvent processes. Discussion in this section on vapor-liquid systems is focused on the choices of operating conditions and solvent to solute ratios for product separation and solvent recovery.

The second major section of the paper deals with the liquid-liquid equilibrium (LLE) behavior that occurs in the vicinity of the solvent critical point. The methane + n-hexane system studied experimentally by Lin, et al. (ref.10) is used in this part of the study. The effects of temperature and pressure on equilibrium compositions are discussed.

VAPOR-LIQUID SYSTEMS

Phase diagram

Pressure-temperature projections of the multidimensional pressure-temperature-composition surface at constant compositions of methane + n-butane + n-decane mixtures were constructed using equation of state methods; an example is shown in Figure 1. Because of the availability of extensive experimental data (ref.8) for this ternary system, the behavior shown in Figure 1 is known to approximate the experimental behavior. The mole ratio of methane to butane plus decane, hereafter referred to as solvent to solute ratio, is 49 to 1 for the mixture in Figure 1. The mole fraction of butane for this mixture is 0.09262. Because of the high concentration of methane, the critical point of the mixture, estimated to be 191 K and 48 atm, is very close to that of pure methane (T_c= 190.5 K, P_c=45.4 atm). Lines of constant fraction vaporized (V/F, where V is moles of vapor and F is moles of feed) are also shown. The feed mixture composition remains constant throughout the diagram, but the compositions of the equilibrium vapor and liquid phases change as functions of V/F, T, and P. Note the very large region in which there is retrograde behavior both isothermally and isobarically, a region which is bounded by the dew line (V/F=1.0) and the locus of temperature maxima for the V/F curves. Within this retrograde region, extraction, product separation, or solvent recovery processes can be carried out. Some general features of the retrograde region are mentioned here.

Starting at Point A on Figure 1, which is on the dew line (V/F=1.0) at 200 K, if the pressure is decreased isothermally, the dew-point fluid initially condenses (i.e., V/F decreases). Condensation continues as the pressure is reduced isothermally to point B. As the pressure is further reduced below point B, V/F increases until it reaches the value of unity at the lower dew point (not shown). The variation of the fraction vaporized (V/F) with temperature along an isobar in the retrograde region is similar to the behavior of V/F with pressure.

Product separation and solvent recovery

For consideration of operating conditions for extraction or solvent recovery, plots of the vapor mole fraction of the heavy components along constant pressure traverses (Figure 3) and constant temperature traverses (Figure 4) are valuable. The 50 atm isobar of Figure 3 has the most striking feature, the steepness of the left-hand-side of the curve. With a small increase in temperature, the mole fraction of butane plus decane in the vapor phase can be reduced by an order of magnitude. At higher pressures, the drop in composition is not as drastic. The 30 atm isobar does not show the same behavior as the higher pressure ones because it lies below the critical pressure of the mixture and therefore the low temperature phase boundary is a bubble point rather than a dew point. This demonstrates that a large change in solubility of heavy components in a solvent

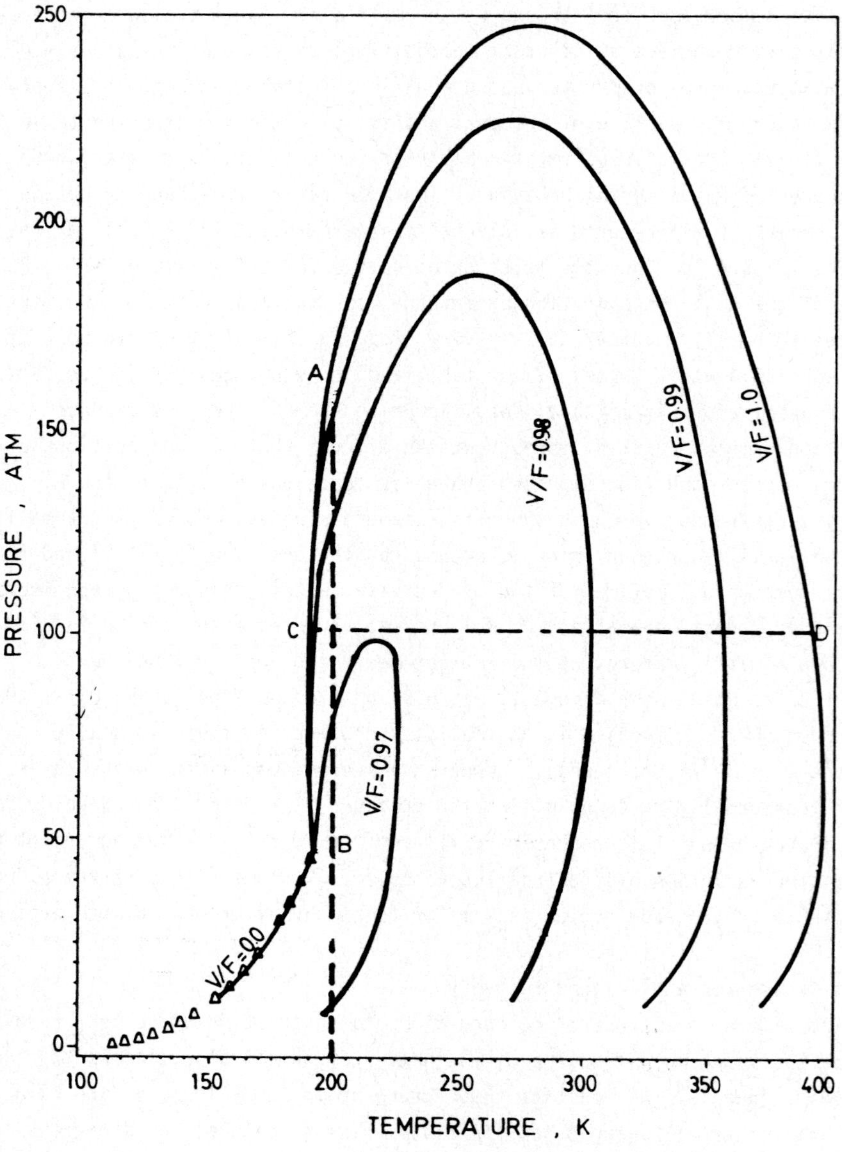

Figure 1. Pressure Temperature Phase Diagram at Constant Composition for the Methane + n-Butane + n-Decane System (Δ is Methane Vapor Pressure)

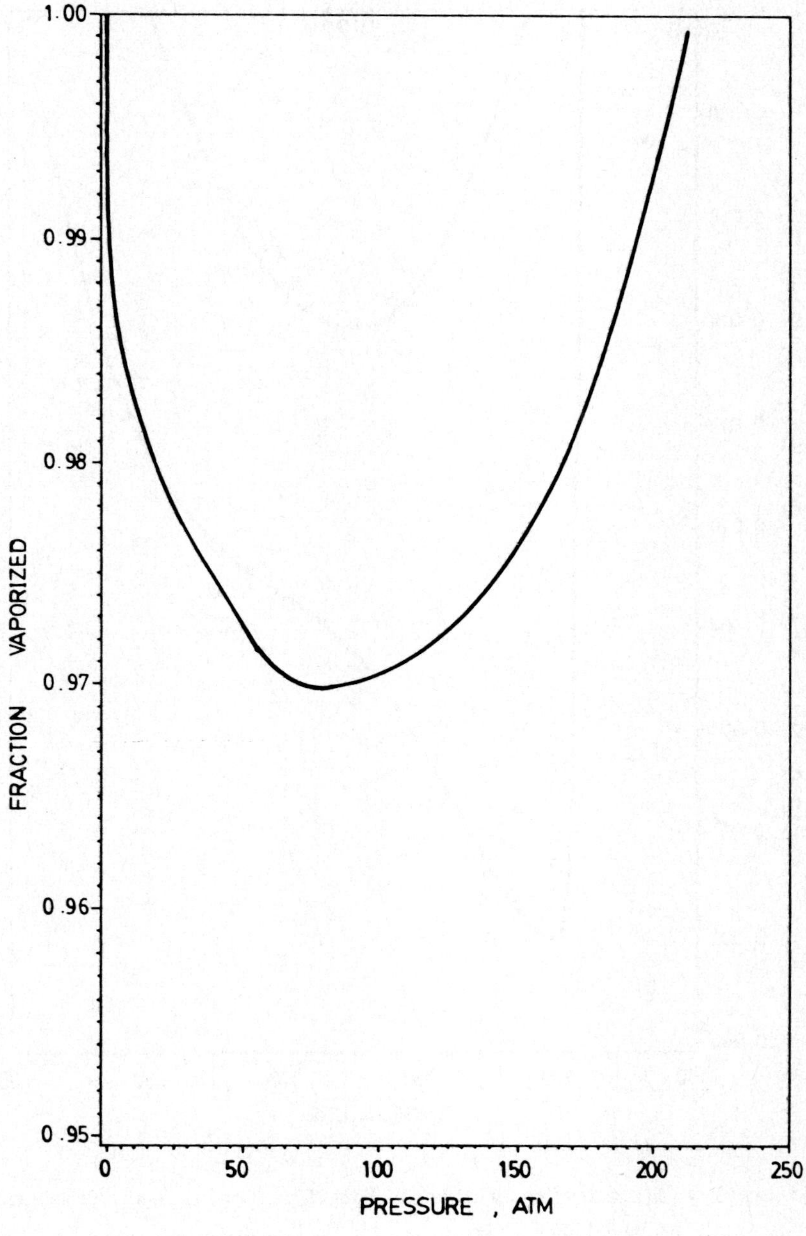

Figure 2. Behavior of Fraction Vaporized (V/F) in the Retrograde Region at 200 K.

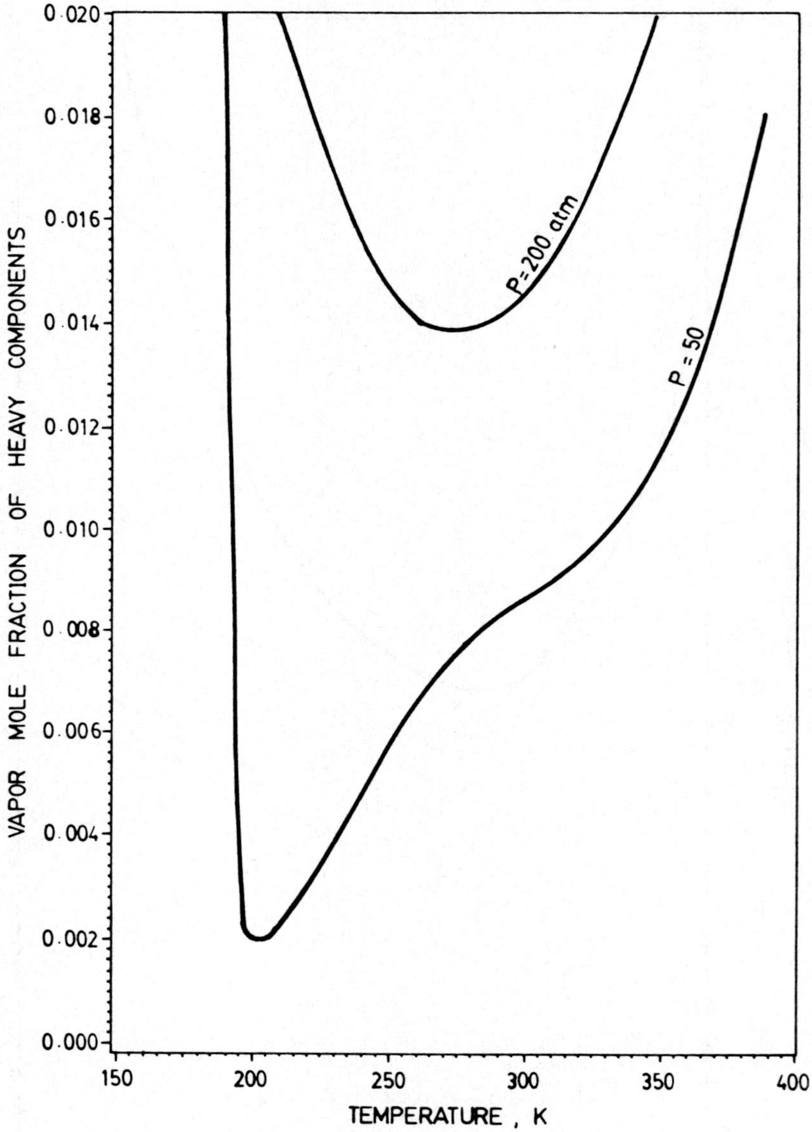

Figure 3. Effect of Temperature on Mole Fraction of Heavy Components in the Vapor Phase.

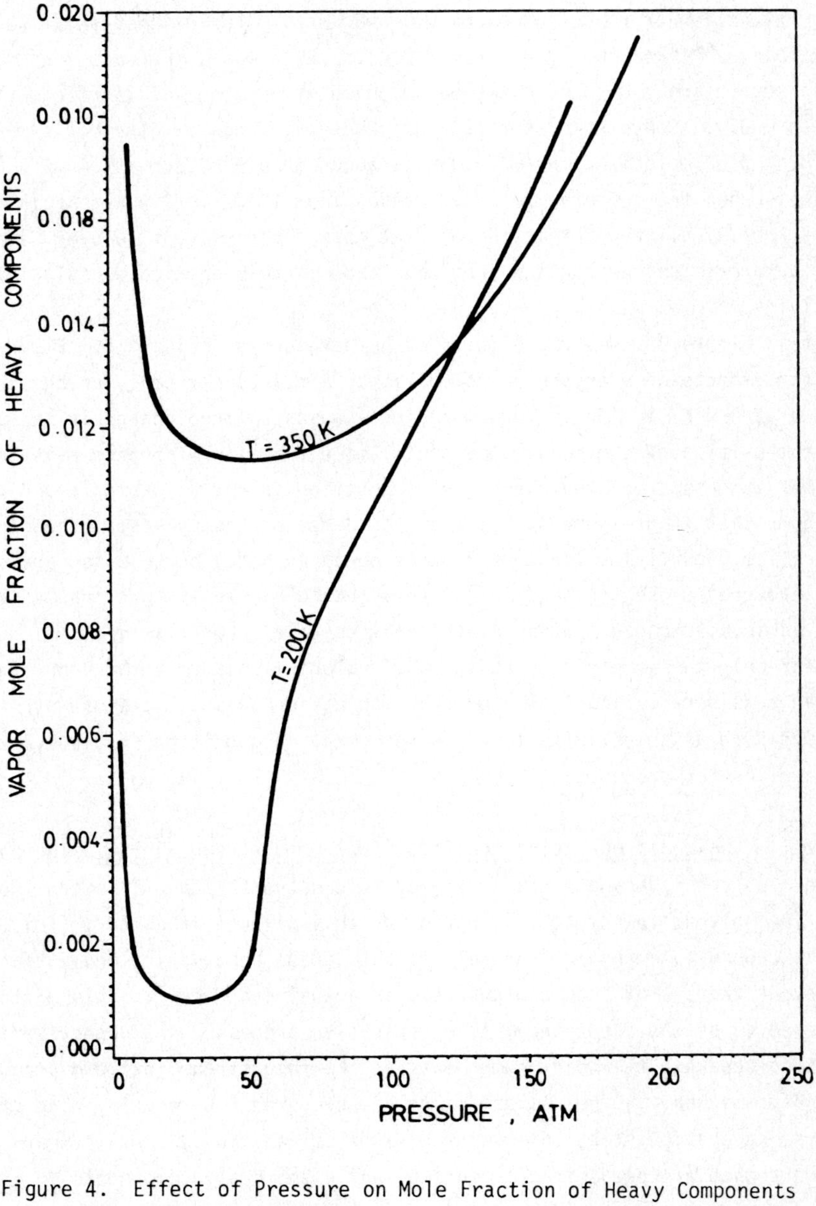

Figure 4. Effect of Pressure on Mole Fraction of Heavy Components in the Vapor Phase.

can be achieved by a small change in temperature at conditions close to the critical point of the mixture. This solubility sensitivity can be utilized for either extraction or rejection of heavy components. For extraction at 50 atm, for example, a temperature drop from 200 K to 190 K would increase the mole fraction of butane + decane in the vapor phase by an order of magnitude (from 0.002 to 0.02). Conversely, for solvent recovery, a dew point vapor stream containing 0.02 mole fraction of butane + decane would reject 90% of its heavy components when the temperature is increased from 190 K to 200 K. Brulé and Corbett (ref.9) point out that the product extraction/solvent recovery steps of actual supercritical extraction processes exploit this particular solubility sensitivity.

Just as Figure 3 shows the effects of temperature on solubility, Figure 4 shows the effects of pressure on solubility. A relatively small change in pressure at 200 K (in the 50-70 atm region) causes a large change in the vapor phase composition of the heavy components (mole fraction of butane plus decane). At higher temperatures, the change in composition is not as dramatic. Figure 4 also shows that if pressure swings are considered as the process methodology to recover (i.e., purify) solvent vapor that contains heavy components, operating at temperatures close to the critical temperature of the mixture results in the lowest achievable concentration of the heavies (i.e., purer solvent).

So far only the combined composition of butane + decane in the vapor phase has been considered. The following four subsections discuss the use of other variables such as selectivity in choosing operating conditions and solvent to solute ratios for processes.

Case 1. Choice of operating conditions for product separation. The same methane + butane + decane mixture as used to construct Figure 1 is considered here. The solvent to solute mole ratio for this mixture is 49 to 1 (this high ratio is chosen to emphasize the role of retrograde behavior; economic factors may prevent the use of such a high ratio in actual processes). This mixture is considered to be the feed stream to a separation process. The objective in this case is to separate the heavy components in the feed stream into two products, one light, one heavy. The solvent is to be recovered for recycle. Therefore, after the separation step, the overhead vapor stream should contain heavy components that are predominantly butane, while the heavy components in the bottoms liquid stream should be mainly decane.

The degree of separation of the solute into a butane enriched overhead stream and a decane enriched bottoms stream can be examined in terms of the selectivity defined by the relation

$$\text{selectivity} = \{x(2)/[x(2) + x(3)]\}\,100\%$$

where x(2) is the butane mole fraction and x(3) is the decane mole fraction. The objective is to choose operating conditions such that the vapor phase selectivity is large and the liquid phase selectivity is small. Calculations were performed to study the effects of variation of operating pressure at a temperature of 200 K; the results are given in Figure 5 and Table 1.

Figure 5 shows that at 200 K the selectivity of the overhead product increases as the operating pressure decreases. At a pressure of 55 atm, the overhead product, i.e., the solute in the overhead stream, is almost pure butane. However, at this pressure, the selectivity of the bottoms product increases rapidly with decreasing pressure and has already reached 36%. The curve labeled overhead product in Figure 5 shows that at 55 atm, only about 17% of the solute is in the overhead stream. This means that at 55 atm the heavy components in the feed can be separated into two products; 17% goes to the light overhead product which is almost pure butane, the other 83% goes to the heavy bottoms product which contains about 36% n-butane. In this case, although a very pure light product is obtained, its total amount is relatively small. Furthermore, the heavy product contains almost as much light component as the feed (36% vs. 46%), so the goal of extraction of the light component from the solute is not achieved.

The amount of the light product produced as well as the degree of separation can be controlled by judiciously choosing the operating pressure. For example, as can be noted in Table 1, a change in operating pressure from 55 atm to 65 atm increases the overhead product from 17% to 34% of the solute in the feed while the liquid phase selectivity decreases from 36% to 22% and the vapor phase selectivity decreases only slightly, from 99% to 95%. This means that 70% of the butane in the solute is extracted in the overhead stream (0.95x0.34/0.46 = 0.70). The operating conditions chosen should, of course, depend on the overall process objectives and constraints such as product purity requirements and number of stages available. In the examples above there is a trade-off between increased extraction and decreased overhead product purity as the pressure is increased. Operating at even higher pressures, however, may not be very beneficial. At 75 atm, the overhead product contains 40% of the solute in feed (an increase of 5%), but the vapor phase selectivity decreases to 89% (Table 1). Note that the best operating conditions exploit the steepness of the curves depicted in Figure 5. The general conclusion is that product separation is maximized by operation at temperatures and pressures which both are just slightly above the mixture critical point.

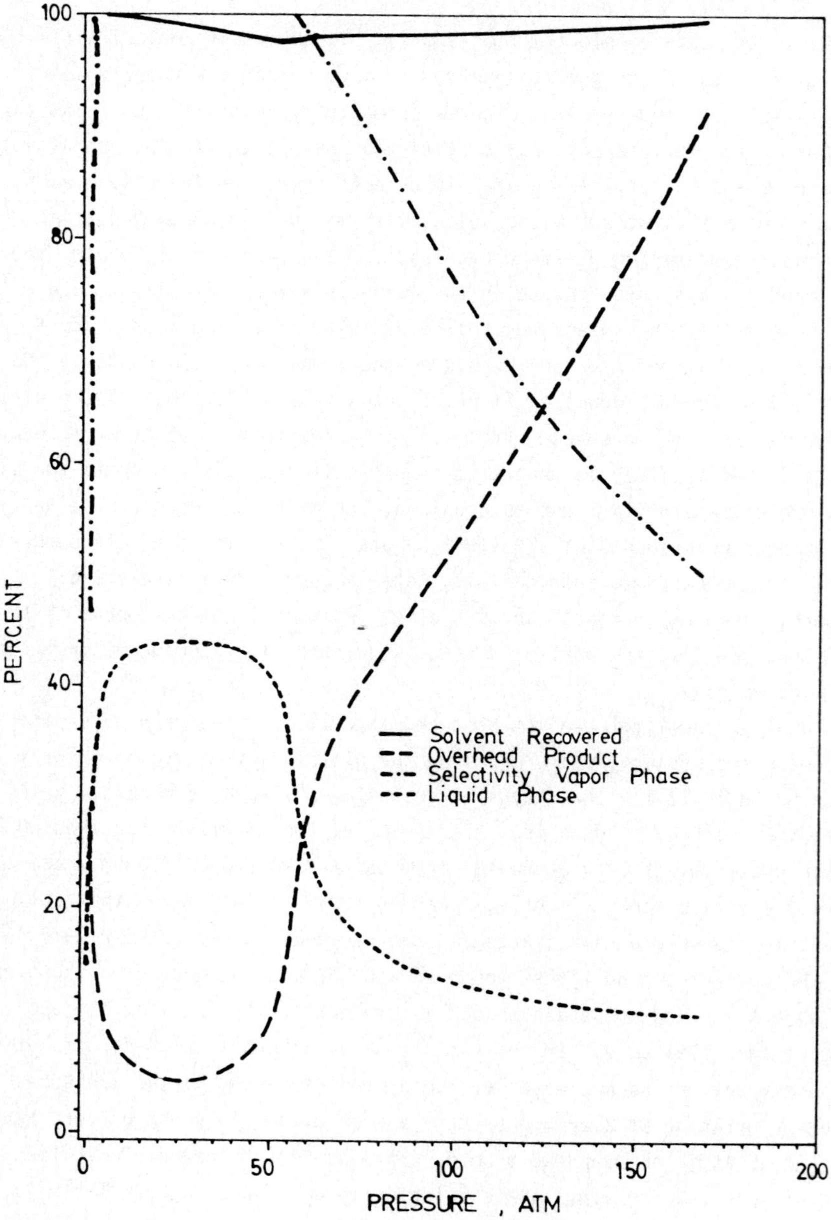

Figure 5. Effect of Variation of Operating Pressure on Product Separation at 200 K.

Case 2. Choice of operating conditions for solvent recovery. Consider a pressure swing recovery process in which the feed to the process and the operating temperature are the same as in Case 1. The objective in this case is to separate solvent from the heavy components (i.e., solvent recovery, not product separation). Figure 6 shows that if the operating pressure is 50 atm, the overhead stream will have 0.002 mole fraction of heavy components. The percent solvent recovered at this pressure is 98% (see Figure 5). The product loss is less than 10% (overhead product curve is Figure 5), so that although 10% of the product gets recycled, 90% of the product is obtained in the bottoms stream (with only 2% of the solvent).

Case 3. Choice of solvent to solute ratio for product separation. The effects of varying the solvent to solute ratio on product separation are examined in this case. Calculations were made at 200 K for solvent to solute ratios of 49:1, 24:1, and 9:1 at varying pressures. The percent of n-butane in the solute feed is 46% for all three cases. Figure 7 shows that at the lower solvent to solute ratios, the selectivities in both the vapor and the liquid phases are higher at any given pressure. The higher selectivity is desirable for the vapor phase but is undesirable for the liquid phase. Figure 8 shows that at a fixed temperature and pressure, say 200 K and 65 atm, more of the solute in the feed can be obtained in the overhead product at the higher solvent to solute ratios. Furthermore, for the lower solvent to solute ratios, the increase in the overhead product with increasing pressure in the 50 to 60 atm range is less pronounced. The fact that at fixed temperature and pressure, increasing solvent to solute ratio causes the overhead product in supercritical extraction processes to increase also has been reported by Brulé and Corbett (ref.9). It is concluded that the highest solvent to solute ratios which are feasible within design and operating constraints should be used in supercritical extraction processes.

Case 4. Choice of solvent to solute ratios for solvent recovery. The effects of solvent to solute ratio on solvent recovery are considered. Figure 6 shows that the same solvent purity can be achieved at 50 atm and below for all three solvent to solute ratios considered. However, from Figure 9, operating the process at the lower solvent to solute ratios results in much lower solvent recovery (98% for the ratio 49:1 vs 84% for the ratio 9:1, both at P=50 atm). The product loss in the overhead stream is smaller, of course, for the smaller solvent to solute ratio (see Figure 8).

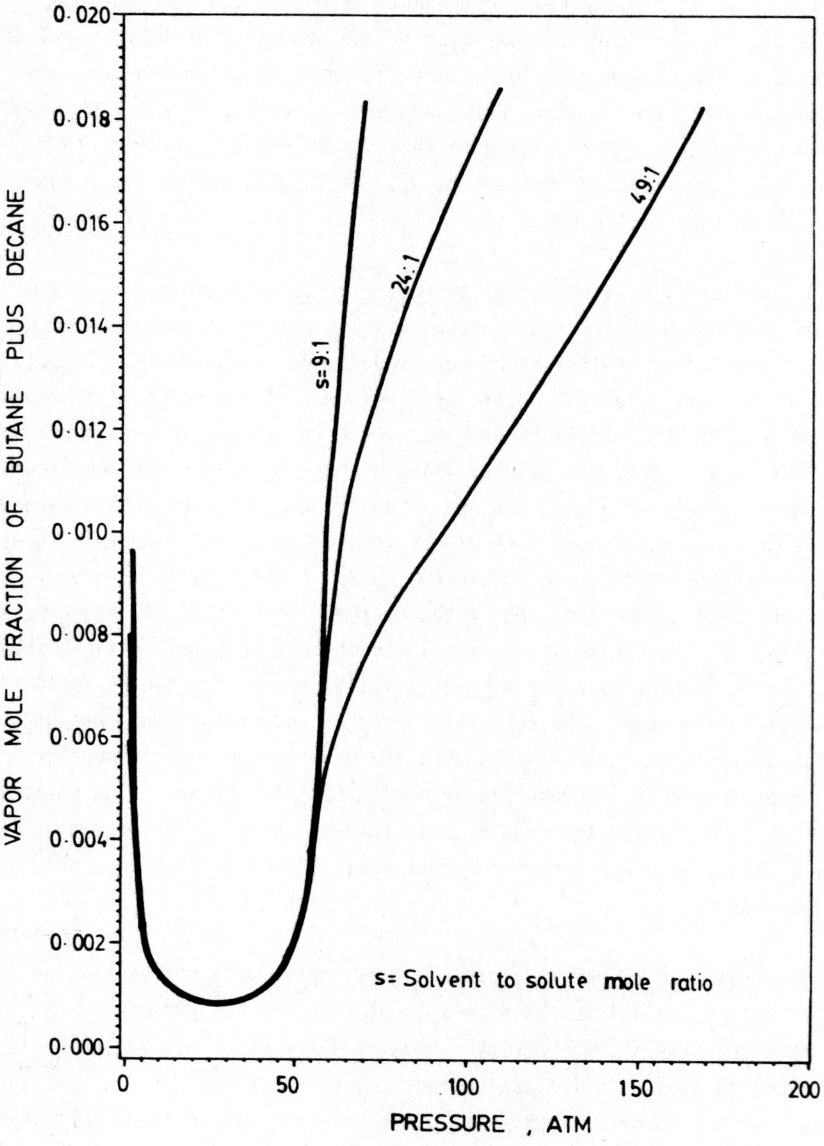

Figure 6. Effect of Solvent to Solute Ratio on Solvent Purity.

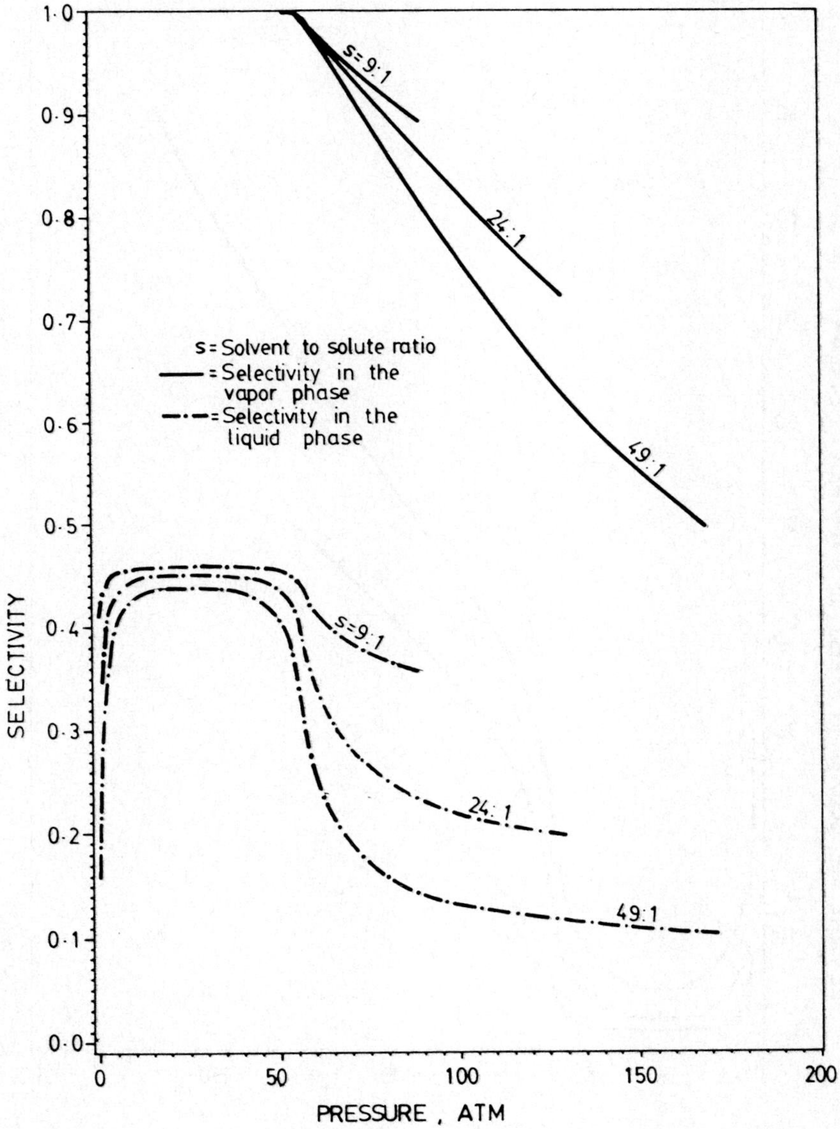

Figure 7. Effect of Solvent to Solute Mole Ratio on Vapor and Liquid Phase Selectivities.

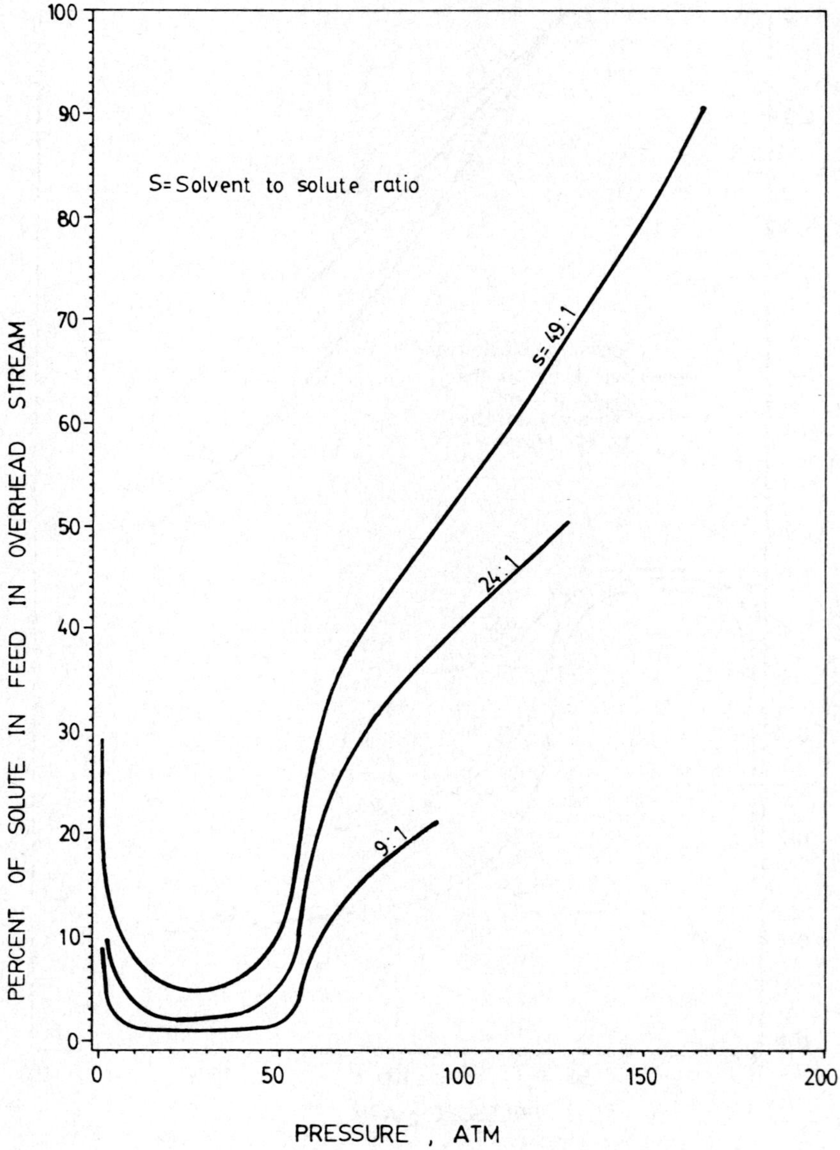

Figure 8. Effect of Solvent to Solute Ratio on Overhead Project.

LIQUID-LIQUID SYSTEMS

The liquid-liquid system chosen for study is the methane + n-hexane binary studied experimentally by Lin, et al. (ref.10). A ternary or multicomponent system would be more desirable for study than a binary system because product selectivity could be studied. However, the paucity of experimental ternary liquid-liquid critical region data would make it difficult to study ternary systems quantitatively. Since a binary system is used, our study will be limited to the distribution of n-hexane in the liquid phases at different conditions.

The methane + n-hexane system is a class V mixture according to the Van Konynenburg and Scott (ref.11) classification scheme. The P-T projection for this class of mixtures is shown in Figure 10. The region of interest here lies between the upper critical solute temperature (UCST), where the light liquid (L1) is equal in composition to the vapor phase (G), and the lower critical solution temperature (LCST), where the liquid liquid (L1) and the heavy liquid (L2) compositions are equal. As shown in Figure 10, at any temperature between UCST and LCST there exists a fixed pressure where all three phases (G, L1, L2) coexist. At pressures greater than the three-phase pressure, at most two phases are possible, G-L1 or L1-L2.

For the methane + n-hexane system, the UCST is at 195.91 K and 51.4 atm and the LCST is at 182.46 K and 33.7 atm. The equilibrium liquid-liquid isotherm compositions are shown versus pressure in Figure 11, while equilibrium liquid-liquid isobar compositions are shown versus temperature in Figure 12. The light liquid (L1) becomes heavier (contains more n-hexane) as the temperature is decreased, while the heavy liquid (L2) becomes lighter. To more clearly illustrate the effect of temperature on extraction, the product yield is plotted in Figure 13 for the case of a feed which is 90 mole percent methane and 10 mole percent n-hexane. Product yield is defined as the percentage of n-hexane in the feed that goes into L1. As the temperature is decreased from 195.91 K to 190 K, the product yield increases dramatically from 0% to 70%. Note that the 195.91 K, 51.4 atm condition corresponds to the UCST and the pressure of 51.4 atm also corresponds closely to the critical pressure of the 190 K isotherm. To study the effect of pressure on extraction, calculations were made at 47.1 atm using the same feed compositions (i.e., 90% methane, 10% n-hexane). Figure 13 shows that in the liquid-liquid critical region, increasing pressure toward the critical pressures (at constant temperature) increases product yield. Similarly, it can be concluded that if the liquid-liquid critical temperature is approached via isobaric steps in the two phase region from higher temperatures, then the highest product yield occurs at temperatures just above the critical temperature. As was noted for vapor-liquid systems, conditions very near to the

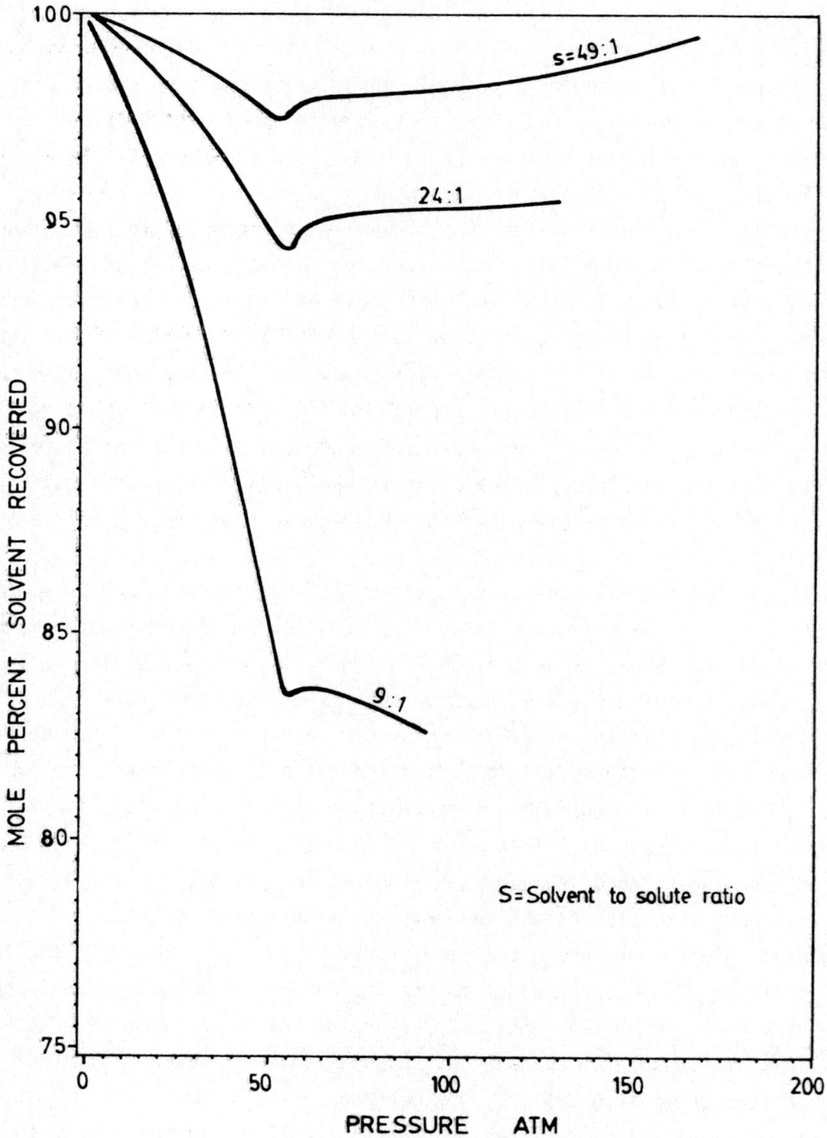

Figure 9. Effect of Solvent to Solute Ratio on Solvent Recovery.

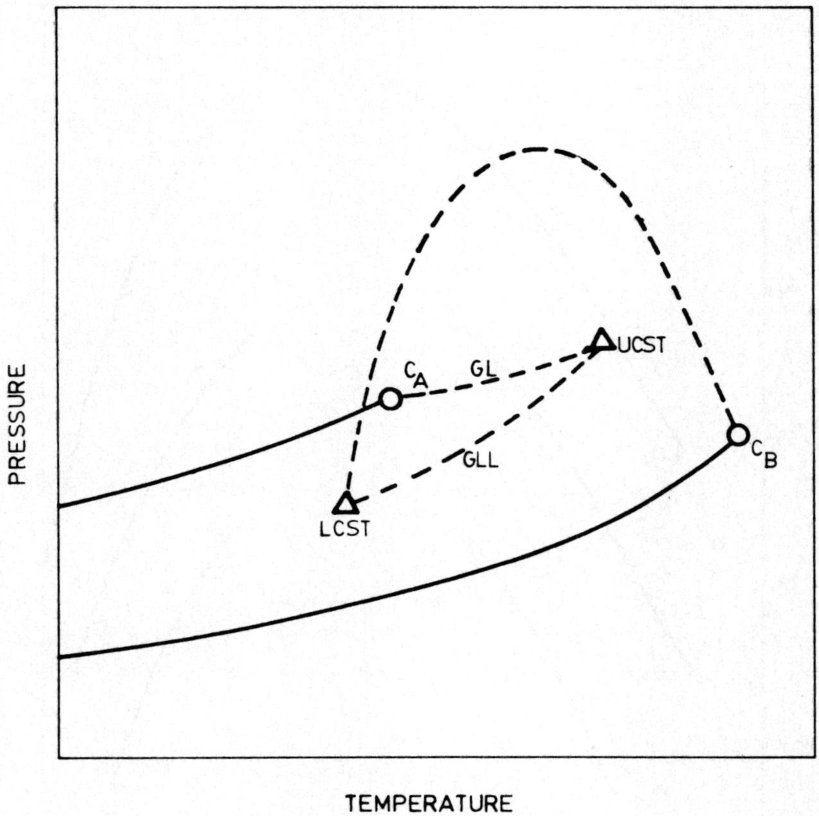

Figure 10. Pressure-Temperature Projection for a Class V Mixture.

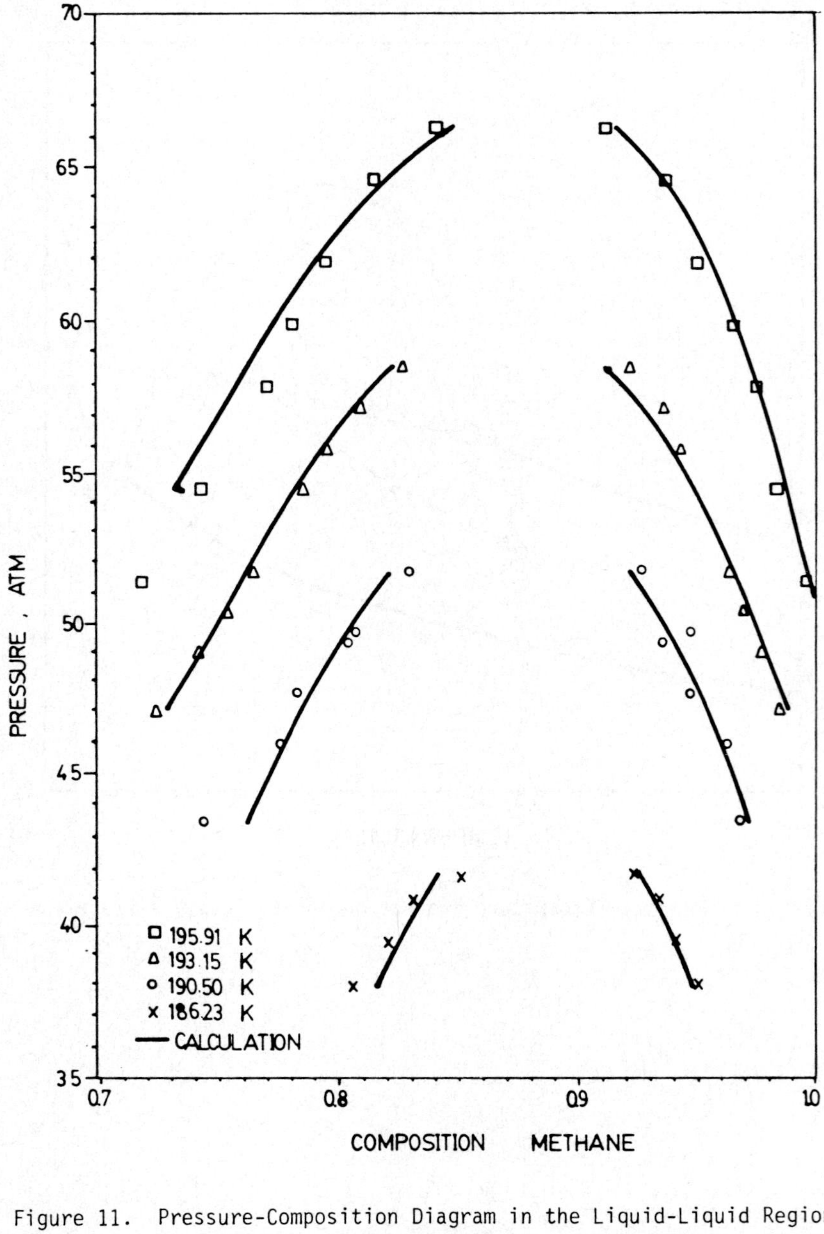

Figure 11. Pressure-Composition Diagram in the Liquid-Liquid Region for the Methane-n-Hexane System (Symbols are the Experimental Data of Lin, et al., 1977).

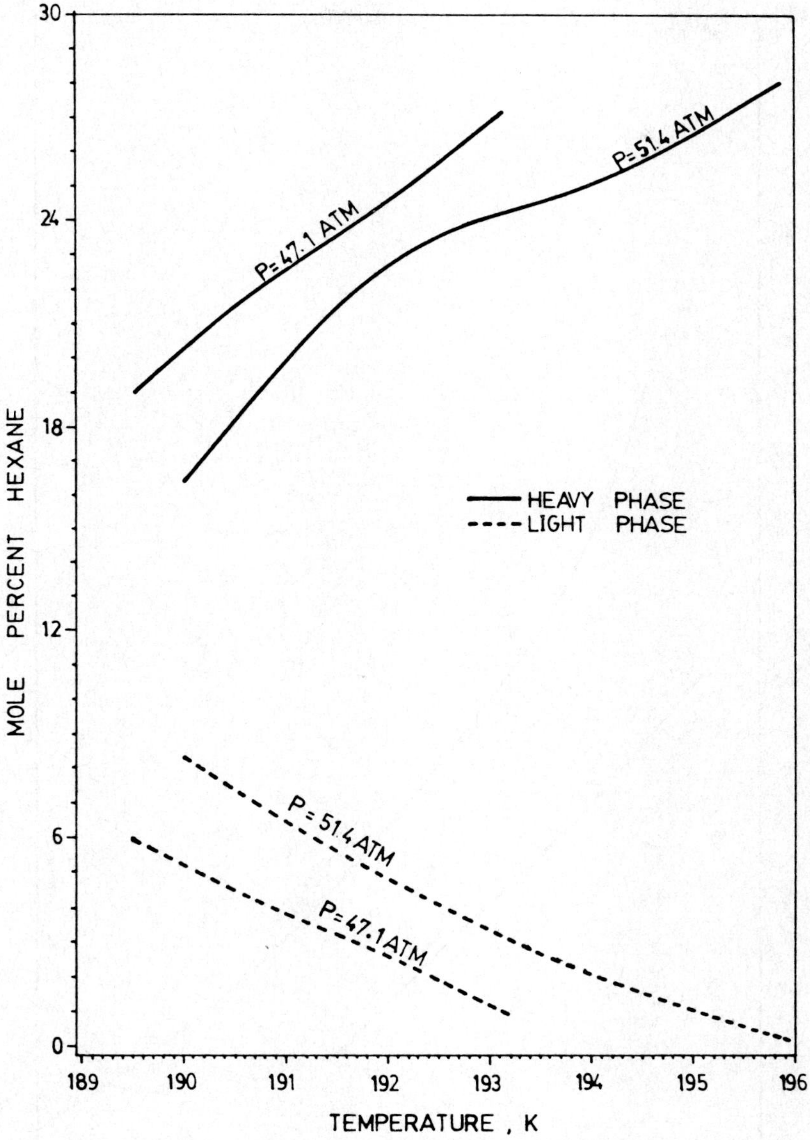

Figure 12. Compositions of Liquid Phases at 51.4 Atm and 47.1 Atm as Functions of Temperature

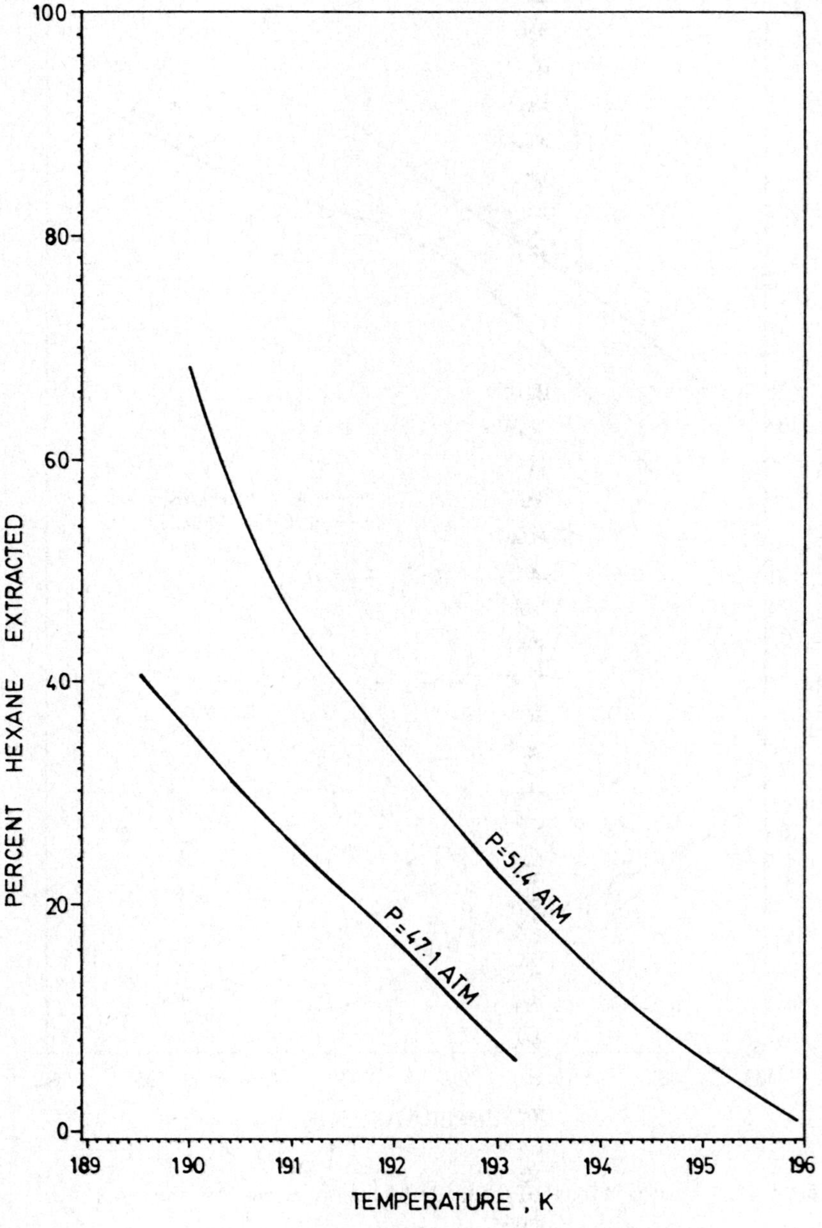

Figure 13. Product Yield for a Solvent to Solute Mole Ratio of 9:1.

mixture critical point seem to be the best operating conditions for liquid-liquid systems.

Figure 14 shows the effect of solvent to solute ratio on product yield at 195.91, 193.15, and 192.90 K and at constant pressure of 51.4 atm. At constant temperature and pressure, increasing solvent to solute ratio increases product yield. At constant solvent to solute ratio, decreasing temperature (toward the critical) increases product yield. These results show that for the type of behavior of the methane plus n-hexane system, choosing operating temperatures near the mixture LCST (e.g., at 190 K) and operating pressures near the mixture critical pressure (e.g., at 51.4 atm for T=190 K) yields the best light-phase product yield.

DISCUSSION AND CONCLUSIONS

The information presented herein demonstrates the value of using fundamental thermodynamics and particularly classical phase behavior information in understanding supercritical extraction phenomena and in designing processes utilizing supercritical extraction. It obviously is not enough to know that a process is being operated at supercritical conditions with respect to the solvent if a supercritical extraction process is to be well designed. Herein we have utilized real system phase behavior to illustrate points of interest with respect to both fundamental fluid mixture behavior and process behavior. Because the approach to equilibrium is relatively rapid in fluid systems near critical points, only equilibrium systems have been considered. The effects of operating conditions on both product separation and solvent recovery for vapor-liquid systems are clearly seen from examples given for the methane + n-butane + n-decane system. The effects of solvent to solute ratio also are amply illustrated for this system. In addition, the methane + n-hexane system is used to illustrate behavior which can be encountered for systems which can form two liquid phases at temperatures near or above the critical temperature of the solvent. Again, more judicious choices for operating conditions and solvent to solute ratio are possible when the phase behavior for the system is established. For most supercritical extraction processes, the best operating conditions generally are as near as possible to the critical point of the mixture and are supercritical with respect to the mixture critical point, irrespective of the solvent critical point. It is also generally true that for vapor-liquid systems both product separation and solvent recovery are enhanced by increased solvent to solute ratios. The design of any process must take economics into account, but it is believed that the technical basis for the design would be enhanced by the type of information discussed in this paper.

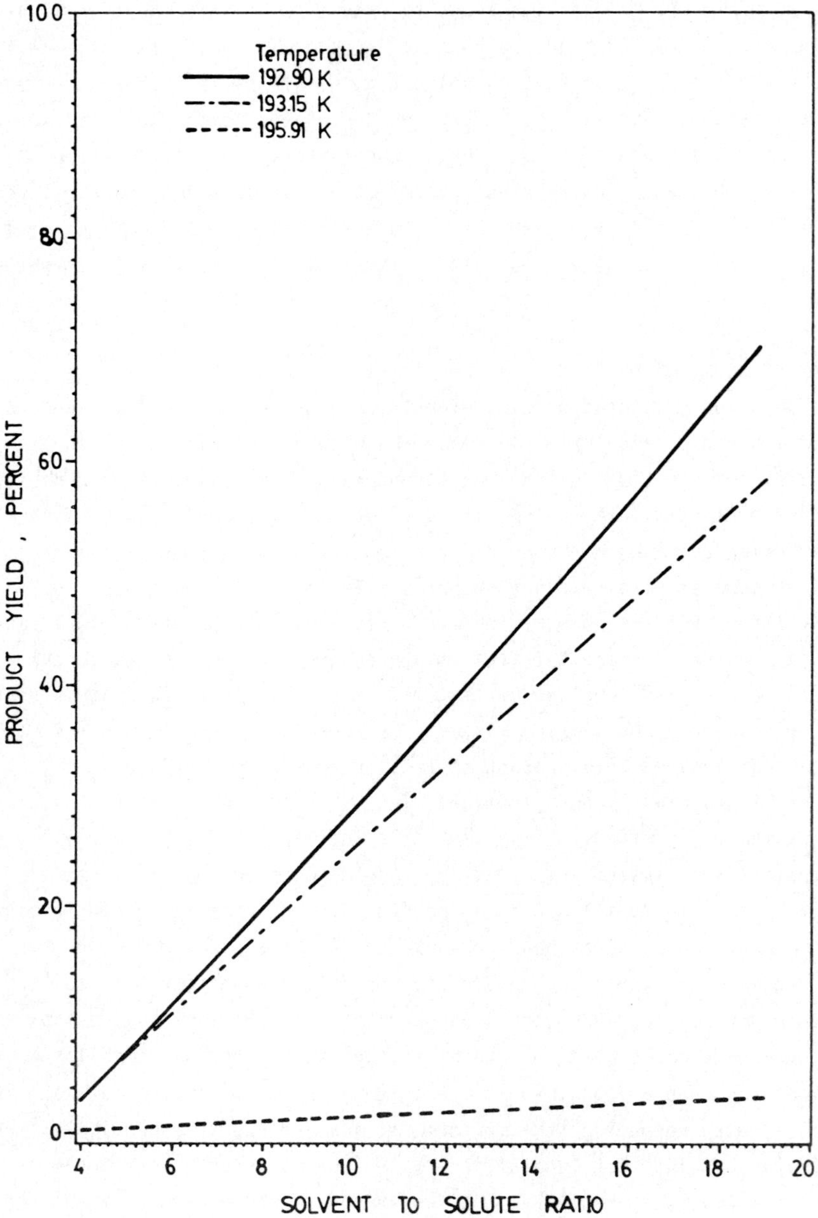

Figure 14. Effects of Solvent to Solute Ratio on Product Yield at 51.4 Atm.

Table 1.
Effects of Variation of Operating Pressure on Product Separation at 200K

System Operating Pressure Atm	Selectivity, percent Feed	Overhead Product	Bottoms Product	Percent of Solute in Overhead Product	Percent of n-Butane in Solute Extracted
55	46	99	36	17	37
65	46	95	22	34	70
75	46	89	18	40	77

ACKNOWLEDGEMENT

This work was supported by the University of Oklahoma and the Department of Energy Pittsburgh Energy Technology Center under contract no. DE-FG22-83PC60804.

LITERATURE CITED

1 Waals, Van der, Arch. Nederland Sci. Ser. 2, Vol. 15 126 (1910).
2 Kalichevsky, V.A., Refining Lubricating Oils, Reinhold Pub. Corp. (1938).
3 Godlewicz, M. and S. Pilat, The Oil and Gas Journal, 34(32), 76 (1933).
4 Vink, D.J., A.M. Ames, R.A. David, and D.L. Katz, 'Multiple-Phase Hydrocarbon Systems,' The Oil and Gas Journal 24,(Nov. 1940).
5 Kuenen, J.P., 'On retrograde condensation and the critical phenomena of two substances,' Commun. Phys. Lab. University Leiden, 4(7) (1892).
6 Patten, V.F.L. and D.C. Ivey, 'Phase Equilibria in High Pressure Condensate Wells,' Oil Weekly, 20 (Dec. 1938).
7 Lacey, W.N., API, Bulletin 210, 65 (1932).
8 Reamer, H.H., B.H. Sage, and W.N. Lacey, 'Volumetric and Phase Behavior of the Methane-n-Butane-n-Decane System,' Ind. Eng. Chem., 39, 77 (1947); 43, 1436 (1951); 44, 1671 (1952).
9 Brulé, M.R. and R.W. Corbett, 'What makes critical-solvent processes work?,' Hydrocarbon Processing, (June 1984).
10 Lin, Y.N., R.J.J. Chen, P.S. Chappelear, and R.K. Kobayashi, 'Vapor-Liquid Equilibrium of Methane-n-Hexane System at Low Temperature,' J. Chem. Eng. Data, 22(4) (1977).
11 Van Konynenburg, P.H. and R.L. Scott, 'Critical Lines and Phase Equilibria in Binary van der Waals Mixtures,' Phil. Trans. Roy. Soc., 298, 495 (Dec. 1980).

Supercritical Fluid Technology, edited by J.M.L. Penninger, M. Radosz, M.A. McHugh and V.J. Krukonis
1985 Elsevier Science Publishers B.V., Amsterdam — Printed in The Netherlands

THERMODYNAMIC ANOMALIES IN SUPERCRITICAL FLUID MIXTURES

G. MORRISON, J.M.H. LEVELT SENGERS, R.F. CHANG and J.J. CHRISTENSEN[1]
Thermophysics Division, National Bureau of Standards, Gaithersburg, MD 20899
[1]Department of Chemical Engineering Sciences, Brigham Young University, Provo Utah 84602

ABSTRACT

We use thermodynamic stability considerations, the Gibbs-Konowalow rule and properties of dilute near-critical mixtures to explain the striking features of the excess enthalpy versus composition curves studied along isotherm-isobars by Christensen and coworkers for a number of supercritical binaries. We show that knowledge of the supercritical enthalpy is essential for an understanding of supercritical solubility.

INTRODUCTION

One of us, Christensen, with coworkers, measured excess enthalpies of binary mixtures at pressures usually exceeding the critical pressures of both components and over temperature ranges often spanning the critical temperatures of both components. In many cases, these excess enthalpies show very striking behavior. Typically, the excess enthalpy H^E may drop to a minimum of the order of -1000 J/mol as the composition x departs from zero; then, the excess enthalpy may rise linearly to a maximum of similar size before returning to zero at x=1. See refs.1-7 for examples in the systems of carbon dioxide and hydrocarbons. The general appearance of these H^E - x curves can be readily understood from the behavior of the H - x diagram at fixed pressure, as schematically indicated in Fig. 1. The addition of hexane to slightly supercritical carbon dioxide, for instance, induces a phase separation which is the cause of the large enthalpy effects. The excess enthalpy, which is the difference of the actual enthalpy and the mole average of the pure-component enthalpies, shows the maximum, minimum and straight tie line encountered in many of the graphs published by Christensen and his group.

On somewhat closer inspection, however, experimental features are seen that are not immediately clear from a schematic plot like Fig. 1. Thus, some isotherm-isobars do not seem to have both a maximum and a minimum. Some isotherm-isobars do not traverse a two-phase region but show large excess enthalpies anyway. Also, the initial slopes of the excess curves, both near x = 0 and x = 1, can have either sign, depending on the temperature. It is these more delicate

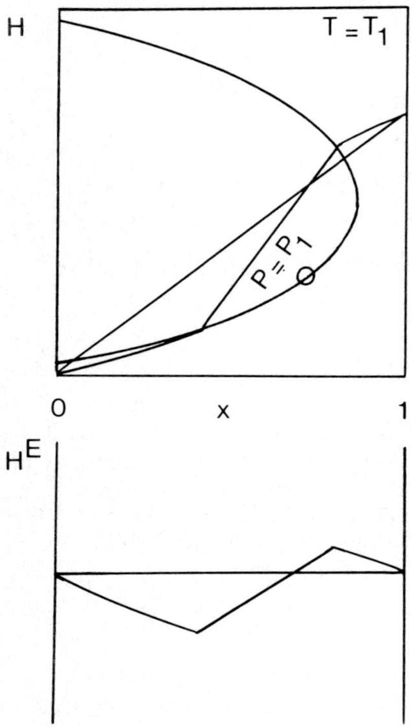

Fig. 1. The enthalpy and excess enthalpy of a binary mixture at constant temperature T_1 slightly above the critical temperature of the most volatile component (x=1). The addition of the nonvolatile component induces a phase separation. As a consequence, isobars at pressures P_1 below the critical pressure of the most volatile component traverse a two-phase region, beginning in a vapor - like state on the right and ending in a compressed liquid on the left. The excess enthalpy shows the maximum, minimum and straight portion typical of many of the excess enthalpy plots published by Christensen and coworkers (ref. 1-7).

features that we also want to understand. As a specific example, we will examine the data for the mixture carbon dioxide - toluene, for which the location of the critical line is known, and for which a large amount of data is available, namely fifteen different H^E isotherm-isobars (ref.8). The critical line passes through a substantial maximum in P-T space (Fig. 2). As a consequence, even at pressures considerably higher that the critical pressures of both pure components, the mixture can enter the two-phase region along isobars. At constant pressure this mixture may show a closed-loop coexistence curve with a lower and an upper critical solution temperature, an LCST and an UCST, respectively, analogous to those encountered in partially miscible binary liquid mixtures.

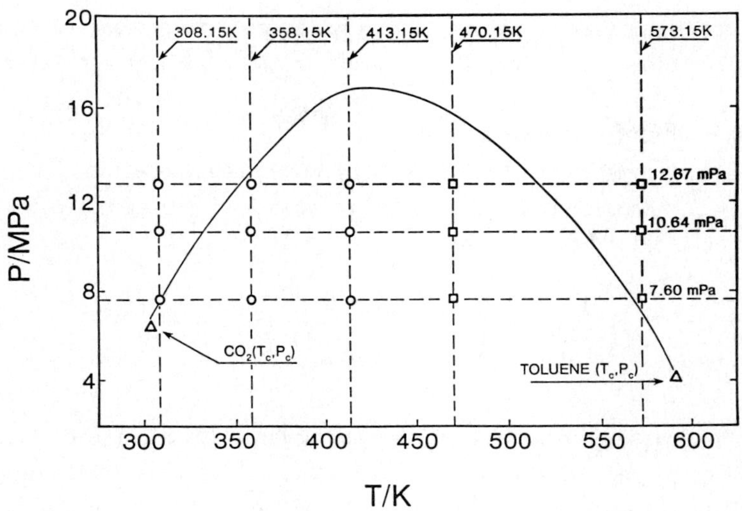

Fig. 2. Location of the experimental isotherm-isobars of the system carbon dioxide -toluene, and their relation to the critical line (ref.8).

In the next sections, we sketch the principles necessary to understand this extensive data set with such unusual features. Then we proceed with a detailed analysis and a discussion of some of the quantitative aspects of those phenomena. The kinds of behavior seen in these enthalpy data and in similar measurements of volumes near critical points are related to two important practical considerations in working with supercritical fluids as solvents. The first, as we discuss in the final section of this paper, is that the temperature and pressure dependence of the solubility of otherwise inert solids in super-critical phases depends on the partial molar properties of the extracted components in the supercritical phase. The nearness of critical lines and points and the relation of the extracting conditions to nearby phase boundaries can dramatically affect not only the size of the partial molar quantities but even their signs. The second, a point that we will note briefly, is that many of the enthalpies of mixing reported are of the same order as enthalpies of phase transitions; indeed, there is a close connection. Thus, temperature -- and likewise pressure -- response to mixing in these kinds of systems needn't be small and may make an important contribution to the energy balances in the design of supercritical extraction processes. The purpose of this paper is thus twofold: first, to explain the intricacies of a fascinating set of data; and, second, to present an overview of the connection between a variety of phenomena in the region of a critical point.

STABILITY AND THE GIBBS-KONOWALOW RULE

In a recent Comment (ref.9), one of us (G.M.) showed how the $(\partial^2G/\partial x^2)_{PT} \geq 0$, imposes conditions on the curvature of the excess-enthalpy curves near critical lines. Since

$$\frac{\partial}{\partial T}\;\frac{1}{T}\left(\frac{\partial^2 G}{\partial x^2}\right)_{TP} = -\;\frac{1}{T^2}\left(\frac{\partial^2 H}{\partial x^2}\right)_{TP} = -\;\frac{1}{T^2}\left(\frac{\partial^2 H^E}{\partial x^2}\right)_{TP} \quad , \tag{1}$$

and $(\partial^2G/\partial x^2)_{TP}$ must increase when the system moves away from the critical line into the one-phase region, the curvature of H^E is always negative at an upper critical solution temperature (where $\partial^2G/\partial x^2$ increases at T <u>inc</u>reases) and always positive at a lower critical solution temperature (where $\partial^2G/\partial x^2$ increases as T <u>de</u>creases). Thus, the stability criterion determines the curvature of H-x curves in the one-phase region near a critical point, a fact also discussed in several textbook on the thermodynamics of mixtures[10, 11].

The Gibbs-Konowalow rule [Appendix] imposes a second condition, pertaining to the change in slope of an isotherm-isobar as it enters the two-phase region. For the enthalpy, this rule states that at any point on the phase boundary of the binary mixture:

$$\left.\frac{dT}{dx}\right|_{P,\sigma} = -\;\frac{T\,(\partial^2G/\partial x^2)_{TP}}{\Delta H/\Delta x - (\partial H/\partial x)_{TP}} \quad . \tag{2}$$

Here $dT/dx|_{P,\sigma}$ is the slope of the isobaric dew-bubble curve and $\Delta H/\Delta x$ the slope of the tie line. We explain the implications of this rule schematically in Figs. 3 and 4.

In Fig. 3 we show the P-T projection and critical line of a binary mixture for three cases, a critical line with (a) positive slope, (b) negative slope and (c) a maximum. In all three cases we consider an isobaric path that intersects the critical line. In Fig. 4 we plot the T-x isobaric coexistence loop, the isobaric H-x diagram and the isobaric H^E-x diagram for the cases corresponding to Figs. 3a, b and c. In case (a) we encounter a lower critical solution temperature, the mixture splitting into two phases as the temperature is raised. We draw attention to the right extremum of this loop, marked by a black dot. This extremum corresponds with a similar extremum on the H-x plot. Since at this point $\partial T/\partial x|_{P,\sigma}$ is infinite, but $(\partial^2G/\partial x^2)_{PT}$ is finite and positive, Eq. (2) implies that $\Delta H/\Delta x = (\partial H/\partial x)_{PT}$ so that tie line and isotherm-isobar are collinear at the phase boundary. The isotherm labeled (3) in the H-x plot of Fig. 4b enters the two-phase region without a change of slope. Below this extremum, $(\partial T/\partial x)_{P,\sigma}$ becomes positive and since $(\partial^2 G/\partial x^2)_{PT} > 0$, $\Delta H/\Delta x < (\partial H/\partial x)_{PT}$. The converse is true above the right extremum where $(dT/dx)_{P,\sigma}$ is negative. In the one-phase region near the LCST, the curvature of H-x isotherm-isobars needs to be positive, and therefore the Gibbs-Konowalow rule forces the near-critical isotherm-isobars to have the same convexity as the ones in the nearby one-phase region.

In the case of Fig. 4b, where the critical line has a negative slope, an upper critical solution temperature is reached. Again, the Gibbs-Konowalow rule requires the isotherm-isobar passing through the left extremum to enter the two-phase region without a change of slope. On opposite sides of the left extremum, isotherm-isobars undergo opposite changes in slope and near-critical isotherm - isobars assume the convexity of the critical iostherm-isobar. By definition of the excess property, the H^E vs x curves can now be constructed (Figs. 4 a, b, bottom). The isotherms are numbered in order of ascending temperature. In the transformation from H to H^E, the information on slopes is modified, but all information regarding convexity and changes of slopes at phase boundaries is retained.

In Fig. 4c, we sketch the situation that is of most interest, namely, that of a maximum in the P-T critical line, such as occurs in the system carbon dioxide and toluene. It is now possible to enter and exit a two-phase region at a pressure exceeding <u>both</u> critical pressures. Therefore, in T-x space, a closed-loop coexistence curve results, as sketched in Fig. 4c, with a lower (LCST) and an upper (UCST) critical solution temperature as indicated. The H-x diagram likewise contains a closed - loop coexistence region. We now construct the isotherm-isobars passing through the left and right extrema without change of slope, and adjacent isotherm-isobars, as before. Again, isotherms are numbered in order of ascending temperature. In the excess enthalpy, we find all features noted in the work of the Christensen group: the curves passing through the middle portion of the loop display a maximum and a minimum connected by a straight portion; the initial slopes on either side can have either sign; and some isotherm-isobars near the critical points do not have both a maximum and a minumum.

BEHAVIOR OF THE DILUTE MIXTURE

There is one additional feature of the H-x diagram that we need to examine. Suppose we draw the Figures 4 for the case where the two-phase region extends almost over the entire x region, that is, where one of the two components is very near its critical point. It can be inferred from the Figures 4 that in

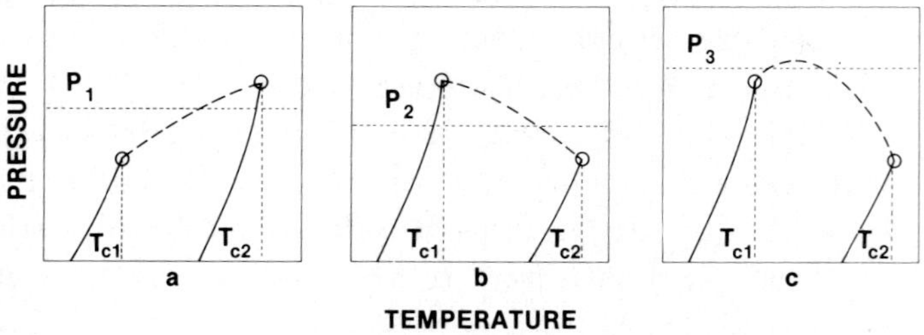

Fig. 3. P-T diagrams of three binary mixtures: (a) the critical line slopes upwards, (b) the critical line slopes downwards, and (c) the critical line passes through a maximum. Isobaric paths intersecting the critical line are indicated. The critical points are indicated by o, the vapor pressure curves of the pure components by full curves, the critical line by a dashed curve.

Fig. 4. Top row: T-x loops for the isobars indicated in Fig. 3 a, b and c. The dashed curve is the critical line. In case a, we have a lower critical solution temperature (LCST), in b, an upper critical solution temperature (UCST) and in c, both an LCST and an UCST. The critical points are indicated by o and the left and right extrema of the T-x loops by ●.

Middle row: H-x diagrams for the isobars indicated in Fig. 3a, b and c. The isobar passes through the left and right extrema without a change of slope. On opposite sides of the extremum, the convexity of the isobar as it enters the two-phase region is opposite. Near an LCST, $\partial^2H/\partial^2 > 0$ while near an UCST, $\partial^2H/x^2 < 0$. Isotherms are numbered in order of ascending temperature.

Bottom row; H^E plots constructed from the H plots in the middle row.

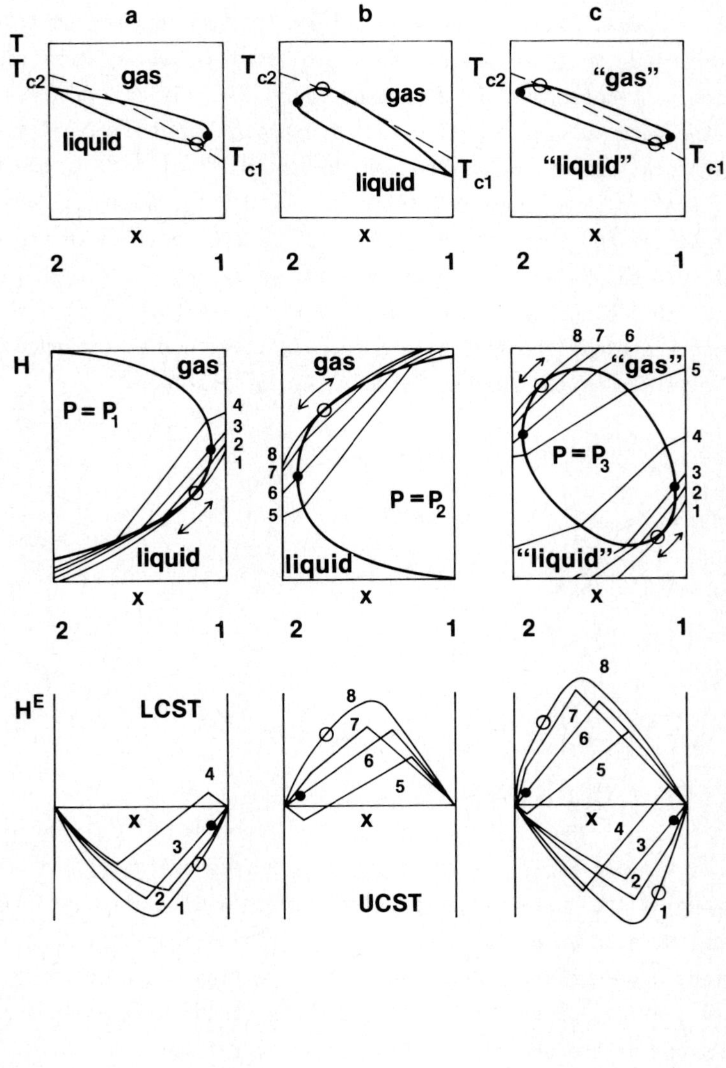
a
b
c
T
T_{c2}
gas
liquid
T_{c1}
"gas"
"liquid"
x
2
1
H
$P = P_1$
$P = P_2$
$P = P_3$
H^E
LCST
UCST
1
2
3
4
5
6
7
8

that case the mixture's critical point will move towards the extremum, which itself moves towards the nearby pure-component critical enthalpy value. The isotherm-isobars between the two-phase boundary and the pure-fluid axis thus will become very steep. The convexity imposed by the Gibbs-Konowalow rule will further accentuate this steepness. In fact, both $(\partial H/\partial x)_{PT}$ and $(\partial V/\partial x)_{PT}$ diverge at the pure-fluid critical point in a manner that depends on the path of approach. At infinite dilution, they diverge strongly, as the isothermal compressibility. This leads to the very interesting consequence that the partial molar volume and enthalpy of the solute diverge, while those of the solvent become path-dependent and do not approach the critical volume and enthalpy of the pure solvent on all paths of approach to the solvent's critical point. In a recent Letter, we worked out the mathematical form of the behavior of $(\partial V/\partial x)_{PT}$ (ref.12). Here we will derive how $(\partial H/\partial x)_{PT}$ diverges for classical equations of state, and indicate how these results are modified in the case of nonclassical critical behavior. To begin with, we express $(\partial H/\partial x)_{PT}$ in terms of derivatives of the Helmholtz free energy $a(V, x, T) + RT[x\ln x + (1 - x)\ln(1 - x)]$. (We have to separate out the mixing term because it cannot be expanded at x=0). We denote derivatives with repeated subscripts, for instance

$$\text{entropy } S = -a_T \qquad \text{specific heat } C_V = -Ta_{TT}$$
$$\text{pressure } P = -a_V \qquad \text{compressibility } K_{Tx} = (Va_{VV})^{-1} \qquad \text{etc.} \tag{3}$$

Thus we find readily that

$$(\partial V/\partial x)_{TP} = -(\partial V/\partial P)_{xT}\,(\partial P/\partial x)_{VT} = -a_{Vx}/a_{VV} \tag{4}$$

and

$$(\partial H/\partial x)_{TP} = a_x - Ta_{Tx} - V\,a_{Vx} + T\,a_{VT}\,a_{Vx}/a_{VV} \tag{5}$$
$$= a_x - Ta_{Tx} + T(\partial P/\partial T)_{Vx}\,(\partial V/\partial x)_{TP}$$

For classical, van der Waals - like behavior, the free energy $a(V, x, T)$ can be expanded at the pure-solvent critical point (x=0). All coefficients in the expansion, such as a^c_x, a^c_{Tx}, a^c_{Vx}, a^c_{VT} are finite, but a^c_{VV}, $a^c_{VVV} = 0$. The coefficient a^c_{VT} equals the negative of the limiting slope of the vapor pressure curve, $dP/dT|^c_{CXC}$ and the coefficient a^c_{Vx} can be written in terms of the initial slope of the critical line as shown in ref.12:

$$a^c_{Vx} = -\,dP/dx|^c_{CRL} + dP/dT|^c_{CXC}\,dT/dx|^c_{CRL} \tag{6}$$

All unusual behavior of $(\partial V/\partial x)_{TP}$ and $(\partial H/\partial x)_{TP}$ orginates from the fact that

$a^c_{VV} = 0$. The mathematical behavior of these two derivatives is inferred by studying how $(a_{VV})^{-1}$, in Equations (4) and (5), diverges on various paths of approach. If the limit $x \to 0$ is taken first, $(Va_{VV})^{-1}$ will become equal to the compressibility of the pure solvent, and therefore both $(\partial H/\partial x)_{TP}$ and $(\partial V/\partial x)_{TP}$ will diverge strongly, that is as the solvent's compressibility, with coefficients that can be predicted from the slopes of the vapor pressure curve and the critical line at the critical point. It is the tail of this divergence of the initial slope that we observe in Figs. 4. If, however, the limit $x \to 0$ is not taken first, the behavior of $(a_{VV})^{-1}$ must be analyzed on the path in question. If we expand a_{VV}, we obtain

$$a_{VV} = 0 + a^c_{VVT}\ \delta T + a^c_{VVx}\ x + a^c_{VVVV}\ (\delta V)^2 \qquad (7)$$

with

$$\delta T = T - T_c,\ \delta V = V - V_c\ .$$

On the critical line, δT and δV are linear in x and $(a_{VV})^{-1}$ will diverge as x^{-1}. On the critical isotherm-isobar, $(\delta V)^3$ varies as x and $(a_{VV})^{-1}$ diverges as $x^{-2/3}$. As a consequence, on the critical line the partial molar volume and enthalpy of the solute diverge as x^{-1}, and the partials of the solvent, though remaining finite, do not approach the critical volume or enthalpy of the pure solvent. On the critical isotherm-isobar, the partial molar volume and enthalpy of the solute diverge as $x^{-2/3}$, while on this path the partials of the solvent do approach the critical values V_c, H_c of the pure solvent.

It is well known that fluids and fluid mixtures near their critical point do not behave as the classical model predicts, and it is therefore necessary to ask how the above results must be modified in order to accommodate nonclassical critical behavior. Many of the above conclusions are carried over to the nonclassical model with only minor, albeit essential modifications (ref. 12). Thus, if the limit $x \to 0$ is taken first, the partials again diverge strongly, as the solvent's compressibility, and with the same prefactor; this compressibility, however, diverges as $|\Delta T|^{-\gamma}$, with $\gamma = 1.24$, rather than with $\gamma = 1$ in the classical case. On the critical line, a_{Vx} and a_{VV} have weak zeroes; nevertheless, the partial molar volume and enthalpy remain finite for nonzero x, and they diverge as 1/x at the pure-solvent critical point. On the critical isotherm-isobar, the partials diverge as $1/x^{(1 - 1/\delta)} \sim 1/x^{0.8}$, with the critical exponent $\delta = 4.8$ instead of 3 in the classical case.

Since the divergence of these partial molar quantities is strong, it will be felt in a large range around the solvent's critical point. It therefore needs to be thoroughly grasped in cases where dilute near-critical mixtures are studied, such as in solubility near the lower critical endpoint.

THE EXPERIMENTAL EXCESS ENTHALPIES

Let us consider in some detail the H^E-x data obtained for the system CO_2 - toluene (Figs. 2, 5-8). Fig. 2 shows the location of the experimental isotherm-isobars with respect to the critical line. Fig. 5 shows a typical set of

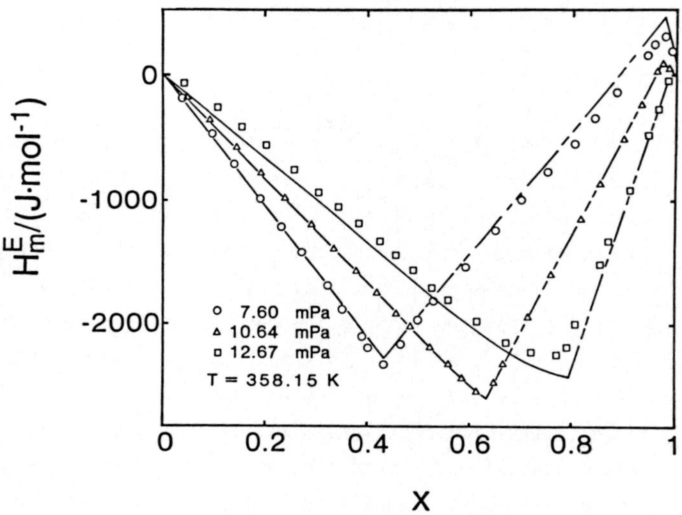

Fig. 5. Typical data set for three isobars of carbon dioxide and toluene at 358.15 K (ref.8).

experimental data: three different isobars at a fixed temperature. In Fig. 6-8 we have regrouped the data, so that five isotherms are collected at each pressure. We will first look at the H^E-x behavior on the 10.64 MPa isobar (Fig. 6). At 308.15 K, both components are in the liquid phase and the excess enthalpy is small. On the right hand side, however, H^E is already in the process of developing the curvature appropriate to the vicinity of an LCST, cf. Fig. 4a. At 358.15 K, the isotherm-isobar traverses the two-phase region. The incipient small upward slope of H^E, just before the steeply descending tie line develops near x=1, is evidence that this isotherm-isobar enters the two-phase region at negative $dT/dx|_{P,\sigma}$, i.e. on the other side of the right extremum in Fig. 4a

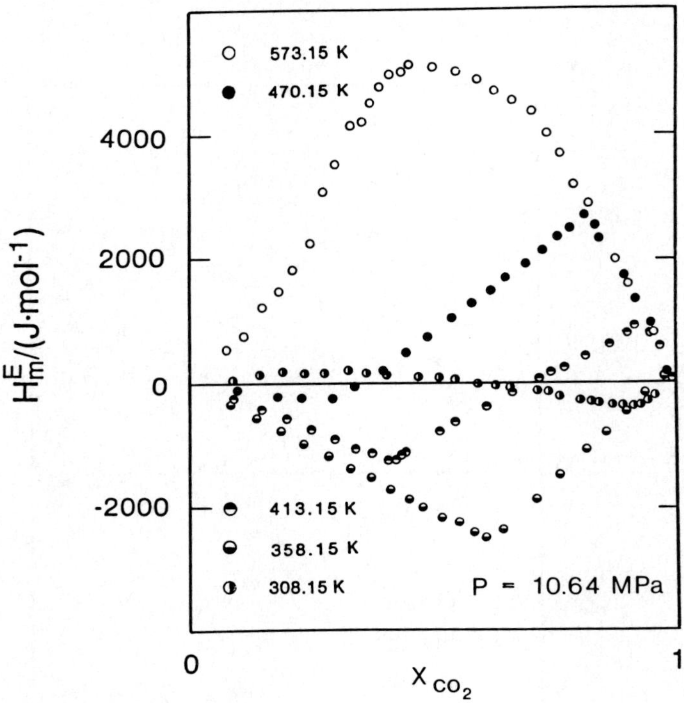

Fig. 6. Composite drawing of five H^E-x isotherms at the pressure of 10.64 MPa.

than the critical point. At 413.15 K and 470.15 K, the two-phase region is traversed in the middle of the x range, and displays the characteristic maximum, minimum and tie line. At 573.15 K, however, no two-phase region is entered. At this temperature and pressure, however, CO_2 is a supercritical fluid of fairly low density, while toluene is a liquid-like near-critical fluid. Thus, there is still a very large excess enthalpy mimicking a true phase transition. The curvature of the excess enthalpy plot is that appropriate to nearness to an UCST.

The isotherms at 12.67 MPa show similar features (Fig. 7). An exception is the 358.15 K isotherm, which has a slope opposite to that at 10.64 MPa near x=1. This isotherm-isobar enters the two-phase region without noticeable change of slope, which indicates that it enters near the right extremum in the

T-x loop of Fig. 3a, 4a.

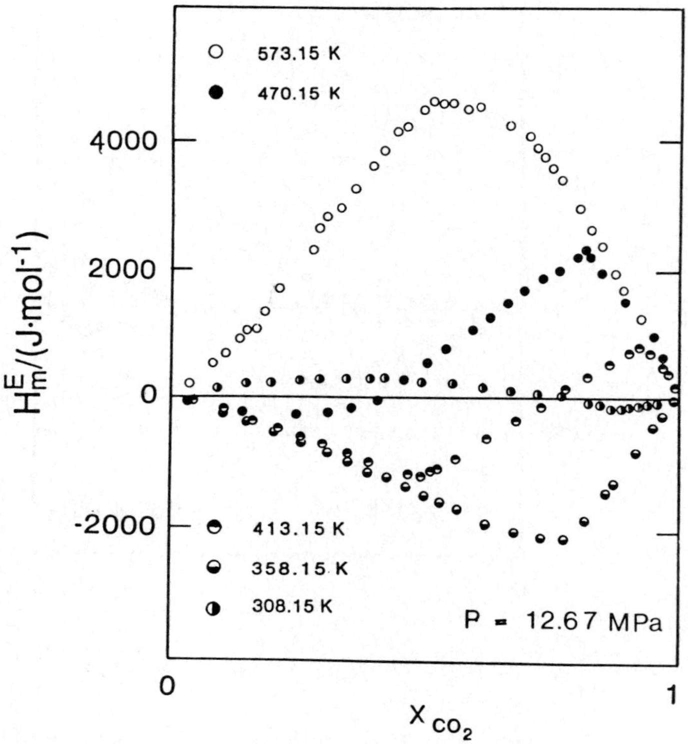

Fig. 7. Composite drawing of five H^E-x isotherms at the pressure of 12.67 MPa.

Let us now turn to the 7.60 MPa isobar, Fig. 8, which is slightly above the critical pressure of carbon dioxide (Fig. 2). At 308.15 K this isobar passes through the mixture critical point, showing the curvature appropriate to an LCST and no straight portion. At 358.15 K, 413.15 K and 470.15 K, the isobars are uneventful, traversing the two-phase region at intermediate x values and showing by the way they enter the two-phase region that they are in regions of negative $dT/dx|_{P,\sigma}$ on both sides. Finally, the one-phase 573.15 K isobar shows clear evidence of the enormous effect of low-density carbon dioxide mixed with liquid toluene. The convexity of this curve indicates nearness to an UCST. On

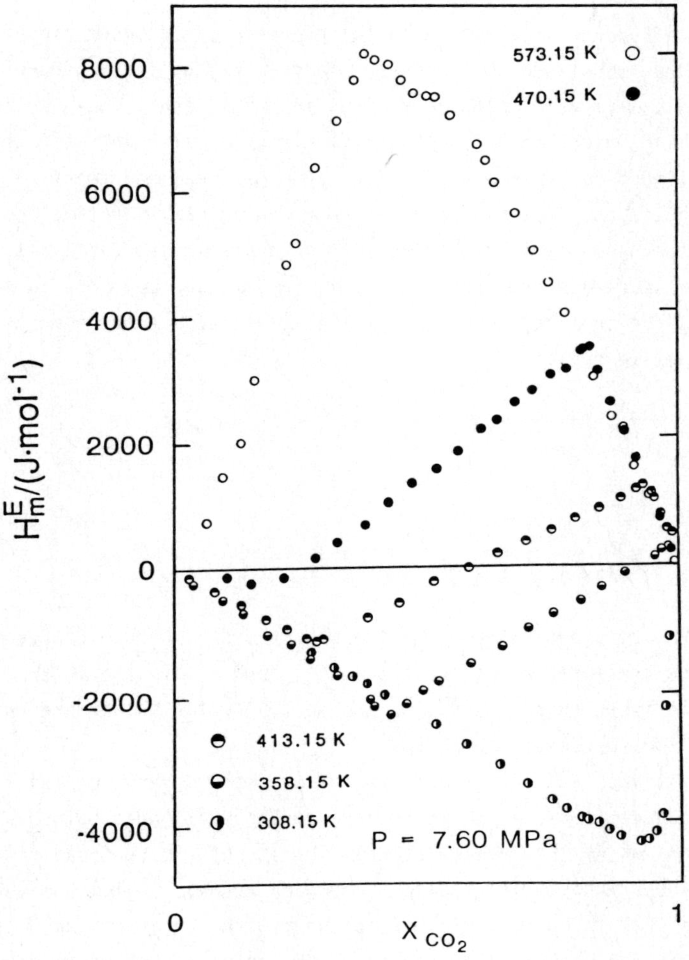

Fig. 8. Composite drawing of five H^E-x isotherms at the pressure of 7.60 MPa.

the toluene - rich side (near x=0) we also note the increasing steepness of the initial slope as the temperature increases. This initial slope diverges as the compressibility of pure toluene. In this case, where a volatile solute has been added, the divergence is to + ∞.

SUPERCRITICAL SOLUBILITY

Consider a supercritical fluid in the presence of an inert liquid or solid phase. The inert substance will dissolve to some extent in the supercritical phase and attain a certain mole fraction x_2 in this phase. This implies that the critical line which begins at the solvent's critical point is interrupted by the vapor-liquid-solid three-phase line at a critical end point. (It usually reemerges at a second critical end point near the solute's critical point). Near the critical end point, the solubility of the solute becomes a strong function of pressure and temperature. It is easy to see why this is so. By requiring that the chemical potential of the solute remain the same as that in the inert phase, we obtain

$$(\partial x_2/\partial P)_{T,\sigma} = \frac{V_s - \overline{V}_2}{(\partial^2 G/\partial x_2^2)_{TP}\ (1-x_2)} \tag{8}$$

and

$$(\partial x_2/\partial T)_{P,\sigma} = \frac{(\overline{H}_2 - H_s)/T}{(\partial^2 G/\partial x_2^2)_{TP}\ (1-x_2)} \tag{9}$$

These equations are again examples of the Gibbs-Konowalow rule [Appendix]. Here σ indicates the presence of the inert phase; V_s and H_s are the molar volume and enthalpy of the inert phase and $\overline{V}_2$, $\overline{H}_2$ the partial molar volume and enthalpy of the solute in the supercritical phase.

The derivatives $(\partial x_2/\partial P)_{T,\sigma}$ and $(\partial x_2/\partial T)_{P,\sigma}$ will diverge at the critical end points because $(\partial^2 G/\partial x^2)_{TP}$ goes to zero strongly at these points. Because $(\partial^2 G/\partial x^2)_{TP}$ behaves as the inverse compressibility of a pure fluid, it will, according to the principles of scaling, approach zero as $|T-T_c|^\varepsilon$ on the critical isobar, and as $|P-P_c|^\varepsilon$ on the critical isotherm with $\varepsilon = \gamma/\beta\delta$ and γ, β and δ the usual critical exponents characterizing the critical behavior of compressibility, coexistence curve and critical isotherm, respectively. For our classical model, ε=2/3, while fluids, which behave nonclassically, have $\varepsilon \sim 0.8$. Scaling also implies that on the critical isotherm

$$P-P_c \propto |x-x_c|^\delta \tag{10}$$

and on the critical isobar

$$T-T_c \propto |x-x_c|^\delta \tag{11}$$

with $\delta = 3$ for classical and $\delta = 4.8$ for the nonclassical behavior characteristic of fluids.

The above consideractions give the general scaling features of the solubility near a critical end point as far as they are caused by the divergence of $(\partial^2G/\partial x^2)^{-1}_{PT}$ in (8) or (9). The relevance of our preceding discussion of the behavior of the enthalpy of supercritical fluids to the problem of supercritcial solubility becomes clear when we consider the numerators in (8) and (9). We have concluded that the excess enthalpies, and therefore also the partials, may vary strongly with composition in the supercritical regime. These variations will distort the simple power-law dependence of (10) and (11). This distortion becomes acute in the case of low solubility, when the critical endpoint is close to the solvent's critical point, at which point, as we have seen, the partials diverge to $-\infty$. In the presence of an excess of inert phase, x is always finite and the infinity of the partials is never reached. In actual situations, however, such as the classic case of naphtalene in ethylene, the critical solubility may be less than 1 mol % and large negative partials are to be expected. Eqns. (8) and (9) imply that the enhancement of the solubility with increasing pressure or decreasing temperature, caused by the denominator approaching zero at the critical endpoint, is reinforced by the near-divergence of the numerator. Thus, in the general supercritical regime, and particularly in the case of low solubility at the critical endpoint, understanding the behavior of excess enthalpies and volumes is a key to understanding supercritical solubility.

CONCLUSIONS

For an understanding of supercritical solubility it is important to know the behavior of the partial molar volume and enthalpy of the solute. In the supercritical regime the excess enthalpies of a binary mixture are often large and vary strongly with composition. Thermodynamic stability considerations, however, dictate the convexity of the enthalpy and excess enthalpy versus composition in binary mixtures, and the manner in which phase boundaries are crossed. These considerations, and the need for $(\partial H/\partial x)_{PT}$ to diverge at the pure-component's critical points, explain all qualitative features of excess enthalpies such as those reported by Christensen et al. We have made this point for the specific example of the system carbon dioxide - toluene for which the critical line is known and the data set is extensive.

ACKNOWLEDGMENTS

The experimental work at Brigham Young University has been supported by the U.S. Department of Energy, Office of Basic Energy Sciences, Contract # DE-AC02-82ER13024, and by the donors of the Petroleum Research Fund administered by the American Chemical Society. Our attention was drawn to the key for interpreting the data, the Gibbs-Konowalow rule, by a reviewer of an earlier paper [12]. We received helpful comments and criticism from Prof. D. Zebolsky.

APPENDIX

The Gibbs-Konowalow rule relates changes of temperature, pressure and composition of coexisting phases to the slope of the phase boundary. Derivations are given in many textbooks (ref.10, 14). Here, we present a general argument, inspired by an observation of Griffiths in an Appendix to ref.15.

Consider a function C(X, Y) which is continuous across a phase boundary, but whose derivatives are not. The boundary curve separates region I and II, each of which, in principle, might be a multiphase region. Along this boundary curve, we have

$$dC^{I} = \left(\frac{\partial C}{\partial X}\right)^{I} dX + \left(\frac{\partial C}{\partial Y}\right)^{I} dY$$

and (A1)

$$dC^{II} = \left(\frac{\partial C}{\partial X}\right)^{II} dX + \left(\frac{\partial C}{\partial Y}\right)^{II} dY$$

Since C is continuous, we have $dC^{I} = dC^{II}$ and therefore

$$\delta \left(\frac{\partial C}{\partial X}\right) = - \frac{dY}{dX}\bigg|_{\sigma} \delta \left(\frac{\partial C}{\partial Y}\right) \qquad (A2)$$

where σ denotes the boundary curve and δ the difference between derivatives in region I and in region II at the boundary curve.

As an example of the usefulness of (A2), consider for a one-component fluid $\mu(P, T)$ as the function C and the vapor pressure curve P(T) as the boundary curve. Then

$$\delta \left(\frac{\partial \mu}{\partial P}\right)_{T} = - \frac{dT}{dP}\bigg|_{\sigma} \delta \left(\frac{\partial \mu}{\partial T}\right)_{P} . \qquad (A3)$$

Since $(\partial\mu/\partial P)_T = V$ and $(\partial\mu/\partial T)_P = -S$, we obtain Clapeyron's equation, $(dP/dT)_{\sigma} = \Delta S/\Delta V$.

We will now prove the Gibbs-Konowalow rule, Eq. (2) of the paper. To this end, we take for C(X,Y) the function $\frac{\mu_1}{T}(T, x_2)$ at constant pressure, and as the boundary curve the dew-bubble curve in T-x space where I is a one-phase, II a two-phase region. Applying (A2) we obtain

$$\delta \left(\frac{\partial\mu_1/T}{\partial T}\right)_{Px} = - \left.\frac{dx}{dT}\right|_{P,\sigma} \delta \left(\frac{\partial\mu_1/T}{\partial x}\right)_{TP} \quad \text{(A4)}$$

Since

$$\frac{\partial\mu_1/T}{\partial T} = - \frac{\bar{H}_1}{T^2}$$

and (A5)

$$\frac{\partial\mu_1/T}{\partial x_2} = - \frac{x_2}{T} \left(\frac{\partial^2 G}{\partial x^2}\right)_{TP} ,$$

we immediately derive that

$$\left.\frac{dx_2}{dT}\right|_{P,\sigma} = - \frac{\bar{H}_1^I - \bar{H}_1^{II}}{T\, x_2\, (\partial^2 G/\partial x_2^2)_{TP}^I} \quad . \quad \text{(A6)}$$

because $(\partial^2 G/\partial x_2^2)_{TP} = 0$ in the two-phase region. Since $\bar{H}_1 = H - x_2\,(\partial H/\partial x_2)_{TP}$ and x_2, H are continuous across the phase boundary, we find, with $x_2 = x$, Eq. (2), namely

$$\left.\frac{dT}{dx}\right|_{P,\sigma} = - \frac{T\,(\partial^2 G/\partial x^2)_{TP}^I}{\Delta H/\Delta x - (\partial H/\partial x)_{TP}} \quad . \quad \text{(A7)}$$

We have made use of the fact that in the two-phase region $(\partial H/\partial x)_{PT} = \Delta H/\Delta x$, the slope of the tie line. By choosing for C(X, Y) the function $\mu_1(P, x_2)$ and for the boundary curve the P-x dew-bubble curve at constant temperature we derive the expression analogous to (A7) for $dP/dx|_{T,\sigma}$:

$$\left.\frac{dP}{dx}\right|_{T,\sigma} = \frac{(\partial^2 G/\partial x^2)_{PT}^I}{\Delta V/\Delta x - (\partial V/\partial x)_{TP}} \quad , \quad \text{(A8)}$$

We now derive Eq. (9) by applying Eq. (A1) to the case of the supercritical solvent, component 1, in the presence of an inert solid phase, component 2. We consider μ_2 (T, x_2) at constant pressure as the continuous function and the boundary curve that at which the solid phase first appears. Then

$$\delta \left(\frac{\partial\mu_2}{\partial T}\right)_{Px} = - \left.\frac{dx_2}{dT}\right|_{P,\sigma} \delta \left(\frac{\partial\mu_2}{\partial x_2}\right)_{TP} \quad \text{(A9)}$$

and since $(\partial\mu_2/\partial x_2)_{PT} = (1 - x_2)\,(\partial^2G/\partial x^2)_{PT}$, we derive, in analogy with (A3) - (A6)

$$\left.\frac{dx_2}{dT}\right|_{P,\sigma} = \frac{\bar{H}_2^{I} - \bar{H}_2^{II}}{T(1-x_2)(\partial^2G/\partial x_2^2)_{TP}^{I}} \qquad \text{(A10)}$$

Since the solid phase is presumed inert, $\bar{H}_2^{II} = H_s$, the molar enthalpy of the solid, so that

$$\left.\frac{dx_2}{dT}\right|_{P,\sigma} = \frac{\bar{H}_2^{I} - H_s}{T(1-x_2)(\partial^2G/\partial x_2^2)_{TP}^{I}} \qquad \text{(A11)}$$

which is Eq. (9). Eq. (8) is derived in the same way, starting with μ_2 (P,x) at the function C(X,Y) and with the P-x dew-bubble curve at constant T as the boundary curve.

The same device that we used to derive the Gibbs-Konowalow rule was used by Ehrenfest (ref.16) to derive the relations, named after him, between jumps in second derivatives along the critical line. Take, for instance, the volume V(P, T) along the dew-bubble curve in P-T space at constant x.

$$\delta\left(\frac{\partial V}{\partial T}\right)_{Px} = -\left.\frac{dP}{dT}\right|_{x,\sigma}\,\delta\left(\frac{\partial V}{\partial P}\right)_{Tx} \;. \qquad \text{(A12)}$$

Since $(\partial V/\partial T)_{Px} = V\alpha_{Px}$, the expansion coefficient, and $(\partial V/\partial P)_{Tx} = -VK_{Tx}$, the compressibility, we have

$$\left.\frac{dP}{dT}\right|_{x,\sigma} = \frac{\alpha_{Px}^{II} - \alpha_{Px}^{I}}{K_{Tx}^{II} - K_{Tx}^{I}} \;. \qquad \text{(A13)}$$

In the P-T plane, however, the dew-bubble curve and the critical line are tangent at the critical point, so that

$$\left.\frac{dP}{dT}\right|_{CRL} = \frac{\delta\,\alpha_{Px}}{\delta\,K_{Tx}} \;, \qquad \text{(A14)}$$

one of the Ehrenfest relations (ref.16).

REFERENCES

1 C. Pando, J.A.R. Renuncio, T.A. McFall, R.M. Izatt and J.J. Christensen, J. Chem. Thermodynamics, 15 (1983), 173-180.

2 C. Pando, J.A.R. Renuncio, R.M. Izatt and J.J. Christensen, J. Chem. Thermodynamics, 15 (1983) 231-235.
3 C. Pando, J.A.R. Renuncio, R.M. Izatt and J.J. Christensen, J. Chem. Thermodynamics, 15 (1983) 259-266.
4 C. Pando, J.A.R. Renuncio, R.S. Schofield, R.M. Izatt and J.J. Christensen, J. Chem. Thermodynamics, 15 (1983) 747-755.
5 J.J. Christensen, S.P. Christensen, R.S. Schofield, P.W. Faux, P.R. Harding and R.M. Izatt, J. Chem. Thermodynamics, 15 (1983) 115.
6 J.J. Christensen, S.P. Christensen, R.S. Schofield, P.W. Faux, P.R. Harding, and R.M. Izatt, J. Chem. Thermodynamics, 16 (1984) 249-255.
7 J.J. Christensen, T.A.C. Walker, R.S. Schofield, P.W. Faux, P.R. Harding and R.M. Izatt, J. Chem. Thermodynamics, 16 (1984) 445-453.
8 J.J. Christensen, Private Communication, April (1984).
9 G. Morrison, J. Chem. Phys., 78 (1983) 4790. See also J.L. Copp and D.H. Everett, Trans. Faraday Soc., 53, (1975) 9-18, for discussion of stability conditions.
10 J.S. Rowlinson and F.L. Swinton, "Liquids and Liquid Mixtures", Third Edition, Ch. 4, Butterworth Scientific, London 1982.
11 A. Williamson, "An Introduction to Non-Electrolyte Solutions", Wiley, New York, 1960, pp. 89-93.
12 R.F. Chang, G. Morrison and J.M.H. Levelt Sengers, J. Phys. Chem. Letter, 88 (1984) 3389-3391.
13 J.M.H. Levelt Sengers, G. Morrison and R.F. Chang, Fluid Phase Equilibria 14 (1983) 19-44.
14 J.D. v.d. Waals and P.H. Kohnstamm, "Lehrbuch der Thermodynamik" Vol. 2 Ch. 4. Ed. Leipzig, J.A. Barth 1912.
15 T. Doiron, R.P. Behringer and H. Meyer, J. Low Temp. Phys., 24 (1975) pp. 345-363.
16 P. Ehrenfest Proc. Acad. Sci. Amsterdam, 36 (1933) 153-157.

Supercritical Fluid Technology, **edited by J.M.L. Penninger, M. Radosz, M.A. McHugh and V.J. Krukonis**
1985 Elsevier Science Publishers B.V., Amsterdam — Printed in The Netherlands

THEORY OF THE PRESSURE EFFECT IN DENSE GAS EXTRACTION

S. KIM, J. M. WONG, and K. P. JOHNSTON
Department of Chemical Engineering, University of Texas, Austin, Texas 78712

ABSTRACT

Several general aspects of the pressure effect on supercritical fluid phase behavior are identified and characterized quantitatively for dilute solutions. These result from the critical behavior of the solvent and are independent of the strength of the solute-solvent interaction. Solubility versus pressure isotherms, which depend upon specific as well as general aspects of the pressure effect, are predicted for nonpolar systems using no adjustable parameters. A "chemical effect" is superimposed on the "pressure effect" by doping CO_2 with a small amount of SO_2.

INTRODUCTION

In dense gas or supercritical fluid extraction (SFE), the yield and the selectivity are often extremely sensitive to pressure. This pressure effect is a basis for emerging and existing SFE applications that include food and pharmaceutical processing, specialty chemicals separation, heavy oils processing, polymer processing, environmental control, and synfuels production (refs. 1, 2, 3).

The development of SFE processes has been hindered by the limited understanding of the phase behavior. There are two kinds of phenomena which must be understood in order to predict the phase behavior relevant to SFE. The general type is a result of the proximity to the critical point of the solvent and is unaffected by the properties of the solute. The specific type depends upon the particular force constants of both the solute and solvent. In the section , "Universal behavior in dilute supercritical fluid solutions," the general kinds of phenomena are described in terms of k_T, the isothermal compressibility of the solvent. The next section, "Prediction of solubility versus pressure isotherms of nonpolar systems," describes a method to calculate solute-solvent interaction energies from the molecular volumes of the components. In the final section, "Effects of co-solvents in supercritical fluid extraction," a "chemical effect" is superimposed on the "pressure effect" to improve several important properties of the solvent.

UNIVERSAL BEHAVIOR IN DILUTE SUPERCRITICAL FLUID SOLUTIONS

The partial molar volume is the key thermodynamic property that describes the pressure effect on supercritical fluid phase behavior. For example, the fugacity coefficient ϕ, which is used to calculate the solubility of solids in supercritical solvents, is actually a pressure integral of the partial molar volume. The partial molar volume is a differential quantity that describes the solution behavior at a particular pressure, and is therefore useful for the development and testing of models for the supercritical phase. It can be used to determine maxima and minima in the solubility versus pressure isotherms (ref. 4), as well as a region of practical interest where the solubility is extremely sensitive to pressure. Unfortunately, the physical understanding of $\bar{v}_2$ has been poor, primarily because of the lack of data.

Eckert et al. (ref. 5) developed a novel experimental technique to obtain $\bar{v}_2$ for a solid in a supercritical fluid within several degrees of T_{C1}, e.g., naphthalene in carbon dioxide at 35°C. These data are interpreted below using classical critical theory. The physical behavior of $\bar{v}_2$ is related to a wider class of critical phenomena. The objective of this section is to identify and to quantify various general aspects of the pressure effect.

Criticality and the effect of pressure on solubility

Solubility maxima and minima. The thermodynamic analysis of the pressure effect on the solubility can be obtained along the equilibrium line as shown by Paulaitis and co-workers (ref. 6)

$$\left(\frac{\partial \ln y_2}{\partial P}\right)_{T,\sigma} \left[1 + \frac{y_2}{RT} \int_0^P \left(\frac{\partial \bar{v}_2}{\partial y_2}\right)_{T,P} dP\right] = \frac{v_2^S - \bar{v}_2}{RT} . \tag{1}$$

An equivalent expression is

$$\left(\frac{\partial y_2}{\partial P}\right)_{T,\sigma} = \frac{v^S - \bar{v}_2}{\left(\frac{\partial \mu_2^G}{\partial y_2}\right)_{T,P}} \tag{2}$$

These relationships describe one of the most important relationships in SFE, i.e. the effect of P on y_2.

At pressures which are significantly lower than the critical pressure, a dilute solution obeys Henry's law and

$$\left(\frac{\partial \mu_2^G}{\partial y_2}\right)_{T,P} = RT/y_2 \tag{3}$$

Substitution of Eq. 3 into Eq. 2 yields

$$(\partial \ln y_2/\partial P)_{T,\sigma} = \frac{v_2^S - \bar{v}_2}{RT} \tag{4}$$

which is equivalent to the expression used by van Wasen and Schneider (ref. 7).

This expression may also be derived by assuming that $\bar{v}_2$ or ϕ_2 is not a function of composition.

At low pressures, $\bar{v}_2 >> v_2^S$ and the solubility decreases with pressure. Furthermore, the pressure derivative of Eq. 4 indicates that the concavity is positive since $(\partial\bar{v}_2/\partial P)_T$ is more negative than $(\partial v_2^S/\partial P)_T$. The absolute value of the latter is small since the solid is essentially incompressible. This region where the solubility decreases is found at a pressure that is lower than the range in Figure 1. The minimum in the solubility occurs at a

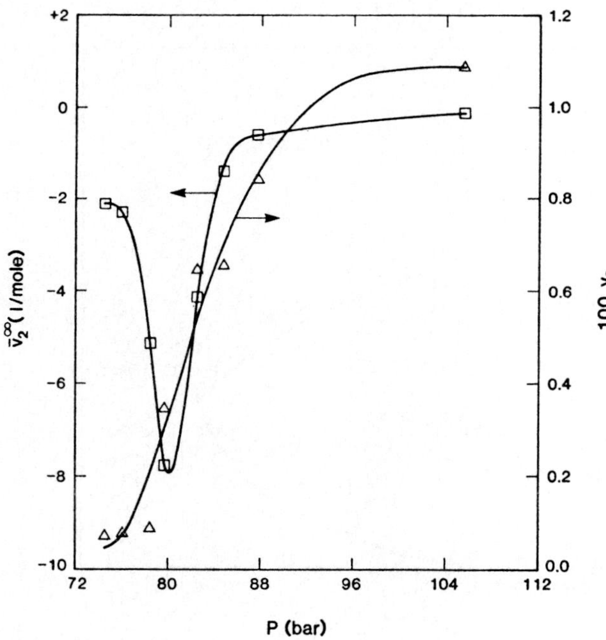

Fig. 1. Solubility and $\bar{v}_2^\infty$ versus pressure for naphthalene in carbon dioxide, T = 35°C, (Solubility data - ref. 20; $\bar{v}_2$ data - ref. 5). The lines are drawn through the data for clarity.

pressure below P_{c1} where $\bar{v}_2 = v_2^S$. Consider the other extreme. At a pressure which is significantly above P_{C1}, $\bar{v}_2$ becomes greater than v_2^S due to repulsive forces and thus the solubility decreases. At the pressure where $\bar{v}_2$ equals v_2^S the solubility is a maximum. The location of this point depends upon the specific chemical properties of the components. This assumes that the denominator in Eq. 2 does not vanish at these points, which was verified by Kurnik and Reid (ref. 4) based on the Peng-Robinson equation of state. The region near the upper critical end point is an exception to this prediction as discussed below.

Complex behavior near the critical end points. In the critical region, the behavior becomes more complex for two reasons:

1) $\bar{v}_2$ is extremely sensitive to pressure, and
2) $(\partial\mu_2^G/\partial y_2)_{T,P}$ vanishes along the critical line.

A P-T projection of the P-T-X diagram is shown in Figure 2 for systems where the triple point of the less volatile component exceeds the critical temperature of the solvent. In this case, the critical line, which connects the pure component critical points, is intersected twice by the three phase solid-liquid-gas(SLG) line. These points of intersection are called the lower critical end point (LCEP) and the upper critical end point (UCEP).

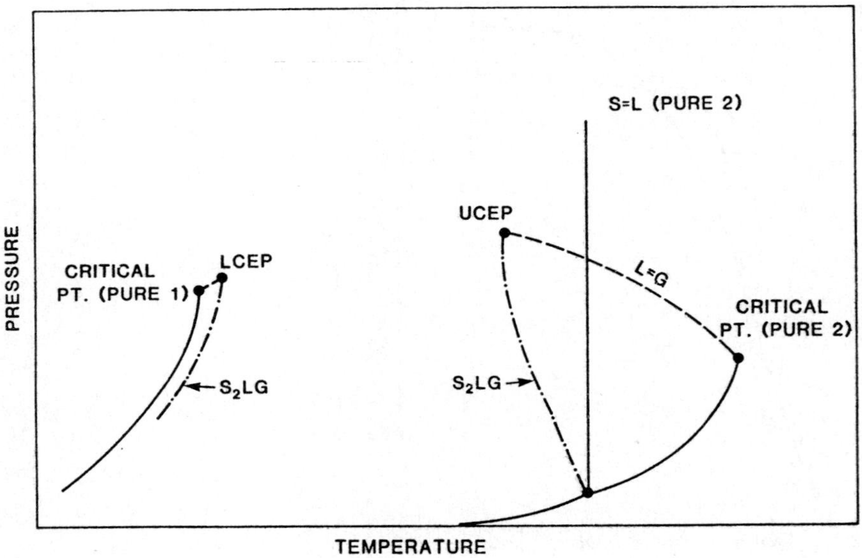

Fig. 2. Pressure-temperature diagram: S_2LG line intersects critical line for systems of the type naphthalene-carbon dioxide.

For systems such as naphthalene, anthracene, m-dinitrobenzene, hexaethylbenzene, hexachloroethane and stilbene in ethylene, the LCEP is close to the critical temperature of ethylene. The quantity $(T_{LCEP} - T_{c1})/T_{c1}$ was less than 0.01 for all of these systems. The proximity to T_{c1} has an interesting effect on the LCEP which is not found for the UCEP.

The $\bar{v}_2^\infty$ is related to the isothermal compressibility by

$$\bar{v}_2^\infty = -(\partial V/\partial P)_{T,n}(\partial P/\partial n_2)_{T,V}^\infty \quad . \tag{5}$$

At infinite dilution, the isothermal compressibility of the mixture approaches that of the pure solvent which diverges at P_{c1}. The factor $(\partial P/\partial n_2)_{T,V}$ does not approach zero in the near critical region as demonstrated experimentally by Ziger (ref. 8) and conceptually by Gitterman and Procaccia (ref. 9). Substitution of Eq. 5 into Eq. 2 gives

$$\left(\frac{\partial y_2}{\partial P}\right)_{T,\sigma} = \frac{v_2^S + (\partial V/\partial P)_{T,n}\,(\partial P/\partial n_2)_{T,V}}{(\partial \mu_2^G/\partial y_2)_{T,P}} \quad . \tag{6}$$

The analysis will be restricted to a solid condensed phase, where there are only two points on the critical locus, i.e. the LCEP and UCEP. At the LCEP, $(\partial y_2/\partial P)_{T,\sigma}$ can become quite large for two reasons, namely k_T becomes large and $(\partial \mu_2^G/\partial y_2)_{T,P}$ vanishes. Thus SFE is an interesting example in which the associated thermodynamics exhibit a strong anomaly which is a result of the presence of a solid phase of relatively constant chemical potential. At the UCEP, $\bar{v}_2$ is not large, and the sensitivity of y_2 to pressure is only due to $(\partial \mu_2^G/\partial y_2)_{T,P}$. The UCEP is described by Paulaitis et al. (ref. 1) using a P versus y_2 diagram, which has characteristics that are similar to a μ_2^G versus y_2 diagram.

The solubility is infinitely sensitive to pressure at both the LCEP and UCEP. There are interesting opportunities for supercritical fluid extraction in these regions (ref. 10). The location of the UCEP depends upon the specific chemical nature of the solute.

Extrema in $\partial y_2/\partial P$ and $\bar{v}_2$ near the critical isochore of the solvent. This section identifies the extrema in $\partial y_2/\partial P$ that occur at temperatures that lie between the LCEP and UCEP. This point where the effect of pressure on solubility is a maximum or the second derivative equals zero is general in that it does not depend upon the properties of the solute. The derivative of Eq. 6 with respect to pressure at constant temperature gives

$$\left(\frac{\partial^2 \ln y_2}{\partial P^2}\right)_{T,\sigma} = - \frac{\left(\frac{\partial \bar{v}_2}{\partial P}\right)_{T,\sigma} + \left(\frac{\partial \ln y_2}{\partial P}\right)_{T,\sigma}\left[\frac{\partial}{\partial \ln y_2}\left(\frac{\partial \mu_2^G}{\partial P}\right)_{T,\sigma}\right]_{T,P}}{(\partial \mu_2^G/\partial \ln y_2)_{T,P}} \tag{7}$$

where it is assumed that the solid is incompressible. This equation will be used to demonstrate that a maximum in the slope of the solubility occurs at the same pressure where $\bar{v}_2$ is a minimum. At this pressure, the second term on the right-hand side vanishes for the following reasons. At saturation,

$$\mu_2^G = \mu_2^S \tag{8}$$

and the derivative $(\partial \mu_2^G/\partial P)_{T,\sigma}$ is equal to

$$(\partial\mu_2{}^S/\partial P)_{T,\sigma} = v_2{}^S \tag{9}$$

for a pure solid phase. Since v_2^S is not a function of composition, this term vanishes. Finally, the denominator on the right-hand side does not vanish, except on the critical line.

Now that the maximum in the slope of the solubility has been related to $(\partial\bar{v}_2/\partial P)_T$, the next step is to express the latter in terms of $(\partial k_T/\partial P)_T$ for the case of infinite dilution. The data reported by Ziger (ref. 8) show that $(\partial P/\partial n_2)^\infty$ is fairly constant for a reduced density range of 0.8 to 1.8. Differentiation of Eq. 5 gives an approximate relationship for this region

$$\left(\frac{\partial\bar{v}_2{}^\infty}{\partial P}\right)_T = -\frac{\partial}{\partial P}\left(\frac{\partial V}{\partial P}\right)_{T,n}\cdot\left(\frac{\partial P}{\partial n_2}\right)_{T,V}^\infty + (\partial V/\partial P)_{T,n}\cdot\frac{\partial}{\partial P}\left(\frac{\partial P}{\partial n_2}\right)_{T,V}^\infty \tag{10}$$

where the second term is insignificant compared to the first one since $\partial P/\partial n_2$ is a weak function of pressure. Therefore, $\bar{v}_2{}^\infty$ reaches a minimum at the point where the isothermal compressibility of the pure fluid reaches a maximum, regardless of the solute (see Fig. 3).

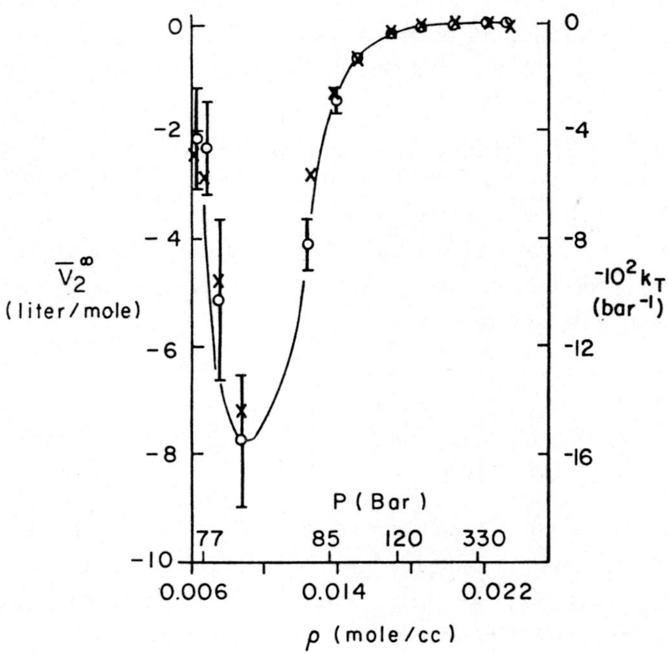

Fig. 3. $\bar{v}_2{}^\infty$ for naphthalene in carbon dioxide cross plotted against the isothermal compressibility of pure carbon dioxide, T = 35°C (Data of Eckert et al., ref. 5) (o: $\bar{v}_2$; x: k_T)

The relationship between $\bar{v}_2$ and $(\partial v/\partial P)_T$ may be explained physically. The introduction of a solute molecule into a reservoir containing a highly compressible fluid condenses the solvent in a manner which is similar to the effect of the external pressure. This presumes that the unlike pair forces are attractive. This may be explained as a two-step process. At constant temperature and volume, the addition of a solute molecule causes a decrease in the pressure because of the attractive forces, and $n(\partial P/\partial n_2)_{T,V,n_1}$ is negative. Next, the pressure is restored to the original pressure. This requires a pronounced decrease in the volume since the compressibility is large. Therefore, the solute functions as a test molecule which probes the nature of the compressibility of the solvent. Wheeler (ref. 11) used a decorated lattice gas model to predict that $\bar{v}_2^{\infty}$ diverges to negative infinity at the critical point of the solvent. If the solute-solvent interaction is repulsive, $\bar{v}_2^{\infty}$ diverges to positive infinity, although this has not been demonstrated experimentally.

It is well known that the pure fluid vapor pressure curve may be extended linearly beyond T_c along the critical isochore. This line is defined by

$$\Delta P^* = B\Delta T^* \tag{11}$$

where

$$\Delta P^* = P/P_c - 1 \text{ and } \Delta T^* = T/T_c - 1. \tag{12}$$

The experimental constant B is 7.3 for SF_6 (ref. 12), 7.4 for CO_2 (ref. 13), and 6.6 for ethylene (ref. 14). Furthermore, it is known that the maximum value of k_T is located on the critical isochore for a given temperature above T_c (e.g., ref. 14). Table 1 lists the values of the slope for dimensionless plots of ΔP^* versus ΔT^* for several different properties.

TABLE 1

Universal behavior on the ΔP^* versus ΔT^* locus

Property	Slope = $\Delta P^*/\Delta T^*$
1. critical isochore of pure ethylene	6.6
2. critical isochore of pure carbon dioxide	7.4
3. maximum in k_T of pure ethylene	5.7
4. minimum in $\bar{v}_2^{\infty}$ for naphthalene in ethylene or in carbon dioxide	5.5
5. minimum in $\bar{v}_2^{sat}$ for naphthalene in ethylene or in carbon dioxide	5.7
6. maximum in $(\partial y_2/\partial P)_T$ for naphthalene in ethylene, in carbon dioxide, and diphenylamine in carbon dioxide (ref. 9)	7.6

Notice that the slope of ΔP^* versus ΔT^* is about 15% smaller for the maximum in the compressibility than for the critical isochore for ethylene. This is due to the experimental uncertainty in k_T. The slope is about the same for the maximum in the compressibility of ethylene and the minimum in the partial molar volume of naphthalene in ethylene, respectively (see Fig. 4). The slope in ΔP^*

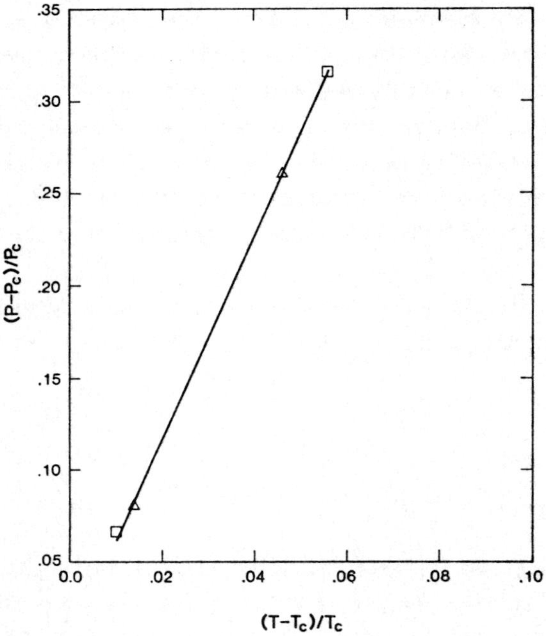

Fig. 4. ΔP^* versus ΔT^* locus of points where $\bar{v}_2^{\infty}$ is a minimum (naphthalene in carbon dioxide: △; naphthalene in ethylene: □)

versus ΔT^* is 7.6 for $(\partial^2 y/\partial P^2)_{T,\sigma} = 0$ as reported by Gitterman and Procaccia (ref. 9). The difference between this value and the one for the maximum in the compressibility was unexpected. One cause may be the large range for the former as it extends to the UCEP. For example, the slope appears to be less than 7.6 for naphthalene in ethylene for the two points where $(T-T_c)/T_c$ is less than 0.06. Furthermore, the second derivative in the y_2 versus P data cannot be located accurately, especially when the difference of two similar numbers is involved, i.e., $T-T_c$. It appears that the slope in ΔP^* versus ΔT^* where $(\partial^2/\partial P^2)_{T,\sigma}$ is zero should be constant for a variety of solids in a fluid near the LCEP; however, it will vary perhaps 25% for different solvents as does the value of B. At the UCEP and LCEP, a separate consideration is involved, i.e. $(\partial \mu_2{}^G/\partial y_2)_{T,\sigma}$ vanishes.

Classical Critical Phenomena and the Partial Molar Volume

The apparently complex behavior of $\bar{v}_2^\infty$ (e.g., see Figure 1) can be represented in a simple manner using classical critical theory. At infinite dilution, $\bar{v}_2^\infty$ and vk_T are related by the factor $n(\partial P/\partial n_2)_{T,V}^\infty$ as shown in Eq. 5. In the near critical region where ρ_r is between 0.8 and 1.8, the experimental value of $n(\partial P/\partial n_2)_{T,V}^\infty$ is relatively constant. In the limit of infinite dilution, it becomes a constant at the critical point independently of path (refs. 12, 15). As a result, the critical relationships which are applicable to the compressibility can also be used for $\bar{v}_2^\infty$. The remainder of this section discusses these relationships which are summarized in Table 2. All of the exponents were obtained using a linear least squares regression on a double logarithmic relationship.

TABLE 2

Critical exponents for the partial molar volume relations

	Relation	Path	System	Temp. (°C)	Exponent
1	$\bar{v}_2^\infty \propto \lvert P-P_o\rvert^{-\varepsilon}$	T_{c1}	Naphthalene-Carbon dioxide	35 45	0.9 1.0
			Naphthalene-Ethylene	12 25	0.8 1.0
2	$\bar{v}_2^\infty \propto \lvert T-T_c\rvert^{-\gamma}$	ρ_{c1}	Naphthalene-CO_2		1.0
			Naphthalene-Ethylene		0.9
3	$\bar{v}_2^\infty \propto \lvert P-P_c\rvert^{-\gamma}$	ρ_{c1}	Naphthalene-CO_2		1.2
			Naphthalene-Ethylene		0.9
4	$\bar{v}_2 \propto x^{-1}$	crit. locus	n-Heptane-Ethane		1.00
5	$\bar{v}_2 \propto x^{-2/3}$	$T_{c1}-P_{c1}$	not available		

1. Classical exponents: $\gamma=1$, $\delta=3$, $\varepsilon=2/3$
2. Nonclassical exponents: $\gamma=1.24$, $\delta=4.8$
3. Relations 1-3 are based on the analogous ones for $(\partial v/\partial P)_T$

The Helmholtz free energy, the pressure, and the chemical potential can be expanded at the critical point as a Taylor series in density and temperature, e.g., as seen in the van der Waals equation of state. The expansion may be scaled (refs. 16, 17) to yield

$$\Delta P^* = P_{30}(\Delta\rho^*)^3 \{1 + (P_{11}/P_{30}) \Delta T^* /(\Delta\rho^*)^2\} \tag{13}$$

where the P_{ij}'s are constants related to derivatives at the critical point. In this scaled equation, $\Delta P^*/(\Delta\rho^*)^3$ is a function of only one variable, namely $\Delta T^*/(\Delta\rho^*)^2$ or the scaled temperature. Several asymptotic power laws may be derived for k_T, $(\partial\rho/\partial P)_T$ and $(\partial V/\partial P)_T$. On the critical isochore where $\Delta\rho^* = 0$, differentiation of Eq. 13 yields

$$(\partial\rho/\partial P)_T \propto |\Delta T^*|^{-\gamma} \tag{14}$$

with γ=1 which is a strong divergence.
Alternatively,

$$k_T \propto |P - P_c|^{-\gamma} \tag{15}$$

on the critical isochore, and

$$k_T \propto |P - P_c|^{-\varepsilon} \tag{16}$$

on the critical isotherm as discussed by Griffiths and Wheeler (ref. 18). Here $\varepsilon = 1 - 1/\delta$ is equal to 2/3 for a van der Waals gas.

In Figure 5, an approximate correlation is presented for log $|\bar{v}_2^\infty|$ for naphthalene in carbon dioxide at 35°C versus log $|P-P_o|$. For comparison, log $|\partial V/\partial P|_T$ of the pure solvent is also plotted. The reference pressure P_o is chosen where k_T is a maximum and $\bar{v}_2^\infty$ is a minimum. For the system naphthalene-CO_2 at 35°C, this pressure is 79.7 bar. The experimental exponent (see Eq. 16) is 0.9 for $\bar{v}_2^\infty$ compared with the predicted value of 2/3 (see Table 2). This exponent is empirical since T_r=1.01, and the classical expression is strictly valid only for the asymptotic condition $T=T_c$. At T_r=1, $(\partial V/\partial P)_T$ is totally symmetric with respect to $P-P_c$. For T_r>1 we choose to replace P_c with P_o, and $(\partial V/\partial P)_T$ is not fully symmetric with respect to $|P-P_o|$. This asymmetry is evident in the supercritical isotherms for a pure fluid. For this reason, the data points at log $|P-P_o|$ = 0.57 and 0.71 (76.0 and 74.6 bar) are above the line in Figure 5. The linearity in Figure 5 is sensitive to the location of P_o, particularly for small values of the abscissa. Although this is an approximate approach, the linear representation of the data is simplified compared to Figure 1. A complex mixture property, $\bar{v}_2^\infty$, has been related to $(\partial V/\partial P)_T$ of a pure fluid.

The effect of temperature on $\bar{v}_2$ along the critical isochore, where the pressure effect is a maximum, is shown in Table 2. The $\bar{v}_2^\infty$ versus $T-T_c$ plot gives γ=0.97 compared with the critical value of unity. The effect of pressure on $\bar{v}_2$ is analogous to that of temperature along the critical isochore which is evident from Eq. 11 (see Table 2).

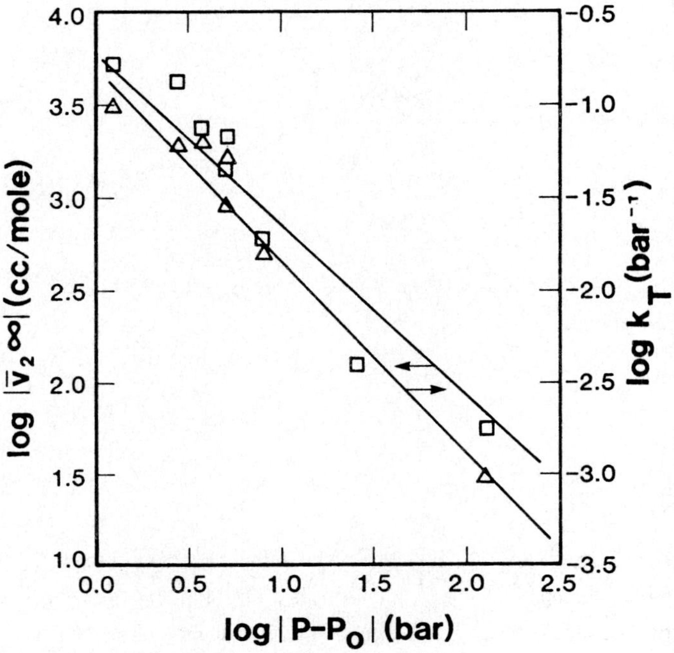

Fig. 5. $\bar{v}_2^\infty$ and k_T versus $P_\infty - P_0$ at 35C for the system naphthalene-CO_2 and pure CO_2, respectively. (□: $\bar{v}_2^\infty$; △: k_T)

The relationship between $\bar{v}_2$ and y_2 may be obtained along the critical locus or on the critical isotherm/isobar using an expansion in the Helmholtz free energy (ref. 17). Expansion about V_c, T_c and x=0 ($x \equiv y_2$) gives

$$A = A^C + [A_V^C + A_{VT}(\delta T)]\,\delta V + [A_X^C + A_{XT}(\delta T)]\,x$$
$$+ \tfrac{1}{2}[A_{VV}^C + A_{VVT}^C \delta T](\delta V)^2 + [A_{VX}^C + A_{VXT}^C\,\delta T](\delta V)x + \tfrac{1}{2}[A_{XX}^C + A_{XXT}^C\,\delta T]x^2$$
$$+ \tfrac{1}{2}[A_{VVX}{}^C + A_{VVXT}^C \delta T](\delta V)^2 x + RT\,x \ln x \qquad (17)$$

where the subscripts denote partial derivatives. The terms of interest in Eqs. 19, 20 and 22 are the only ones which have been retained. The relationship

$$(\partial V/\partial x)_{PT} = \frac{-(\partial P/\partial x)_{VT}}{(\partial P/\partial V)_{XT}} \qquad (18)$$

is used to obtain $\bar{v}_2$ where Eq. 17 gives

$$-(\partial P/\partial x)_{VT} = A_{VX} = A_{VX}^C + \text{higher order terms} \qquad (19)$$

and

$$- (\partial P/\partial V)_{XT} = A_{VV} = A^C_{VVT}\ \delta T + A^C_{VVX} x \ . \tag{20}$$

The stability determinant

$$\text{Det} \begin{vmatrix} A_{VV} & A_{VX} \\ A_{VX} & A_{XX} \end{vmatrix} = 0 \tag{21}$$

defines the critical line. Two of the elements are already listed in Eqs. 19-20, while Eq. 17 also gives

$$A_{XX} = A^C_{XX} + RT/x \quad . \tag{22}$$

Combining Eqs. 19-22 yields

$$A^C_{VVT}\ (\delta T) = [(A^C_{VX})^2 - A^C_{VVX}\ RT]x/RT \ . \tag{23}$$

Along the critical line, ΔT is linear in x. Near the pure fluid critical point, A_{VV} approaches zero strongly while A_{XX} diverges, but the product is finite. Substitution of Eq. 23 into Eq. 20, then Eq. 19 and Eq. 20 into Eq. 18 gives

$$(\partial V/\partial x)_{PT} = -RT/(xA^C_{VX}) \tag{24}$$

The partial molar volume is defined by

$$\bar{v}_2 = V + (1 - x)(\partial V/\partial x)_{P,T} \quad . \tag{25}$$

Substitution of Eq. 24 into 25 gives the desired result

$$\bar{v}_2 = V_c - \frac{RT}{A^C_{VX}} \cdot \frac{1}{x} \tag{26}$$

along the critical line. Along the critical isotherm/isobar, an alternate result is obtained

$$\bar{v}_2 \propto |x|^{-2/3} \quad , \tag{27}$$

thus, the approach of $\bar{v}_2$ to negative infinity is a path function as noted by Wheeler (ref. 11) using a nonclassical model.

The prediction of the theory is accurate for the partial molar volume data of Wu and Ehrlich (ref. 19) along the critical line. The data give $\bar{v}_2 \propto x^{-1.00}$ which is identical to Eq. 26 (see Figure 6 and Table 2). Given these examples of the general aspects of the pressure effect, the remainder of the paper will discuss the specific aspects.

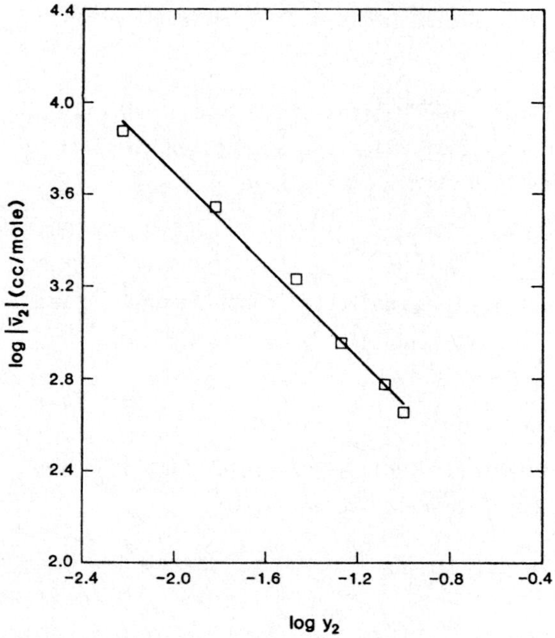

Fig. 6. $|\bar{v}_2|$ versus y_2 along the critical locus for ethane in n-heptane. (Data of Wu and Ehrlich, ref. 19).

PREDICTION OF SOLUBILITY VERSUS PRESSURE ISOTHERMS OF NONPOLAR SYSTEMS

In order to focus on the supercritical phase, we restrict the following discussion to binary solid-fluid equilibria (SFE) in which the solid phase is essentially pure. This region lies between the lower and upper critical end points in a class III system (see Figure 2). It is convenient to define a near supercritical region for a reduced density range of 1 to 1.3 where the compressibility of the solution and the partial molar volume of the solute are quite large. In the dense supercritical region where the reduced density is greater than 1.3, the magnitude of these properties is considerably more moderate. The scope of this investigation includes only the dense supercritical region; while further complexities that arise in the near supercritical region are described elsewhere (refs. 15, 21, 22, 23).

In the dense gas approach, the solubility is obtained using an equation of state. The solubility of a solute, y_2, in a supercritical fluid is

$$y_2 = \frac{P_2^{Sat} \exp \{v_2(P - P_2^{Sat})/RT\}}{\phi_2 P} \quad (28)$$

where the fugacity coefficient is obtained from an equation of state

$$\ell n\phi_2 = \int_0^P \left(\frac{\bar{v}_2}{RT} - \frac{1}{P}\right) dP \quad . \tag{29}$$

Recent investigators have used various modifications of the van der Waals equation of state to correlate SFE data (1,2,15,23). The attraction constant is obtained using a van der Waals one-fluid theory mixing rule

$$a = \sum_i \sum_j y_i y_j a_{ij} \quad . \tag{30}$$

The unlike-pair attraction parameter a_{12} is an universal function of $T_{C_{12}}$ and $P_{C_{12}}$ where

$$T_{C_{12}} = (T_{C_1} T_{C_2})^{\frac{1}{2}} (1 - k_{12}) \tag{31}$$

and k_{12} is a binary interaction parameter. Typically, a solubility isotherm can be correlated with an average absolute relative deviation

$$AARD = \frac{1}{N} \sum_{i=1}^{N} \left| \frac{y_{2i}^{calc} - y_{2i}^{exp}}{y_{2i}^{exp}} \right| \tag{32}$$

of 5-20% using only one binary interaction constant k_{12}. For a series of solids in a given supercritical solvent, the variation in k_{12} is often unpredictable. This problem is a result of the extreme difference in the critical temperatures of the solute and solvent. The concept of corresponding states based on critical properties (see eq. (31)) is limited for these highly asymmetric mixtures. Although this approach is quite useful for correlating complex SFE isotherms, an alternative treatment is required for the more ambitious goal of predicting the behavior.

For nonpolar systems, the unlike pair interaction energy is a function of the polarizabilities and molecular volumes of the components. This direct functionality was used for these highly asymmetric mixtures instead of eq. 31. The hard sphere - van der Waals equation of state (HS-VDW EOS) was chosen for use in eq. (29). It combines an accurate expression for the hard sphere repulsive forces (ref. 26) with the standard van der Waals attractive term. The fugacity coefficient is given by

$$\ell n(\phi_2 z) = \frac{\partial}{\partial n_2} \left(\frac{A-A^o}{RT}\right)^{HS}_{T,V,n_1} - 2(y_1 a_{12} + y_2 a_{22}) \rho/RT \tag{33}$$

where z is the compressibility factor, and HS refers to hard sphere.

Wong and co-workers (ref. 27) derived an expression for the van der Waals unlike pair attraction constant, a_{12}, for nonpolar molecules

$$a_{12} = \frac{4}{3} \pi^2 \alpha_1 I_1 I_2/(I_1 + I_2)\theta \tag{34}$$

where α is the polarizability, I is the ionization potential and

$$\theta = \frac{A_2 \, B_2^{2/3}}{(b_1^{1/3} + b_2^{1/3})^3} \tag{35}$$

where b is the molecular volume and A is the atomic number. (The numerator is related linearly to α_2.) The a_{22} is obtained in a similar fashion.

The HS-VDW equation of state was used to correlate the solubility data with an average deviation of 11% for the 44 isotherms shown in Table 3.

TABLE 3

Comparison of Predictions of the HS-VDW Equation of State with Experimental Results

Solute	T °C	$10^{-7} a_{12}$ [1] (bar-cm^6)	%AARD[2] [3]	%AARD[4]	ref.
		carbon dioxide			
naphthalene (n)	35	1.38	5.9	6.0	20
	45	1.36	11.6	13.0	
	55	1.33	19.0	27.9	
phenanthrene (ph)	50	1.79	8.9	8.9	23
	70	1.74	7.1	12.0	
anthracene (a)	50	1.79	12.6	13.7	23
	70	1.74	13.1	35.0	
hexamethylbenzene (h)	50	1.72	7.3	7.3	23
	70	1.69	4.4	14.2	
fluorene (f)	35	1.73	9.6	14.3	23
	50	1.70	1.7	12.5	
	70	1.66	17.2	8.5	
pyrene (py)	35	2.01	15.5	27.2	23
	50	1.99	11.9	21.5	
	70	1.95	3.5	3.9	
2,3-dimethyl-naphthalene (2,3-dmn)	35	1.66	7.1	7.5	24
	45	1.63	1.8	5.7	
	55	1.62	6.3	18.8	
2,6-dimethyl-naphthalene (2,6-dmn)	35	1.66	6.5	13.6	24
	45	1.63	3.0	3.0	
	55	1.62	6.1	21.1	
		ethylene			
naphthalene (n)	25	1.72	13.7	13.7	22,28
	35	1.71	19.9	19.9	
	45	1.70	12.7	12.7	
	50	1.70	18.5	18.5	

Solute	T °C	$10^{-7}a_{12}$ [1] (bar-cm^6)	%AARD[2] [3]	%AARD[4]	ref.
			ethylene		
phenanthrene (ph)	25	2.21	12.7	17.6	22
	45	2.19	15.0	15.0	
	70	2.17	15.5	15.5	
anthracene (a)	50	2.18	8.7	33.0	22
	70	2.16	9.8	14.1	
	85	2.15	10.1	11.0	
hexamethylbenzene (h)	25	2.14	17.9	17.9	23
	45	2.12	18.0	18.1	
	70	2.08	20.7	25.7	
fluorene (f)	25	2.10	24.9	24.9	23
	45	2.08	16.6	16.6	
	70	2.06	20.1	20.1	
pyrene (py)	45	2.42	22.6	22.6	23
	75	2.40	20.4	20.4	
2,3-dimethyl-naphthalene (2,3-dmn)	35	2.02	3.6	3.8	24
	45	2.01	2.2	2.4	
	55	2.00	9.4	9.4	
2,6-dimethyl-naphthalene (2,6-dmn)	45	2.01	1.5	16.8	24
	55	2.00	3.1	8.7	

[1] Predicted value using eqs. (35)-(38).
[2] Average absolute relative deviation is defined by eq. (32).
[3] Result of correlation when a_{12} is regressed from each solubility isotherm.
[4] AARD when a_{12} is predicted using eqs. (35)-(38).

The optimized values of a_{12} are related linearly to θ as shown in Figure 7. The results of a linear regression are

$$(a_{12})_{CO_2} = (2.74 \times 10^6) + (2.61 \times 10^6)\,\theta \tag{36}$$

and

$$(a_{12})_{C_2H_4} = (4.80 \times 10^6) + (3.26 \times 10^6)\theta \tag{37}$$

with a correlation coefficient of 0.99 for both solvents. The units of a_{12} and θ are $(cm^3/gmol)^2$ bar and $(cm^3/gmol)^{-1/3}$, respectively. The a_{12} exhibited a slight temperature dependence which was approximated by

$$a_{12}(T)/a_{12}(T{=}323°K) = 1.0 + C_1(T-323) \tag{38}$$

where $C_1 = -1.13 \times 10^{-3}$ °K^{-1} and -3.72×10^{-4} °K^{-1} for systems in carbon

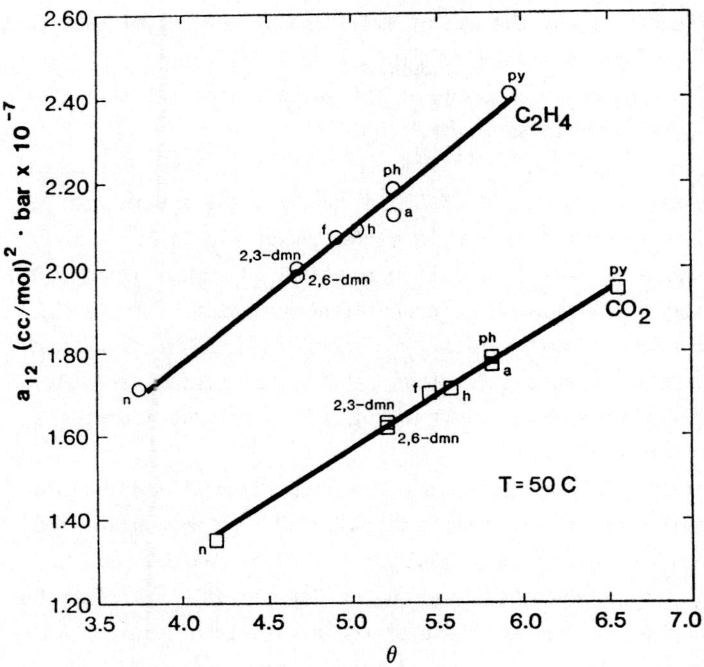

Fig. 7. Unlike pair attraction constant a_{12} versus θ which is a function of the molecular volumes of the solvent and solute molecules (code given in Table 3).

dioxide and ethylene, respectively. As expected, the attraction constants are greater for the more polarizable solvent, ethylene.

The isotherms in Table 3 were predicted without adjustable parameters with an AARD of 16% using Eqs. 35-38 to calculate a_{12}. The accuracy is impressive since it is comparable with that of other approaches which use an adjustable binary interaction parameter, k_{12}. The only input data for this predictive model are the sizes of the components which are calculated easily. This represents a major departure from the conventional approach where k_{12}'s are used to compensate for the inadequacies of the geometric mean combining rule for $T_{C_{12}}$. Instead, the interaction constant a_{12} is calculated directly from the relevant properties of the components, i.e., the polarizabilities and molecular volumes.

EFFECTS OF CO-SOLVENTS IN SUPERCRITICAL FLUID EXTRACTION

The approach is to exploit an additional dimension of supercritical fluid extraction in which substances may be separated based on polarity and specific forces such as electron donor-acceptor complexation. The thermodynamic analysis is focused on the supercritical phase by choosing a pure solid for the condensed phase and examining solid-fluid equilibria. A "chemical effect" is superimposed on top of the "pressure effect" by the use of small amounts of co-solvents that contain polar and in some cases acidic or basic functional groups. Carbon dioxide was chosen as the primary component of the solvent since it is nontoxic, nonflammable and has a critical temperature, of 304°K.

The polarizability of CO_2 is 26.5×10^{-25} cm³ which is less than that of all of the hydrocarbons except methane. At 35°C and 200 bar, the square root of the cohesive energy density or the solubility parameter of CO_2 is 6.5 $(cal/cc)^{\frac{1}{2}}$ which is nearly comparable to that of liquid isopentane. At these conditions, the solubility of aromatic solids such as phenanthrene or benzoic acid in CO_2 is several orders of magnitude below the ideal solubility of a solid in a liquid (ref. 30). Since the benzoic acid (2)-carbon dioxide (1) system is highly nonideal ($\gamma_2 > 100$), the addition of a small amount of co-solvent promotes a significant solubility enhancement.

The literature contains only a few examples of experimental solid-fluid equilibria data for solid-co-solvent-supercritical fluid systems where the volatility of the co-solvent is between that of the other two components. Kurnik and Reid (ref. 31) demonstrated that the solubility of a solid may be enhanced in a supercritical solvent by the presence of a second solute. A new interpretation of this phenomenon is presented below. Irani and Funk (ref. 32) showed that the introduction of benzene induces miscibility in the carbon dioxide-decalin system. Peter and Brunner (ref. 33) increased the concentration of glycerides in CO_2 by adding small amounts of acetone. Bruner (ref. 33) also studied the effect of co-solvents on the separation factor of hexadecanol and octadecane in CO_2 and in N_2O. The composition of the liquid phase was not measured and consequently it is difficult to determine how much the co-solvent affected the supercritical phase relative to the liquid phase.

Co-solvents are useful in supercritical fluid chromatography (SFC) as they often promote large modifications in the retention times of compounds such as purines (refs. 34, 35). These data represent the combined effects of the co-solvent in the supercritical as well as in the stationary phase. Unfortunately, it is difficult if not impossible to use these data to determine the effects of the co-solvent in the supercritical phase alone. Joshi and Prausnitz (ref. 36) calculated the effects of co-solvents on the solubility of benzoic acid using the Redlich-Kwong equation of state.

Solubility versus pressure isotherms were measured for the solid benzoic acid in pure CO_2 and in CO_2 doped with SO_2 (ref. 29). The density of the solvent mixture was measured to $\pm 1.0 \times 10^{-4}$ g/cc using a Mettler-Paar DMA-512 vibrating tube densitometer. Figure 8 shows solubility isotherms at 35°C for benzoic acid for various mole percent concentrations of SO_2 in CO_2. At 45°C, the binary benzoic acid - CO_2 data (not shown) agree with the results of Kurnik

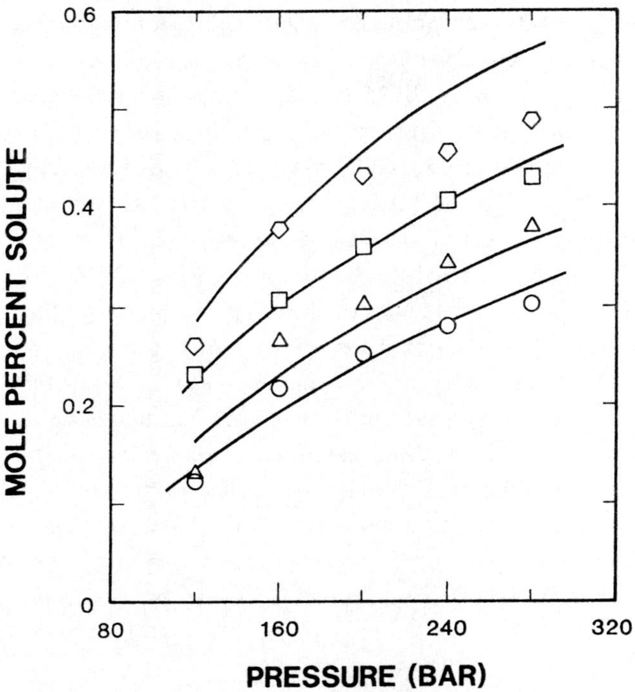

Fig. 8. Solubility of benzoic acid in carbon dioxide with sulfur dioxide co-solvent (○ - pure CO_2, △ - 1.86 mol% SO_2, □ - 4.35 mol% SO_2, ⬠ - 5.85 mol% SO_2, — calculated using a_{23} from 3.24% SO_2 curve and HS-VDW EOS).

et al. (ref. 37) to within 2 percent. The solubility at 120 bar with 5.8 mole percent SO_2 is comparable to the solubility in pure CO_2 at 280 bar. This example demonstrates that the yield may be maintained at a greatly reduced pressure by adding only a small amount of co-solvent.

The shape of each ternary isotherm is similar to the binary one such that the ratio of y_2 in the ternary system (CO_2 + co-solvent) over y_2 in the binary system (pure CO_2) is relatively invariant with respect to pressure. This serendipitous discovery simplifies the development of the models for the phase equilibria. The "pressure effect" is about the same in the mixed solvent as in pure CO_2, but in the former case the "chemical effect" shifts the isotherm upward by a relatively constant factor.

The solubility isotherms were correlated using Eq. 29 and

$$\ell n(\phi_2 z) = \frac{\partial}{\partial n_2}\left(\frac{A-A^\circ}{RT}\right)^{\text{hard sphere}}_{T,V,n_1,n_3} - 2(y_1 a_{12} + y_2 a_{22} + y_3 a_{23})\rho/RT. \qquad (39)$$

This expression is an extension of Eq. 33 to the ternary mixture. The key parameters in this expression are the binary attraction constants a_{12} and a_{23}.

The former was regressed from the binary data, and the latter from the ternary data. Since most equations of state do not predict the density of solutions accurately in the critical region, the experimental value was used.

The solubility isotherms of the benzoic acid-SO_2-CO_2 system are predicted with an AARD of 8% for the co-solvent concentrations shown in Figure 8. The a_{23} was regressed from the isotherm where x_3=4.35 mol%, and was used to calculate the isotherms at x_3 = 1.86, 3.24, and 5.85%. This demonstrates that the model can predict the co-solvent induced solubility enhancement as a function of the concentration of co-solvent. The molecular volume of SO_2 is 1.46 times that of CO_2. The ratio of a_{23}/a_{12} is 1.46 which is enough to increase the solubility of benzoic acid even though SO_2 is a larger molecule than CO_2. The increase in attraction due to SO_2 predominates over the increase in repulsion.

This work is being continued using strongly polar but small co-solvents such as methanol. The rational utilization of these co-solvents could improve the economics of existing or proposed processes through reduction in the pressure and/or the recycle ratio. It could lead to a new generation of extractions for components which have similar volatility but different types of chemical forces and for compounds which have extremely limited solubilities in pure fluids, e.g. biomolecules.

ACKNOWLEDGEMENT

We are grateful to Prof. C. A. Eckert and Dr. J.M.H. Levelt Sengers, whose comments were invaluable at the start of this project.

This material is based on work supported by the National Science Foundation under Grant No. CPE-8306327. Any opinions, findings, and conclusions or recommendations expressed in this publication do not necessarily reflect the views of the National Science Foundation. We also acknowledge the Separations Research Program at The University of Texas and the Atlantic Richfield Foundation.

Acknowledgment is made to the Donors of the Petroleum Research Fund, administered by the American Chemical Society, for partial support of this work. Further support is acknowledged from the Dow Chemical Company Foundation and The University of Texas Bureau of Engineering Research.

NOTATION

f_i	fugacity of i'th component in solution
k_T	isothermal compressibility = $-1/V\ (\partial V/\partial P)_T$
M	molecular weight
ΔP^*	reduced pressure difference = $(P-P_c)/P_c$
P_{ij}	derivatives at the critical point
ΔT^*	reduced temperature difference = $(T-T_c)/T_c$

$\bar{v}_i$	partial molar volume of i'th component
x,y	mole fraction

Subscripts

1	solvent
2	solute
3	co-solvent
c	critical
i,j	component designation
σ	along phase boundary

Superscripts

G	gas
L	liquid
S	solid
sat	saturation
o	reference

Greek

γ	activity coefficient or critical exponent
ε	critical exponent defined by Eq. 16

REFERENCES

1. Paulaitis, M. E., McHugh, M. A. and C. P. Chai, "Solid Solubilities in Supercritical Fluids at Elevated Pressure," in Chemical Engineering at Supercritical Fluid Conditions, Ed. by M. E. Paulaitis et al., Ann Arbor Science (1983).

2. McHugh, M. A., "Extraction with Supercritical Fluids," Recent Developments in Separation Science, Ed. by N. N. Li, J. M. Calo, V. IX, CRC Press (1984).

3. Johnston, K. P., "Supercritical Fluids," Encyclopedia of Chemical Technology, 3rd Ed., Wiley, NY (1984).

4. Kurnik, R. T. and R. C. Reid, A.I.Ch.E. J., 27, 861 (1981).

5. Eckert, C. A., Ziger, D. H., Johnston, K. P., and T. K. Ellison, Fluid Phase Equilib, 14, 167 (1983).

6. Paulaitis, M. E., Johnston, K. P. and C. A. Eckert, J. Phys. Chem., 85, 1770 (1981).

7. van Waser, U., and G. Schneider, J. Phys. Chem., 84, 229 (1980).

8. Ziger, D. H., Ph.D. Dissertation, Univ. of Illinois (1983).

9. Gitterman, M. and I. Procaccia, J. Chem. Phys., 78, 2648 (1983).

10. Paulaitis, M. E., Krukonis, V. J., Kurnik, R. T., and R. C. Reid, Rev. Chem. Eng., 1, 179 (1982).

11. Wheeler, J. C., Ber. Bunsenges. Phys. Chem., 76, 308 (1972).

12. Krichevskii, I. R., Russ. J. Phys. Chem., 41, 1332 (1967).

13. Krichevskii, I. R., Khusainova, S. R., and L. A. Makarevich, Dokl. Phys. Chem., 206(6), 1388-90 (1972).

14. Douslin, D. R. and R. H. Harrison, J. Chem. Thermodynamics, 8, 301 (1976).

15. Chang, R. F., Morrison, G., and J.M.H. Levelt-Sengers, J. Phys. Chem., 88, 3389 (1984).

16. Levelt Sengers, J.M.H., "Critical Behavior in Fluids," in High Pressure Technology, v. II, Ed. by I. L. Spain, Dekker, Inc., NY (1977).

17. Levelt Sengers, J.M.H., Morrison, G., and R. F. Chang, Fluid Phase Equilibria, Fluid Phase Equilib., 14, 19-44 (1983).

18. Griffiths, R. B. and J. C. Wheeler, Phys. Rev. A2, 1047 (1970).

19. Wu, P. C. and P. Ehrlich, A.I.Ch.E. J., 19, 533 (1973).

20. Tsekhanskaya, Y. V., M. B. Iomtev, and E. V. Mushkina, Russ. J. Phys. Chem., 38, 1173-1176 (1965).

21. Kim, S. and K. P. Johnston, Presented at AIChE Annual Meeting, San Francisco (1984).

22. Johnston, K. P. and C. A. Eckert, AIChE J. 1981, 27(5), 773-79.

23. Johnston, K. P., Ziger, D. H., and C. A. Eckert, Ind. Eng. Chem. Fundam., 21, 191 (1982).

24. Kurnik, R. T.; Holla, S. J.; Reid, R. C., J. Chem. Eng. Data 1981, 26, 47-51.

25. Mackay, M. E.; Paulaitis, M. E., Ind. Eng. Chem. Fundam. 1979, 18(2), 149-53.

26. Mansoori, G. A.; Leland, T. W., J. Chem. Soc. Faraday Trans II, 1972, 68, 320-44.

27. Wong, J. M., Pearlman, R. S., Johnston, K. P., submitted to J. Phys. Chem., 1985.

28. Diepen, E. A.; Scheffer, F.E.C., J. Phys. Chem. 1953, 57, 575-577.

29. Lahiere, R. J., M.S. Thesis, University of Texas, Austin, TX (1984).

30. Prausnitz, J. M., "Molecular Thermodynamics of Fluid-Phase Equilibria," pp. 422-429, Prentice-Hall Inc., Englewood Cliffs, NJ (1969).

31. Kurnik, R. T.; Reid, R. C., Fluid Phase Equilibria 8, 1982, 93-105.

32. Irani, C. A.; Funk, W. W., Recent Developments in Separation Science 1977, 3, Part A, 171-92.

33. Peter, S.; Brunner, G., Angew Chem Int. Ed. Engl. 1978, 17, 746-50.

34. Randall, L. G., "Supercritical Fluid Chromatography," Hewlett Packard Technical Paper No. 102.

35. Gere, D. R., Science 1983, 222, 253-9.

36. Joshi, D. K.; Prausnitz, J. M., AIChE J. 1984, 30, 522-5.

37. Kurnik, R. T.; Holla, S. J.; Reid, R. C., Chem. Engr. Data 1981, 26, 47-51.

Supercritical Fluid Technology, edited by J.M.L. Penninger, M. Radosz, M.A. McHugh and V.J. Krukonis
1985 Elsevier Science Publishers B.V., Amsterdam — Printed in The Netherlands

A MODIFICATION OF THE RANDOM-FLUID APPROXIMATION FOR SOLID-SUPERCRITICAL FLUID EQUILIBRIA

Charles J. Mart and Kyriakos D. Papadopoulos
Department of Chemical Engineering, Tulane University
New Orleans, Louisiana 70118 USA

ABSTRACT

The role of the random-fluid approximation in predicting the solubility of hydrocarbon solids in a supercritical fluid is examined and a largely empirical modification to this approximation is proposed. The modification expresses the temperature and density dependence of the binary interaction coefficient in the van der Waals mixing rule. Using this approach, solubility predictions are greatly improved over the case where the random-fluid (van der Waals) approximation is used.

INTRODUCTION

The concept of supercritical fluid extraction is not new and actually was first recognized over a hundred years ago. Hannay and Hogarth (ref.1) discovered that solid compounds could be dissolved in supercritical fluids which possessed densities similar to those of liquids. In recent years, the utilization of supercritical fluids in a gaseous extraction process has received much interest as an alternative to distillation and liquid extraction by the food, petroleum, metallurgical, and synthetic fuel industries. The application of supercritical extraction processes in these industries has been reviewed in some detail by Williams (ref.2). Some applications are: the removal of vegetable oils from foods, the deasphalting of petroleum fractions, the extraction of ozocerites (high-melting parafins) from mineral ores, and the extraction of liquid fuels from coal. In all of these instances, the solvent power of the supercritical fluid can be varied over a wide range by manipulating its temperature and pressure. As the density of the solvent fluid increases, the fluid is able to dissolve considerable quantities of non-volatile compounds that would be barely soluble in the same fluid in its ordinary gaseous state.

The supercritical fluid extraction process is related to both distillation and liquid extraction since enhancement of the solute's vapor pressure and phase separation are both involved. Supercritical fluid extraction is more advantageous than distillation since the separation can be accomplished at low temperatures. In comparison to liquid extraction, the supercritical fluid has higher diffusivities and lower viscosities than a liquid solvent, allowing for more rapid extraction and phase separation. Furthermore, the supercritical fluid extraction is reversible, in the sense that the solute and the solvent can be separated by reducing the solvent density. This reduction in density can be accomplished by altering the pressure and/or the temperature of the solvent. The reversibility of the process also allows for purer products of separation since the problem of residual solvent is greatly diminished.

Over the last few years, a number of investigators have studied the solubility of hydrocarbon solids in supercritical fluids. Mackay and Paulaitis (ref.3) treated the supercritical fluid as an expanded liquid and utilized activity coefficients at infinite dilution in the equilibrium calculations. Prausnitz (ref.4) and Kurnik et. al. (refs.5,6) have used the semi-empirical, cubic equations of state to estimate the fugacity of the heavy component in the fluid phase. Eckert et. al. (ref.7) also have illustrated the use of cubic equations of state to calculate the partial molar volume of a non-volatile solute in supercritical fluids. The hard sphere (Carnahan-Starling) equations were used by Johnston and Eckert (ref.8) to model the fugacity of the heavy component in the fluid phase. This method avoids the corresponding states treatment based on the critical properties of each component. In order to obtain the solubility of the heavy component near the critical point of the solvent, Johnston et. al. (ref.9) proposed an equation based on the Augmented van der Waals theory. Modeling of the solubility near the critical point of the solvent is important, since, in this region, the solute will precipitate from the solvent with a small decrease in pressure. The cubic equation models fail to represent the solubility data over the full range of pressures for an isotherm. In this paper, a popular, semi-empirical, cubic equation of state (ref.10), which has one adjustable parameter, namely the interaction parameter k_{12}, is used. By using physical arguments, a modified interaction parameter, ℓ_{12} is proposed. This parameter consists of an adjustable term, ℓ'_{12} and a term obtained from a recipe which is based on physical but also largely empirical arguments.

THERMODYNAMIC MODELING FOR EQUILIBRIA

In order to calculate the solubility of a non-volatile hydrocarbon solid in the supercritical solvent, the fugacity of the heavy component in the solid phase and in the fluid phase is equated:

$$\hat{f}_2^S = \hat{f}_2^F \tag{1}$$

With this criterion for equilibrium, classical thermodynamic relationships are used to calculate the fugacities of the two phases in equilibrium. In the case of solid-supercritical fluid equilibria, the solubility of the solvent in the solid phase can be considered negligible and the solid phase can be assumed to be pure. Therefore, all non-ideal behavior exhibited by the system is due to the behavior of the fluid phase. According to Prausnitz (ref.4), the fugacity of a pure solid at a given temperature and pressure is given by

$$\hat{f}_2^S = p_2^{sat}\ \hat{\phi}_2^{sat}\ \exp\left[\int_{p_2^{sat}}^{p} \frac{v_2^S}{RT}\, dp\right] \tag{2}$$

At conditions removed from its critical point, the solid phase can be considered incompressible and, therefore, the volume could be independent of temperature and pressure. Because the solid is considered to be non-volatile, the fugacity coefficient for the solid at saturation conditions is unity. Equation (2) therefore simplifies to

$$\hat{f}_2^S = p_2^{sat}\ \exp\left[\frac{v_2^S\,(p - p_2^{sat})}{RT}\right] \tag{3}$$

The supercritical fluid phase is treated as a dense or highly compressed gas, and the fugacity of the heavy component in the fluid phase can be calculated from

$$\hat{f}_2^F = y_2 \hat{\phi}_2^F \, p \tag{4}$$

Substituting equations (3) and (4) into equation (1) gives the solubility of the heavy component in the fluid phase as

$$y_2 = \frac{p_2^{sat}}{p} \, \frac{\exp\left[\dfrac{v_2^s \, (p - p_2^{sat})}{RT}\right]}{\hat{\phi}_2^F} \tag{5}$$

The non-ideal behavior of the fluid system is characterized by the fugacity coefficient, $\hat{\phi}^F$, which is evaluated using an equation of state and thermodynamic relationships. The accuracy of the determination of the solubility is dependent upon the accuracy of the determination of the fugacity coefficient. For this reason, equations of state which are capable of predicting phase behavior in all regions are important for solubility determinations. The Peng-Robinson (ref.10) equation of state (see Table 1), is used in the calculation of the fugacity coefficient for the heavy component in the fluid phase. This equation is chosen because it offers better predictions for liquid densities than other van der Waals-based, cubic equations. The attractive parameter, a, can be regarded as a measure of the intermolecular attractive force, while the covolume parameter, b, is related to the size of hard spheres for representation of the molecule. The component attractive and covolume parameters are calculated using a corresponding states theory, based on the critical properties of the component. The relationships for these parameters, a_i and b_i, are also contained in Table 1. In order to calculate the attractive and covolume parameters for the mixture, a mixing rule which is concentration-dependent is used.

Table 1

Equation of State and Fugacity Coefficient for the Peng-Robinson Equation

Equation of State: $P = \frac{RT}{v-b} - \frac{a(T)}{v(v+b) + b(v-b)}$

where $a = 0.45724 \frac{R^2T_c^2}{P_c}\alpha$; $\alpha = \left[1 + \kappa(1-T_r^{\frac{1}{2}})\right]^2$

$\kappa = 0.37464 + 1.54226\omega - 0.26992\omega^2$; $b = .0778 \frac{RT_c}{P_c}$

Fugacity Coefficient: $\ln \hat{\phi}_k = \frac{1}{b}\left[\frac{\partial(nb)}{\partial n_k}\right](Z-1) - \ln(Z - \frac{bP}{RT})$

$$- \frac{a}{2\sqrt{2}bRT}\left[\frac{1}{na}\frac{\partial(n^2a)}{\partial n_k} - \frac{1}{b}\frac{\partial(nb)}{\partial n_k}\right]\ln\left[\frac{Z+2.414\frac{bP}{RT}}{Z-0.414\frac{bP}{RT}}\right]$$

MIXING RULES

In order to solve the equation of state and compute the expression for the fugacity coefficient, an assumption about the behavior of the fluid is made and, therefore, a mixing rule is obtained. For the mixture parameters, four mixing rules were used - van der Waals one fluid, van der Waals with surface fractions, conformal solution, and conformal solution with surface fractions. All of these mixing rules assume a random fluid model and they are shown in Table 2. Using the van der Waals mixing rule, the covolume parameter for the mixture is determined by equation (1') of Table 2. To relax the geometric mean assumption of the attractive parameter, Chueh and Prausnitz (ref.11) introduced a binary interaction coefficient, k_{ij}. The form for the attractive parameter is given in equations (2') and (3'). Equations (4') through (6') give the expressions for a van der Waals mixing rule where surface fractions have replaced mole fractions. The mixing rule proposed by Radosz et al (ref.12), equations (7') through (10') of Table 2, are obtained from an extension of conformal solution theory. Kreglewski and Kay (ref.13) proposed a mixing rule for asymmetric systems based on defining the size effects in terms of surface fractions rather than mole fractions. Equations (11') through (16') of Table 2 illustrate this mixing rule and its derivatives. The binary interaction coefficient, k_{ij}, in all these mixing rules is an empirically-obtained constant which characterizes the deviation of the mixture attractive parameter from the geometric mean of the component i - component j pair attractive parameter. Donohue and Prausnitz (ref.14) have shown that this interaction parameter is not just an extra parameter used to fit experimental data, but that it is actually related to the attractions of the molecules in the system. In this way, the interaction parameter can be used to reflect temperature and density effects on the i-j pair attractive parameter.

Table 2
Random-Fluid Mixing Rules

van der Waals	van der Waals with surface fractions	conformal solution	conformal solution w/ surface fractions
$b = \sum_i y_i b_i$ (1')	$b = \sum_i y_i b_i$ (4')	$b = \sum_i \sum_j y_i y_i b_{ij}$ (7')	$b = \sum_i \sum_j y_i y_i b_{ij}$ (11')
		$b_{ij} = \frac{1}{8} \left\{ b_{ii}^{1/3} + b_{jj}^{1/3} \right\}^3$ (8')	$b_{ij} = \frac{1}{8} \left\{ b_{ii}^{1/3} + b_{jj}^{1/3} \right\}^3$ (12')
$a = \sum_i \sum_j y_i y_j a_{ij}$ (2')	$a = \sum_i \sum_j s_i s_j (1-k_{ij}) \sqrt{a_{ii} a_{jj}}$ (5')	$a = \sum_i \sum_j y_i y_j b^{0.25} \frac{a_{ij}}{b_{ij}^{0.25}}$ (9')	$a = \sum_i \sum_j s_i s_j b^{0.33} \frac{a_{ij}}{b_{ij}^{0.33}}$ (13')
$a_{ij} = (1-k_{ij}) \sqrt{a_i a_j}$ (3')	$s_i = \frac{y_i b_i^{2/3}}{\sum_k y_k b_k^{2/3}}$ (6')	$a_{ij} = (1-k_{ij}) \sqrt{a_{ii} a_{jj}}$ (10')	$a_{ij} = (1-k_{ij}) \sqrt{a_{ii} a_{jj}}$ (14')
			$s_i = \frac{y_i b_i^{2/3}}{\sum_k y_k b_k^{2/3}}$ (15')
			$\delta_{ij} = \frac{a_{ij}}{b_{ij}^{0.33}}$ (16')

INTERACTION COEFFICIENT MODIFICATION

By utilizing a generalized mixing rule to determine the empirical constants of a pure fluid equation of state, an equation of state for a mixture can be obtained. Snider and Herrington (ref.15) have demonstrated, using the Longuet-Higgins model for a hard-sphere fluid, that the equation of state for the mixture has a different dependence on temperature and density than the pure fluid equation. Furthermore, at higher densities, the chemical potential of the fluid becomes extremely sensitive to variations in the density and to the form of the mixing rule. For these two reasons, at high densities the form of the mixing rule must be able to account for its increased dependence upon temperature and density, as well as for non-random structures which tend to form. The covolume parameter of cubic equations of state depends very little upon the configuration of the molecules and is not affected by density and temperature changes. The average attractive energy between molecules, however, depends highly upon the configuration of the system, and the attractive parameter warrants the modification. Therefore, modification to the random-fluid mixing rule for the attractive parameter should be in terms of density and temperature variables. Vera and

Prausnitz (ref.16) have suggested a similar dependence for the attractive parameter, where the temperature and density dependence is obtained from a sixteen-constant, empirical equation.

In the dilute gas or at low densities, molecules of the fluid are randomly arranged and pair interactions predominate. As the density increases, molecules tend to arrange themselves in non-random groups and the average energy between two molecules in the group is greater than the energy of interaction for an isolated pair. In a highly dense fluid or liquid, molecules are extremely close together so that the repulsive forces form a structure which will not be changed to any great extent by the attractive interactions. Therefore, at moderate densities or in the supercritical state, this "rigid" structure is not present and interactions between pairs of "isolated" molecules and between molecules in a non-random group are important to the properties of the fluid. The van der Waals-type equations are unable to account for the structural effects created by the attractive forces of molecules in a non-random group. To date, very little work has been done on determining the magnitude of these multiple body forces. Barker et. al. (ref.17), Rowlinson et. al. (refs.18,19) and Singh (ref.20) have performed calculations for the effect of three-body forces on the effective pair potentials of fluids. The modification of the random fluid assumption is based on the work of Sherwood and Prausnitz (ref.21) and Deiters (ref.22). Sherwood and Prausnitz's method corrects the third virial coefficient for three-body effects, while Deiters' method corrects a semi-empirical equation of state for three-body effects. The derivation presented here uses a probabilistic approach analogous to that of Deiters'.

For a dense gas system, the assumption of pair-wise additivity of the intermolecular potentials have proven to be only a fair approximation of the true potential. Consider a non-random group of n molecules. The total energy of interactions in this group is equal to the sum of all pair interactions plus a term which accounts for the multi-body interactions

$$U_n = \sum_{i,j} u(i,j) + \Delta u_n \tag{6}$$

Multi-body interactions become important as the density of the fluid increases, since at higher densities more molecules are involved in some sort of "grouping". In order to obtain the corrected value to the interaction energy for a molecular pair due to the non-random group, the dynamic situation of a molecular pair entering and leaving the non-random group is considered. The molecular density of a non-random group is assumed to be the close-packed density and this density is, therefore, defined as the reference density

$$\rho^{ref} = \frac{1 \text{ molecule}}{b} \tag{7}$$

Since the number of molecular pairs to exit a non-random group is dependent upon density, the energy associated with the pair leaving the non-random group is equal to $\rho^* \Delta u_n$. The term ρ^* is a reduced density with respect to the reference density. For a pair of molecules to exit the non-random group, it must possess an energy equal to or greater than $\rho^* \Delta u_n$. In this way, the probability that a pair of molecules will exit the non-random group is given by

$$P_2 = \exp\left[-\frac{\Delta u_n}{kT}\rho^*\right] \tag{8}$$

Conversely, the probability that the pair will remain in the non-random group is given by

$$P_n = 1 - \exp\left[-\frac{\Delta u_n}{kT}\rho^*\right] \tag{9}$$

To put this derivation in a more useful form, the fluid is assumed to behave according to a square-well potential, so that the energies in the derivation can be written in terms of the depth of potential-well parameter, ε. Then the corrected value of the pair interaction energy becomes a correctd value for the well-depth parameter. This corrected value to account for the pair of molecules entering and leaving a non-random group is expressed as the expectation value

$$\varepsilon_{eff,n} = \varepsilon^{(2)} P_2 + \varepsilon^{(n)} P_n \tag{10}$$

where $\varepsilon^{(2)}$ is the two-body potential well-depth and $\varepsilon^{(n)}$ is the effective well-depth for a two-body interaction when account is taken of multiple-body effects. The influence of multiple-body forces on a pair of molecules is to decrease the absolute value of the pair-potential well-depth parameter by some increment of $\Delta\varepsilon_n$. Since n pair-interactions are present in the non-random group, the average well-depth for the energy of interaction between two molecules in that group is given by

$$\varepsilon^{(n)} = \varepsilon^{(2)} - \frac{\Delta\varepsilon_n}{n} \tag{11}$$

Substituting equations (8), (9) and (11) into equation (10) and exchanging the interaction energy for the well-depth parameter, the effective well-depth for the pair will be given by

$$\varepsilon_{eff,n} = \varepsilon^{(2)} \exp\left[\frac{-\Delta\varepsilon_n}{kT}\rho^*\right] + \left(\varepsilon^{(2)} - \frac{\Delta\varepsilon_n}{n}\right)\left\{1 - \exp\left[\frac{-\Delta\varepsilon_n}{kT}\rho^*\right]\right\} \tag{12}$$

Writing the exponential functions in equation (12) in terms of a Taylor series and truncating after the first order term, the effective well-depth for the pair simplifies to

$$\varepsilon_{eff,n} = \varepsilon^{(2)}\left\{1 - \frac{(\Delta\varepsilon_n)^2}{\varepsilon^{(2)}nkT}\rho^*\right\} \tag{13}$$

Up to this point, the derivation has followed an extention of Deiters' arguments to n-body interactions. To extend this type of analysis to a dense fluid in which all sorts of groupings are possible, consider an average well-depth over all the effective well-depths in the fluid

$$\varepsilon_{ave} = \sum_{n=3}^{\infty} \alpha_n \varepsilon_{eff,n} \tag{14}$$

In the infinite series, the coefficients α_n are weighting factors which depend on the number of non-random groups of n molecules. By substitution of the effective well-depth, equation (13) into equation (14), the average well-depth becomes

$$\varepsilon_{ave} = \varepsilon^{(2)}\left\{1 - \sum_{n=3}^{\infty} \frac{\alpha_n}{n} \frac{(\Delta\varepsilon_n)^2}{\varepsilon^{(2)}kT}\rho^*\right\} \tag{15}$$

In van der Waals-type equations of state, the attractive parameter, a, of the equation can be linked to the well-depth parameter for the potential by

$$a = \frac{2\pi N_A \varepsilon \sigma^3}{3} \tag{16}$$

Equation (15) can, therefore, be used to calculate the correlation to the attractive parameter due to multiple-body effects. When dealing with supercritical fluid extraction, non-random groups of dissimilar-sized molecules will have the greatest impact on the behavior of the fluid. The existence of non-random groups consisting of one large molecule of the solute and several solvent molecules introduces the asymmetry and non-randomness into the fluid, and thus the mixing rule. Replacing the a_{12} parameter for the parameter in equation (15) yields

$$a_{12'ave} = a_{12} \left\{ 1 - \sum_{n=3}^{\infty} \frac{\alpha_n}{n} \frac{2\pi\sigma_{12}^3 (N_A \Delta\varepsilon_n)^2}{3a_{12}kT} \rho^* \right\} \tag{17}$$

Substituting the equation

$$a_{12} = (1 - k_{12}) \sqrt{a_{11}\, a_{22}} \tag{18}$$

on the right hand side of equation (17) yields an expression for the modified interaction parameter ℓ_{12}

$$\begin{aligned} \ell_{12} &= k_{12} + (1-k_{12}) \sum_{n=3}^{\infty} \frac{\alpha_n}{n} \frac{2\pi\sigma_{12}^3 (N_A \Delta\varepsilon_n)^2}{3a_{12}kT} \rho^* \\ &= \ell'_{12} + \sum_{n=3}^{\infty} \frac{\alpha_n}{n} \frac{2\pi\sigma_{12}^3 (N_A \Delta\varepsilon_n)^2}{3\sqrt{a_{11}a_{22}}\,kT} \rho^* \end{aligned} \tag{19}$$

or

$$\ell_{12} = \ell'_{12} + \frac{2\pi\sigma_{12}^{3} N_A^{2}}{3kT\sqrt{a_{11}a_{22}}} \rho^* \sum_{n=3}^{\infty} \frac{\alpha_n \Delta\varepsilon_n}{n} \tag{20}$$

In order to preserve the number of adjustable parameters to one, an attempt was made to find an equation for the infinite sum in equation (20) that is general and applicable to all systems. The equation is developed using dimensional analysis arguments, by a counting and weighting of the interactions between the solvent and solute molecules, and above all, by empiricism. This summation is expressed as a function of the attractive parameter divided by the b_{12} (covolume) parameter of the equation of state. This function is obtained from an averaging of the attractive parameters, dependent upon the coordination number and temperature. The summation is replaced by

$$\sum_{n=3}^{\infty} \frac{\alpha_n \Delta\varepsilon_n}{n} = \frac{(\Delta\beta)^2}{z'} \tag{21}$$

where

$$\Delta\beta = \frac{\pi T_r \sqrt{z'}}{N_A 6 b_{12}} \left[\frac{(z'+1) a_{11} \sqrt[z']{a_{11}a_{22}}}{z'a_{11} + \sqrt{a_{11}a_{22}}} \right]^{1/z'} \tag{22}$$

In its final form, the modified binary interaction coefficient, ℓ'_{12} becomes

$$\ell_{12} = \ell'_{12} + \left[\frac{2\pi\sigma_{12}^{3} (N_A \Delta\beta)^2 b\rho_c}{3z'kT \sqrt{a_{11}a_{22}}} \right] \rho_{red} \tag{23}$$

with $\Delta\beta$ given by equation (22). The density parameter in equation (23) is defined as a pseudo-reduced mixture density. That is, the mixture density is divided by the critical density of the solvent.

DISCUSSION

Traditional Mixing Rules

The solubility of naphthalene in supercritical ethylene at 35°C was determined using the Peng-Robinson equation with both the van der Waals and conformal solution mixing rules as shown in Figure 1. In these calculations, a constant value for the binary interaction coefficient was used, as in previous investigations. For the van der Waals mixing rule, the binary interaction coefficient equaled 0.018. For the conformal solution mixing rule, the coefficient equaled -0.03. At a pressure of 80 atm, the Peng-Robinson model with the van der Waals mixing rule under-predicts the solubility by approximately 53 per cent. At 270 atm, the model over-predicts the mole fraction of naphthalene in ethylene by approximately 65 per cent. Furthermore, at the higher pressures, the slope of the solubility curve as predicted by the model is greater than the slope of the experimental curve. The deviations between the predicted and experimental solubilities, therefore, will increase as the pressure increases. The Peng-Robinson model with the conformal solution mixing rule exhibits the same solubility behavior as that model with the van der Waals mixing rule, also shown in Figure 1. At the pressures mentioned above, the model with the conformal solution mixing rule shows even larger deviations from the experimental solubility data. These general trends have been illustrated by Kurnik (ref.5,6) in earlier work with the Peng-Robinson equation of state. Manipulation of the binary interaction coefficient can decrease the deviation between the prediction of the model and the experimental data points over certain ranges of the pressure, but there will be an increase in the deviation over other ranges of the pressure.

In order to examine the role of the mixing rule in the model more closely, the binary interaction coefficient in four mixing rules was calculated for each experimental solubility data point on an isotherm. The four mixing rules were: van der Waals, van der Waals with surface fractions, conformal solution, and conformal solution with surface fractions. The results for the ethylene-naphthalene system at 35°C are contained in Figure 2. In general, the interaction coefficient in the four mixing rules all exhibit the same functional behavior with respect to pressure and temperature. The magnitude of the coefficient did differ, however, for the four mixing rules. Similar trends were observed for the same system at different temperatures and for other systems (i.e., carbon dioxide - naphthalene). The discrepancies in accurately predicting the solubility, therefore, seem to be due to the random fluid assumption and not the form of the particular mixing rule.

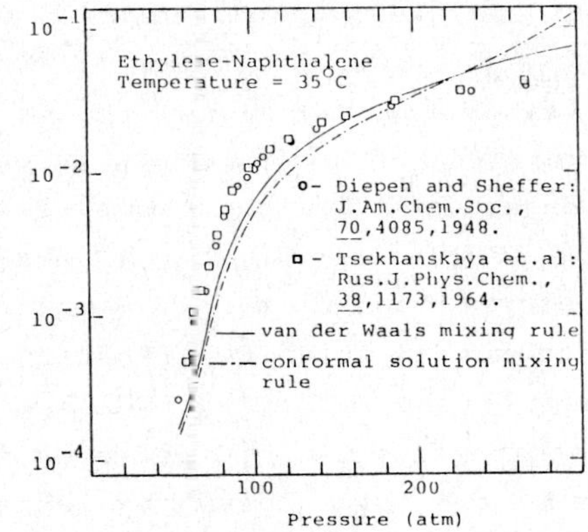

Figure 1. Solubility (p-y) isotherm of naphthalene in ethylene determined by Peng-Robinson model with two mixing rules.

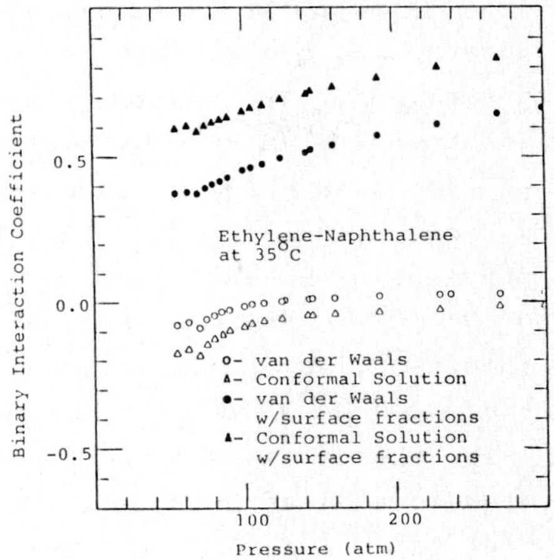

Figure 2. Binary interaction coefficient versus pressure obtained from experimental solubility data for the ethylene-naphthalene system.

Analysis of Interaction Coefficient Data

As suggested by the modification of the binary interaction coefficient, the coefficient should be some function of reduced density rather than pressure. In Figure 3, the binary interaction coefficient is plotted against reduced solvent density for the ethylene-naphthalene system. Examination of the binary interaction coefficients for a temperature of 35°C suggested that a linear function was appropriate, especially at the higher densities. The data from the 45°C solubility isotherm supports a linear function of the interaction coefficient with reduced solvent density. The data for the carbon dioxide-naphthalene system (Figure 4) illustrates a similar behavior.

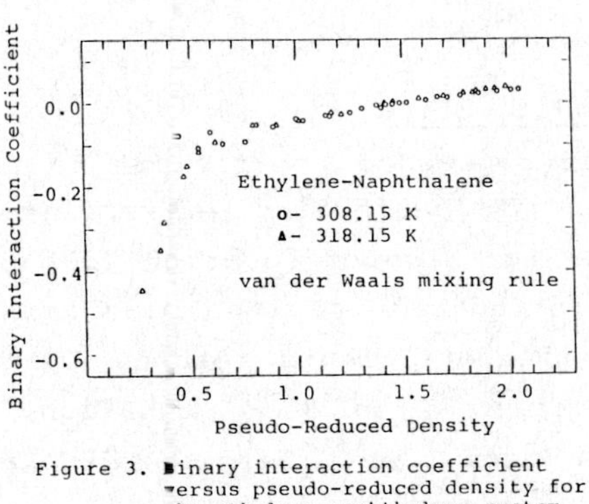

Figure 3. Binary interaction coefficient versus pseudo-reduced density for the ethylene-naphthalene system.

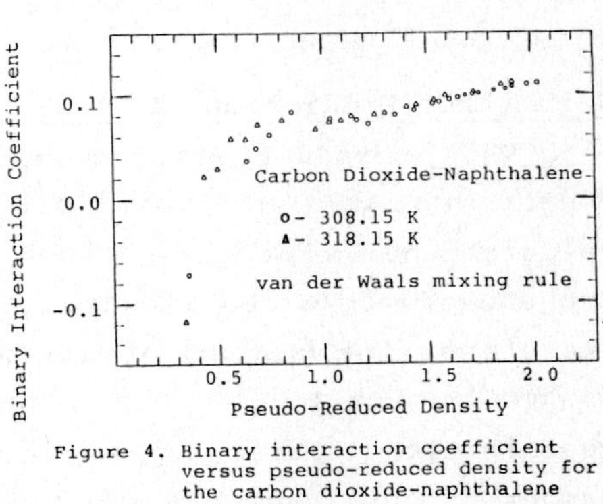

Figure 4. Binary interaction coefficient versus pseudo-reduced density for the carbon dioxide-naphthalene system.

Using a least squares method, the slope of the line for the interaction coefficient at reduced densities greater than 0.5 was determined. These lines all possessed a correlation coefficient in excess of 0.90. Using these values for the slope, the value of $\Delta\beta$, the average non-additive contribution to the depth of well parameter, in equation (23) was determined for the two systems, ethylene-naphthalene and carbon dioxide-naphthalene. A summary of the non-additive energy is contained in Table 3. Since the interest lies in the effects of the non-additive contribution on the i-j pair depth of well parameter, the percentage of ε_{12} is calculated. These percentages, ranging from 10 - 50 per cent for coordination numbers, are similar to the values given by Deiters (ref.22) and Kohler (ref.23) for the contribution of non-additive effects. This fact lends credibility to the density-functionality obtained for the binary interaction coefficient.

Table 3

Summary of Non-Additive Energy Contribution to

Depth of Well Parameter for i-j Pair

Ethylene-Naphthalene

Temperature (K)	Coordination Number	$\Delta\beta$ (ergs)	Percent of ε_{12}
308.15	6	1.676×10^{-13}	37.69
	8	1.934×10^{-13}	43.50
	12	2.370×10^{-13}	53.29
318.15	6	1.916×10^{-13}	43.74
	8	2.212×10^{-13}	50.50
	12	2.710×10^{-13}	61.86

Carbon Dioxide-Naphthalene

Temperature (K)	Coordination Number	$\Delta\beta$ (ergs)	Percent of ε_{12}
308.15	6	1.326×10^{-13}	29.26
	8	1.532×10^{-13}	33.80
	12	1.876×10^{-13}	41.38
318.15	6	1.411×10^{-13}	31.71
	8	1.629×10^{-13}	36.60
	12	1.996×10^{-13}	44.84

Solubility Predictions with the Modification

The introduction of a second regressed coefficient in the model increases the difficulties in performing extrapolative calculations. For this reason the function of equation (23) for the non-additive contribution is introduced into the model, and, therefore, the number of regressed coefficients returns to the original number for the model, one. In this equation, a coordination number of eight is used, and this coordination number is obtained from implied values for supercritical or dense fluids.

Using equations (22) and (23) in the model greatly improves the correlation between experimental and predicted solubilities. Not only is the deviation between the two solubilities lessened, but the slope of the predicted isotherm now resembles that of the experimental isotherm. Figure 5 illustrates the solubility isotherm predicted by the Peng-Robinson model with the modification of the binary interaction coefficient. There is good agreement between experimental and predicted solubilities over the full pressure range. For comparative purposes, the predicted isotherms for the carbon dioxide-naphthalene system with a constant binary interaction coefficient and with the coefficient modification are illustrated in Figure 6. There is again improvement in the solubility predictions with the coefficient linearization. Similar results are obtained for both systems at 318.15K. Figures 5 and 6 illustrate the improvement in solubility predictions using the binary interaction coefficient modification over that of the constant coefficient. The coefficients in equation (23), however, were obtained from an analysis of the data for these two systems at 35 and 45°C. Therefore, a more stringent test of the modification is to utilize it at a different temperature and system.

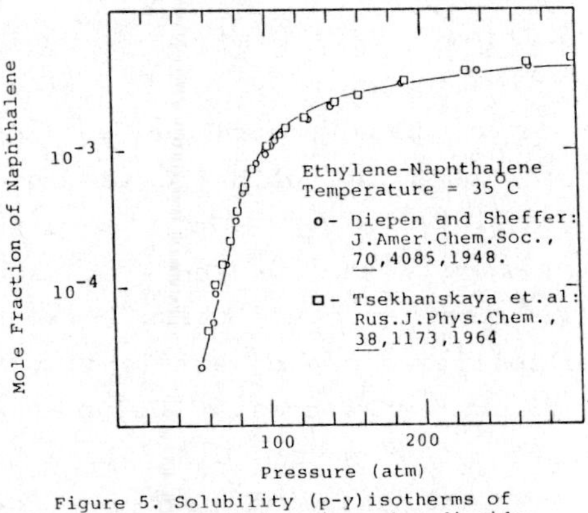

Figure 5. Solubility (p-y) isotherms of naphthalene in carbon dioxide determined with the binary interaction coefficient modification.

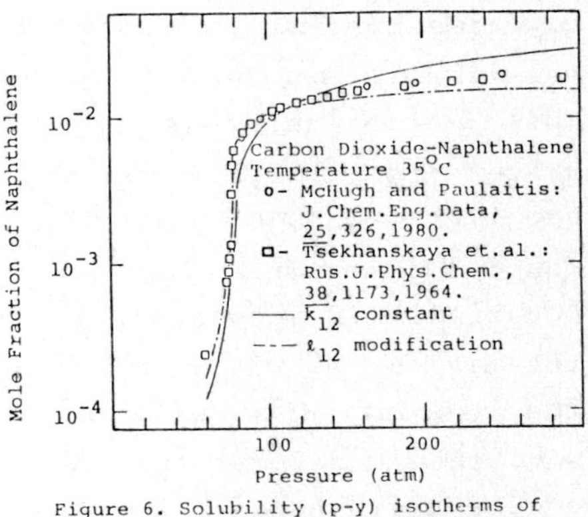

Figure 6. Solubility (p-y) isotherms of naphthalene in carbon dioxide determined with the binary interaction coefficient modification and a constant coefficient.

The solubility isotherms for the ethylene-naphthalene system at 25°C, obtained using the proposed modification and a constant coefficient, are compared in Figure 7. As in Figures 5 and 6, the modification of the binary interaction coefficient improves the correlation between experimental and predicted solubilities.

As further support, the solubility isotherms for the carbon dioxide-naphthalene system at 55°C is illustrated in Figure 8.

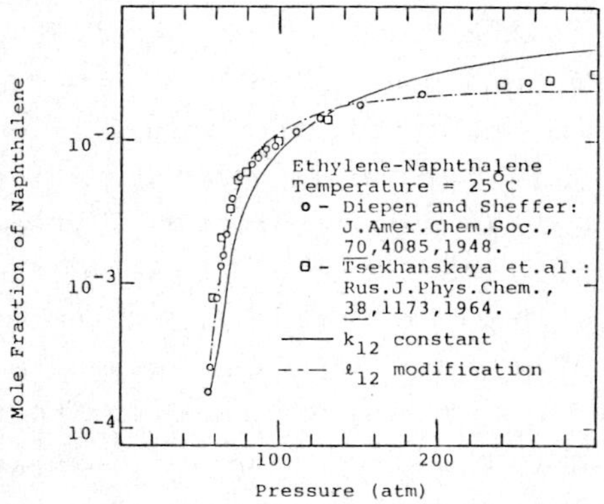

Figure 7. Solubility (p-y) isotherms of naphthalene in ethylene determined with the binary interaction coefficient modification and a constant coefficient.

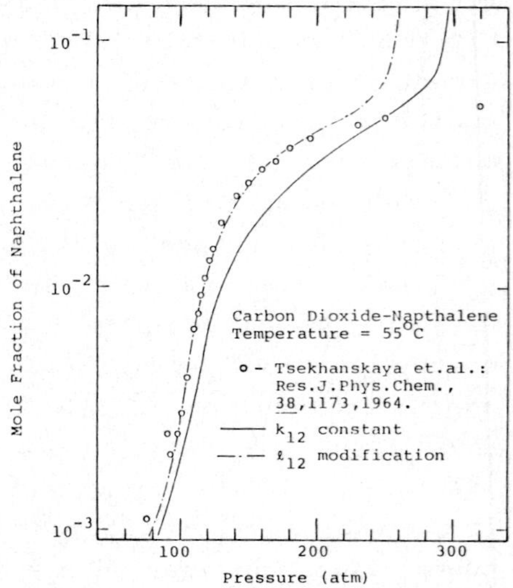

Figure 8. Solubility (p-y) isotherm of naphthalene in carbon dioxide determined with the binary interaction coefficient modification and a constant coefficient.

As in the previous examples, the modification of the binary interaction coefficient improves the correlation between the experimental and calculated solubilities. This case is unusual, however, due to the discontinuity in the calculated solubilities. This discontinuity is indicative of a phase transition of the system from solid-fluid coexisting phases to solid-liquid-fluid coexisting phases. Modell, Hong, and Heiba (ref.24) have also found similar behavior for this system with the Peng-Robinson equation. If a constant k_{12} is used, better correlation between the calculated and experimental solubilities can be obtained, but the discontinuity will occur at lower pressures. Using the suggested modification, the pressure at which the discontinuity occurs corresponds to the upper critical endpoint pressure for this system. Figures 9 and 10 compare the solubilities of anthracene in ethylene and carbon dioxide as predicted with the binary interaction coefficient modification and a constant coefficient. At pressures less than 200 atm, there is little difference between the two methods. This is probably due to the small amount of anthracene in the fluid phase. However, as the pressure increases, introducing the modification yields closer agreement between the calculated solubilities and the experimental.

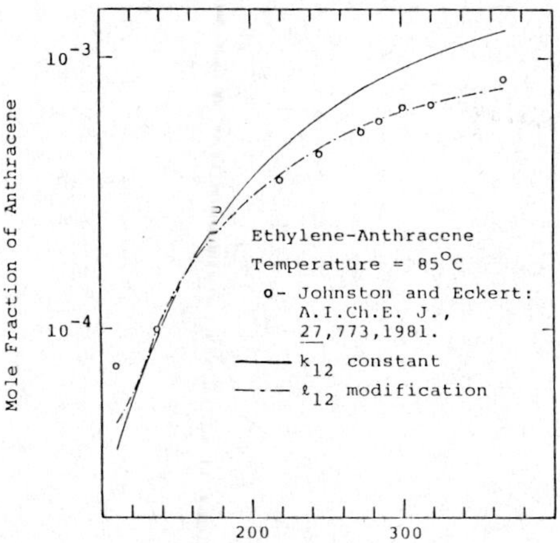

Figure 9. Solubility (p-y) isotherm of anthracene in ethylene determined with the binary interaction coefficient modification and a constant coefficient.

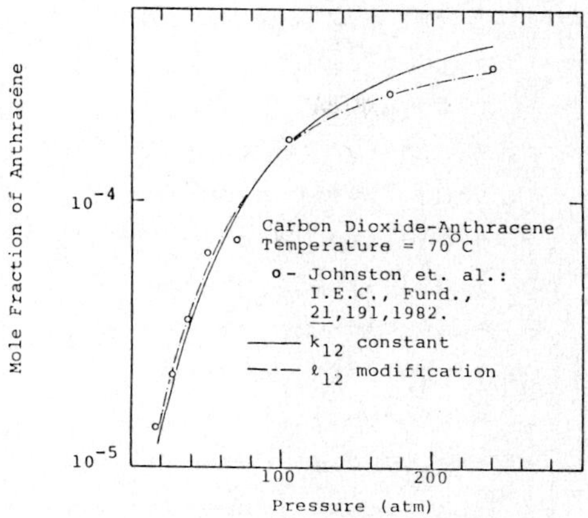

Figure 10. Solubility (p-y) isotherm of anthracene in carbon dioxide determined with the binary interaction coefficient modification and a constant coefficient.

CONCLUSIONS

The work performed indicates that the random-fluid approximation is inadequate for the calculation of solid-supercritical fluid equilibria. Models should incorporate the localized fluctuations of the concentration due to the molecular-size difference of the system and multiple-body interactions in order to adequately predict equilibria. Cubic equations can provide good predictions of the solubility of hydrocarbon solids in supercritical fluids if the proposed modification of the binary interaction coefficient is used. This modification takes into account the multiple-body interactions that occur at higher densities. It is not based on a theoretical derivation, but rather on a series of largely empirical arguments that nevertheless derive from the physics of asymmetrical mixtures. Prediction of the solubility using this modification is in excellent agreement with experimental data. This is achieved while the number of adjustable parameters in the equation of state remains one.

ACKNOWLEDGMENTS

The authors express their appreciation to Raymond V. Bailey, Vijay T. John, Blaine S. Fury and to the reviewers of this paper for helpful suggestions and comments. Judith Lozano typed the manuscript.

NOTATION

a = attractive parameter

b = covolume parameter

f = fugacity

k_{ij} = binary interaction coefficient

k = Boltzmann constant

ℓ_{12} = modified binary interaction coefficient

ℓ'_{12} = adjustable parameter

N_A = Avogadro's number

p = pressure

P = probability

R = universal gas constant

T = absolute temperature

T_r = reduced temperature

u_{ij} = intermolecular potential function

U = total system energy

Δu = non-additive energy for multi-body interactions

V = volume

v = molar volume

y = fluid phase mole fraction

z' = coordination number

Z = compressibility

GREEK SYMBOLS

ε = depth of potential well

$\Delta\varepsilon$ = non-additive energy for multi-body interaction with square-well potential

ρ = density

ρ^* = $b\rho$

ϕ = fugacity coefficient

σ = hard core diameter

ω = accentric factor

κ, α = weighting factors

SUPERSCRIPTS

s solid phase

F fluid phase

^ mixture property

eff effective value

(n) number of bodies interacting

* reduced property with respect to covolume parameter

sat saturation conditions

References

1. Hannay, J.B., and J. Hogarth, "On the Solubility of Solids in Gases," Proc. Roy. Soc., London, A29, p. 324 (1879).

2. Williams, D.F.: "Extraction with Supercritical Gases," Chem. Eng. Sci., 36, 1769 (1981).

3. Mackay, M.E., and M.E. Paulaitis, "Solid Solubilities of Heavy Hydrocarbons in Supercritical Solvents," I&EC Fund., 18, 149 (1979).

4. Prausnitz, J.M., Molecular Thermodynamics of Fluid-Phase Equilibria, Prentice-Hall, Englewood Cliffs, New Jersey, (1969).

5. Kurnik, R.T., S.J. Holla, and R.C. Reid, "Solubility of Solids in Supercritical Carbon Dioxide and Ethylene," J. Chem. Eng. Data, 26, 47 (1981).

6. Kurnik, R.T., and R.C. Reid, "Solubility Extremes in Solid-Fluid Equilibria," AIChE J., 27, 861 (1981b).

7. Eckert, C.A., D.H. Ziger, K.P. Johnston and T.K. Ellison: "The Use of Partial Molal Volume Data to Evaluate Equations of State for Supercritical Fluid Mixtures," Fluid Phase Equil., 14, 167 (1983).

8. Johnston, K.P., and C.A. Eckert, "An Analytical Carnahan - Starling-van der Waals Model for Solubility of Hydrocarbon Solids in Supercritical Fluids," AIChE J., 27, 773 (1981).

9. Johnston, K.P., D.H. Ziger, and C.A. Eckert, "Solubilities of Hydrocarbon Solids in Supercritical Fluids. The augmented van der Waals Treatment," I&EC Fund., 21, 191 (1982).

10. Peng, D.Y., and D.B. Robinson, "A New Two-Constant Equation of State," I&EC Fund., 15, 59 (1976).

11. Chuen, P.L., and J.M. Prausnitz, "Vapor-Liquid Equilibria at High Pressures. Vapor-Phase Fugacity Coefficients in Nonpolar and Quantum-Gas Mixtures," I&EC Fund., 6, 492 (1967).

12. Radosz, M., H.M. Lin, and K.C. Chao, "High-Pressure Vapor-Liquid Equilibria in Asymmetric Mixtures Using New Mixing Rules," I&EC Proc. Des. Dev., 21, 653 (1982).

13. Kreglewski, A., and W.B. Kay, "The Critical Constants of Conformal Mixtures," J. Phys. Chem., 73, 3359 (1969).

14. Donohue, M.D., and J.M. Prausnitz, "Perturbed Hard Chain Theory for Fluid Mixtures: Thermodynamic Properties for Mixtures in Natural Gas and Petroleum Technology," AIChE J., 24, 849 (1978).

15. Snider, N.S., and T.M. Herrington, "Hard-Sphere Model of Binary Liquid Mixtures," J. Chem. Phys., 47, 2248 (1967).

16. Vera, J.H., and J.M. Prausnitz, "Generalized van der Waals Theory for Dense Fluids," Chem. Eng. J., 3, p. 1 (1972).

17. Barker, J.A., D. Henderson, and W.R. Smith, "Pair and Triplet Interactions in Argon," Mol. Phys., 17, 579 (1969).

18. Casanova, G., R.J. Dulla, D.A. Jonah, J.S. Rowlinson, and G. Saville, "Effective Pair Potentials in Fluids in the Presence of Three-Body Forces," Mol. Phys., 18, 589 (1970).

19. Dulla, R.J., J.S. Rowlinson, and W.R. Smith, "Effective Pair Potentials in Fluids in the Presence of Three-Body Forces. II," Mol. Phys., 21, 299 (1971).

20. Singh, V., "Perturbation Theory for Fluids of Non-Spherical Molecules in the Presence of Three-Body Forces. I" Mol. Phys., 29, 155 (1975).

21. Sherwood, A.E., and J.M. Prausnitz, "Third Virial Coefficient for the Kihara, Exp-6, and Square-Well Potentials," J. Chem. Phys., 41, 413 (1964).

22. Deiters, U., "A New Semiempirical Equation of State for Fluids-I.," Chem. Eng. Sci., 36, 1139 (1981).

23. Kohler, F.: The Liquid State, Monographs in Modern Chemistry, Volume 1, Verlag Chernie GmbH, Weinheim, Germany, p. 79, 1972.

24. Modell, M., G. Hong, and A. Heiba: "The Use of the Peng-Robinson Equation in Determining Supercritical Behavior," A.I.Ch.E. Annual Meeting, San Francisco, CA, 1979.

Supercritical Fluid Technology, edited by J.M.L. Penninger, M. Radosz, M.A. McHugh and V.J. Krukonis
1985 Elsevier Science Publishers B.V., Amsterdam — Printed in The Netherlands

APPLICATION OF THE GROUP CONTRIBUTION EQUATION OF STATE TO SUPERCRITICAL FLUID EXTRACTION

E.A. BRIGNOLE, S. SKJOLD-JØRGENSEN and AA. FREDENSLUND
Instituttet for Kemiteknik, The Technical University of Denmark, DK-2800 Lyngby, Denmark

ABSTRACT

A group-contribution equation of state and new parameter table are used to design near critical fluid (NCF) extraction processes for separating alcohols from water. A new energy-efficient process is proposed for producing pure ethanol from fermentation broth, which uses propane or butane as solvents. This process requires only 3400 kJ/kg of absolute ethanol compared with 10000 kJ/kg of conventional azeotropic distillation.

INTRODUCTION

Near critical fluid (NCF) extraction for separating alcohols from water has several attractive features: high selectivity, low solubility of nonpolar NCF in the aqueous phase, ease of solvent recovery and low NCF heats of vaporization. However , the search for suitable solvents and process conditions is hindered by the scarcity of experimental data and difficulties involved in predicting phase equilibria in mixtures of polar and nonpolar components in the critical region.

Table 1 which shows the available experimental data on phase equilibria for alcohol-water-NCF ternary systems indicates that carbon dioxide has been the most frequently studied solvent for separating ethanol from water. However, it has been shown that using supercritical or liquid carbon dioxide as a solvent cannot compete economically with conventional distillation (Brignole et al (1984)). Therefore, other solvents with higher critical temperatures, such as propane or butane, have been proposed to obtain better separation of ethanol from water. In addition, it was shown that propane and butane could produce absolute ethanol whereas carbon dioxide could not.

Brignole et al (1984) have demonstrated the applicability of a group contribution equation of state (GC-EOS) developed by Skjold-Jørgensen (1984) to the prediction of phase equilibria of ethanol-water mixtures with liquid or supercritical carbon dioxide. The GC-EOS was initially developed for gas solubility predictions up to 25 MPa in polar and nonpolar solvents and their mixtures. The horizontally shaded area in Fig. 1 indicates that data from low concentration high-density region were used to determine the GC-EOS parameters. On the other hand, NCF extraction involves high NCF concentration and intermediate phase densities. Thus, considerable extrapolation is needed to bridge these regions.

Previously published parameters for gas-solvent interactions were obtained mainly from gas solubility data (Skjold-Jørgensen (1984)). Also, since the available group parameters are valid only for primary alcohols, an extended data base was used to establish new parameters for primary, secondary and tertiary alcohols, water, carbon dioxide, propane and other paraffinic groups.

GROUP CONTRIBUTION EQUATION OF STATE

In the derivation of the GC-EOS, Skjold-Jørgensen (1984) assumed that the residual Helmholtz function is the sum of two contributions, repulsive and attractive,

$$A^R = A^R_{rep} + A^R_{attr} \tag{1}$$

The repulsive term is that for hard spheres derived by Mansoori and Leland (1972). The attractive term is based on the van der Waals equation

$$\frac{A^R_{attr}}{N} = - a\rho \tag{2}$$

where the energy parameter has been defined as:

$$a = \frac{z}{2} g q^2 \tag{3}$$

to make it amenable for group contribution treatment. The parameter g is the characteristic attractive energy per segment, and q is the surface area parameter as defined in the UNIFAC method (Fredenslund et al (1975)).

The computation of A^R_{attr} for mixtures is based on an expression similar to the NRTL equation (Renon and Prausnitz (1968)), but written in a group contribution form:

$$A^R_{attr} = -\frac{z}{2} \sum_i^{NC} N_i \sum_j^{NG} \upsilon_j^i q_j \sum_k^{NG} \theta_k (g_{kj} \tilde{q}\tau_{kj} \rho) / \sum_\ell^{NG} \theta_\ell \tau_{\ell j} \tag{4}$$

where

$$\tau_{kj} = \exp(\alpha_{kj}(g_{kj}-g_{jj})\tilde{q}\rho/RT)$$

TABLE I

Experimental Data on Phase Equilibria of Alcohol-Water-NCF Ternary Systems.

Alcohol	NCF	Temperature range, K	Pressure range, MPa	Authors and Comments
Ethanol	CO_2	313 - 333	7.5-20.0	Kuk and Montagna (1983), ethanol loadings int the NCF at given ethanol/water feed ratios.
2-Propanol	CO_2	313	10.3	Ibid
Ethanol	CO_2	308 - 338	10.3-17.7	Paulaitis et al (1981), tie lines.
Ethanol	Ethylene	308	10.3-17.7	Ibid
Ethanol	Ethane	313 - 323	5.0 - 8.0	McHugh et al (1983), bubble and dew points.
Ethanol	CO_2	298	6.5	Moses et al (1982), average distribution coefficient (liquid CO_2)
2-Propanol	CO_2	333	10.4	Radosz (1984), graphical tie lines
2-Propanol	CO_2	313 - 333	10.0-13.2	Paulaitis et al (1984), graphical tie lines.

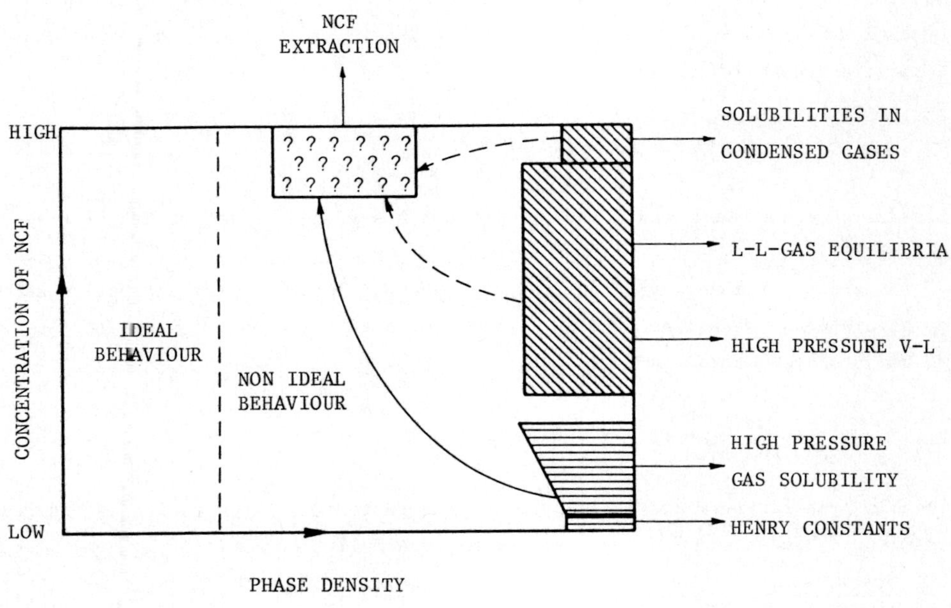

Fig. 1 Experimental Basis of the GC-EOS.

and

$$\tilde{q} = \sum_{i}^{NC} N_i \sum_{j}^{NG} \upsilon_j^i \, q_j$$

$$\theta_j = (q_j/\tilde{q}) \sum_{i}^{NC} N_i \, \upsilon_j^i$$

As in the UNIFAC equation the interactions are considered to take place through the surface of the segments, and correspondingly surface fractions are used instead of mole fractions.

The total Helmholtz function is obtained by

$$A = A^R_{rep} + A^R_{attr} + A^{ideal} \tag{5}$$

and the equation of state is obtained by differentiation with respect to volume:

$$P = (\frac{\partial A}{\partial V})_{T,N} \tag{6}$$

GC-EOS Parameters

The repulsive part of the GC-EOS requires only pure-component parameters, namely the hard-sphere diamaters. These are obtained from acentric factors, normal boiling points or particular vapor pressures. The hard-sphere diameter is assumed temperature-dependent following the generalized expression proposed by Fermeglia and Mollerup (1984):

$$d = d_c(1.- 0.12 \exp(- \frac{2}{3}(T_c/T)) \; 1.065655 \tag{7}$$

Besides pure-component parameters the GC-EOS has group parameters and binary parameters.

The group parameters are the group surface parameter and the attractive energy parameter between like segments (g_{jj}). The attractive energy parameter is assumed to be temperature-dependent:

$$g_{jj} = g^*_{jj}(1.+ g'_{jj}(\frac{T}{T^*_j} - 1) + g''_{jj} \; \ell n(\frac{T}{T^*_j})) \tag{8}$$

The binary parameters are the binary damping factors (α_{ij}), which are considered asymmetric, and the attractive energies between unlike segments (g_{ij}), which are defined as:

$$g_{ij} = k_{ij} \sqrt{g_{ii} g_{jj}} \tag{9}$$

where k_{ij} is a symmetric binary parameter, which is temperature dependent:

$$k_{ij} = k^*_{ij}(1+k'_{ij}\ \ell n(\frac{T}{T^*_{ij}})) \qquad (10)$$

where

$$T^*_{ij} = (T^*_i + T^*_j)/2 \qquad (11)$$

PARAMETER ESTIMATION

The group parameters were estimated in the way explained by Skjold-Jørgensen (1984). The surface area parameters for the secondary and tertiary alcohol groups were treated as adjustable parameters. This way, it was possible to obtain the same group parameters for all alcohol-alkane interactions. The q values obtained are only slightly different from the values based on the Bondi surface areas.

Furthermore a distinction is made between the paraffinic groups CH_3 and WS-CH_3 for nonpolar (paraffinic or aromatic) and polar compounds that are water soluble (WS) like alcohols or ketones. Besides additional groups (different surface areas) are introduced when paraffinic groups are linked to a quaternary carbon, because there is a reduction in the group surface area that is attributable to a steric effect (CH_3(B), CH_2(B), WS-CH_3(B) and WS-CH_2(B)).

The pure group parameters are shown in Table II, the binary parameters k_{ij} are shown in Table III, and the binary damping factors are shown in Table IV. It should be noted that the binary interaction parameters for the alkane-alcohol and alkane-water groups have been assumed temperature-independent. References for the experimental data used to estimate the parameters are available as suplementary material from the authors.

CORRELATION AND PREDICTION OF BINARY PHASE EQUILIBRIA

The GC-EOS capabilities to represent highly asymmetric systems have been tested on mutual solubilities of water and n-alkanes. The comparison with experimental values is shown in Fig.2; the experimental values are the selected values recommended by Sørensen and Arlt (1979). It can be seen that the solubilities are predicted in both phases with acceptable accuracy using a single set of temperature-independent group interaction paramenters.

Simultaneous correlation of vapor-liquid equilibria and liquid-liquid equilibria is another difficult test for the GC-EOS. Fig.3 shows predicted mutual solubilities of n-alcohols and water compared with smoothed experimental data recommended by Sørensen and Arlt (1979). Fig.4 and Fig.5 show predicted vapor-liquid equilibria for n-butanol - water and ethanol - water mixtures compared

TABLE II

Pure Group Parameters for the GC-EOS

Group	No.	T /K	q	g^*_{jj}	g'_{jj}	g''_{jj}
CH_3	1	600.0	0.848	316910.	-0.9274	0.00
CH_2	2	600.0	0.540	356080.	-0.8755	0.00
CH	3	600.0	0.228	356080.	-0.8755	0.00
CH_3(B)	4	600.0	0.789	316910.	-0.9274	0.00
CH_2(B)	5	600.0	0.502	355130.	-0.8755	0.00
WS-CH_3	6	600.0	0.848	316910.	-0.9274	0.00
WS-CH_2	7	600.0	0.540	356080.	-0.8755	0.00
WS-CH	8	600.0	0.228	356080.	-0.8755	0.00
WS-CH_3(B)	9	600.0	0.789	316910.	-0.9274	0.00
WS-CH_2(B)	10	600.0	0.502	355130.	-0.8755	0.00
PROPANE	11	369.8	2.236	436890.	-0.4630	0.00
CH_2OH	12	512.6	1.124	1207500.	-0.6502	0.00
CHOH	13	512.6	0.908	1207500.	-0.6502	0.00
COH	14	512.6	0.714	1207500.	-0.6502	0.00
H_2O	15	647.3	0.866	1697000.	-0.6707	0.00
CO_2	16	304.2	1.261	531890.	-0.5780	0.00

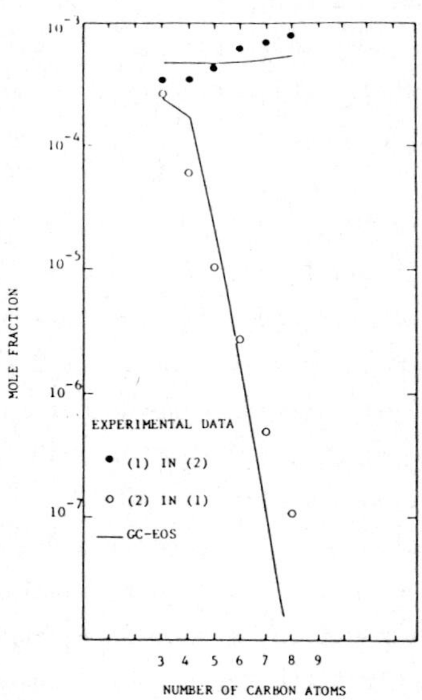

Fig. 2 Binary Mutual Solubilities: Water(1)- N-Alkanes(2), T=298 K

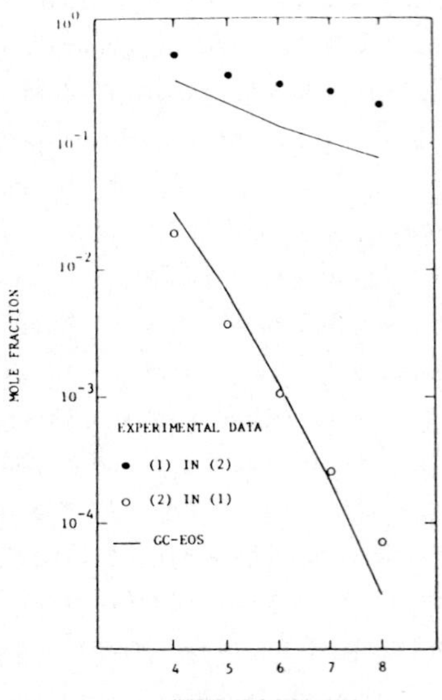

Fig.3 Binary Mutual Solubilities: Water(1)- N-Alkanols(2),T=298 K

TABLE III

Binary k^*_{ij} and k'_{ij} parameters for the GC-EOS. The k^*_{ij} values are found above the diagonal

No.	1	2	3	4	5	6	7	8
1	x	1.000	1.000	1.000	1.000	1.000	1.000	1.000
2	0.000	x	1.000	1.000	1.000	1.000	1.000	1.000
3	0.000	0.000	x	1.000	1.000	1.000	1.000	1.000
4	0.000	0.000	0.000	x	1.000	1.000	1.000	1.000
5	0.000	0.000	0.000	0.000	x	1.000	1.000	1.000
6	0.000	0.000	0.000	0.000	0.000	x	1.000	1.000
7	0.000	0.000	0.000	0.000	0.000	0.000	x	0.000
8	0.000	0.000	0.000	0.000	0.000	0.000	0.000	x
9	0.000	0.000	0.000	0.000	0.000	0.000	0.000	0.000
10	0.000	0.000	0.000	0.000	0.000	0.000	0.000	0.000
11	0.000	0.000	0.000	0.000	0.000	0.000	0.000	0.000
12	0.000	0.000	0.000	0.000	0.000	0.000	0.000	0.000
14	0.000	0.000	0.000	0.000	0.000	0.000	0.000	0.000
15	0.000	0.000	0.000	0.000	0.000	0.000	0.000	0.000
16	0.000	0.000	0.000	0.000	0.000	0.000	0.000	0.000

No.	9	10	11	12	13	14	15	16
1	1.000	1.000	1.000	0.715	0.715	0.715	0.560	0.892
2	1.000	1.000	1.000	0.682	0.682	0.682	0.560	0.814
3	1.000	1.000	1.000	0.682	0.682	0.682	0.560	0.814
4	1.000	1.000	1.000	0.715	0.715	0.715	0.560	0.892
5	1.000	1.000	1.000	0.682	0.682	0.682	0.560	0.814
6	1.000	1.000	1.000	0.715	0.715	0.715	0.560	0.892
7	1.000	1.000	1.000	0.682	0.682	0.682	0.560	0.814
8	1.000	1.000	1.000	0.682	0.682	0.682	0.560	0.814
9	x	1.000	1.000	0.715	0.715	0.715	0.560	0.892
10	0.000	x	1.000	0.682	0.682	0.682	0.560	0.814
11	0.000	0.000	x	0.800	0.800	0.800	0.560	0.875
12	0.000	0.000	0.231	x	1.000	1.000	0.949	0.985
13	0.000	0.000	0.231	0.000	x	1.000	0.949	0.985
14	0.000	0.000	0.231	0.000	0.000	x	0.949	0.985
15	0.000	0.000	0.000	-0.191	-0.191	-0.191	x	0.923
16	0.000	0.000	0.000	0.000	0.000	0.000	0.122	x

TABLE IV

Binary damping factors for the GC-EOS

No.	1	2	3	4	5	6	7	8
1	0.000	0.000	0.000	0.000	0.000	0.000	0.000	0.000
2	0.000	0.000	0.000	0.000	0.000	0.000	0.000	0.000
3	0.000	0.000	0.000	0.000	0.000	0.000	0.000	0.000
4	0.000	0.000	0.000	0.000	0.000	0.000	0.000	0.000
5	0.000	0.000	0.000	0.000	0.000	0.000	0.000	0.000
6	0.000	0.000	0.000	0.000	0.000	0.000	0.000	0.000
7	0.000	0.000	0.000	0.000	0.000	0.000	0.000	0.000
8	0.000	0.000	0.000	0.000	0.000	0.000	0.000	0.000
9	0.000	0.000	0.000	0.000	0.000	0.000	0.000	0.000
10	0.000	0.000	0.000	0.000	0.000	0.000	0.000	0.000
11	0.000	0.000	0.000	0.000	0.000	0.000	0.000	0.000
12	10.220	10.220	10.220	10.220	10.220	10.220	10.220	10.220
13	10.220	10.220	10.220	10.220	10.220	10.220	10.220	10.220
14	10.220	10.220	10.220	10.220	10.220	10.220	10.220	10.220
15	0.494	0.494	0.494	0.494	0.494	1.229	0.637	0.637
16	3.369	3.369	3.369	3.369	3.369	3.369	3.369	3.369

No.	9	10	11	12	13	14	15	16
1	0.000	0.000	0.000	1.471	1.471	1.471	0.494	3.369
2	0.000	0.000	0.000	1.471	1.471	1.471	0.494	3.369
3	0.000	0.000	0.000	1.471	1.471	1.471	0.494	3.369
4	0.000	0.000	0.000	1.471	1.471	1.471	0.494	3.369
5	0.000	0.000	0.000	1.471	1.471	1.471	0.494	3.369
6	0.000	0.000	0.000	1.471	1.471	1.471	1.229	3.369
7	0.000	0.000	0.000	1.471	1.471	1.471	0.637	3.369
8	0.000	0.000	0.000	1.471	1.471	1.471	0.637	3.369
9	0.000	0.000	0.000	1.471	1.471	1.471	1.229	3.369
10	0.000	0.000	0.000	1.471	1.471	1.471	0.637	3.369
11	0.000	0.000	0.000	1.051	1.051	1.051	0.435	3.369
12	10.220	10.220	10.010	0.000	0.000	0.000	-1.370	-0.390
13	10.220	10.220	10.010	0.000	0.000	0.000	-1.933	-0.390
14	10.220	10.220	10.010	0.000	0.000	0.000	-2.188	-0.390
15	1.229	0.637	0.434	-1.370	-1.933	-2.188	0.000	-0.393
16	3.369	3.369	3.369	0.468	0.468	0.468	0.236	0.000

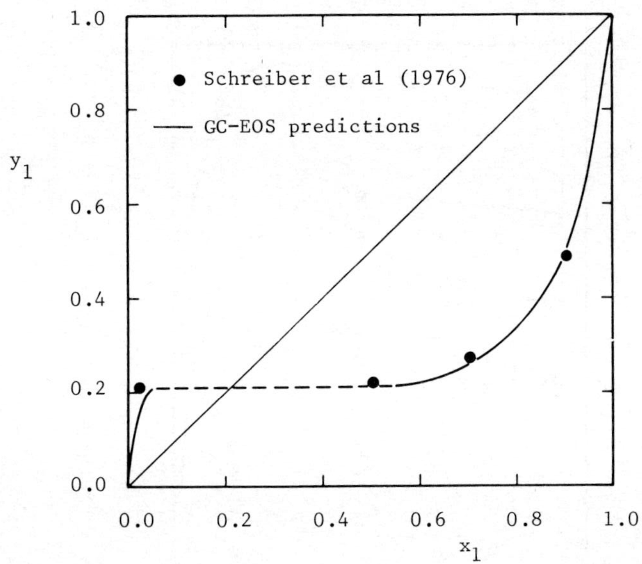

Fig. 4 Vapor Liquid Equilibria, System: N-Butanol(1)-Water(2), T=333 K.

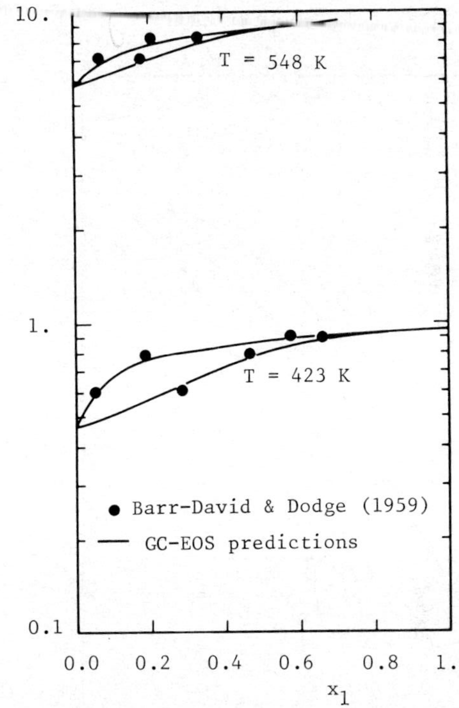

Fig. 5 Vapor Liquid Equilibria, System:Ethanol(1)-Water(2).

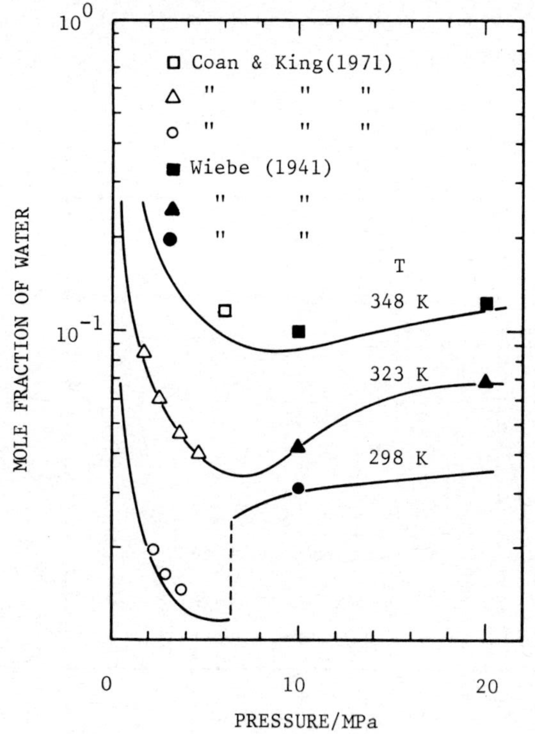

Fig. 6 Water Solubility in Compressed and Liquid CO_2.

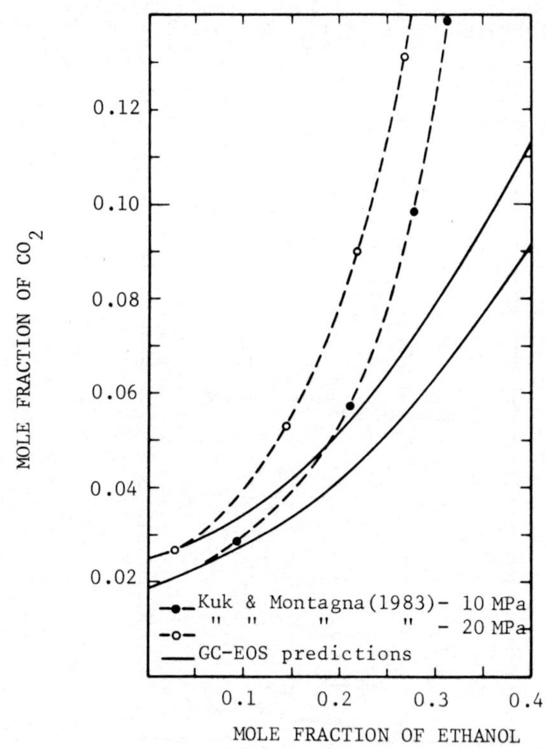

Fig. 7 Solubility of CO_2 in ethanol-water mixtures at 333 K.

with experimental data. The agreement between experimental and predicted values is quite good considering the wide range of pressures, temperatures and alcohol molecular weights covered with a single set of binary group interaction parameters. Also, water solubility in compressed and liquid carbon dioxide, as shown in Fig. 6 is well represented by the GC-EOS.

The NCF extraction of liquids differs from the NCF extraction of solids in that NCF solvents are, to some extent, soluble in the liquid phase. While the solubility of gases in mixed solvents is within the predictive capabilities of the GC-EOS, the predicted solubilities in aqueous solutions may be off by 50 % (Skjold-Jørgensen (1984)). Fig. 7 shows that, for mole fractions of ethanol in the aqueous phase greater than 0.2, discrepancies between the calculated and experimental mole fractions, measured by Kuk and Montagna(1983), become significant. This limits, in the case of carbon dioxide the applicability of the GC-EOS to ethanol concentrations in water lower than 0.15 mole fraction. The same limitation applies to isopropanol- and n-propanol-water mixtures.

PREDICTION OF NCF PHASE EQUILIBRIA FOR TERNARY SYSTEMS

Water-ethanol-carbon dioxide experimental phase equilibrium data (Kuk and Montagna (1983)) are compared with GC-EOS predictions in Fig. 8. The predicted solubilities of water and ethanol in the carbon dioxide phase are in good agreement with experiment. However, for ethanol mole fractions in the liquid phase greater than 0.2, the predictions of ethanol solubilities are underestimated by 30%. Also, it can be seen that the increase in solubility with pressure, above the critical pressure of carbon dioxide, is much greater for ethanol than for water. Table V shows another comparison between calculated and measured (Paulaitis et al (1981) tie lines for this system. Even though reasonable predictions are obtained at low alcohol concentrations, the GC-EOS breaks down when the alcohol concentration in the liquid phase is greater than 0.2 mole fraction. The most important and difficult property to predict is the alcohol distribution coefficient because it requires accurate prediction of the alcohol fugacity coefficient in both polar and nonpolar phases. Therefore, predictions of the GC-EOS should be taken as a first approximation because errors of the order of 30 % can be expected.

However, the effects of pressure and temperature on the distribution coefficient are correctly represented. This can be seen in Fig. 9 and 10 where predictions for the system water-isopropanol-carbon dioxide are shown in comparison with experimental data. Paulaitis et al (1984) have shown that and increase in pressure from 10. MPa to 13 MPa at 333 K has a pronounced effect on the value of the distribution coefficient of isopropanol. Fig. 9 shows that the isopropanol distribution coefficient can increase by a factor of four when the pre-

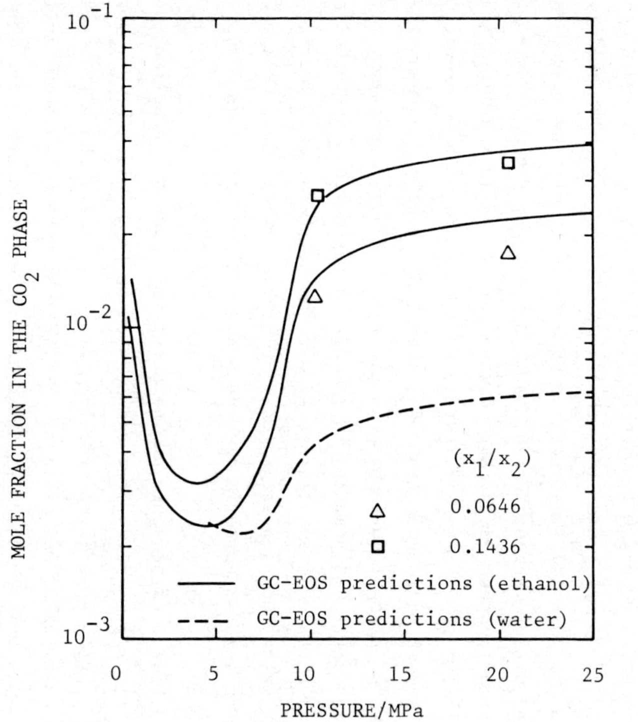

Fig. 8 Ethanol(1)-Water(2)-CO_2(3) Phase Equilibria at 313 K. Comparison of Experimental Data of Kuk and Montagna (1983) with GC-EOS predictions.

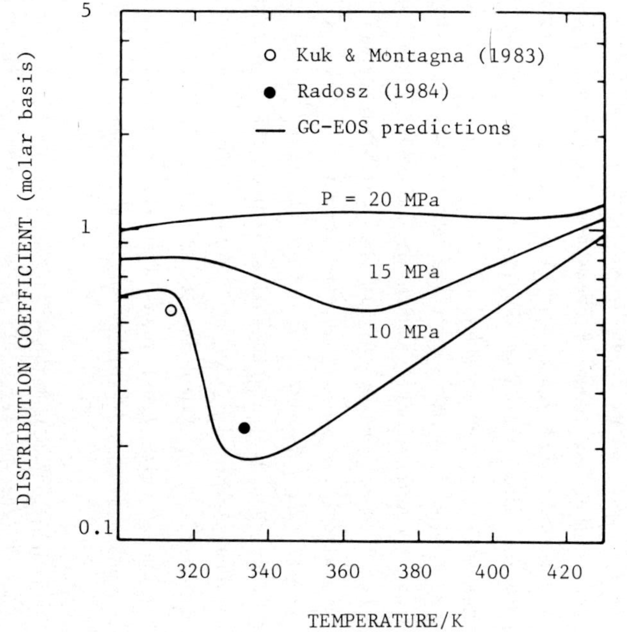

Fig. 9 Distribution Coefficients of Isopropanol between Water and CO_2. Isopropanol mole fraction in the aqueous phase: 0.114 (free of CO_2 basis)

TABLE V

Comparison of Calculated and Experimental Tie Lines for Water(1), ethanol(2) and CO_2(3). Data of Paulaitis et al(1981) at 308 K.

P/MPa		mole fractions					
		y_1	y_2	y_3	x_1	x_2	x_3
10.3	exp	0.0040	0.005	0.991	0.944	0.032	0.024
	cal	0.0038	0.0077	0.989	0.945	0.033	0.022
	exp	0.007	0.0210	0.972	0.852	0.118	0.030
	cal	0.0041	0.0226	0.973	0.855	0.118	0.027
	exp	0.0120	0.0440	0.944	0.678	0.249	0.074
	cal	0.0042	0.0320	0.964	0.696	0.255	0.048
13.8	exp	0.0040	0.005	0.991	0.943	0.032	0.025
	cal	0.0047	0.011	0.984	0.943	0.033	0.024
	exp	0.006	0.020	0.974	0.851	0.118	0.030
	cal	0.0048	0.028	0.967	0.854	0.117	0.028
	exp	0.015	0.054	0.931	0.691	0.238	0.071
	cal	0.005	0.040	0.955	0.694	0.255	0.051
17.2	exp	0.004	0.005	0.991	0.942	0.032	0.026
	cal	0.0047	0.011	0.984	0.943	0.033	0.024
	exp	0.009	0.025	0.966	0.851	0.118	0.031
	cal	0.0052	0.032	0.962	0.853	0.117	0.029
	exp	0.015	0.053	0.932	0.683	0.242	0.075
	cal	0.0056	0.046	0.948	0.693	0.254	0.053

TABLE VI

Predicted Distribution Coefficients of Alcohols at Infinite Dilution between Water and N-Butane at 5 MPa. Mole fraction basis.

Alcohol	T = 298 K	T = 363 K	T = 403 K
Ethanol	0.23	0.83	1.61
N-Propanol	0.90	2.69	4.48
2-Propanol	0.69	2.00	3.4
N-Butanol	3.06	7.72	11.1
Iso-Butanol	2.69	4.64	6.5
2-Butanol	5.04	9.96	13.2
T-Butanol	1.27	3.08	4.9
N-Pentanol	12.8	25.3	30.8

TABLE VII

Predicted Distribution Coefficients at Infinite Dilution of Alcohols between Water and Carbon Dioxide at 298 K and 7.6 MPa

Alcohol	Distribution Coefficient
Ethanol	0.190
N-Propanol	0.500
2-Propanol	0.470
N-Butanol	1.11
Isobutanol	1.73
2-Butanol	2.35
T-Butanol	0.72
N-Pentanol	3.02

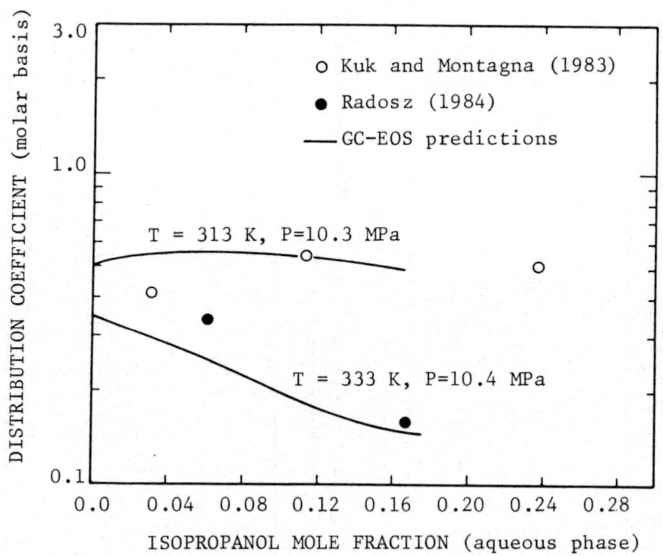

Fig. 10 Distribution Coefficients of Isopropanol between Water and CO_2.

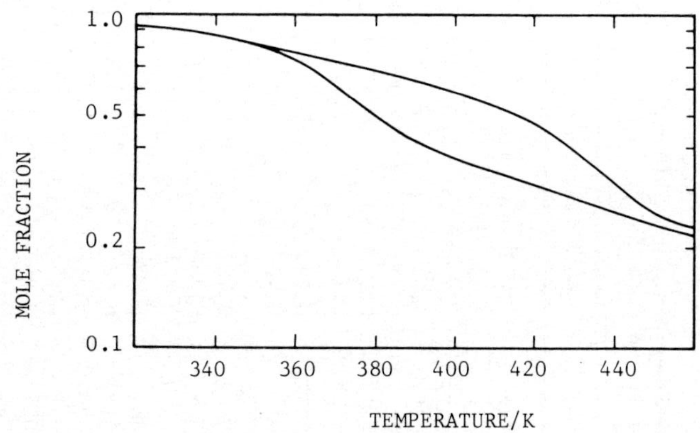

Fig. 11 Ethanol mole fraction in the extract (free of CO_2 basis). The ethanol mole fraction in the feed to the extractor is 0.043.

ssure is increased from 10 to 15 MPa. This is attributable to large variations in density with pressure and temperature near the critical region.

Also, when the mole fraction of IPA in the aqueous phase is greater than 0.15, the existance of three phases, vapor-liquid-liquid, is predicted at 333 K and 10.4 MPa. This is consistent with the findings of Paulaitis et al (1984) and Radosz (1984) who found three phases near these conditions.

NCF EXTRACTION WITH LIGHT HYDROCARBONS

A major motivation for the revision of the GC-EOS parameter table was to obtain more reliable predictions for ethanol-water mixtures with light hydrocarbons near their critical conditions. Such solvents allow for operation at higher temperatures and, therefore, lead to higher distribution coefficients, as shown in Table VI.

Another important factor, selectivity of a NCF solvent, can be quantified by the ethanol concentration in the extract on a solvent-free basis. This concentration decreases with increasing extraction temperatures, Fig.11. Therefore, an extract less rich in ethanol is obtained at higher temperatures. However this loss can be compensated if the NCF solvent exhibits a water entrainment effect. By this effect, water is removed from ethanol while recovering the solvent.

Water can be removed from ethanol in the solvent recovery tower if its relative volatility with respect to the solvent is greater than one. The predicted relative volatility of water with respect to propane and butane is greater than one for high solvent concentrations in the extract phase, as shown in Fig.12 and discussed in Appendix. However, the predicted relative volatily of water with respect to carbon dioxide is less than one, which is also shown in Fig.12. Therefore, propane and butane, but not carbon dioxide can produce pure ethanol as the bottoms from the solvent recovery tower. This has also been demonstrated by simulation of the extract distillation process using a rigorous multi-stage distillation program coupled with the GC-EOS predicted fugacities.

To sum up, even though the distribution coefficients of n-propanol, isopropanol and higher alcohols between water and carbon dioxide are reasonably high, carbon dioxide extraction can not produce pure alcohols. The use of light hydrocarbons, as dual effect solvents (extraction and entrainment), results in both high distribution coefficients (Table VI) and the possibility to produce pure alcohols.

One process alternative utilizing NCF extraction solvents is that based on a heat pump scheme (Brignole et al (1984)). This way significant energy savings, compared to conventional distillation were demonstrated for propane- or butane-based ethanol-from-water extraction cycles. However the use of a heat pump results in greater capital cost and limits precess conditions. For instance, the

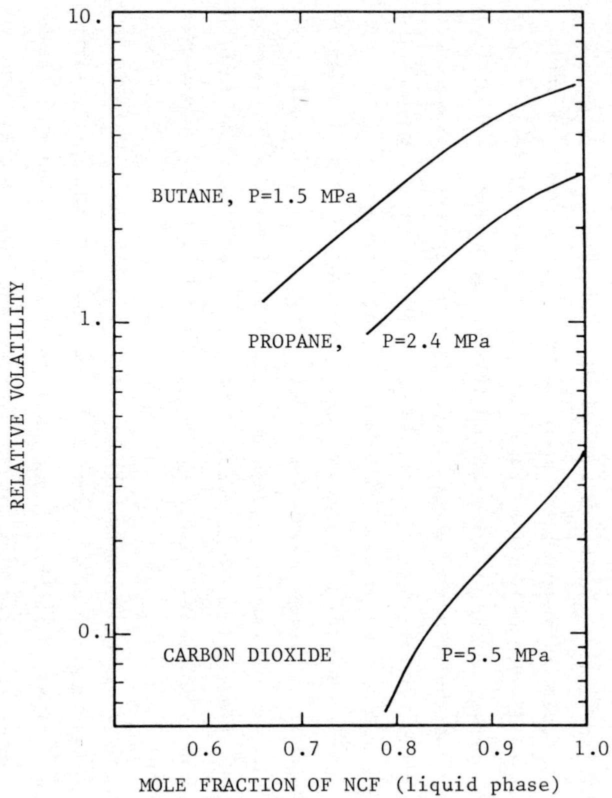

Fig.12 Relative Volatilities of Water with respect to the NCF, in the NCF Recovery Column. The ethanol/water molar ratio in the liquid phase is 10.

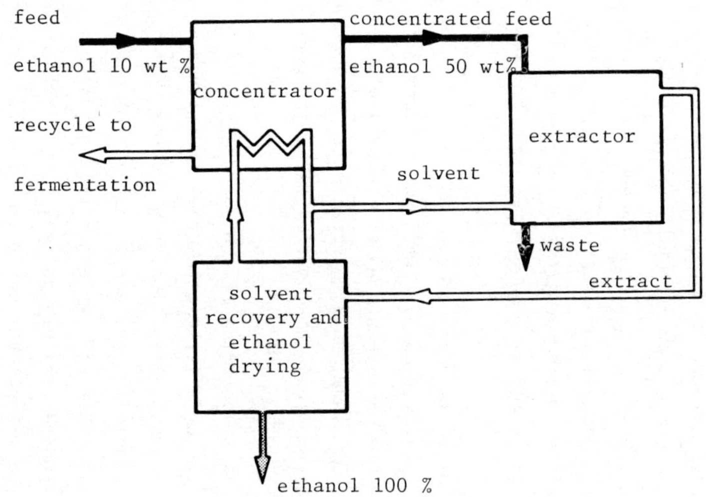

Fig. 13 NCF Extraction cyle with Preconcentrator.

NCF concentrations in the solvent recovery column have to be high to keep a low temperature difference between the column top and bottom.

Another alternative is to preconcentrate the feed by distillation at reduced pressures to increase the alcohol concentration to about 25 mole %. Subsequently, the mixture is extracted with hot propane or butane, and the solvent is recovered by distillation in a third column. The condenser of this column is integrated with the reboiler of the preconcentrator as shown in Fig.13. Reduced pressure in the preconcentrator is needed to enable the heat integration. Furthermore, milder temperature conditions in the preconcentrator may be beneficial when the feed is a broth that should be recycled to the fermentation process. This processing scheme consumes 3400 kJ/kg, on a absolute ethanol basis, when the feed is a 10 wt % ethanol in water solution. This is far below the energy consumption of conventional distillation coupled with azeotropic distillation which is close to 10000 kJ/kg (Black (1980)).

CONCLUSIONS

The GC-EOS gives reasonable predictions of phase equilibria in alcohol-water-NCF mixtures. With the revised parameters for alcohol, water and paraffinic groups, it is possible to correlate and predict vapor-liquid and liquid-liquid equilibria for alcohol-water-alkane mixtures over a wide ranges of pressure, temperature and carbon number. However, for carbon dioxide the GC-EOS application is limited to alcohol concentrations in the water phase of up to 0.15 mole fraction.

The phase equilibrium predictions for alcohol-water-NCF systems have confirmed that propane or butane are promising solvents for the separations of alcohols by NCF extraction. An energy efficient cyclic process for producing pure alcohols is proposed, which consumes three times less energy than conventional azeotropic distillation.

ACKNOWLEDGEMENT

One of us, E.A.B. gratefully acknowledges the Consejo Nacional de Investigaciones Científicas y Técnicas de la República Argentina award of a fellowship.

APPENDIX

THE ENTRAINMENT EFFECT

Water is removed from the liquid phase in the section of the column below the feed. A material balance for water (component 1) in this section of the column indicates that

$$\ell_{1,f} = v_{1,f-1} + b_1 \tag{A.1}$$

where f is the feed plate and b_1 is the water flow rate in the bottoms product. If there is a water entrainment effect in this section

$$\ell_{1,f} >> b_1$$

and

$$\ell_{1,f} \simeq v_{1,f} \qquad (A.2)$$

using the stripping factor definition

$$S_{1,n} = \frac{K_{1,n} V}{L}$$

where V and L are the liquid and vapor molar low rates, equation (A.2) can be written as:

$$\ell_{1,f} = S_{1,f-1}\, \ell_{1,f-1} \qquad (A.3)$$

If water is being stripped out of the liquid phase, then

$$\frac{\ell_{1,f}}{\ell_{1,f-1}} > 1$$

or

$$S_{1,f-1} > 1$$

in general

$$S_{1,n} > 1$$

If we take into account that ethanol and water are rather dilute in the feed to the column, then

$$(L/V) \simeq 1$$

and

$$K_{1,n} > 1$$

For the same reason the equilibrium ratio or the NCF (K_2) is close to unity.

Therefore

$$\alpha_{12,n} = \frac{K_{1,n}}{K_{2,n}} > 1$$

is required in the solvent recovery column to obtain the required water entrainment effect. It is not sufficient for the water to be more volatile than ethanol in this section. If the solvent is removed first, then the ethanol - water mixture ends up at the binary azeotropic point.

REFERENCES

Barr-David and B.F.Dodge, "Vapor Liquid Equilibrium at High Pressures", J.Chem. Eng.Data 4 107 (1959).

Black, C., "Distillation Modeling of Ethanol Recovery and Dehydration Process for Ethanol and Gasohol", C.E.P., September, 78 (1980).

Brignole, E.A., S.Skjold-Jørgensen and Aa.Fredenslund, "Application of a Local Composition Equation of State to Supercritical Fluid Phase Equilibrium Problems"' Ber.Bunsenges.Phys.Chem. 88 801 (1984).

Coan, C.R. and A.D.King, J.Am.Chem.Soc. 98 1857(1971)

Fermeglia, M. and J.Møllerup, "Phase Equilibria Calculations by Means of a Modified Carnahan-Starling-van der Waals Equation of State", Paper D14, CHISA,Prague, 1984.

Fredenslund, Aa., R.L.Jones and J.M.Prausnitz, "Group-Contribution Estimation of Activity Coefficients in Nonideal Liquid Mixtures", A.I.Ch.E.J. 21 1086 (1975).

Kuk, M.S. and J.C. Montagna, "Solubility of Oxygenated Hydrocarbons in Supercritical Carbon Dioxide", Chapter 4 in "Chemical Engineering at Supercritical Fluid Conditions", Eds.M.E.Paulaitis et al., An Arbor Science pub. (1983).

Mansoori, G.A. and T.W.Leland, "Statistical Thermodynamics of Mixtures", J.Chem. Soc.Faraday Trans. II 68 320 (1972).

Mc Hugh, M.A., M.W.Mallet and J.P.Kohn, "High Pressure Fluid Phase Equilibria of Alcohol-Water-Supercritical Solvent Mixtures", Chapter 5 in "Chemical Engineering at Supercritical Fluid Conditions", Eds. M.E.Paulaitis et al., An Arber Science pub. (1983)

Moses, J.M., K.E.Gocklen and R.P. De Filippi, "Pilot Plant Critical-Fluid Extraction of Organics from Water", A.I.Ch.E. Annual Meeting, Paper No. 127c, Los Angeles, 1982.

Paulaitis, M.E., R.G.Kander and J.R.Diandreth,"Phase Equilibria Related to Supercritical-Fluid Solvent Extractions", Ber.Bunsenges. Phys.Chem. 88 869 (1984).

Paulaitis,M.E., M.L.Gilbert and C.A. Nash, presented at the 2nd World Congress of Chemical Engineering, Montreal, 1981

Radosz, M., "Variable-Volume Circulation Apparatus for Measuring High-Pressure Fluid Phase Equilibria"., Ber.Bunsenges.Phys.Chem. 88 859 (1984)

Renon, H. and J.M.Prausnitz, "Local Compositions in Thermodynamic Excess Functions for Liquid Mixtures", A.Ch.E.J. 14 135 (1968).

Schreiber, E., E. Schettau, D.Rant and H.Schubert, Z.Phys.Chem.(Leipzig) 247 23 (1971).

Skjold-Jørgensen, S., "Gas Solubility Calculations II: Application of a New Group Contribution Equation of State", Fluid Phase Equilibria 16 317 (1984)

Sørensen, J.M. and W.Arlt, "Liquid-Liquid Equilibrium Data Collection-Binary Systems", DECHEMA Chemistry Data Series, Vol. V, Part 1 (1979)

Wiebe, R., Chem.Rev. 29 475 (1941)

LIST OF SYMBOLS

A^R	Residual Helmholtz function
a	Energy parameter in van der Waal equation
b_i	Component flow rate in the bottoms product
d	Hard sphere diameter
d_c	Hard sphere critical diameter
g_{ji}	Attractive energy parameter for group j
g^*_{ji}	Reference value of the attractive energy parameter for group j
g'_{ji}, g''_{jj}	Coefficients in eq. (8)
k_{ij}	Binary interaction parameter
K_i	Distribution coefficient of component i between a NCF phase and aqueous phase in molar fraction basis (y_i/x_i)
n	Plate number
N	Number of moles
N_i	Number of moles of component i
NC	Number of components
NG	Number of groups
P	Pressure
q_j	Surface area parameter of group j
R	Gas constant
S	Stripping factor
T	Absolute temperature
T_c	Critical temperature
T^*_j	Reference temperature for group j
V	Molar volume
x	Mole fraction in the liquid phase
y	Mole fraction in the vapor or NCF phase
α_{ij}	Binary damping factor
ρ	Molar density
θ_j	Surface fraction of group j
ν^i_j	Number of groups j in molecule i

Supercritical Fluid Technology, edited by J.M.L. Penninger, M. Radosz, M.A. McHugh and V.J. Krukonis
1985 Elsevier Science Publishers B.V., Amsterdam — Printed in The Netherlands

DESIGN OF SUPERCRITICAL-FLUID-EXTRACTION PROCESSES USING CONTINUOUS THERMODYNAMICS

Ronald L. Cotterman, Dimitrios Dimitrelis and John M. Prausnitz

Materials and Molecular Research Division, Lawrence Berkeley Laboratory, and Chemical Engineering Department, University of California, Berkeley, California 94720

ABSTRACT

We present a new molecular-thermodynamic framework which is particularly useful for describing thermodynamic properties of mixtures containing very many components. This framework, called continuous thermodynamics, is combined with an equation of state to calculate phase equilibria as required for computer-aided design of supercritical-fluid-extraction processes. To illustrate this method, we consider a process for upgrading a heavy-crude residuum with a supercritical solvent. Continuous thermodynamics provides an efficient tool for determining optimum operating conditions.

INTRODUCTION

In recent years, increasing attention has been given to the use of supercritical fluids for separation and purification processes (ref. 1-5). The design of such processes requires knowledge of phase behavior to evaluate solvent selectivity and capacity and to calculate process yields. This work presents modeling techniques necessary for computer-aided design of supercritical-fluid-extraction processes.

Practical supercritical-fluid processes are often concerned with mixtures that contain a very large number of different components (such as fossil fuels or vegetable oils) (ref. 6-9). However, almost all proposed models for supercritical-extraction processes have been confined to binary and ternary systems. In this work, we present a new molecular-thermodynamic framework developed for many-component mixtures, called continuous thermodynamics (ref. 10). This framework is combined with an equation of state to calculate phase equilibria for design of a process to upgrade a heavy crude residuum with a supercritical solvent. An important advantage of the continuous thermodynamics framework is that it avoids arbitrary lumping of the residuum composition into pseudocomponents.

PROCESS DESCRIPTION

Figure 1 shows schematically a process for the separation of a heavy crude residuum into asphaltenes, resins and oils. This process is similar to the Kerr-McGee ROSE process described by Gearhart and Garwin (ref. 11) and by Brulé et al. (ref. 12,13). To illustrate the calculation procedure, we choose propane as the solvent. The critical temperature and the critical pressure of propane are 369.8 K and 42.5 bar, respectively.

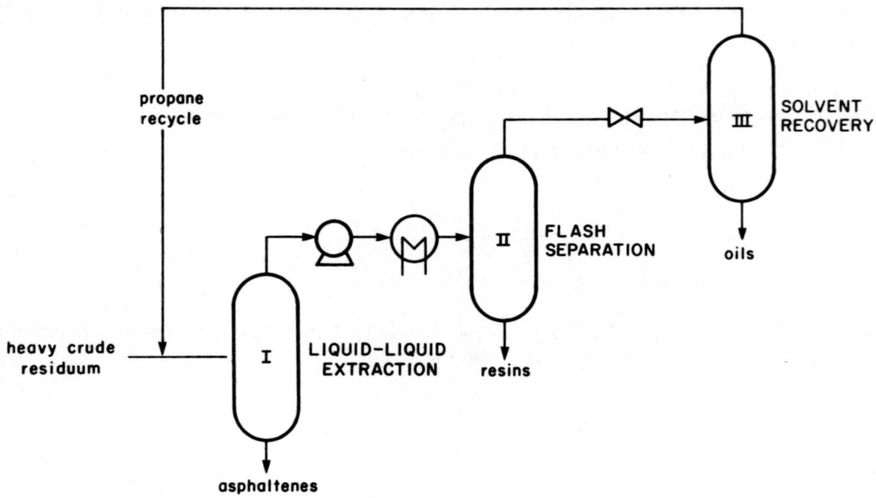

Fig. 1. Three-stage extraction for high-boiling petroleum fractions using propane

In the first stage of the process, the heavy crude residuum is contacted with solvent (propane) in a multi-stage liquid-liquid extraction. Here, the highest molecular-weight portion of the feed, the asphaltene fraction, is removed. This separation is typically carried out at temperatures below the solvent critical temperature, at moderate pressures and at high solvent-to-feed ratios (ref. 14).

The upper phase leaving Stage I contains the intermediate and the light molecular-weight portions of the feed, i.e., the resins and the oils. In Stage II of the process, we would like to separate the oils from the resins. If the pressure in Stage II is too low, resins and oils are separated from the solvent, but not from each other. If the pressure is too high, there is no significant separation of propane or oils from resins. To avoid very high temperatures, it is advantageous to operate Stage II in the mixture retrograde region where the solvent is supercritical and where large changes in solubility occur with small changes in temperature or pressure (or both). The retrograde region is attained in Stage II through a single-stage vapor-liquid flash of the outlet stream from Stage I. The liquid outlet stream from Stage II is enriched in resins and the vapor outlet stream contains most of the oils.

In a final stage, pressure is reduced; the oils are removed as a liquid and the solvent is removed as a vapor. Solvent is compressed and recycled to Stage I.

We are here concerned with the phase behavior of the solvent-oils-resins mixture to determine desirable operating conditions for Stage II. We discuss how the phase behavior in the retrograde region may be used to effect a separation between oils and resins.

CONTINUOUS THERMODYNAMICS FRAMEWORK

To represent the phase behavior of the solvent-oils-resins mixture, we need to account for the disperse nature of the resins and the oils fractions; we do this using continuous thermodynamics. Continuous thermodynamics provides a framework wherein the composition of a mixture of very many components is described by a continuous distribution function $F(I)$ of some characterizing property I (ref. 10,15-19). In this work, we use molecular weight as characterizing property I.

The continuous distribution function is normalized such that

$$\int_I F(I)\, dI = 1 \tag{1}$$

For many mixtures (e.g. when a solvent is present in excess amount), it is advantageous to use a semicontinuous description. In semicontinuous mixtures, the composition of a portion of the mixture is described by one or more discrete mole fractions (such as that of the solvent) whereas the remainder is described by one or more continuous distributions (such as one for the resins and another for the oils). The composition of each discrete component is given by a mole fraction x_i and the composition of each continuous fraction in the mixture is weighted with mole fraction η_j. These mole fractions are normalized for a given phase; for s discrete components and t continuous fractions we write

$$\sum_i^s x_i + \sum_j^t \eta_j \int_I F_j(I)\, dI = 1 \tag{2}$$

where $F_j(I)$ is the distribution function for fraction j.

FLASH-CALCULATION PROCEDURES

Stage II is considered to be a single-stage isothermal flash. Figure 2 shows schematically a flash separation for a semicontinuous mixture with one discrete component and one continuous fraction. Temperature, pressure and feed composition are specified; the compositions and relative amounts of the outlet streams are to be calculated. At the temperatures and pressures of interest, the feed to Stage II separates into vapor and liquid phases.

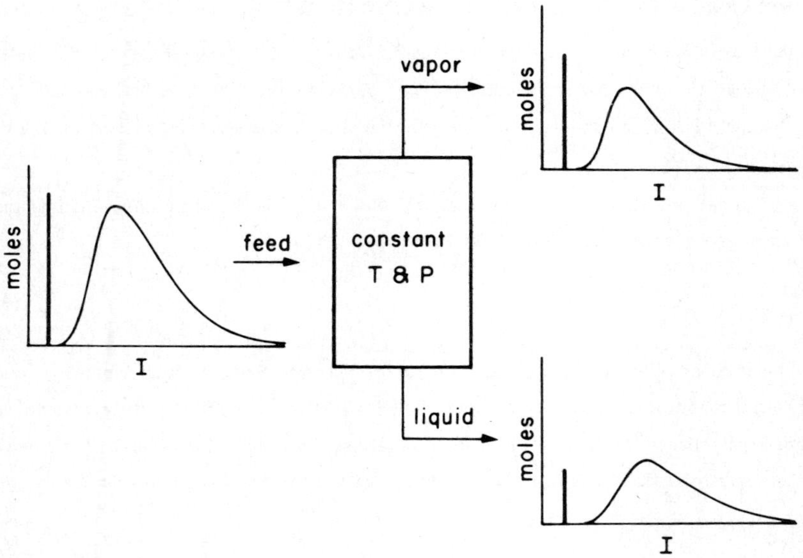

Fig. 2. Flash calculation for a semicontinuous mixture.

The feed and outlet streams are related through material balances; the two outlet streams are assumed to be in thermodynamic equilibrium.

We define the fraction vaporized ξ as the ratio of moles of vapor to moles of feed. For every discrete component, the material balance is

$$z_i = \xi\, y_i + (1-\xi)\, x_i \tag{3}$$

where z_i, y_i and x_i are feed, vapor and liquid mole fractions for component i, respectively.

For every continuous fraction, we introduce a distribution function $F_j(I)$ and a mole fraction η_j for each phase. Thus, we write

$$\eta_j^F F_j^F(I) = \xi \eta_j^V F_j^V(I) + (1-\xi)\eta_j^L F_j^L(I) \tag{4}$$

where superscripts F, V and L designate feed, vapor and liquid, respectively.

Since the effluent streams are at equilibrium, we equate the chemical potential μ of each component in each stream. For every discrete component i,

$$\mu_i^V = \mu_i^L \tag{5}$$

and for every continuous fraction j, we equate chemical potentials throughout the range of index I

$$\mu_j^V(I) = \mu_j^L(I) \tag{6}$$

Equations (3)-(6) form a set of non-linear, integral-algebraic equations. Regardless of the form we choose for the distribution functions $F(I)$, it is not possible to solve these equations analytically. As discussed elsewhere (ref. 20), the procedure for solving the flash-calculation equations in semicontinuous mixtures introduces numerical integration by Gaussian quadrature.

PHASE EQUILIBRIA FROM PERTURBED-HARD-CHAIN THEORY

Chemical potentials in semicontinuous fluid mixtures are found from an expression for the molar residual Helmholtz energy $\mathbf{a}^r$. The residual Helmoltz energy represents the difference between the total Helmholtz energy for the mixture and an ideal-gas mixture at the same temperature, volume and composition.

For every discrete component i, the residual chemical potential μ_i^r is found by differentiating the residual Helmholtz energy $n\,\mathbf{a}^r$ with respect to mole number

$$\mu_i^r = \left(\frac{\partial n\,\mathbf{a}^r}{\partial n_i}\right)_{T,V,n_k} \tag{7}$$

where n is the total number of moles in the phase of interest. Temperature T, total volume V and number of moles of component k (n_k) are held constant. Following Salacuse and Stell (ref. 16), we define the residual chemical potential $\mu_j^r(I)$ for a continuous fraction using a funtional derivative of $n\,\mathbf{a}^r$ with respect to the extensive distribution function $n_j F(I)$ at a fixed value of the distributed variable, $I = I^+$

$$\mu_j^r(I) = \left(\frac{\delta n\,\mathbf{a}^r}{\delta n_j F(I^+)}\right)_{T,V,I^+=I} \tag{8}$$

The residual chemical potential $\mu_j^r(I)$ is a continuous function of the distributed variable I over the entire range of I.

To illustrate, we use the perturbed-hard-chain theory (ref. 21,22) to derive an expression for the molar residual Helmholtz energy of pure hydrocarbon components and their mixtures. Details are given elsewhere (ref. 23).

The molar residual Helmholtz energy $\mathbf{a}^r$ is expressed as the sum of repulsive and attractive contributions:

$$\mathbf{a}^r = \mathbf{a}^r(repulsive) - \mathbf{a}^r(attractive) \tag{9}$$

For the repulsive contribution, we use a generalized Carnahan-Starling (ref. 24) term as a function of reduced molar volume $\tilde{v}=v/v^*$ and Prigogine's parameter c (ref. 25)

$$\frac{\mathbf{a}^r}{RT}(repulsive) = f_{rep}(c,\tilde{v}) \tag{10}$$

The expression for the attractive term is determined from pure-fluid and computer-simulation data as a function of reduced temperature $\tilde{T}=T/T^*$, reduced molar volume $\tilde{v}$ and parameter c

$$\frac{\mathbf{a}^r}{RT}(attractive) = f_{att}(c,\tilde{v},\tilde{T}) \tag{11}$$

There are three parameters for a pure fluid, v^*, $\epsilon q/k$ and c, which are characteristic of molar hard-core volume, molecular potential energy and the number of external degrees of freedom of the molecule, respectively. We define a characteristic temperature $T^* \equiv \epsilon q/ck$ where ϵ is the potential energy per unit area, q is the external molecular surface area and k is Boltzmann's constant.

The expression for the Helmholtz energy is extended to mixtures using suitable mixing rules to relate mixture parameters to composition and to pure-component parameters. The pressure-explicit equation of state is obtained by differentiating the residual Helmholtz energy with respect to volume

$$P = \frac{nRT}{V} - \left(\frac{\partial n\mathbf{a}^r}{\partial V}\right)_{T,\text{all } n} \tag{12}$$

Pure-component parameters v^*, $\epsilon q/k$ $(=cT^*)$ and c are fit to vapor-pressure and liquid-density data. These parameters are then correlated as functions of molecular weight for various hydrocarbon classes (ref. 23). Figure 3 schematically shows the correlations for alkanes, alkylcyclohexanes and alkylbenzenes. The molecular significances of v^* and cT^* indicate that (for a class of compounds) they should be linear functions of molecular weight, as observed. At high values of molecular weight, since cT^* and c are nearly proportional to molecular weight, T^* approaches a constant as molecular weight becomes very large. Therefore, the plots shown in Figure 3 can be extrapolated with confidence to higher molecular weights, where experimental data are scarce.

PHASE-EQUILIBRIUM CALCULATIONS FOR STAGE II FLASH

We represent the feed to Stage II as a semicontinuous mixture of solvent (here, propane) and two continuous distributions (one for the oils and another for the resins). The overall feed composition contains 91 wt % propane, 6 wt % oils and 3 wt % resins. Figure 4 shows the sum of the molar distributions for the resins and oils. The molar distribution $(\eta F(I))$ for each continuous fraction represents the distribution of moles for that fraction per one mole of total feed. For simplicity, the molar distributions for both the resins and the oils are chosen to be analytic functions; we use beta distributions (ref. 26) which dramatically emphasize the difference between oils and resins. The oils cover the molecular-weight range 200-500 with mean 325 and variance 2500. The resins cover the range 450-850 with mean 650 and variance 6000.

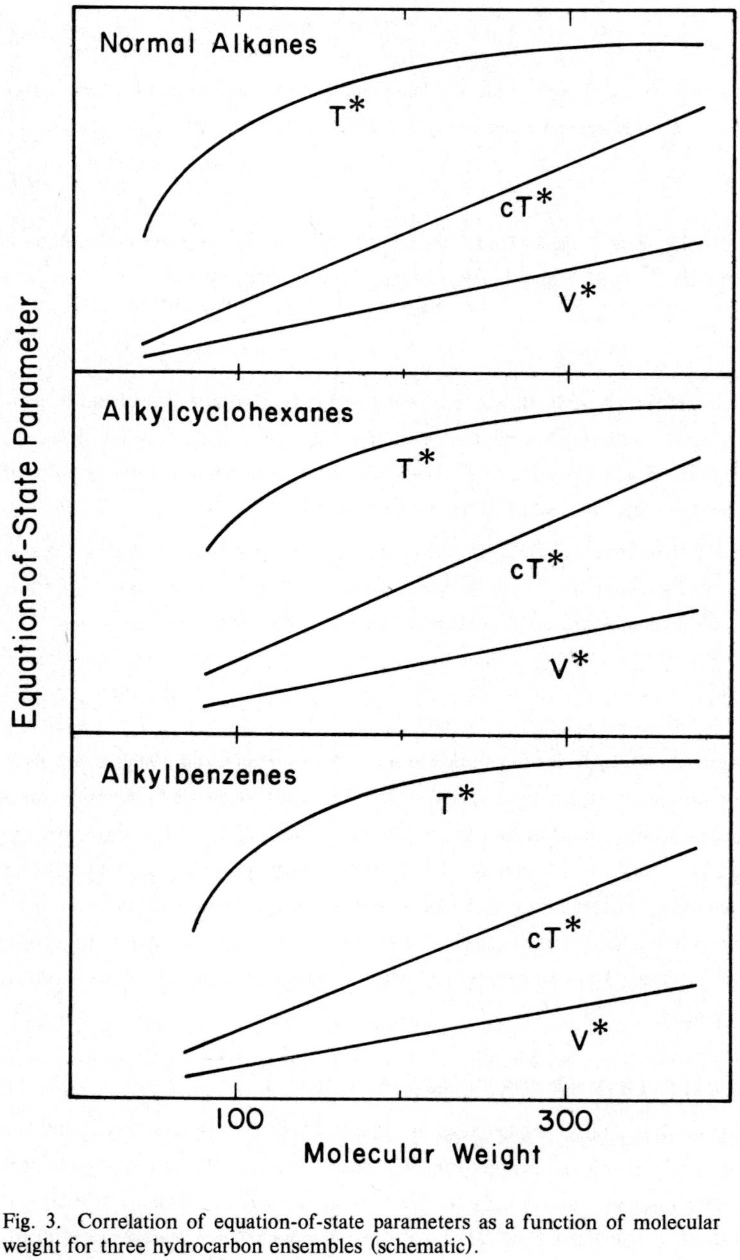

Fig. 3. Correlation of equation-of-state parameters as a function of molecular weight for three hydrocarbon ensembles (schematic).

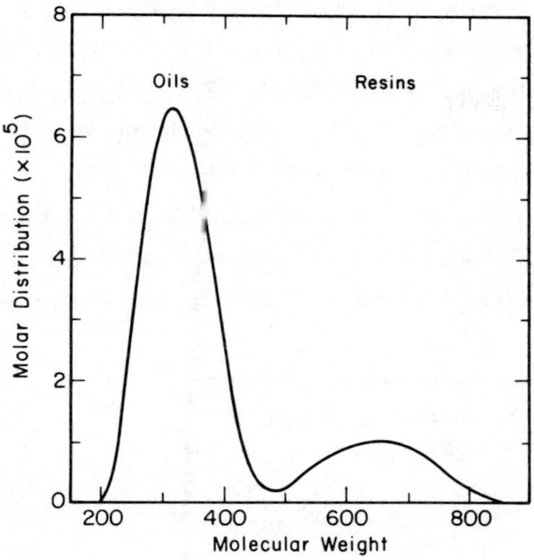

Fig. 4. Sum of feed distributions for resins and oils to Stage II flash (feed composition: 91 wt% propane, 6 wt% oils and 3 wt% resins).

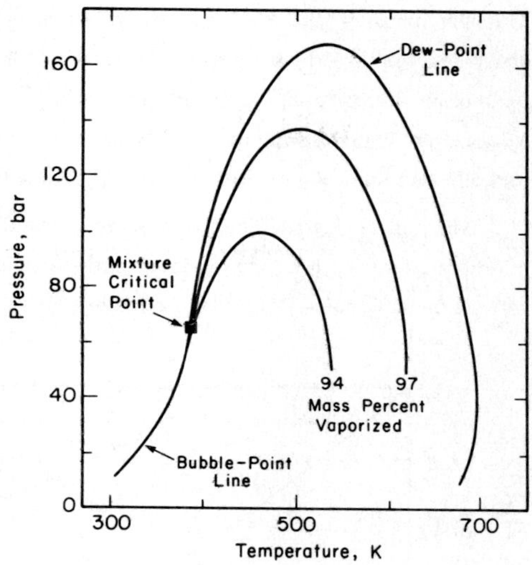

Fig. 5. Pressure-temperature diagram for the propane-oils-resins mixture entering Stage II (10:1 solvent-to-feed mass ratio).

Thermodynamic properties are given by the perturbed-hard-chain theory using alkane correlations for pure-component parameters. Figure 5 shows the calculated pressure-temperature diagram for the semicontinuous mixture. Bubble-point, dew-point and constant percent-vaporized lines end at the mixture critical point. The area inside the bubble-point and the dew-point curves is a two-phase vapor-liquid region; outside, the fluid mixture is a homogeneous single phase.

Phase-equilibrium calculations are generally very difficult near the mixture critical point; the solution procedure is sensitive to initialization of variables and convergence is slow. To overcome these difficulties, we use a tracing technique described by Michelsen (ref. 27). Phase-equilibrium calculations are initiated at low pressures (where convergence methods are very stable) and then progress to higher pressures by stepping along lines of constant fraction vaporized.

The area of the pressure-temperature diagram above the mixture critical point within the two-phase envelope is called the retrograde region (ref. 28). As discussed earlier (ref. 29), retrograde behavior is typical for propane/heavy-hydrocarbon mixtures with a large excess of propane. In the retrograde region, condensation occurs by increasing temperature (at constant pressure) from the left-hand side of the dew-point line. This retrograde region establishes operating conditions for Stage II. The relative amounts of the oils and the resins in the vapor and liquid phases can be altered by changing the temperature and pressure. Optimum temperature and pressure are chosen where the greatest separation between resins and oils is effected.

Figure 6 shows yields of oils and resins in the outlet streams of the Stage II flash. The yield for a component is defined

$$\text{Yield} \equiv \frac{\text{mass in outlet stream}}{\text{mass in feed stream}}$$

The upper diagram (Fig. 6) shows the yield of oils in the vapor outlet stream. For the pressures which are above the mixture critical pressure (80 bar and 100 bar), vapor yields are unity at the dew-point of the mixture and decrease with rising temperature in the retrograde region. Yields eventually increase to unity again at very high temperatures. Below the mixture critical pressure (50 bar), vapor yields increase monotonically with temperature from zero at the mixture bubble point to unity at the mixture dew point.

The middle diagram (Fig. 6) shows the yield of resins in the liquid outlet stream. Below the mixture critical pressure (50 bar), liquid yields decrease with rising temperature from unity at the bubble point to zero at the dew point. Above the mixture critical pressure (80 bar and 100 bar), liquid yields increase

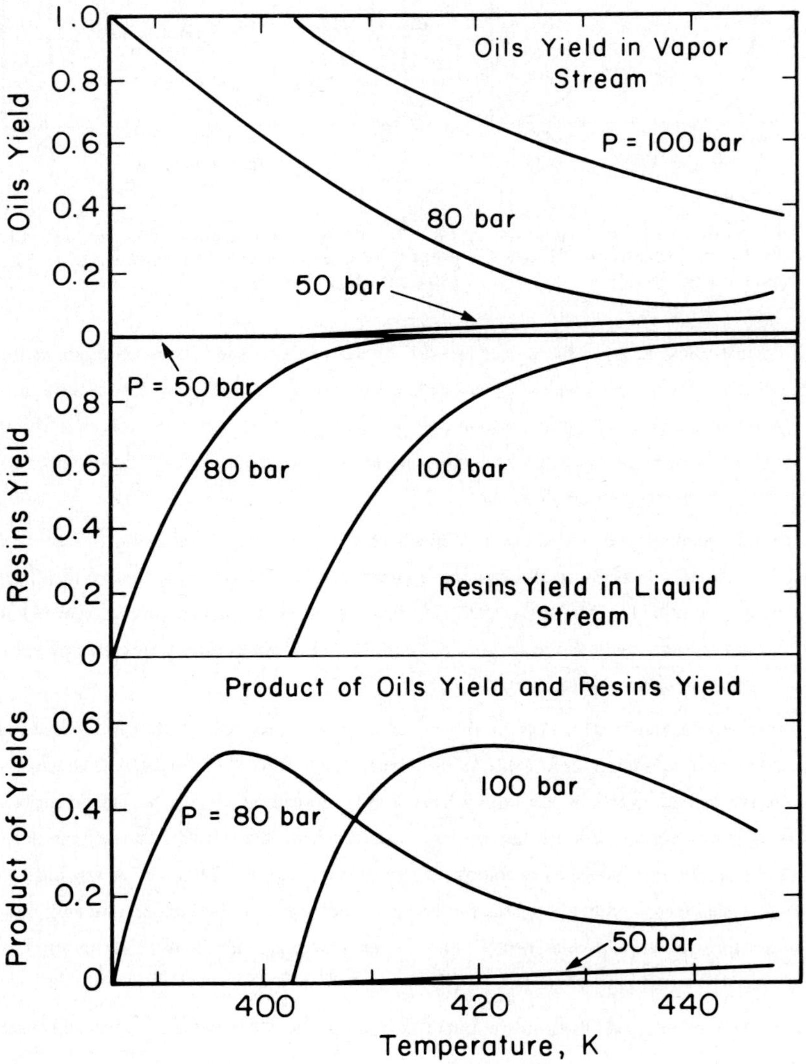

Fig. 6. Yields of oils and resins in outlet streams of Stage II flash.

from zero at the dew point and pass through a maximum with rising temperature.

For an optimum design, we would like to maximize both the oils yield in the vapor and the resins yield in the liquid. The bottom diagram of Figure 6 shows the product of these yields. Above the mixture critical pressure, the yields product shows a distinct maximum as temperature rises. At higher pressures, the dome of this maximum becomes broader and moves to higher temperatures. Below the mixture critical pressure, the yields product shows a very broad maximum at high temperatures.

Based on the analysis of the product of the oils vapor yield and the resins liquid yield, optimal operating conditions for the second-stage flash are chosen to be 80 bar and 398 K. At these conditions, Table 1 gives calculated outlet-stream compositions and Figure 7 shows calculated outlet-stream molecular-weight distributions for the continuous fractions; the molar fraction vaporized is 0.963. The molar distributions in Figure 7 represent the relative amount of the total feed distribution in each outlet stream.

TABLE 1

Calculated optimum outlet-stream compositions from Stage II flash at 398 K and 80 bar.

	Composition, wt %		
	Feed	Vapor	Liquid
Propane	91	94.87	53.56
Oils	6	4.43	21.19
Resins	3	0.70	25.25

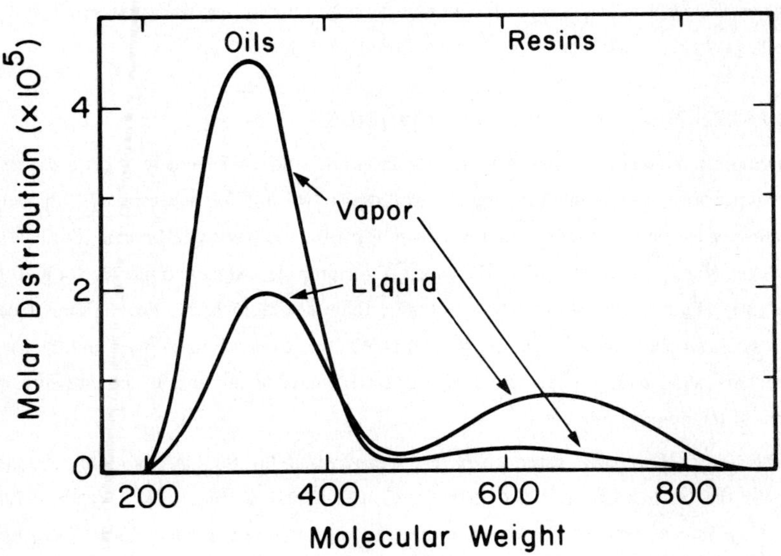

Fig. 7. Outlet-stream molecular-weight distributions for the continuous fractions from Stage II flash at 398 K and 80 bar.

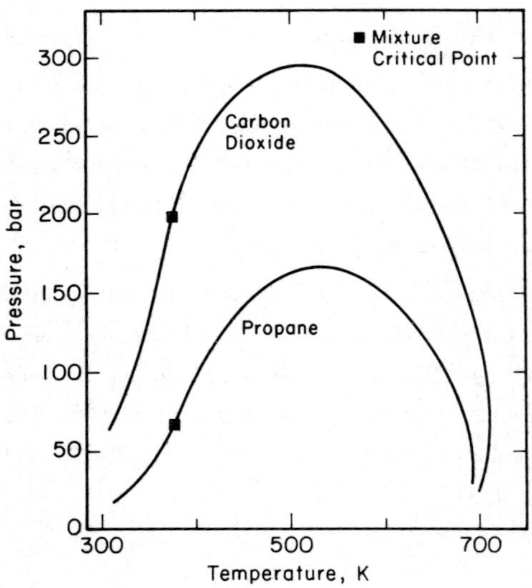

Fig. 8. Pressure-temperature diagram for solvent-oils-resins mixtures using either propane or carbon dioxide (10:1 solvent-to-feed mass ratio).

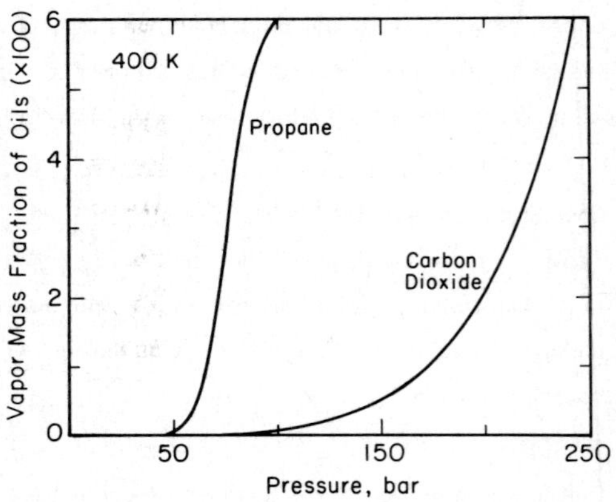

Fig. 9. Comparison of oils capacity in two solvents for Stage II vapor outlet stream (feed composition: 91 wt% solvent, 6 wt% oils and 3 wt% resins).

Using continuous thermodynamics, we are able to calculate not only the relative amounts of the oils and the resins in each outlet stream, but we are also able to describe the shape of the molecular-weight distribution in each phase. Figure 7 shows that the vapor stream is enriched in the oils fraction and the liquid stream contains most of the resins. The liquid distributions are more disperse than those in the vapor; this indicates that the propane in the vapor preferentially extracts the lighter-molecular-weight compounds (or, propane has a greater selectivity for the lower molecular weights).

COMPARISON BETWEEN PROPANE AND CARBON DIOXIDE

Continuous thermodynamics can be used to screen for the most efficient solvent to effect a desired separation. To illustrate, since carbon dioxide is the most popular solvent for supercritical-extraction processes, we compare its possible use and advantage versus propane for the separation process considered here. Figure 8 shows the phase diagram for the feed to Stage II for propane and carbon dioxide for a 10:1 solvent-to-feed mass ratio. Both the propane mixture and the carbon dioxide mixture have the same feed composition but the calculated phase behavior is very different. Whereas the propane mixture and the carbon dioxide mixture have similar critical temperatures, the carbon dioxide mixture critical pressure is much higher than that of the propane mixture.

As Figure 9 shows, to achieve with carbon dioxide the same solubility (or capacity) of oils in the vapor outlet stream as that with propane, much higher pressures would be required. Therefore, for separation of heavy from intermediate hydrocarbons, propane is the preferred solvent. Carbon dioxide may have possible advantages over propane in terms of selectivity; however, the higher operating pressures would most likely negate any significant process advantages.

PROPANE RECOVERY IN STAGE III

As a final example, we consider the recovery of solvent from Stage II for recycle to the process inlet. The majority of the solvent from Stage II exits in the vapor stream with the oils. This vapor stream is fed to a third stage for solvent recovery. The pressure-temperature diagram for the feed to Stage III is similar to that for the feed to Stage II, shown in Figure 5.

Figure 10 shows the yields of propane and oils in the outlet streams of Stage III. As in the Stage II analysis, at 50 bar (below the mixture critical pressure), the propane vapor yield and oils liquid yield increase from zero at the bubble-point of the mixture with rising temperature. Alternately, above the mixture critical point, at 100 bar, the yield curves decrease with rising temperature from unity at the dew point. Temperature and pressure in Stage III should be selected to maximize both the propane vapor yield and the oils liquid yield. As indicated by Figure 10, this maximum occurs at low pressures and tempera-

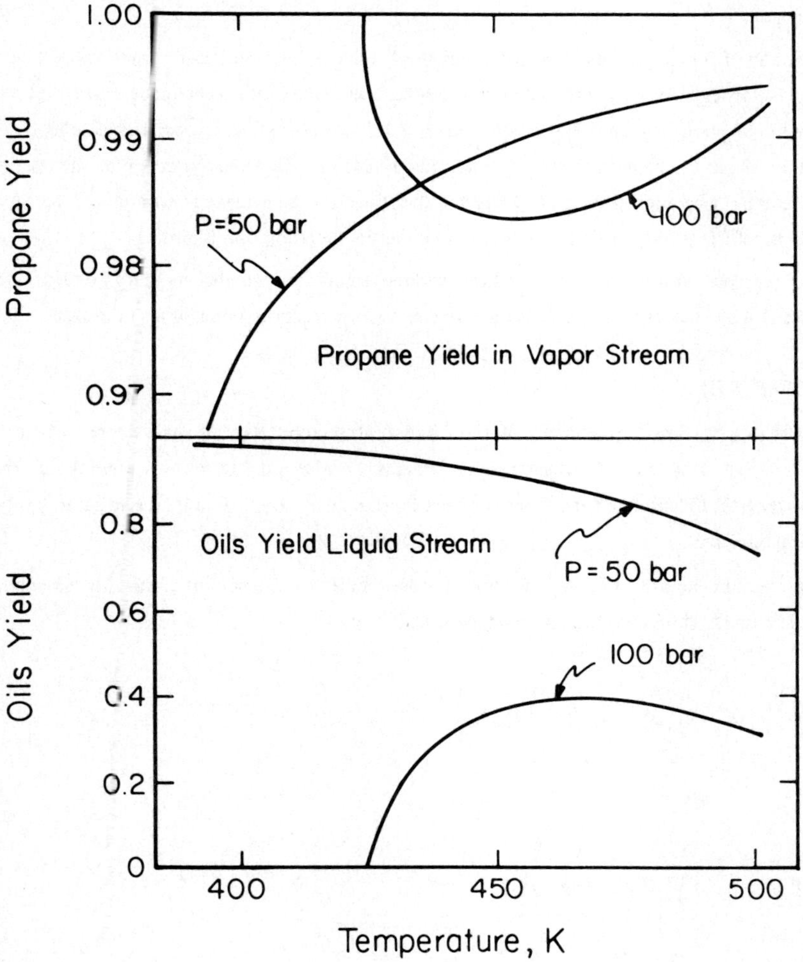

Fig. 10. Propane recovery in Stage III. Yields of oils and propane in outlet streams.

tures near the mixture dew point. However, since this is not a difficult separation, pressure should be maintained high enough so that recovered propane need not be recompressed before it is recycled to Stage I. Thus the actual operating temperature and pressure depend on the conditions chosen for the first stage.

The other process outlet streams which contain the asphaltenes and the resins have recoverable propane. Based on the amount of propane in these streams and the process economics, it may be necessary to recycle propane from these streams to Stage I.

CONCLUSIONS

Continuous thermodynamics is a useful tool for process-design calculations. For systems which contain a large number of components, continuous thermodynamics accounts for the disperse nature of the molecular-weight distributions. Continuous thermodynamics may be combined with the perturbed-hard-chain theory to calculate high-pressure phase equilibria as needed for design of supercritical-fluid-extraction processes.

As an application of these methods, we have outlined calculations for a process for upgrading a heavy crude residuum using propane as a solvent. The crude can be fractionated into asphaltenes, resins and oils using a three-stage extraction process. The second stage of this process utilizes supercritical propane to separate resins from oils in the retrograde region of the phase diagram. By proper selection of the operating conditions, optimum separation between the oils and the resins can be achieved. Continuous thermodynamics provides an efficient tool for determining these optimum operating conditions.

Continuous thermodynamics may be introduced without significant modification into conventional numerical algorithms and molecular-thermodynamic models used for computer-aided process design.

ACKNOWLEDGEMENTS

This work was supported by the Director, Office of Energy Research, Office of Basic Energy Sciences, Chemical Sciences Division of the U.S. Department of Energy. Additional support was provided by the Fannie and John Hertz Foundation and the Donors of the Petroleum Research Fund, administered by the American Chemical Society.

The authors are also grateful to Barry J. Schwarz for assistance in the equation-of-state development and to Robert J. Topliss for computer-programming assistance.

NOTATION

Symbols

a	=	molar Helmholtz energy
c	=	external-degrees-of-freedom parameter
$F(I)$	=	molar distribution function
I	=	distributed variable (taken here as molecular weight)
I^+	=	dummy variable for functional differentiation
k	=	Boltzmann's constant
n	=	number of moles
P	=	absolute pressure
q	=	external molecular surface area
R	=	gas constant
s	=	number of discrete components
t	=	number of continuous components
T	=	absolute temperature
$\tilde{T}$	=	reduced temperature, $\tilde{T}=T/T^*$
T^*	=	characteristic temperature, $T^*=\epsilon q/ck$
v	=	molar volume
$\tilde{v}$	=	reduced molar volume, $\tilde{v}=v/v^*$
v^*	=	hard-core molar-volume parameter
V	=	total volume
x	=	discrete-component liquid mole fraction
y	=	discrete-component vapor mole fraction
z	=	discrete-component feed mole fraction

Subscripts

i	=	discrete components
j	=	continuous fraction

Superscripts

F	=	feed stream
L	=	liquid stream
V	=	vapor stream
r	=	residual property

Greek Symbols

μ	=	chemical potential
ξ	=	fraction of feed vaporized
η	=	continuous-fraction mole fraction
ϵ	=	potential energy per unit surface area

LITERATURE CITED

1 E.U. Franck and G.M. Schneider (Eds.), "Supercritical Fluid Solvents", *Ber. Bunsen. Physik. Chem.* **9** (1984).

2 M.E. Paulaitis, V.J. Krukonis, R.T. Kurnik, and R.C. Reid, *Rev. Chem. Eng.,* **1** , 179 (1983).

3 H. Renon (Ed.), "Supercritical Fluids: Their Chemistry and Applications", *Fluid Phase Equil.* **10**: 2-3 (1983).

4 M.E. Paulaitis, et al., (Eds.), "Chemical Engineering at Supercritical Fluid Conditions", Ann Arbor Science Publishers, Ann Arbor, Michigan 1983.

5 G.M. Schneider, E. Stahl and G. Wilke (Eds.), "Extraction with Supercritical Gases", Verlag Chemie, Weinheim, West Germany 1980.

6 D.F. Williams, *Chem. Eng. Sci.* **36** (1981) 1769.

7 N. Gangoli and G. Thodos, *IEC Prod. Res. Dev.* **16** (1977) 208.

8 C.A. Irani and E.W. Funk, "Separations Using Supercritical Gases" from "CRC Handbook, Recent Developments in Separation Science", Chapter 3, Volume 3, (N. Li, Ed.), CRC Press 1977.

9 R.M. Adams, A.H. Knebel and D.E. Rhodes, *Chem. Eng. Prog.* **75** (1979) 44.

10 R.L. Cotterman, R. Bender and J.M. Prausnitz, *IEC Proc. Des. Dev.,* **24** (1985) 194.

11 J.A. Gearhart and L. Garwin, *Hydrocarbon Proc.* May (1976) 125.

12 M.R. Brulé, R.W. Corbett and S. Watanasiri, *Energy Prog.* **3** (1983) 54.

13 M.R. Brulé and R.W. Corbett, *Hydrocarbon Proc.* June (1984) 73.

14 S.A. Berridge, "Finishing Processes" from "Modern Petroleum Technology", 4th ed. (G.D. Hobson and W. Pohl, Eds.), Applied Science Publishers, Ltd., 1973.

15 J.A. Gualtieri, J.M. Kincaid and G. Morrison, *J. Chem. Phys.,* **77** (1982) 521.

16 J.J. Salacuse and G.Stell, *J. Chem. Phys.,* **77** (1982) 3714.

17 J.M. Prausnitz, *Fluid Phase Equil.* **14** (1983) 1.

18 J.G. Briano and E.D. Glandt, *Fluid Phase Equil.,* **14** (1983) 91.

19 M.T. Rätzsch and H. Kehlen, *Fluid Phase Equil.* **14** (1983) 225.

20 R.L. Cotterman and J.M. Prausnitz, *IEC Proc. Des. Dev.,* **24** (1985) 434.

21 S. Beret and J.M. Prausnitz, *AIChE J.* **21** (1975) 1123.

22 M.D. Donohue and J.M. Prausnitz, *AIChE J.* **24** (1978) 849.

23 R.L. Cotterman, PhD Thesis, University of California, Berkeley, 1985.

24 N.F. Carnahan and K.E. Starling, *AIChE J.* **18** (1972) 1184.

25 I. Prigogine, "The Molecular Theory of Solutions", North-Holland, Amsterdam, 1957.

26 N.L. Johnson and S. Kotz, "Continuous Univariate Distributions-1", Houghton Miffin Company, Boston, MA, 1970.

27 M. Michelsen, *Fluid Phase Equil.* **4** (1980) 4.

28 J.S. Rowlinson and F.L. Swinton, "Liquids and Liquid Mixtures", 3rd ed., Butterworths, London, 1982.

29 R.L. Cotterman, D. Dimitrelis and J.M. Prausnitz, *Ber. Bunsen. Physik. Chem.* **87** (1984) 796.

PART II

Experimental Data & Techniques

Supercritical Fluid Technology, edited by J.M.L. Penninger, M. Radosz, M.A. McHugh and V.J. Krukonis
1985 Elsevier Science Publishers B.V., Amsterdam — Printed in The Netherlands

THE INFLUENCE OF THE SOLVENT GAS ON SOLUBILITY AND SELECTIVITY IN SUPERCRITICAL EXTRACTION

W.J. SCHMITT[1] and R.C. REID[2]

[1]The Upjohn Company, Kalamazoo, MI 49001
[2]Dept. of Chem. Eng., Mass. Inst. Tech. Cambridge, MA 02139

ABSTRACT

The solubilities of three organic solids were measured in carbon dioxide, ethane, fluoroform, and chlorotrifluoromethane. The solutes were selected so that each contained a single organic functional group affixed to a parent bi- or tricyclic aromatic ring. The solvents were chosen on the basis of their chemical dissimilarities yet nearly equal critical temperatures. This restriction makes it possible to treat the experimental binary solubility data by a generalized corresponding states analysis. This analysis permits the physical solvolysis effects of solvent density and solute vapor pressure to be decoupled from the observed solubilities, which in turn enables the more subtle effect of chemical solvent-solute interaction to be quantitatively examined.

It was found that the choice of solvent gas can significantly affect the equilibrium solubility of a particular solute species in that gas, and different solvents either enhance or depress solubility in conjunction with the particular functionality of the solute molecule. Acid-base and solvent-solute association effects appear to play a role in some specific cases.

The experimental data were correlated with a single adjustable parameter Peng-Robinson equation, and regressed binary interaction coefficients (k_{ij}'s) are tabulated for all of the systems studied. In addition, a new empirical correlation of the experimental data is proposed which models the solid-fluid solubility isotherms well. This correlation is useful when precise interpolation of available data to other temperatures and pressures is required and when solubilities must be predicted based on a limited amount of available experimental data.

INTRODUCTION

In the past several years a number of investigators have published equilibrium solubility data for various solids in supercritical fluids. There have, however, been two distinct and separate classes of solute molecules studied. At one extreme there now exists an abundance of solubility data for simple polycyclic hydrocarbons containing no chemical functionality (ref.'s 1,2,3), and, at the other extreme, solubility data for chemically complex solutes, usually natural products, containing several functional groups per molecule (ref.'s 4,5,6). It is the purpose of this investigation to bridge the gap between the two existing sets of data by investigating systematically the solubilities of several monofunctional solid solutes closely related to the simple aromatic hydrocarbons in structure. In this paper we present the solubilities of three solutes (naphthalene, 1,4-naphthoquinone, and acridine) in the solvents carbon dioxide, ethane, chlorotrifluoromethane [R-13], and fluoroform [R-23]. A later paper will expand the study to include several more related solutes.

For the solutes chosen, naphthalene was selected as the parent hydrocarbon possesing no functional groups. This compound has also been studied extensively by others, therefore providing an opportunity to compare our results with previous research in some cases. 1,4-naphthoquinone is an example of a diketone having symmetric carbonyl groups attached to a naphthalenic skeleton. Acridine is an organic base with a heterocyclic nitrogen in the central ring. The parent hydrocarbon of acridine is anthracene, rather than naphthalene. The heterocyclic nitrogen analog of naphthalene (quinoline) is a liquid at room temperature, and the comparison of solubilities between liquid and solid solutes is complicated by the fact that gas readily dissolves in the liquid (solute) phase. The chemical structures and pertinent physical properties of the three solutes are given in Table 1.

Our choice of the supercritical solvents was made so that all had essentially the same critical temperature (see Table 2). Eventually we wished to compare the different supercritical fluids both qualitatively (as being a good or poor solvent) as well as quantitatively with precise equilibrium solubility data. Qualitative statements concerning good and bad solvents

TABLE 1

Physical properties of the solutes

COMPOUND	STRUCTURE	T_c (°C)	P_c (bar)	ω	SOL VOL (cm^3/mol)	A	B
Naphthalene	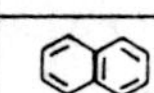	475.	40.5	0.302	110.3	11.458	3733.9
1,4-Naphthoquinone	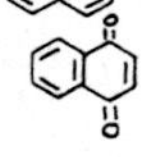	519.	41.2	0.575	111.2	12.610	4739.4
Acridine	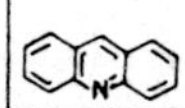	610.	31.9	0.498	178.3	11.596	4740.1

Critical properties were estimated with a factor analysis method (ref. 7).
Vapor pressure constants in $\log_{10}P^{vap}(torr) = A - B/T$ (K)
Naphthalene (ref. 8), 1,4-naphthoquinone (9), Acridine (10).

(for any particular solute) are difficult to defend since, in supercritical extraction, there is little agreement under what conditions comparisons should be made. According to the law of corresponding states, different fluids will behave similarly (with respect to their P-V-T properties) if they are compared at equivalent reduced temperature, pressure, and volume. However, in supercritical extraction, solubility is a strong function of the vapor pressure of the solute, and the vapor pressure is determined by the temperature of the system, not the reduced temperature of the solvent. In order to remove the effect of vapor pressure from comparisons of a solute in a series of different solvents, the comparisons must be made at the same absolute temperature. But in order to apply the principle of corresponding states to the dilute solvent phase, comparisons should be made at the same reduced temperature. Only if the different solvent gases have the same critical temperature can both conditions be simultaneously fulfilled. Within realistic limits, we have succeeded in finding four chemically diverse supercritical solvents that possess critical temperatures within a narrow range of temperatures 29.±3°C. The solvents include carbon dioxide and ethane, which are practical and inexpensive supercritical solvents. These solvents have received considerable attention by previous researchers.

TABLE 2

Solvent properties

Solvent Gas	Formula	T_c(°C)	P_c(bar)	ρ_c ($\frac{mol}{cm^3}$)	ω
Carbon Dioxide	CO_2	31.1	73.8	0.01059	0.225
Ethane	C_2H_6	32.2	48.8	0.0067	0.098
Chlorotri-fluoromethane	$CClF_3$	28.8	39.2	0.0055	0.180
Fluoroform	CHF_3	26.0	48.8	0.0075	0.275

Fluoroform is interesting because of its analogous structure to chloroform, which is a good organic solvent and is capable of hydrogen bonding, as in the system chloroform/acetone. Chlorotrifluoromethane is a completely halogenated methane, and was therefore postulated to be relatively inert chemically, similar to carbon tetrachloride and sulfur hexafluoride.

Having specified the critical temperature of the solvents, one parameter of the corresponding states analysis has been fixed. Since it is convenient to make solubility comparisons in graphical form, of which then ordinate will be mole fraction solute (y_2) in the fluid phase, there remains one degree of freedom representing the choice of abscissa on a two dimensional plot. Either reduced pressure or reduced volume/density are suggested by corresponding states. We have chosen the later representation, since reduced density results in solubility isotherms that are considerably more linear than pressure coordinates over broad ranges of supercritical fluid conditions. Curves intersect less frequently, and the region of retrograde solubility, (peculiar to y_2-pressure diagrams) does not exist in y_2-ρ_r coordinates.

EXPERIMENTAL

A schematic diagram of the equilibrium solubility experimental apparatus is shown in Figure 1. It is similar to those used by Kurnik et al. (ref. 2) and Johnston and Eckert (ref. 11). To expedite data collection, two extraction columns were used in parallel. These could be operated at different temperatures,

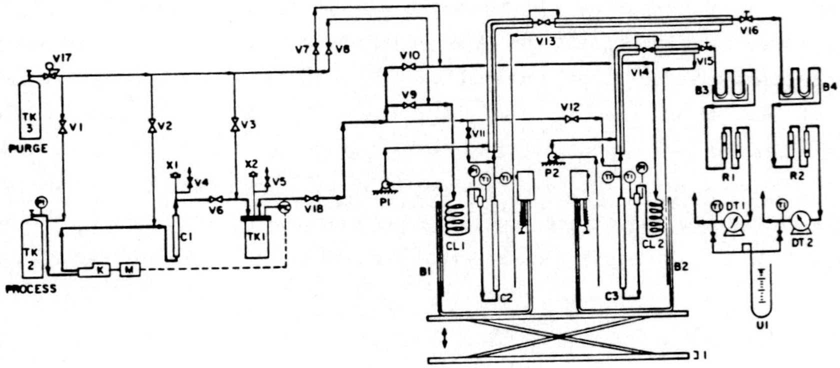

Fig. 1. Supercritical extraction apparatus.

but the pressures in each column were identical since they were fed from a common compressor and surge tank. The system pressure was held to within ±0.3 bar at pressures up to 150 bar and was maintained to within ±1 bar at the highest pressures used (360 bar). Temperatures in the baths were maintained within ±0.1^{o}C and the exiting supercritical fluid from both extraction vessels passed through a line traced with flowing water at the same temperature as in the constant temperature bath. This precaution was necessary to prevent solid precipitation in the exit line under conditions where the fluid may either heat or cool since the solubility of solutes in supercritical fluids may either increase or decrease with temperature depending upon the specific temperature and pressure employed. The only way to avoid this problem is to maintain the test temperature constant up to the let-down valve where the gas is expanded to atmospheric pressure. The precipitated solid was subsequently trapped for weighing and analysis. The operation of the unit is similar to that described earlier (ref. 2).

CALIBRATION AND PROOF TESTS

Preliminary extraction experiments were made with naphthalene and carbon dioxide. The results over a temperature range of 35-45°C and a pressure range of 100 to 300 bar agreed with those of Tsekhanskaya et al. (ref. 12)) to within

±5 mole-% [defined as $(y_{lit} - y_{exp})x100/y_{lit}$]. Also, as will be noted later, there were several instances in our solute-solvent test matrix where our work overlapped that of previous investigators. In all cases except for a single reported 55°C ethane/naphthalene datum point (ref.13), the agreement between our solubility results and those presented in the literature was within about ±5 mole%. Prior to these proof tests with naphthalene and carbon dioxide, we studied the effect of flow rate variations in the extraction cell to insure that we were always operating in a sufficiently low flow regime that equilibrium was attained between the solid and the exit fluid.

RESULTS

With a matrix of three solid solutes and four supercritical fluids, there are twelve binary systems. For each, runs were performed at a minimum of at least two temperatures and over a pressure range that normally extended from about 60 to 365 bar. The experimental solubility data are given in Tables 7 through 18. In each table, the test temperature, pressure, and the equilibrium mole fraction of the solute in the supercritical phase are shown. In addition, the reduced densities and solute enhancement factors are also given. The column defined as reduced density represents the density of the pure supercritical fluid divided by the critical density of the solvent. Data sources for the supercritical fluid densitites at the system conditions were as follows: carbon dioxide (ref.14), ethane (ref. 15), chlorotrifluoromethane (ref. 16), and fluoroform (ref.17). In the last two, experimental density data were not available over the entire temperature and pressure range studied, and, when this was the case, the Peng-Robinson equation of state (ref. 18) was used to predict the pure gas densities from critical properties of the pure solvents as suggested by Peng and Robinson.

Finally, in Tables 7 through 18, there is a column showing the enhancement factor. This factor, termed E, is defined as

$$E = y_2/(P^{vap}/P) \qquad (1)$$

where y_2 is the equilibrium mole fraction of the solute, P the system pressure and P^{vap} the vapor pressure of the pure solute at the system temperature. If the gas phase were ideal and no Poynting correction made, E would then be unity independent of pressure. Thus, the actual solubility is enhanced by a factor E over the value which would have been calculated if gas phase nonidealities and pressure effects on the solid were neglected. Later we shall give an alternate expression for E and indicate how it may be estimated from the thermodynamic properties of the system.

To show the results in graphical form, the equilibrium mole fractions of each solute in the four supercritical fluids have been plotted as a function of reduced density in Figures 2 through 4. The temperature in all of the figures is 55°C. (Note that since the critical temperatures of all four solvents are very similar, then, at equal temperatures, the reduced temperature for the four systems would also be the same)

At this constant temperature, it is seen that the solubilities of all solutes in all supercritical fluids are strong functions of solvent density. The magnitude of the solubilities differ with naphthalene more than the solubilities of the other two solutes. The relative solubilities of the solutes in the four gases are more interesting. Clearly, the lowest extraction occurred with $CClF_3$ (R-13) for all solutes, but the remaining three solvents showed differing abilities to solvate the organic compounds.

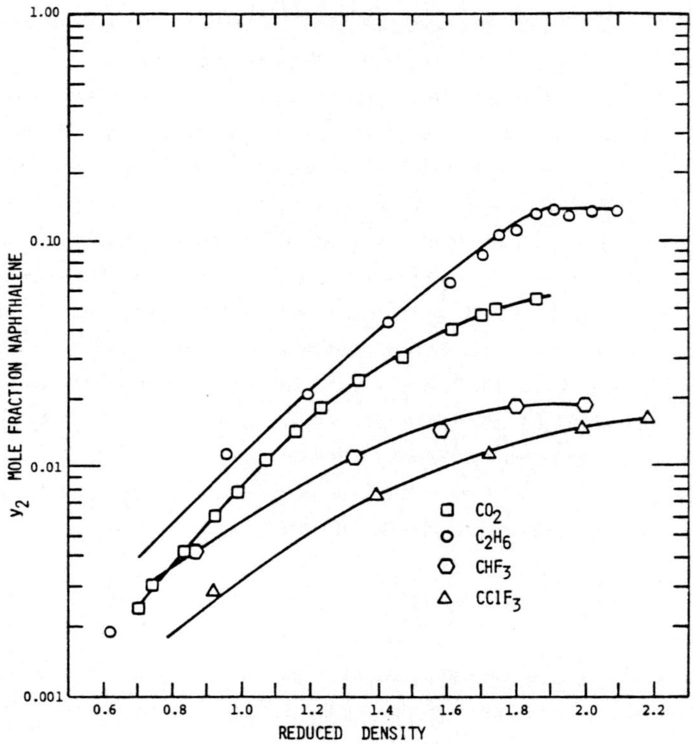

Fig. 2. Solubility of naphthalene in four gases at 55°C

In the case of naphthalene, the highest solubilities were measured in ethane, followed in descending order by carbon dioxide, fluoroform, and finally $CClF_3$ (R-13). The relative difference in the solubility of naphthalene in ethane and in $CClF_3$ is on the order of 350%, measured at a median value of reduced density. It should be reiterated once again that this difference is not caused by differences in the vapor pressure of the solute, but rather appears to be related to the specific nature of the solvent gas. The ethane isotherm is of somewhat peculiar shape at this temperature, reaching a constant value of solubility at a reduced density of approximately 1.9 (corresponding to ~220 bar) and does not increase further. This may be due to the fact that this system is very close to its upper critical endpoint (UCEP) in temperature, known to be 56.6°C and 124 bar (ref. 13). It is known that the solubility of naphthalene changes little with pressure in the vacinity of the UCEP for this system (ref. 13), whereas the change in solubility with temperature is large. For carbon dioxide the UCEP is higher (60.1°C, 256 bar), as reported by McHugh and Yogan (ref.19). The UCEP's of the remaining two systems are not known.

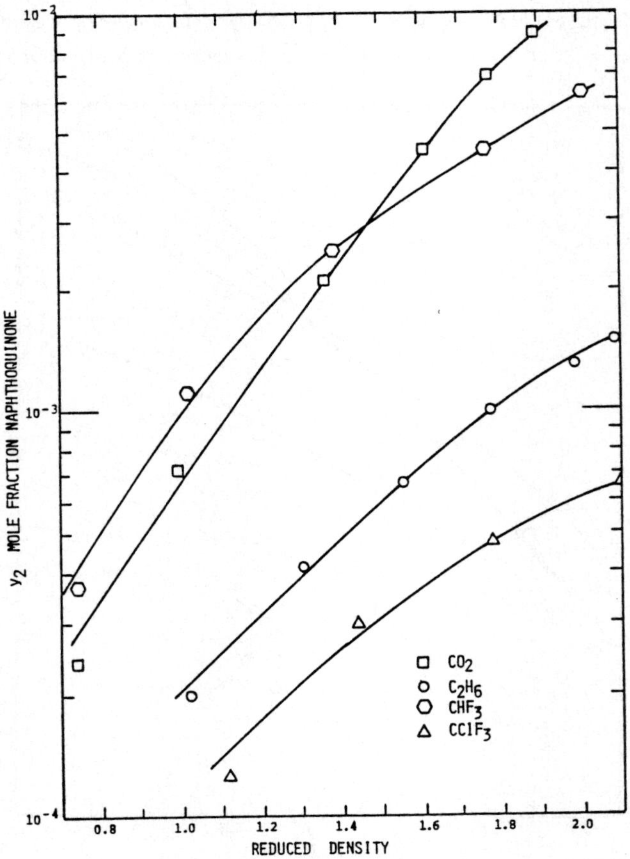

Fig. 3. Solubility of 1,4-naphthoquinone in four gases at 55°C.

The solubilities of 1,4-naphthoquinone in four different gases are shown in Figure 3. Carbon dioxide is an excellent solvent for this compound, as is fluoroform. Indeed, the solubility of naphthoquinone in fluoroform is unexpectedly high and may be associated with partial hydrogen bonding between the acidic proton of fluoroform and the carbonyl groups on the naphthoquinone molecule. Fluoroform appears to be a good solvent for those solutes which are potential electrophillic hydrogen bond acceptors (C=O, CO_2H, and to a lesser degree =N-) but a poor solvent when the solute lacks such functional groups (e.g., naphthalene. Whether or not hydrogen bonding is actually occuring in the supercritical fluid cannot be directly inferred from these data. Spectroscopic information would ultimately be required to prove the existance of hydrogen bonds. Izatt et al. (ref. 20) have postulated that hydrogen bonding may be occurring in the system $CHClF_2$ / N,N-dimethylacetamide at near critical solvent conditions. Their reasoning is based on the relatively large enthalpies of solvation measured for this system but which were not observed in the system $CHClF_2$/ 2,3-dimethylbutane, where hydrogen bonding could not occur.

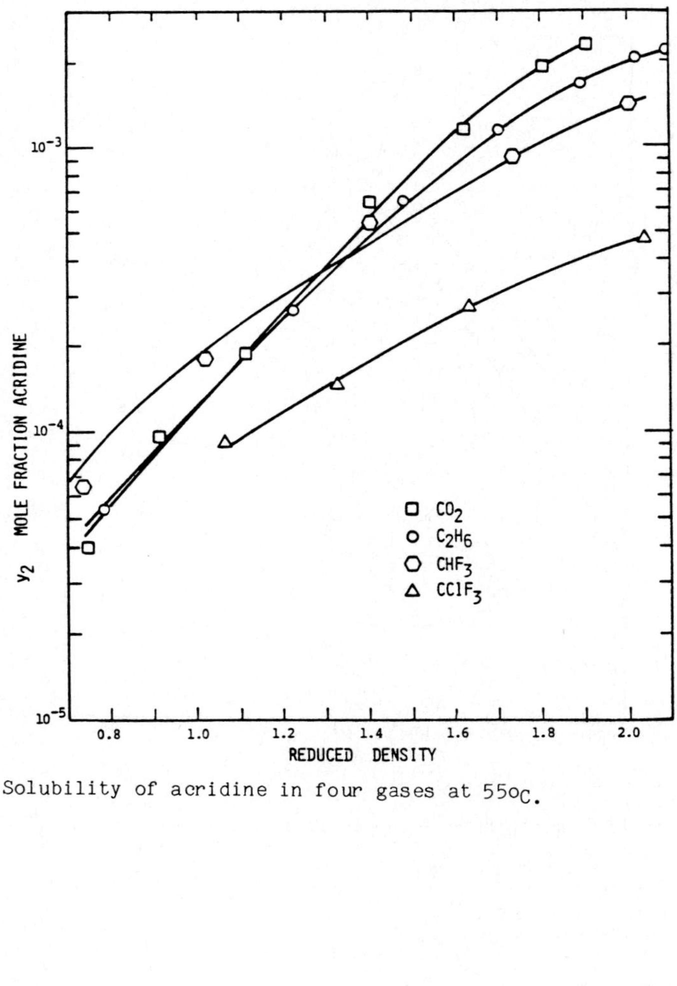

Fig. 4. Solubility of acridine in four gases at 55oC.

With acridine as the solute, carbon dioxide, ethane and fluoroform were all found to be good solvents (see Figure 4). Acridine is actually slightly more soluble in carbon dioxide than it is in ethane, which was unexpected, since anthracene (as well as phenanthrene and naphthalene) is more soluble in ethane than in CO_2 (ref. 1). The only difference between acridine and anthracene is the presence of the heterocyclic nitrogen in acridine, which makes this compound an organic base (pk_a ~5.6). The increased solubility of acridine in carbon dioxide may be due to an acid-base complex which does not occur with the solvent ethane.

DATA CORRELATION

At present, two methods of data correlation have been examined. The first method employs the Peng-Robinson equation of state (ref. 18), and the second is a correlation based on the experimental observation that supercritical solubilities can be linearized over a wide range of fluid conditions if solute enhancement factors are plotted versus the pure solvent densities.

The fundamental equation relating the solubility of a solid in a supercritical solvent is (ref. 21):

$$y_2 = \left(\frac{P^{vap}}{P}\right)\left(\frac{1}{\emptyset_2}\right)\exp\left\{\frac{V_2^s P}{RT}\right\} \qquad (2)$$

Here V_2 is the molar volume of the pure solid solute and $\emptyset_2$ is the fugacity coefficient of the solute in the supercritical phase. The fugacity coefficent of the pure solute, $\emptyset_2^{sat}$ was assumed to be unity; the variation of V_2^s with pressure was neglected; and the approximation $P >> P^{vap}$ was made in the derivation of Eq. 2. The value of $\emptyset_2$ as a function of temperature, pressure, and fluid composition can be modeled with an equation of state as Kurnik et al. (ref. 2) and Paulaitis et al. (ref. 22) have shown with the Peng-Robinson equation. This approach requires that pure component critical constants (T_c, P_c, and the acentric factor) be known for both the solvent and the solute. While this information is readily available for common solvent gases, it is rarely available for solute molecules of interest. Estimation techniques exist which can be used to predict values of T_c and P_c for the solute molecules, as, for example, Lydersen's method (see ref. 23). Our reported acentric factors were derived from vapor pressure measurements of the pure solute compounds near their normal boiling points. Tables 1 and 2 list the solute and solvent pure component properties used in the Peng-Robinson analysis. Besides the pure component properties, a single binary interaction parameter (k_{12}) is required to fit the experimental solubility data. Values of the regressed binary interaction coefficients are given as a function of temperature for each solute/solvent pair in Tables 3-5.

In Figure 6, the Peng-Robinson equation has been used to model the carbon dioxide/acridine system. This model has the ability to reproduce all of the qualitative characteristics of the solubility map as a function of pressure and temperature, including the solubility minimum and the region of retrograde solubility. One problem with this model is that it is often unable to predict quantitatively both low pressure (pressures in the retrograde-solubility region) and high pressure solubility data with a single, constant value of the binary interaction coefficient. This makes it difficult to interpolate and extrapolate accurate experimental supercritical solubility data to temperatures and pressures not actually measured. For this purpose a different method of correlating the data was developed.

The remarkable linearity of the solute enhancement factor (E) with solvent density over a wide range of supercritical conditions has been previously noted (ref. 11). A recent correlation proposed by Ziger and Eckert (ref. 24) uses this linearity as a basis for predicting the solubility of a solid in

TABLE 3

Regressed binary interaction parameters for naphthalene

SOLVENT GAS	Temperature 35°C	45°C	55°C
Carbon Dioxide*	0.0976	0.0973	0.0959
Ethane	0.0469	0.0461	0.0452
Monochlorotri-fluoromethane	0.0879	0.0910	0.0872
Fluoroform	0.0973	0.0972	0.0978

*Data from ref. 12 used to regress for k_{12}.

TABLE 4

Regressed binary interaction coefficients for 1,4-naphthoquinone

SOLVENT GAS	Temperature 35°C	45°C	55°C	70°C
Carbon Dioxide		0.0556	0.0558	0.0529
Ethane	0.114	0.110	0.107	0.106
Monochlorotri-fluoromethane		0.123	0.117	
Fluoroform		0.0222	0.0186	

TABLE 5

Regressed binary interaction coefficients for acridine

SOLVENT GAS	Temperature 35°C	45°C	55°C	70°C
Carbon Dioxide	0.121	0.117	0.116	0.112
Ethane	0.107	0.107	0.095	0.097
Monochlorotri-fluoromethane		0.132	0.126	
Fluoroform		0.108	0.105	

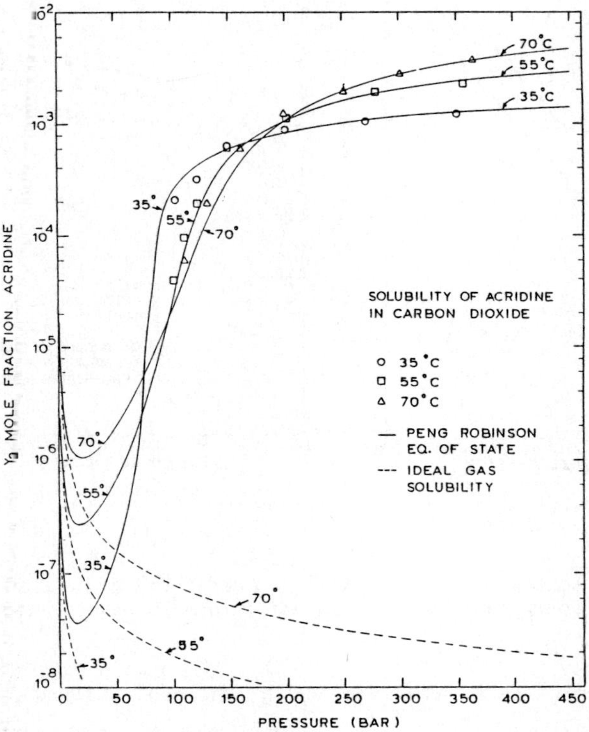

Fig. 5. Correlation of the carbon dioxide/acridine system with the Peng-Robinson equation.

a supercritical fluid. What has apparently not been noted before is that plots of log E versus pure solvent density (or reduced density) exhibit a characteristic fine splitting of the solubility isotherms such that they form a set of closely spaced, yet distinct, parallel lines. This is shown for the naphthalene/fluoroform system in Figure 7. If the individual isotherms could be made to collapse into co-linearity, then the entire system could be modeled by a simple linear relationship. Thus, the following empirical model is proposed:

$$\log E = \alpha\rho_r + \beta + \sigma(T - T_{ref}) \quad (3)$$

where:

α = slope of the reference isotherm
β = y-intercept of the reference isotherm
σ = isotherm spacing constant (K^{-1})

A reference solubility isotherm, onto which the others are to be collapsed,

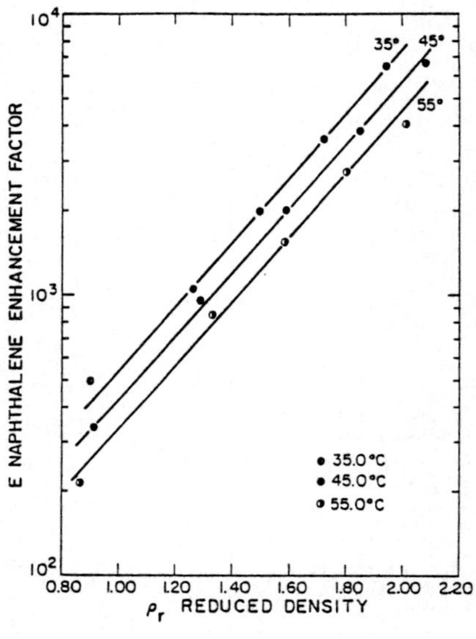

Fig. 6 Enhancement factors of fluoroform in naphthalene

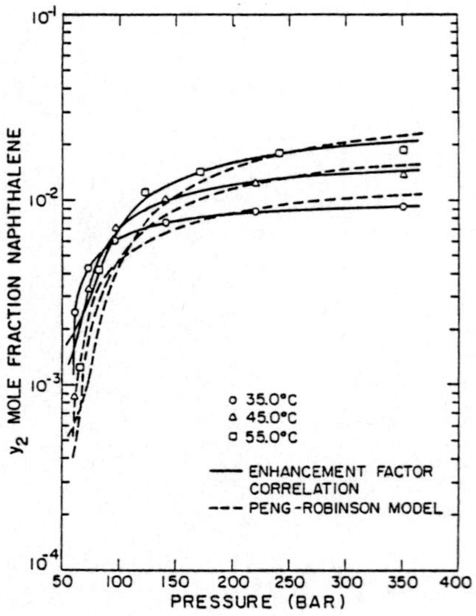

Fig. 7. Correlated solubilities in the fluoroform/naphthalene system system.

is arbitrarily chosen. The parameters α and β can then be found by a linear regression of the experimental log E vs. reduced solvent density data at the reference temperature. To determine σ, all of the remaining isotherms are linearly regressed, and the values of log E at a constant value of the reduced density (such as $\rho_r = 1.5$) are noted. A plot of log $E_{\rho r=1.5}$ versus temperature is a straight line, and the slope of this line is the value of σ for the system, in reciprical temperature. Regressed values of α, β, and σ for the systems presented here are given in Table 6.

Equation (3) can be rewritten in terms of mole fraction solubility as follows:

$$y_2 = (\frac{p^{vap}}{P}) \exp_{10} \{\alpha\rho_r + \beta + \sigma(T-T_{ref})\} \qquad (4)$$

Equation (4) was used to model the solubility isotherms of the naphthalene-fluoroform system, as is shown in Figure 7. The Peng-Robinson (k_{12}) prediction is also shown for comparison. It can be seen that the enhancement factor correlation yields a better fit of the experimental data over the entire range of pressure than does the Peng-Robinson model. The improvement is particularly

TABLE 6

Regressed constants for the enhancement factor correlation

SOLUTE	T_{ref}(°C)	α	β	$\sigma(K^{-1})$
IN CARBON DIOXIDE:				
Naphthalene†	45.0	1.300	1.592	-0.03
Naphthoquinone	55.0	1.800	1.780	-0.017
Acridine	55.0	1.915	1.977	-0.017
IN ETHANE:				
Naphthalene††	45.0	1.680	0.858	-0.006
Naphthoquinone	55.0	1.610	1.110	-0.015
Acridine	55.0	1.801	1.852	-0.015
IN CHLOROTRIFLUOROMETHANE:				
Naphthalene	45.0	1.090	1.201	-0.011
Naphthoquinone	55.0	1.155	1.302	-0.023
Acridine	55.0	1.315	2.100	-0.014
IN FLUOROFORM:				
Naphthalene	45.0	1.115	1.520	-0.011
Naphthoquinone	55.0	1.460	2.163	-0.018
Acridine	55.0	1.570	2.302	-0.016

† Regressed from data of ref. 12.
††Do not use above 50.°C.

prounounced in the low kpressure region including the region of retrograde solubility, where relative errors as large as 75-150% are not uncommon for the Peng-Robinson model in many systems. A detailed mathematical discussion of the enhancement factor correlation is presented elsewhere (ref. 25).

CONCLUSIONS

A generalized corresponding states analysis was used to compare the solubility of three organic model compounds in a series of supercritical fluids at approximately equivalent reduced solvent conditions. It was found that the solubility of a particular solute is significantly affected by the choice of the supercritical solvent used to dissolve the compound. For example, the solubility of naphthalene in ethane is nearly four times greater than it is in chlorotrifluoromethane at all of the temperatures studied. In some cases, chemical solvolysis effects, such as hydrogen bonding (e.g. 1,4-naphthoquinone/CHF_3) and acid/base (acridine/CO_2) interaction were indirectly observed. This type of information could be useful when attempting to optimize the extraction of a desired compound from a solute mixture.

REFERENCES

1 K.P Johnston, D.H. Ziger, and C.A. Eckert, Ind. Eng. Chem. Fundam., 21 (1982) 191.
2 R.T. Kurnik, S.J. Holla, and R.C. Reid, J. Chem. Eng. Data, 26 (1981) 47.
3 M. McHugh, and M.E. Paulaitis, J. Chem. Eng. Data, 25 (1980) 326.
4 E. Stahl and E. willig, Microchim. acta, 2 (1980) 465.
5 V.J. Krukonis, A.R. Branfman, M.G. Broome, and A.T. Sneden, "Supercritical Fluid Extraction of Plant Materials Containing Chemotherapeutic Drugs", Presented at the 87th AIChE National Meeting, Boston, MA (1979).
6 J. Chrastil, J. Phys. Chem., 86 (1982) 3016.
7 K. Joback, S.M. Thesis, Dept. Chem. Eng., Mass. Inst. Tech., Cambridge, MA, Aug. (1984).
8 L. Fowler, W. Trump, and C. Vogler, J. Chem. Eng. Data, 13 (1968) 209.
9 C.G. de Kruif, E.J. Smit, and H.A.J. Govers, J. Chem. Physics 74 (1981) 5838.
10 C.G. de Kruif, C.H.D. van Ginkel, and J. Voogel, "Torsion-Effusion Vapor Pressure Measurements of Organic Compounds",Quatrieme Conference Internationale de Thermodynamique Chemique, 25 au 30, Aout, (1975) Montpellier, France.
11 K.P. Johnston and C.A. Eckert, AICHE J., 27 (1981) 773.
12 Yu V. Tsekhanskaya, M.B. Iomtev, and E.V. Mushkina, Russ. J. Phys. Chem. 38 (1964) 1173.
13 G.S.A. Van Welie, and G.A.M. Diepen, J. Phys. Chem., 67 (1963) 755.
14 I.U.P.A.C., International Thermodynamic Tables of the Fluid State: Carbon Dioxide, ed. by S. Angus, B. Armstrong, and K.M. de Reuck, Permgamon Press (1973).
15 N.B.S., Thermophysical Properties of Ethane, from 90 to 600 K at Pressures to 700 Bar, R.D. Goodwin, H.M. Roder, and G.C. Straty, issued Aug. 1976.
16 A. Michels, T. Wassenaar, G.J. Wolkers, Chr. Prins, and L. v.d.Klundert,, J. Chem. Eng. Data, 11 (1966) 449.
17 K. Hori, S. Okazaki, M. Vematsu, and K. Watanabe, Proc. VIII Symposium on Thermochemical Properties, Vol. II, A.S.M.E., ed. by J. V. Sengers (1981) 380.
18 D.Y. Peng, and D.B. Robinson, Ind. Eng. Chem. Fund, 15 (1976) 59.
19 M.A. McHugh, and T.J. Yogan, J. Chem. Eng. Data, 29 (1984) 112.
21 J.M. Prausnitz, "Solubility of Solids in Dense Gases", Nat. Bureau of Standards Tech., Note 316 (1965).
22 M.E. Paulaitis, M.A. McHugh, and C.P. Chai, Chemical Engineering at Supercritical Fluid Conditions, ed. M.E. Paulaitis, J.M.L. Penninger, R.D. Gray Jr., and P. Davidson, Ann Arbor Science, p.139, (1983).
23 R.C. Reid, J.M. Prausnitz, and T.K. Sherwood, The Properties of Gases and Liquids, 3rd ed., McGraw-Hill, New York, (1977).
24 D.H. Ziger, and C.A. Eckert, Ind. Eng. Chem. Process Des. Dev., 22 (1983)

582.
25 W.J. Schmitt, Doctoral Thesis, Dept. Chem. Eng., Mass. Inst. Tech., Cambridge MA, Dec. (1984)

NOMENCLATURE

E enhancement factor

P_c critical pressure (bar)

P^{vap} vapor pressure (torr)

T_c critical temperature (K)

y_2 solute mole fraction

V_2^s solid volume (cm^3/mol)

Greek

α slope parameter in the Enhancement Factor Correlation

β intercept parameter in the Enhancement Factor Correlation ρ density (mol/cm^3)

ρ_c critical density (mol/cm^3)

ρ_r reduced density

σ splitting constant in the Enhancement Factor Correlation

ω acentric factor

TABLE 7

SOLUBILITY OF NAPHTHALENE IN CARBON DIOXIDE†

Pressure (abs)bar	Mole Fraction	Reduced Density	Enhancement Factor
at 35°C			
79.0	1.37×10^{-3}	0.637	370
79.5	3.01	0.713	818
80.5	4.76	0.897	1,300
87.1	8.10	1.310	2.400
92.2	9.20	1.480	2,900
106.4	11.0	1.565	4,000
122.	12.5	1.640	5,200
142.	13.9	1.715	6,700
162.	15.1	1.781	8,400
192.	16.3	1.847	10,700
223.	17.4	1.904	13,300
294.	18.4	1.989	18,500
334.	18.7	2.026	21,300
at 45°C			
88.2	1.30×10^{-3}	0.730	163
92.2	2.10	0.820	275
95.8	4.30	0.915	586
97.8	5.10	0.980	709
101.	6.90	1.084	990
106.	9.70	1.178	1,460
109.	11.3	1.235	1,750
127.	15.4	1.451	2,780
142.	18.3	1.555	3,690
155.	19.7	1.612	4,340
203.	24.5	1.753	7,070
279.	28.5	1.876	11,300
314.	29.4	1.923	13,100
at 55°C			
94.2	2.00×10^{-3}	0.616	117
99.3	2.40	0.700	149
102.	3.00	0.748	191
106.	4.20	0.838	278
114.	7.70	0.990	548
120.	10.8	1.074	809
126.	14.2	1.159	1,110
132.	18.1	1.235	1,490
143.	23.4	1.348	2,090
162.	30.0	1.470	3,030
198.	39.9	1.612	4,930
233.	45.9	1.706	6,670
253.	48.5	1.753	7,660
324.	55.0	1.866	11,100

† All data from Tsekhanskaya et al. (ref. 12).

TABLE 8

SOLUBILITY OF NAPHTHALENE IN ETHANE

Pressure (abs)bar	Mole Fraction	Reduced Density	Enhancement Factor
at 35°C			
50.9	0.25×10^{-2}	0.926	660
55.6	1.12	1.373	2,090
60.1	1.29	1.482	2,600
65.5	1.42	1.559	3,180
69.1	1.54	1.604	3,600
91.5	1.97	1.742	6,200
99.4	2.08	1.780	6,980
120.5	2.30	1.848	9,450
131.0	2.54	1.897	11,400
150.8	2.56	1.921	13,100
181.0	2.78	1.975	17,200
198.7	3.02	1.204	20,000
250.0	3.24	2.081	27,600
301.0	3.19	2.144	32,700
362.0	3.19	2.196	42,900
at 45°C			
51.0	0.06×10^{-2}	0.517	43
55.5	0.15	0.663	118
65.5	1.36	1.127	1,260
69.1	1.40	1.276	1,380
91.5	2.67	1.584	3,470
120.5	3.46	1.741	5,940
131.0	3.86	1.773	7,130
150.7	4.08	1.845	8,750
199.0	4.74	1.951	13,400
250.0	5.25	2.030	18,600
301.7	5.42	2.090	23,000
364.0	5.85	2.151	29,900
at 55°C			
56.3	0.12×10^{-2}	0.528	42
61.3	0.19	0.625	72
71.5	1.14	0.960	503
80.2	2.09	1.199	1,020
100.8	4.37	1.431	2,750
120.3	6.48	1.613	4,920
164.0	10.7	1.760	10,900
175.0	11.1	1.809	12,100
202.0	13.7	1.872	17,400
206.0	13.0	1.879	17,000
225.0	13.6	1.918	19,100
251.0	12.9	1.959	20,100
302.0	13.6	2.029	25,900
364.0	13.4	2.098	30,400

TABLE 9

SOLUBILITY OF NAPHTHALENE IN FLUROFORM

Pressure (abs)bar	Mole Fraction	Reduced Density	Enhancement Factor
at 35°C			
60.5	$2.46x10^{-3}$	0.903	500
73.3	4.24	1.264	1,050
96.1	6.06	1.498	1,990
141.3	7.49	1.727	3,600
221.0	8.71	1.947	6,500
351.3	9.18	2.155	10,800
at 45°C			
60.5	$0.857x10^{-3}$	0.586	72
73.3	3.30	0.918	338
96.1	7.04	1.287	962
141.3	10.09	1.590	2,000
221.0	12.3	1.850	3,870
351.3	13.4	2.083	6,620
at 55°C			
66.0	$1.24x10^{-3}$	0.570	51
82.8	4.21	0.868	214
123.0	11.1	1.332	850
171.0	14.4	1.587	1,530
240.8	18.4	1.802	2,730
351.4	18.5	2.012	4,050

TABLE 10

SOLUBILITY OF NAPHTHALENE IN MONOCHLOROTRIFLUOROMETHANE

Pressure (abs.)bar	Mole Fraction	Reduced Density	Enhancement Factor
at 35°C			
50.1	2.84×10^{-3}	1.271	478
71.5	4.35	1.662	1,070
98.2	5.41	1.816	1,800
140.9	6.33	2.004	3,000
199.8	7.12	2.171	4,800
281.3	7.66	2.326	7,200
355.3	7.84	2.429	9,500
at 45°C			
50.1	1.17×10^{-3}	0.790	82
71.5	5.16	1.402	525
98.2	7.11	1.666	985
140.9	8.94	1.893	1,760
199.8	10.08	2.083	2,800
281.3	10.35	2.258	4,100
355.3	10.92	2.369	5,500
at 55°C			
51.6	1.08×10^{-3}	0.645	35
61.3	2.82	0.921	108
86.9	7.50	1.394	407
130.3	11.51	1.725	951
200.3	14.71	1.996	1,850
282.0	16.2	2.187	2,900
360.3	15.8	2.314	3,500

TABLE 11

SOLUBILITY OF 1,4-NAPTHOQUINONE IN CARBON DIOXIDE

Pressure (abs.)bar	Mole Fraction	Reduced Density	Enhancement Factor
at 45°C			
101.0	0.932×10^{-3}	1.053	13,400
115.8	1.55	1.325	25,600
145.0	2.62	1.560	53,900
200.5	4.20	1.739	119,000
274.5	5.76	1.871	225,000
363.0	6.97	1.965	359,000
at 55°C			
100.9	0.24×10^{-3}	0.741	1,250
116.0	0.72	0.996	4,270
145.0	2.12	1.363	15,700
200.5	4.59	1.617	46,500
275.0	6.96	1.778	96,500
364.0	8.88	1.899	163,000
at 70°C			
100.8	0.18×10^{-3}	0.538	217
116.0	0.37	0.704	508
145.0	1.67	1.034	2,880
200.5	5.34	1.419	12,800
275.0	9.92	1.636	32,500
364.0	13.5	1.786	58,700

TABLE 12

SOLUBILITY OF 1,4-NAPHTHOQUINONE IN ETHANE

Pressure (abs)bar	Mole Fraction	Reduced Density	Enhancement Factor
at 35°C			
63.0	1.06×10^{-4}	1.519	2,942
74.0	2.02	1.635	6,550
91.0	3.07	1.732	12,350
141.0	4.31	1.888	26,850
202.0	5.03	2.001	45,000
283.	5.24	2.113	65,500
364.	5.37	2.188	85,400
at 45°C			
63.0	0.91×10^{-4}	1.021	820
74.0	1.86	1.385	1,980
91.0	3.54	1.579	4,622
141.	5.97	1.800	12,200
202.	7.65	1.943	22,400
283.	8.99	2.054	36,900
364.	9.35	2.143	49,400
at 55°C			
63.0	0.59×10^{-4}	0.690	190
73.0	2.00	1.021	745
87.0	4.14	1.301	1,840
112.	6.65	1.551	3,800
165.	10.0	1.775	8,430
275.	13.1	1.985	18,500
364.	15.1	2.085	27,900
at 70°C			
63.0	0.76×10^{-4}	0.524	57
73.0	1.39	0.694	121
87.0	4.08	0.949	423
112.	7.30	1.307	976
165.	14.0	1.628	2,760
271.	22.8	1.882	7,360
364.	27.0	2.001	11,700

TABLE 13

SOLUBILITY OF 1,4-NAPHTHOQUINONE IN FLUOROFORM

Pressure (abs)bar	Mole Fraction	Reduced Density	Enhancement Factor
at 45°C			
66.0	0.26×10^{-3}	0.720	2,400
76.0	0.51	0.980	5,580
92.5	0.89	1.245	12,000
131.0	2.57	1.535	48,900
228.0	3.333	1.862	110,200
356.0	4.57	2.085	239,000
at 55°C			
66.0	0.17×10^{-3}	0.569	587
76.0	0.37	0.743	1,450
92.5	0.37	1.020	1,450
131.0	1.12	1.383	5,300
228.0	4.52	1.766	52,600
356.0	6.33	2.014	115,000

TABLE 14

SOLUBILITY OF 1,4-NAPHTHOQUINONE IN MONOCHLOROTRIFLUOROMETHANE

Pressure (abs)bar	Mole Fraction	Reduced Density	Enhancement Factor
at 45°C			
56.3	0.90×10^{-4}	1.061	735
69.2	1.88	1.368	1,890
91.0	2.32	1.611	3,064
141.0	3.22	1.894	6,590
246.0	3.90	2.189	13,780
364.0	4.31	2.380	22,800
at 55°C			
56.3	0.56×10^{-4}	0.775	160
69.2	1.25	1.117	440
91.0	2.99	1.440	1,400
141.0	4.75	1.778	3,420
246.0	6.54	2.112	8,220
364.0	7.56	2.321	14,050

TABLE 15

SOLUBILITY OF ACRIDINE IN CARBON DIOXIDE

Pressure (abs)bar	Mole Fraction	Reduced Density	Enhancement Factor
at 35°C			
102.5	2.06×10^{-4}	1.528	101,000
122.2	3.30	1.666	191,400
151.6	6.38	1.755	431,000
200.0	9.07	1.858	735,000
270.3	10.6	1.963	1,300,000
350.4	12.5	2.044	1,960,000
at 45°C			
101.8	1.18×10^{-4}	1.110	28,000
110.6	2.47	1.315	40,800
122.8	3.26	1.439	60,000
152.1	6.81	1.606	150,000
200.0	11.3	1.744	330,000
270.1	15.1	1.866	605,000
279.4	15.5	1.882	640,000
352.2	17.3	1.967	897,000
at 55°C			
101.6	0.40×10^{-4}	0.749	1,600
110.6	0.96	0.917	5,600
122.5	7.89	1.115	12,300
151.0	6.17	1.408	48,100
201.1	11.65	1.628	121,300
279.6	19.4	1.805	284,000
355.7	22.7	1.911	425,000
at 70°C			
111.2	0.60×10^{-4}	0.652	930
130.1	1.96	0.883	3,100
160.7	6.15	1.182	12,100
199.7	12.0	1.414	29,400
251.4	19.5	1.585	60,500
300.1	28.5	1.691	105,000
364.1	36.6	1.793	164,500

TABLE 16

SOLUBILITY OF ACRIDINE IN ETHANE

Pressure (abs)bar	Mole Fraction	Reduced Density	Enhancement Factor
at 35°C			
61.7	1.22×10^{-4}	1.498	33,700
76.6	3.42	1.658	116,700
100.5	4.91	1.783	221,100
140.6	6.03	1.851	380,200
209.9	6.83	2.014	642,300
283.4	7.12	2.127	905,000
at 45°C			
52.8	0.134×10^{-4}	0.578	1,040
56.1	0.250	0.692	2,090
65.9	0.593	1.147	5,800
75.8	1.86	1.357	20,800
82.4	2.10	1.510	25,500
100.4	5.41	1.649	80,900
140.2	6.53	1.808	138,000
209.8	9.81	1.970	310,000
292.1	11.52	2.081	501,400
361.4	12.78	2.150	696,100
at 55°C			
56.1	0.143×10^{-4}	0.528	422
61.1	0.292	0.630	935
65.9	0.540	0.783	1,800
82.3	2.62	1.224	11,100
100.7	6.41	1.482	33,800
140.4	11.55	1.705	84,600
210.0	16.8	1.893	186,700
279.0	21.1	1.997	309,200
364.0	29.4	2.101	561,200
at 70°C			
61.0	0.299×10^{-4}	0.485	223
82.2	1.78	0.872	1,779
101.3	5.79	1.185	7,200
140.5	15.6	1.526	26,800
210.0	26.7	1.773	68,900
279.0	37.0	1.901	127,900
364.0	48.8	2.027	217,900

TABLE 17

SOLUBILITY OF ACRIDINE IN FLUOROFORM

Pressure (abs)bar	Mole Fraction	Reduced Density	Enhancement Factor
at 45°C			
65.5	0.36×10^{-4}	0.706	3,575
75.5	0.60	0.968	6,860
92.5	1.81	1.245	25,200
134.5	5.67	1.553	115,000
217.0	7.22	1.836	236,100
353.0	10.2	2.081	541,100
at 55°C			
65.5	1.23×10^{-4}	0.561	841
75.5	0.64	0.734	2,580
92.5	1.79	1.020	8,750
143.5	5.39	1.405	38,400
217.0	9.22	1.736	106,000
353.0	14.0	2.009	261,200

TABLE 18

SOLUBILITY OF ACRIDINE IN MONOCHLOROTRIFLUOROMETHANE

Pressure (abs)bar	Mole Fraction	Reduced Density	Enhancement Factor
at 45°C			
67.0	1.05×10^{-4}	1.330	10,500
81.5	1.31	1.524	16,100
115.0	2.01	1.771	34,900
216.0	3.02	2.124	98,300
341.0	4.07	2.350	209,400
at 55°C			
67.0	0.93×10^{-4}	1.068	3,290
81.5	1.47	1.326	6,360
115.	2.75	1.635	16,760
216.	4.78	2.040	54,700
341.	6.41	2.286	115,800

Supercritical Fluid Technology, edited by J.M.L. Penninger, M. Radosz, M.A. McHugh and V.J. Krukonis
1985 Elsevier Science Publishers B.V., Amsterdam — Printed in The Netherlands

AN EXPERIMENTAL STUDY OF PHASE EQUILIBRIA FOR ISOPROPANOL-WATER-CO_2 MIXTURES RELATED TO SUPERCRITICAL-FLUID EXTRACTION OF ORGANIC COMPOUNDS FROM AQUEOUS SOLUTIONS

M.E. PAULAITIS, J.R. DIANDRETH and R.G. KANDER
Dept. of Chemical Engineering, University of Delaware, Newark, Delaware 19716 (USA)

ABSTRACT

The results of an experimental study of fluid-phase equilibria at elevated pressures for the isopropanol-water-carbon dioxide system are presented. Two methods for extracting organic compounds from aqueous solutions using supercritical-fluid solvents are proposed based upon the observed phase behavior.

INTRODUCTION

Supercritical or near-critical fluids can be used as solvents to extract organic constituents from aqueous solutions. A potential application for this extraction technique would be to recover products from fermentation broths. Recovery by conventional methods can be prohibitively expensive since these products are typically present in low concentrations. Furthermore, the chemical nature of these products may preclude the use of conventional separation techniques, such as distillation or extraction with hydrocarbon solvents.

In this paper, we present the results of an experimental study of phase equilibria at elevated pressures for the ternary system: isopropanol-water-carbon dioxide. Our results include equilibrium phase compositions for two, three, and four coexisting fluid phases as a function of temperature and pressure in the vicinity of the critical point of CO_2. Correlation of the experimental data to give a complete description of global phase behavior is accomplished using a thermodynamic model based on a simple cubic equation of state. Two methods for extracting organic solutes from dilute aqueous solutions using supercritical-fluid (SCF) solvents are proposed based upon this equilibrium phase behavior.

Studies of fluid-phase equilibria related to the SCF extraction of organic solutes from aqueous solution are not extensive. Elgin and Weinstock (ref. 1) investigated two- and three-phase equilibria at elevated pressures for several three-component systems consisting of organic-water mixtures with compressed ethylene. Measurements at only one temperature, 15° C, were reported. Two methods for separating organic compounds from water can be identified in their work. The first is a conventional extraction of organic constituents from the aqueous phase with the SCF solvent. This separation is based upon the equilibrium distribution of the organic component between water and compressed ethylene. The second method involves separating organic-water mixtures by forming two liquid phases (one water-rich phase and one organic-rich phase) in the presence of compressed ethylene. This "salting out" process is based upon the higher solubility of ethylene in the organic-rich liquid phase relative to the water-rich liquid phase. Snedeker (ref. 2) investigated similar organic-water mixtures with compressed carbon dioxide.

A number of studies have focused on the extraction of ethanol from water using SCF solvents at elevated pressures. Baker and Anderson (ref. 3) measured the phase boundary for gas-liquid equilibrium for ethanol-water-CO_2 mixtures at temperatures from 10° to 50° C and pressures up to 3000 PSI. These experimental results show that extraction with compressed CO_2 will not give ethanol concentrations greater than that corresponding to the ethanol-water azeotrope. This observation was confirmed by Paulaitis et al. (ref. 4) who measured equilibrium tie-line compositions for this system at similar conditions of temperature and pressure. Kuk and Montagna (ref. 5) also presented a limited number of experimental results for this system. Experimental studies of ethanol-water mixtures with ethylene (ref. 4) and with ethane (ref. 6) at elevated pressures have also been reported. Again, neither compressed gas can extract ethanol to give concentrations greater than the ethanol-water azeotrope. Relatively little work has been done with higher molecular weight alcohols (ref. 5, 7).

Experimental observations of multiple (three or more) equilibrium phases for alcohol-water-SCF systems have also appeared in the literature. Efremova and Shvarts (ref. 8-11) measured critical endpoints (i.e., gas-liquid and liquid-liquid critical points in the presence of a third phase) bounding three-phase, liquid-liquid-gas equilibrium for ethanol-water-CO_2 and methanol-water-CO_2 mixtures at elevated pressures. Equilibrium phase compositions were not measured in these studies. Paulaitis et al. (ref. 4) measured compositions of the three coexisting fluid phases for the ethanol-water-CO_2 system at 35° C and 1045 PSI. Fleck (ref. 12) investigated three- and four-phase equilibria for the n-propanol-water-CO_2 system.

EXPERIMENTAL SECTION

Two experimental techniques were used to measure phase compositions in this study, and are described in detail elsewhere (ref. 13, 14). For two-phase equilibrium, tie-line compositions were measured using a flow technique to facilitate sampling of the phases at elevated pressures. The apparatus was also equipped with a view cell to observe equilibrium phase behavior. For three or more coexisting phases, the equilibrium compositions were measured using a static technique with a variable-volume, view cell. This technique involved calculating phase compositions from component material balances and measured phase volumes at known cell loadings; no sampling was required.

The flow technique and the static technique were used to measure, respectively, two-phase and three-phase equilibrium compositions for the isopropanol-water-CO_2 system at 60° C and 1750 PSI. These results are shown on the triangular phase diagram in Figure 1. Although a direct comparison of measured compositions cannot be made the tie-line compositions can be easily extrapolated to the three-phase compositions to show good agreement between the two experimental techniques. Radosz (ref. 15) also measured three-phase equilibrium compositions at 61° C/1765 PSI and equilibrium tie-line compositions at 60° C/1515 PSI for this system. These results compare favorably with compositions measured at nearly the same conditions in this work.

EQUILIBRIUM PHASE BEHAVIOR

In this section, a detailed description of the equilibrium phase behavior for the isopropanol-water-CO_2 system is given based on experimental measurements and on model-generated results obtained with the Peng-Robinson equation of state (ref. 16). The significant features of this phase behavior are:

(1) For two equilibrium phases, the compositions of the coexisting (water-rich and CO_2-rich) phases can change dramatically with relatively small temperature or pressure changes;

(2) Equilibria involving three and four coexisting fluid phases can be obtained over an extensive range of temperatures and pressures in the vicinity of the critical point of CO_2. At least four <u>different</u> three-phase regions must exist at conditions near those where four phases are in equilibrium.

As a reference, consider the measured compositions for the system ethanol-water-CO_2 plotted on the triangular phase diagrams in Figure 2. Both diagrams are at a temperature of 35° C; the diagram on the left is at a pressure of 1500 PSI and the diagram on the right is at a pressure of 2000 PSI. The

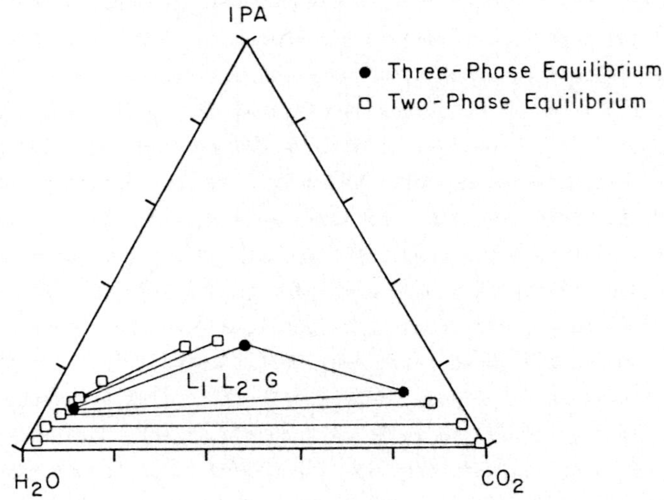

Fig. 1. Measured Tie-Line and Three-Phase Equilibrium Compositions for Isopropanol-Water-CO_2.

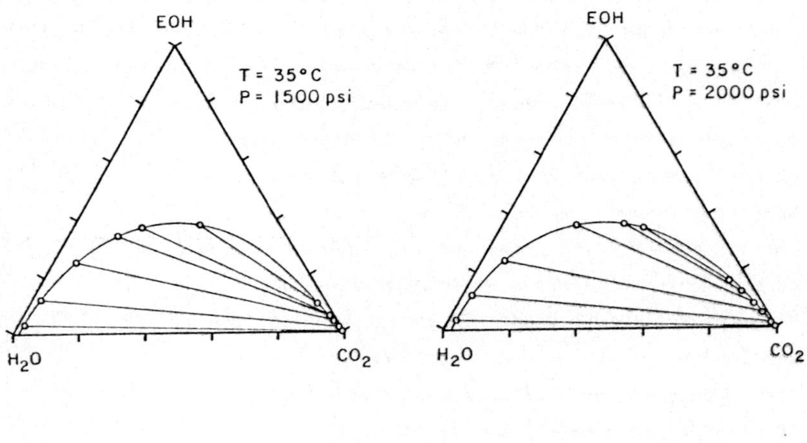

Fig. 2. Measured Equilibrium Tie Lines for Ethanol-Water-CO_2.

pressure change at this temperature causes some shifting of the equilibrium tie-lines to reflect increased SCF (CO_2) solvent capacities and lower alcohol selectivities with higher pressures. However, this pressure effect is a relatively minor one.

Figure 3 shows two similar triangular diagrams for isopropanol-water-CO_2 mixtures which have also been constructed from measured phase compositions. An identical pressure change (from 1500 to 2000 PSI) is shown; however the temperature is now somewhat higher (60^0 C). For this system, the pressure increase produces a much more dramatic shift in tie lines. Again, higher pressures correspond to higher solvent capacities and lower alcohol selectivities. Figure 4 depicts a comparable shift in tie lines with a relatively small temperature change (60^0 to 40^0 C) at a constant pressure of 1500 PSI. These effects can be correlated qualitatively with the density of CO_2 which, at 60^0 C and 1500 PSI, is significantly lower than the CO_2 densities corresponding to the other three sets of temperature and pressure.

The tie lines depicted in Figures 3 and 4 do not, however, shift continuously as either temperature or pressure is varied. At intermediate temperatures and pressures, regions of three or more equilibrium phases can be obtained as shown in Figure 1. The relationship between the observed two- and three-phase equilibria is described with reference to Figure 5. These four phase diagrams depict the transition in phase behavior with increasing pressure at a constant temperature of 60^0 C. At 1500 PSI, two fluid phases exist at equilibrium over the entire composition range for the ternary mixture. As pressure is increased to 1600 PSI, the water-rich liquid phase splits to form two liquid phases in the presence of the CO_2- rich fluid phase. The apexes of the triangle L_1L_2G represent the compositions for these three coexisting phases. The three-phase triangle is also bounded by three two-phase regions which are not shown on this phase diagram. As pressure is increased further (1750 PSI), the composition of the liquid L_2 phase of intermediate density shifts continuously towards the composition of the CO_2-rich fluid phase. Eventually, the L_2 phase and the CO_2-rich G phase merge to form a single phase at pressures just below 2000 PSI. Thus the two-phase region is obtained again; but now the equilibrium tie lines have shifted dramatically as compared to the tie lines at 1500 PSI.

At lower temperatures, a second transition from three to four equilibrium phases will occur with increasing pressure. This transition is depicted on the model-generated triangular phase diagrams in Figure 6. As before, a single three-phase triangle appears with increasing pressure. However, as pressure is increased further (1390 PSI), a second three-phase triangle appears. (Note that the five two-phase regions bounding these three-phase triangles are

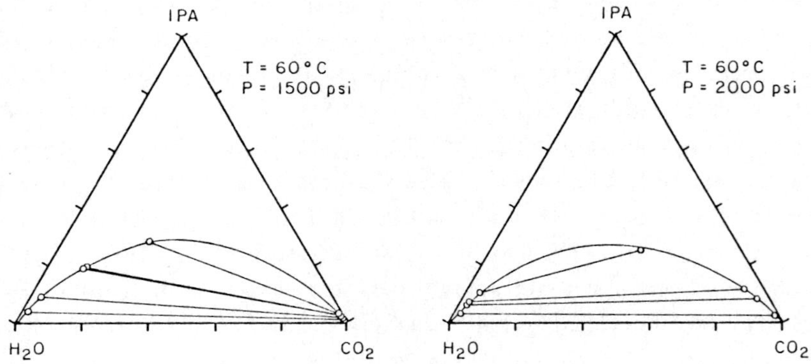

Fig. 3. Measured Equilibrium Tie Lines for Isopropanol-Water-CO_2 at 60^0 C.

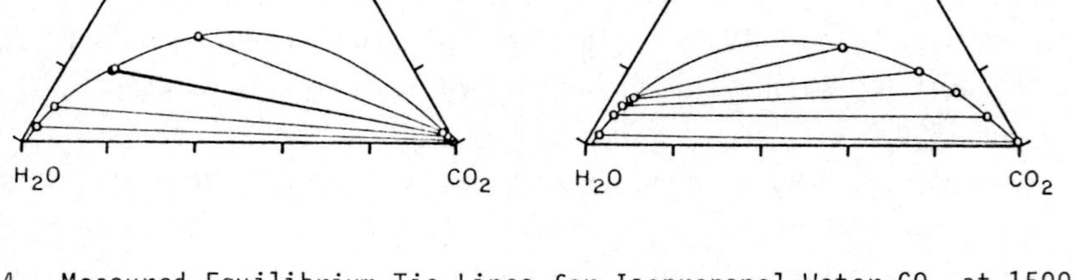

Fig. 4. Measured Equilibrium Tie Lines for Isopropanol-Water-CO_2 at 1500 PSI.

not depicted on the diagram.) At 1405 PSI, the two three-phase triangles merge to form a four-phase quadrilateral. From the phase rule, four equilibrium phases will exist at only one pressure for a three-component system at a specified temperature. At higher pressures (1420 PSI), the four-phase quadrilateral splits to give two different three-phase triangles. (Again, the five two-phase regions that bound the three-phase triangles are not depicted.) One of these three-phase triangles will disappear at higher pressures to recover the transition described in Figure 5. The model-generated phase behavior is in agreement with experimental observations of multi-phase equilibria. Figure 7 shows measured compositions for four-phase equilibrium at 40° C and 1110 PSI which can be compared qualitatively with the four-phase quadrilateral in Figure 6.

Figure 8 shows the model-generated pressure-temperature (PT) projection for the isopropanol-water-CO_2 system, which exhibits the four-phase line (LLLG) for the ternary mixture, the three-phase line (LLG) for the binary water-CO_2 system, the vapor pressure curve for carbon dioxide (CO_2 VP), and the various gas-liquid and liquid-liquid critical lines (dashed lines) which bound the three-phase regions for the ternary mixture. Although these calculated results are not expected to be quantitative, the PT projection does demonstrate the considerable extent of the three-phase behavior and the complexity of the phase behavior in the vicinity of the critical point of CO_2. This second point deserves particular emphasis when interpretting results of simple screening experiments for SCF extractions.

APPLICATIONS

Our experimental results show that carbon dioxide, as a compressed gas or supercritical fluid, will selectively extract isopropanol from aqueous solutions to give alcohol concentrations greater than that corresponding to the isopropanol/water azeotrope. However, the important question is whether this extraction presents an attractive alternative to conventional separation processes. Although an economic comparison of process flowsheets is not within the scope of this paper, several important aspects of potential extraction processes can be addressed by considering the practical implications of the equilibrium phase behavior discussed in the previous section. These are

(1) What are the minimum operating pressures for these extractions?

(2) Can an efficient separation process be devised utilizing the large variations in SCF solvent characteristics that are produced over relatively small temperature and pressure ranges?

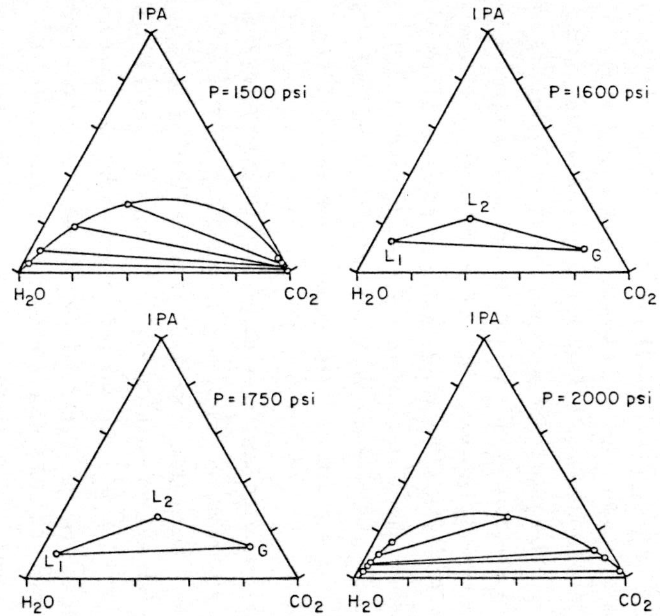

Fig. 5. Observed Two-Phase and Three-Phase Equilibria at 60° C.

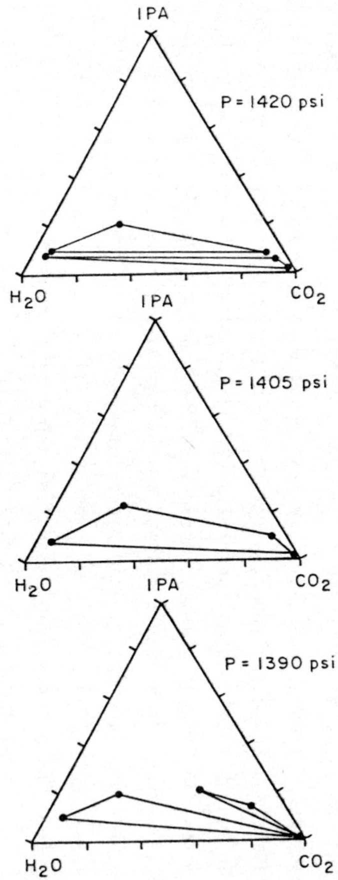

Fig. 6. Model-Generated Three-Phase and Four-Phase Equilibria at 55° C.

(3) Does the observed three-phase equilibrium behavior represent a method for separating isopropanol from water by "salting out" the alcohol?

(4) Can higher molecular weight components be extracted from aqueous solution using dense carbon dioxide and isopropanol as co-solvents?

Minimum operating pressures for SCF extractions are obviously desirable. Although the pressures chosen in this study are reasonable ones, lower bounds on operating pressures can be established based upon the observed phase behavior. It was shown in the previous section that the effect of decreasing pressure at constant temperature is to reduce solvent capacity and increase isopropanol selectivity (see Fig. 3). The effect of increasing temperature at constant pressure is the same; reduced capacity and increased selectivity. Together these effects demonstrate a direct correlation between solvent density and solvent characteristics, such as capacity and selectivity. The lower bound on operating pressure for a particular separation will be determined by the desired solvent characteristics (usually high capacities for SCF solvents) which, in turn, depends on solvent density. Hence, one factor that determines the minimum operating pressure is the minimum practical operating temperature. Another consideration is to avoid conditions at which multiple coexisting phases would be present. From Figure 8, it is obvious that pressures above the L_2L_3-GL_2 critical line are preferred if isopropanol extraction with a "liquid-like" solvent (exhibiting high capacities) is desired. This critical line would therefore establish the relationship between operating temperature and the minimum operating pressure needed to avoid three or more phases in a multistage extraction. For extraction with dense CO_2 at 40° C, Figures 4 and 8 indicate that the minimum operating pressure would be approximately 1300 PSI. Subcritical (relative to CO_2) operating temperatures are also feasible, in which case the minimum operating pressure is determined by the vapor pressure curve for CO_2. Since the critical temperature of CO_2 is just above ambient temperatures (31° C), subambient operating temperatures would be required to significantly reduce the minimum operating pressure below 1300 PSI.

The significance of the observed large variations in solvent properties (i.e., the dramatic shifts in tie lines) with relatively small changes in temperature and pressure is related to solvent/extract separation efficiency. For example, at 60° C and 2000 PSI (see Fig. 3), compressed CO_2 can be used to extract isopropanol to concentrations just above the isopropanol/water azeotrope. A subsequent pressure reduction to 1500 PSI would recover the extract and return most of the CO_2 at relatively high pressures. Temperature adjustments (see Fig. 4) or combined temperature/pressure control can also be used

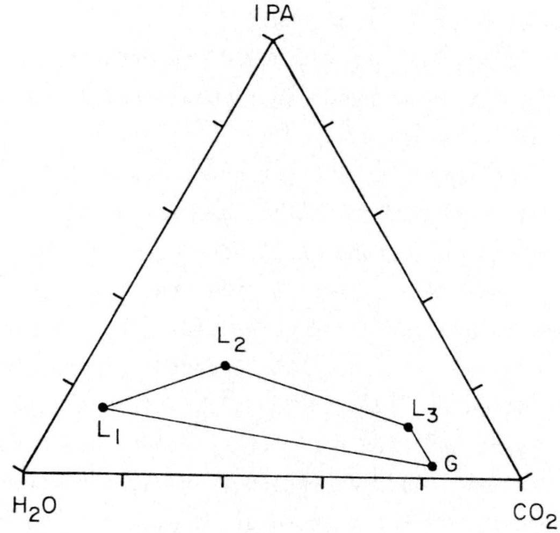

Fig. 7. Measured Four-Phase Equilibrium Compositions at 40° C and 1110 PSI.

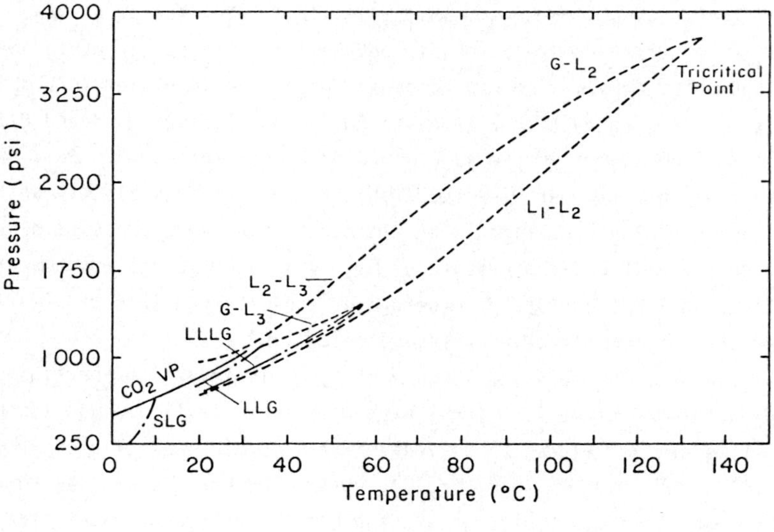

Fig. 8. Model-Generated Pressure-Temperature Projection for Isopropanol-Water-CO_2.

to effect this recovery. In these cases, the goal would be to minimize the required pressure and/or temperature change and thus optimize the extract recovery and solvent recycle processes.

For conventional multi-stage extractions with SCF solvents, it would be desirable to avoid conditions at which three or more equilibrium phases are present. However, the ability to form two liquid phases from a completely miscible mixture by compressing CO_2 above this mixture represents a potential separation method. One advantage of this method would be to lower operating pressures. Another advantage is that the separation is not based upon the ability of the supercritical fluid to dissolve or extract components, but rather relies upon the difference in solubilities of the compressed gas in the various (liquid) components. This advantage becomes important when the mixture constituents are relatively nonvolatile -- e.g., polymeric solutes in dilute aqueous solution.

Perhaps the most interesting application of the phase behavior described in the previous section is to consider isopropanol and CO_2 as co-solvents. (Alternatively isopropanol could be considered as an entrainer.) This solvent mixture could then be used to extract high molecular weight species from dilute aqueous solutions in cases where dense CO_2 alone would not dissolve such species to a practical extent. The phase diagrams for 60° C and 2000 PSI in Figure 3 and for 40° C and 1500 PSI in Figure 4 show fluid-phase mixtures containing appreciable amounts of isopropanol and water. The solvent capacities of these fluids will undoubtedly be much greater than those for dense CO_2. The three-phase equilibrium behavior can be considered in a similar manner (see Fig. 5) where the distribution of high molecular weight species between the L_1 and L_2 liquid phases is used to achieve a desired extraction or separation. Such separation techniques can be particularly important for solutes that are sensitive to temperature and solvent environment.

SUMMARY

The equilibrium behavior that has been observed for the isopropanol-water-CO_2 system is characteristic of phase behavior observed previously for binary solid-supercritical fluid mixtures (ref. 17) where large solubility enhancements and three-phase (solid-liquid-gas) equilibrium occur in the vicinity of the critical point of the supercritical fluid. The phase equilibria can be considerably more complex for the three-component system, however. The complexity for multicomponent systems can be even greater, and hence the importance of understanding and predicting phase equilibria at elevated pressures in the vicinity of the critical point of the SCF solvent is obvious.

The isopropanol-water-CO_2 system serves as two different model systems. First, the observed phase behavior can be related to SCF extraction of organic species, such as alcohols, from dilute aqueous solutions. Second, the isopropanol can be considered as a co-solvent with dense CO_2 and the phase behavior is related to the extraction of high molecular weight components from dilute aqueous solutions. This second extraction method is currently under investigation in our laboratories at the University of Delaware.

ACKNOWLEDGMENTS

Financial support from Exxon Research and Engineering Company is gratefully acknowledged.

REFERENCES

1 J.C. Elgin and J.J. Weinstock, J. Chem. Eng. Data, 4 (1959) 3.
2 R.A. Snedeker, Ph.D. Thesis, Princeton University (1956).
3 L.C.W. Baker and T.F. Anderson, J. Amer. Chem. Soc., 79 (1957) 2071.
4 M.E. Paulaitis, M.L. Gilbert, and C.A. Nash, paper presented at the 2nd World Congress of Chemical Engineering, Montreal, October, 1981.
5 M.S. Kuk and J.C. Montagna, in "Chemical Engineering at Supercritical Fluid Conditions", Ann Arbor Science, Ann Arbor, Michigan, 1983, 101 pp.
6 M.A. McHugh, M.W. Mallett, and J.P. Kohn, in "Chemical Engineering at Supercritical Fluid Conditions", Ann Arbor Science, Ann Arbor, Michigan, 1983, 113 pp.
7 A.W. Francis, J. Phys. Chem., 58 (1954) 1099.
8 G.D. Efremova and A.V. Shvarts, Russ. J. Phys. Chem., 40 (1966) 486.
9 G.D. Efremova and A.V. Shvarts, Russ. J. Phys. Chem., 43 (1969) 968.
10 G.D. Efremova and A.V. Shvarts, Russ. J. Phys. Chem., 46 (1972) 237.
11 A.V. Shvarts and G.D. Efremova, Russ. J. Phys. Chem., 44 (1970) 614.
12 R.E. Fleck, Ph.D. Thesis, University of California - Berkeley (1967).
13 J.R. DiAndreth, Ph.D. Thesis, University of Delaware (1985).
14 R.G. Kander, Ph.D. Thesis, University of Delaware (1985).
15 M. Radosz, Ber. Bunsenges. Phys. Chem., 88 (1984) 859.
16 D.-Y. Peng and D.B. Robinson, IEC Fund., 15 (1976) 59.
17. M.A. McHugh and M.E. Paulaitis, J. Chem. Eng. Data, 25 (1980) 326.

Supercritical Fluid Technology, edited by J.M.L. Penninger, M. Radosz, M.A. McHugh and V.J. Krukonis
1985 Elsevier Science Publishers B.V., Amsterdam — Printed in The Netherlands

POLYMER SOLUTION-SUPERCRITICAL FLUID PHASE BEHAVIOR

A. K. McCLELLAN, E. G. BAUMAN, AND M. A. McHUGH
Department of Chemical Engineering, University of Notre Dame
Notre Dame, Indiana 46556

ABSTRACT

Experimental pressure - temperature (P-T) data are presented for the polystyrene - toluene - carbon dioxide system and the polystyrene - toluene - ethane system. The P-T phase diagrams for these systems exhibit characteristics which are similar to the phase diagrams of simple binary solutions. The addition of supercritical carbon dioxide and supercritical ethane to the polystyrene - toluene solution shifts the lower critical solution temperature curve to lower temperatures.

Polymer - organic solvent - supercritical fluid additive phase behavior is interpreted in terms of Patterson's corresponding states model. Representative calculations using this model are shown for the poly(ethylene-co-propylene) - hexane - supercritical ethane system.

INTRODUCTION

Separating solvent from polymer is one of the final processing stages which occurs during a solution polymerization process. Typically, the solvent is steam stripped from the polymer solution. However, steam stripping is an energy intensive process and, as such, is very costly. It is therefore desirable to invesigate other separation techniques.

Separating the solvent from a polymer solution is also accomplished by forcing the single phase, miscible polymer solution to split into two equilibrium phases: a polymer-rich phase and a solvent-rich phase. It is well known that miscible polymer-solvent solutions exhibit liquid-liquid immiscibility as the solution temperature approaches the critical temperature of the solvent (1). This liquid-liquid immiscibility occurs at the lower critical solution temperature (LCST). The LCST is a function of the chemical nature of the solvent, the polymer molecular weight, and the critical properties of the solvent. For instance, Bardin and Patterson (2) report an LCST to solvent critical temperature (T_c) ratio as low as 0.73 for the polyisobutylene - n-pentane system and as high as 0.98 for the polyisobutylene - cyclooctane system. In general, a polymer with a molecular weight on the order of 10,000 exhibits an LCST which is 20-30 K lower than the solvent T_c while a polymer with a molecular weight of 1,000,000 can have an LCST nearly 100 K below the solvent T_c.

The phase behavior for several polymer - solvent systems are reported in the literature. Freeman and Rowlinson (1) discuss the LCST behavior of hydrocarbon polymers with short chain alkane solvents and their isomers. Baker, et al. (3,4) describe the LCST behavior of polyisobutylene - n-pentane and polyisobutylene - isopentane mixtures. The phase behavior of polystyrene - solvent mixtures is reported by Zeman and Patterson (5), Allen and Baker (6), Saeki, et al. (7), and Cowie and McEwen (8). In each case, the LCST is measured at the vapor pressure of the mixture at various concentrations of polymer. To realize the full potential of this technique it is necessary to understand the phase behavior which can be exhibited by polymer - solvent mixtures at high pressures

A schematic pressure temperature (P-T) diagram for a typical polymer - solvent system is shown in Figure 1a. C_1 is the critical point of the pure solvent and C_2 is the critical point of the polymer, which in this case is probably fictitious, since the polymer will degrade before it exhibits critical phenomena. The dashed lines are the critical mixture curves. The critical mixture curve which starts at C_2 intersects a liquid - liquuid - vapor (LLV) region at the LCST and the branch of the critical mixture curve starting at C_1 ends at an upper critical end point (UCEP). At temperatures below the critical temperature of the solvent a second LLV region appears which ends in a UCEP. The upper critical solution temperature (UCST) curve, in Figure 1a, represents

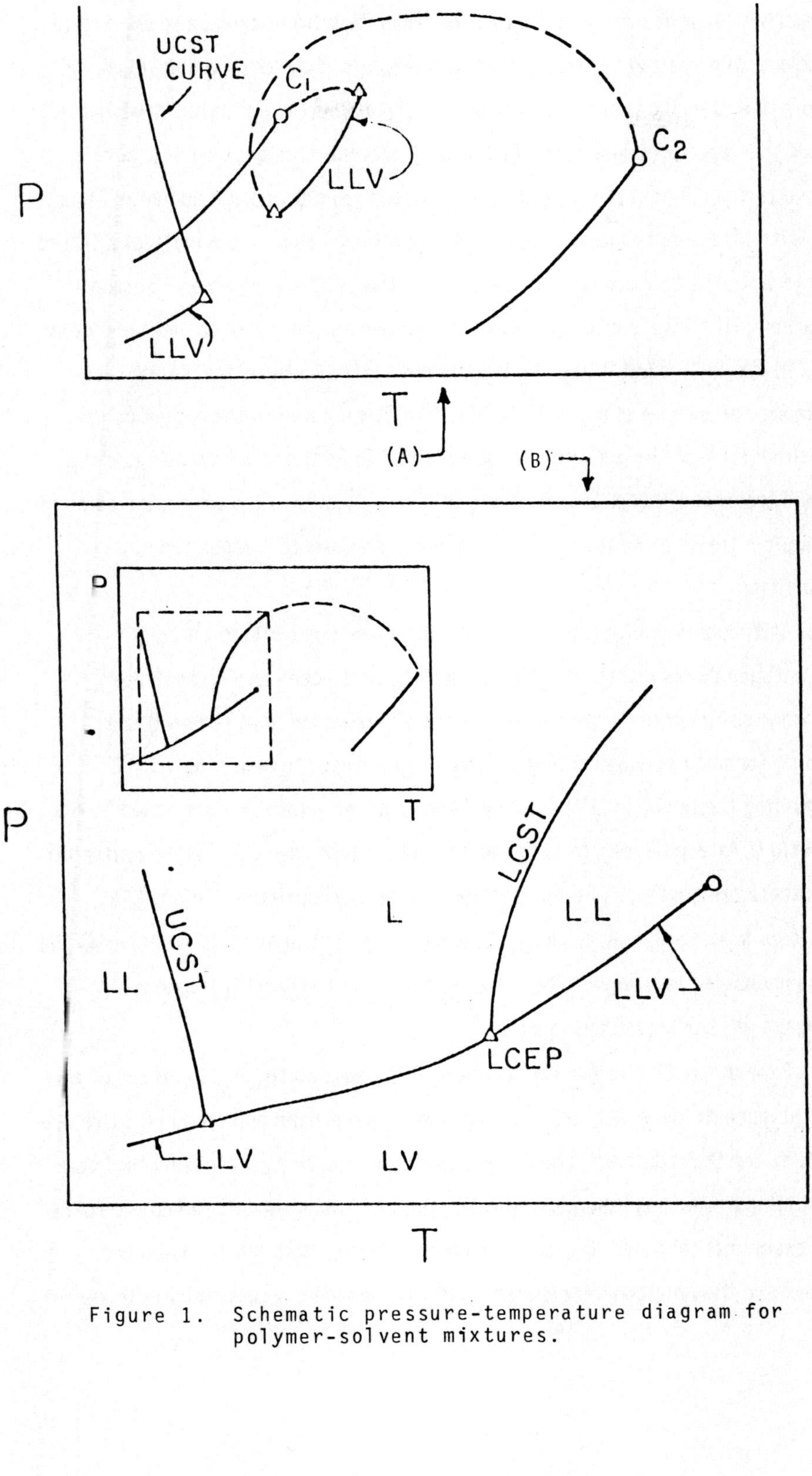

Figure 1. Schematic pressure-temperature diagram for polymer-solvent mixtures.

the effect of pressure on the transition from two liquid phases to one liquid phase as the system temperature is isobarically increased. Baker, et al. (3) show that the LLV curve of a polymer - solvent mixture is indistinguishable from the vapor pressure curve of the pure solvent. The section of Figure 1a close to the critical point of the pure solvent is enlarged and shown in Figure 1b with this simplification. Although the intersection of the critical mixture curve with the LLV curve, is by definition, the LCST for polymer - solvent systems, this intersection is commonly known as the lower critical end point (LCEP) (ie., at the LCEP two liquids become critically identical in the presence of a vapor phase). The critical mixture curve in the region of the critical point of the pure solvent is referred to as the LCST curve since the phase transition which occurs along this curve exhibits the characteristics of a liquid - liquid to fluid transition rather than a liquid - vapor to fluid transition.

Although inducing a polymer-solvent phase split offers certain advantages as compared to steam stripping, it is necessary to heat the polymer solution to temperatures near the solvent critical temperature which, for good polymer solvents, can be very high. To avoid thermally degrading the polymer at these high temperatures inhibitors are added to the solution. Alternatively, Irani, et al. (11) show that the LCST can be shifted to moderate temperatures by introducing a light, supercritical fluid (SCF) additive to the polymer solution. The overall effect of the SCF additive on the phase behavior is to lower the LCST curve without affecting the phase diagram as illustrated in Figure 2.

Experimental data are presented in this paper on the P-T behavior of the polystyrene-toluene-SCF additive system where ethane and carbon dioxide are used as the SCF additive. The experimental procedure for obtaining this data is outlined. The corresponding states model of Patterson is used to interpret the experimental data. Representative calculations using this model are shown for the poly(ethylene-co-propylene) - hexane - supercritical ethane

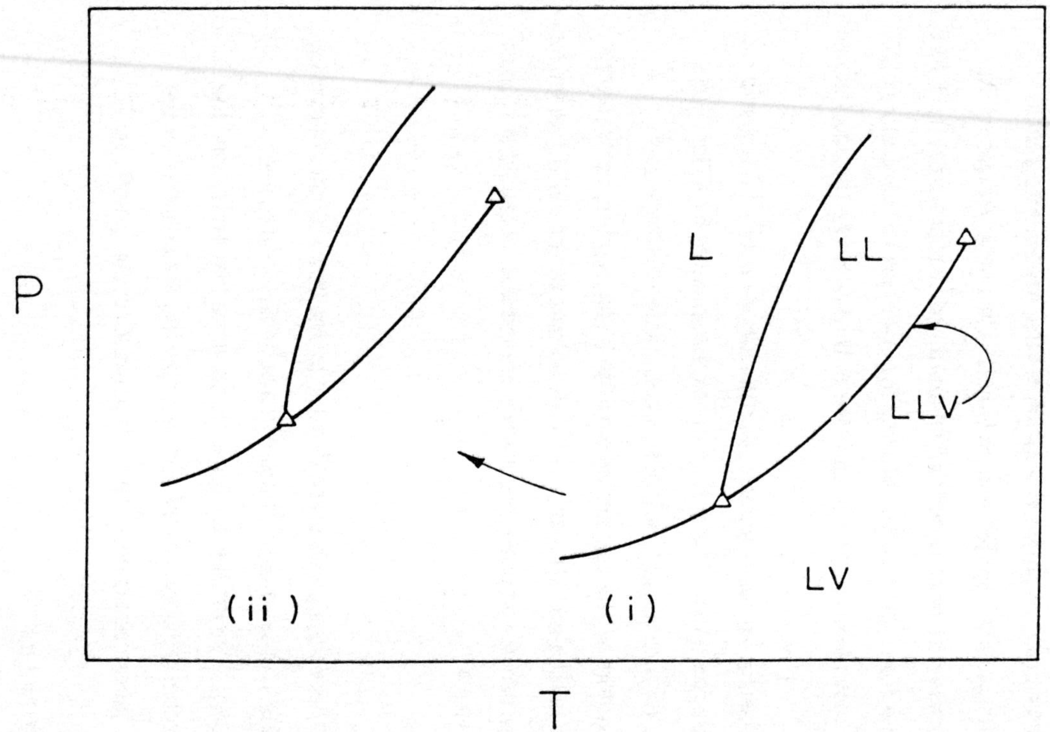

Figure 2. The effect of SCF additive on the polymer-solvent phase behavior.

system.

EXPERIMENTAL

The experimental apparatus and procedure used in this study (see Figure 3) are described elsewhere (13), and therefore, are only briefly described here. A measured amount of polymer solution is first loaded into a high pressure, variable-volume, equilibrium view cell. The concentration of the polymer in the organic solvent is normally maintained at approximately 5 wt% to ensure that the cloud point curve obtained is very close to the actual LCST curve (9,12). A known amount of SCF additive is then added to the equilibrium cell.

The high pressure, variable-volume, equilibrium view cell is designed to operate at 35 MPa and 530 K. The cell contents, illuminated by a fiber light pipe, are viewed through a quartz window, which is secured by a cell end-cap. The contents of the cell are mixed with a stirring bar, activated by a magnet, which is located below the cell. The volume of the cell is varied by a movable piston, hence, the phase boundaries are visually obtained at a fixed overall composition.

MATERIALS

The properties of the polystyrene donated by the Dow Chemical Company are listed in Table 1. The toluene, reagent grade, is supplied by the Sargent-Welch Corp. The ethane, CP grade 99.5% minimum purity, and the carbon dioxide, bone-dry grade 99.8% minimum purity, are supplied by the Linde Company. These components are used without further purification.

RESULTS AND DISCUSSION

The pressure - temperature (P-T) projection of various polystyrene-toluene-SCF transition curves are shown in Figures 4 through 7 for supercritical carbon dioxide and ethane. No attempt is made to locate the

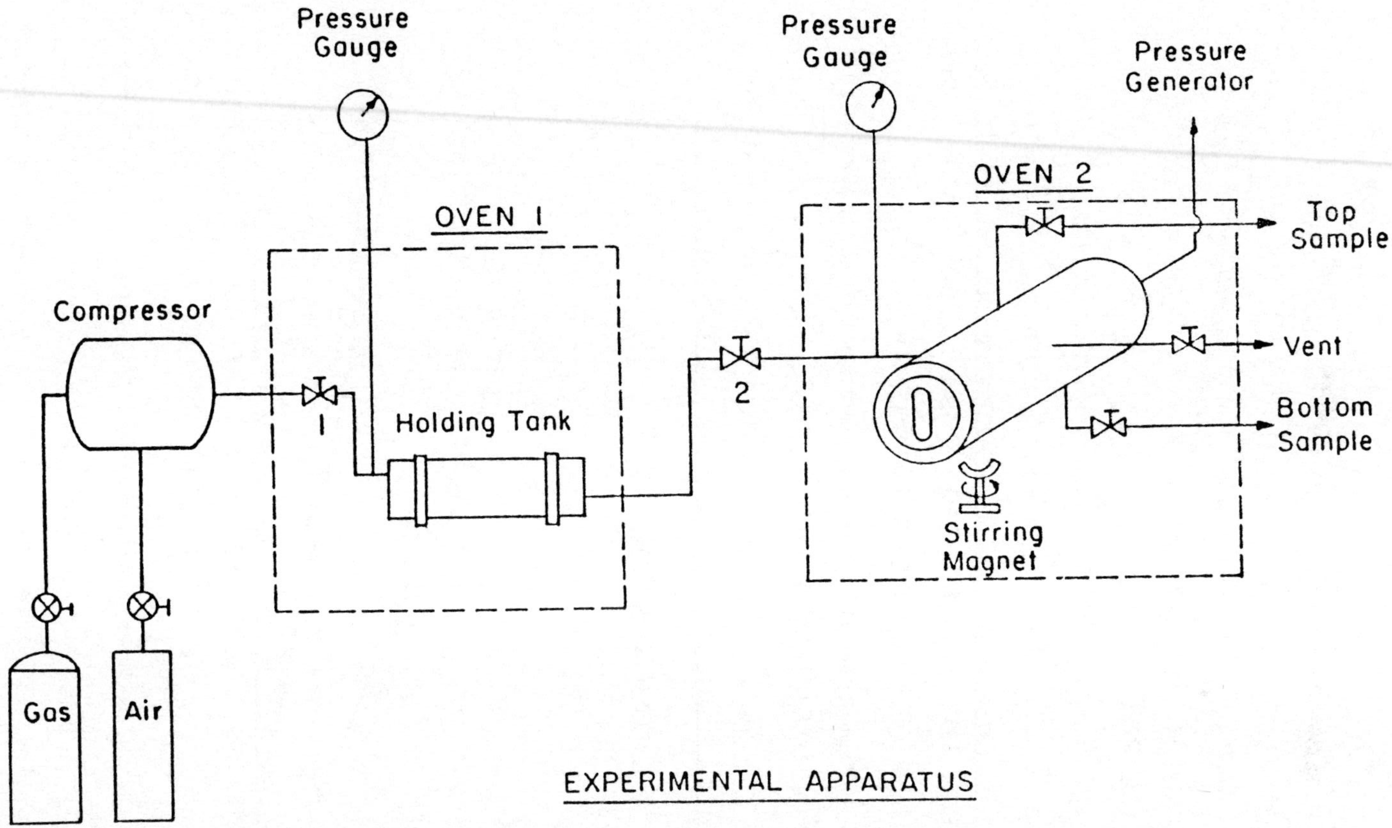

EXPERIMENTAL APPARATUS

Figure 3. Schematic diagram of the experimental apparatus used in this study.

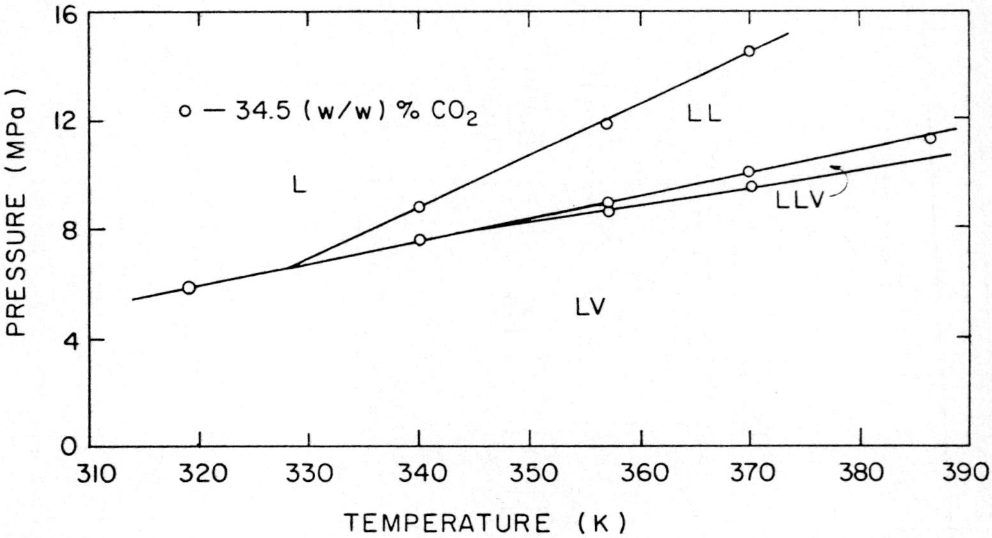

Figure 4. Pressure-temperature projection of the polystyrene-toluene-carbon dioxide system.

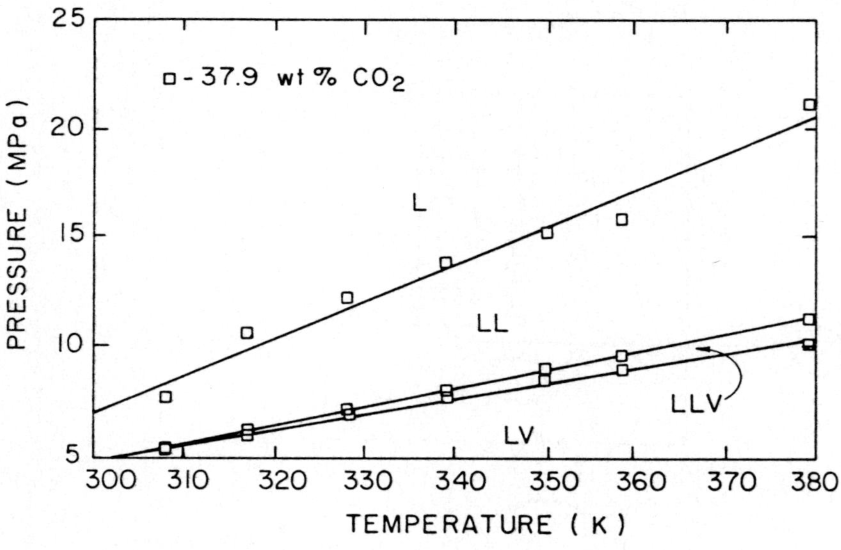

Figure 5. Pressure-temperature projection of the polystyrene-toluene-carbon dioxide system.

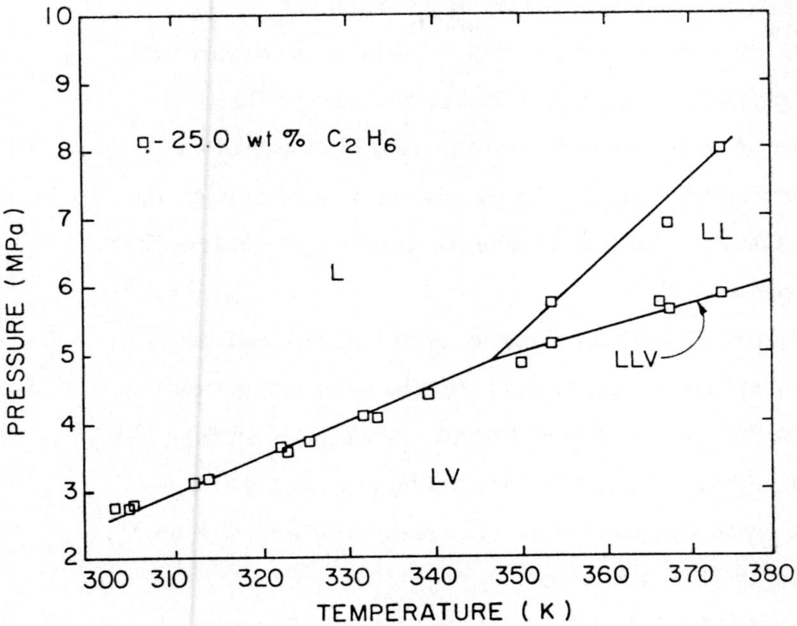

Figure 6. Pressure-temperature projection of the polystyrene-toluene-ethane system.

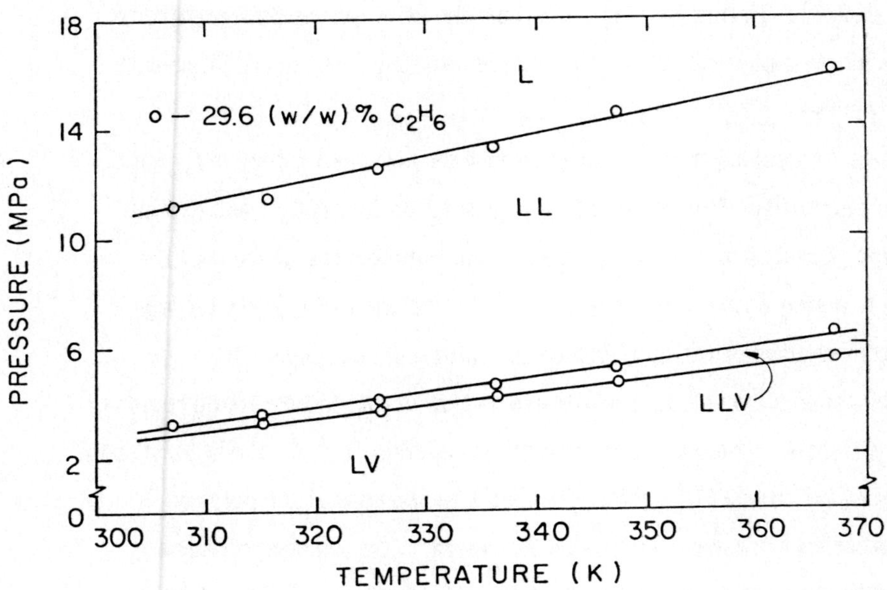

Figure 7. Pressure-temperature projection of the polystyrene-toluene-ethane system.

upper critical solution temperature curve for these systems.

Shown in Figure 4 are the experimental P-T data for the polystyrene - toluene - carbon dioxide system at 34.5 wt% carbon dioxide. The weight percent of carbon dioxide is based on the total weight of the solution. Without any carbon dioxide added to the polystyrene-toluene solution, the phase border curves depicted in this figure are located approximately 220 K higher in temperature (14).

The polystyrene-toluene-carbon dioxide system at 37.9 wt% carbon dioxide is shown in Figure 5. The slope of the LCST curve in this figure has changed considerably from the 34.5 wt% system. The LCEP is not determined for this system, although if the LCST curve and the LLV curve are extrapolated to lower temperatures, the LCEP is approximately 45 K lower than the LCEP for the 34.5 wt% system.

Shown in Figures 6 and 7 are the P-T curves for the polystyrene - toluene - ethane systems for ethane loadings of 25.0 wt% and 29.6 wt%. For the 25 wt% ethane system the LCEP is approximately 347 K. The 29.6 wt% ethane system, shown in Figure 7, exhibits a critical mixture curve that is parallel to the LLV line. In this case, the LCST and the UCST curves have merged to form a single curve and, hence, the critical mixture curve never intersects the LLV curve.

It is interesting to note that Figures 4 through 7 all exhibit a region of LLV immiscibility. This immiscibility behavior is in contrast to the phase behavior exhibited by binary mixtures of small molecules where the LLV region is a single line in P-T space. The LLV regions in Figures 4 through 7 are a consequence of the multicomponent nature of the polymer.

The composition of the equilibrium phases in the liquid - liquid region of the polystyrene - toluene - 29.6 wt% ethane system at 311.1 K and 9.1 MPa is presented in Table 1. This compositional data shows that the polymer concentration increases from 5 wt% to 15 wt% in the polymer-rich phase. Further studies are in progress to expand on this data base and to determine the molecular weight and the molecular weight distribution of the polymer in the equilibrium phases.

Table 1. Composition Analysis for Polystyrene (PS)- Toluene - Ethane (29.6 wt%) System.

T = 311.1 K

P = 9.1 MPa

Component	Light Phase wt %	Heavy Phase wt %
PS	0.3	15.8
Toluene	73.9	64.3
Ethane	25.8	19.9

PS: M_w = 239,000

M_n = 221,000

M_w/M_n = 1.08

MODELLING

To model the phase behavior obtained in this study it is necessary to obtain a value of the Flory - Huggins free energy parameter

$$\chi_c = 1/2(1 + r^{-1/2})^2 \quad (1)$$

where

$$r = V_s^*/V_p^* \quad (2)$$

the superscript * represents a reduction parameter. Using the model of

Patterson and coworkers (15) the critical mixture curve for polymer solutions at low polymer concentrations are calculated with

$$\chi_c(P,T) = c(-\nu^2 U_s/T_s + 1/2C_{p,s}(P_s,T_s) (\tau + P_s V_s^2 \pi/(P_s V_s^2 + 1))^2 \qquad (3)$$

where subscript p represents the polymer and subscript s represents the solvent. P_S, T_S, and V_S are the reduced pressure, temperature, and the volume repectively. The difference in the free volumes of the mixture components is given by

$$\tau = 1 - T_s^*/T_p^* \qquad (4)$$

The ν^2 parameter is a measure of the difference in the chemical nature of the mixture components. In this study ν^2 is fit to experimental data. The parameter π is defined as

$$\pi = P_s^*/P_p^* - 1 \qquad (5)$$

The parameter 3c represents the number of external degrees of freedom of the solvent molecule. This parameter is calculated using

$$c = P^* V^*/RT^* \qquad (6)$$

Finally the parameters U_S and $C_{p,S}$ are the reduced configurational internal energy and the reduced configurational heat capacity of the solvent, respectively. Using a van der Waals type of configurational internal energy,

$$U_s = -V_s^{-1} \qquad (7)$$

the following expression is derived (16)

$$C_{p,s}^{-1} = (1 - 2/3V_s^{-1/3}) - 2(1 - V_s^{-1/3})/(P_sV_s^2 + 1) \qquad (8)$$

and

$$T_s = (P_sV_s + V_s^{-1})(1 - V_s^{-1/3}) \qquad (9)$$

The reduced temperature T_S is defined as

$$T_s = T/T_s^* \qquad (10)$$

The critical mixture curve (i.e., LCST curve) is calculated in the following manner. First the reduction parameters are determined from pure component P-V-T data, (see ref. 15), then the parameters τ, π, and c are determined. The parameter ν^2, which affects the location of the UCST and the LCST curves, is then fitted to the LCEP by solving equations 1 through 10 (5). A non - zero value of ν^2 is needed to predict both an LCST and a UCST curve (17). Once ν^2 is determined, the rest of the critical mixture curve is calculated. The mixing rules for $\langle U^* \rangle$ and $\langle T^* \rangle$ as described by Patterson and Delmas (15) are used for the multicomponent solvents investigated in this study. These mixing rules are

$$\langle U^* \rangle = \Sigma_i x_i U_i^* \qquad (11)$$

where x_i = mole fraction

$$1/\langle T^* \rangle = \Sigma_i X_i/T_i^* \qquad (12)$$

$$X_i = x_i U_i/\langle U^* \rangle \qquad (13)$$

The mixing rule for $\langle P^* \rangle$ is

$$\langle P^* \rangle = \Sigma_i \Sigma_j \; X_i X_j P_{ij}^* \tag{14}$$

$$P_{ij}^* \doteq (P_i^* P_j^*)^{1/2} \tag{15}$$

Based on our results the calculated phase diagram is not extremely sensitive to the form of the mixing rule for $\langle P^* \rangle$.

To test Patterson's model the critical mixture curves for the previously reported poly(ethylene-co-propylene) (EP) - hexane - ethylene system are determined (13). The EP - hexane - ethylene system is modelled since a large body of data are available. Although the EP polymer has a polydispersity of 2.2 and the hexane is actually an isomeric mixture of C_6's, these mixtures exhibit phase behavior which is similar to that of the previously described polystyrene - toluene - SCF mixtures. For these calculations the pure component characteristic parameters for supercritical ethylene are fitted to a function of temperature and pressure using pure component data (i.e., the isothermal compressibility and the thermal expansion coefficient). The modelling results are shown in Figure 8. In general the calculated LCST curves are in agreement with the experimentally determined curves. Notice, however, for the case of 30.0 wt% ethylene it appears that the LCST and UCST curves are merged. If the LCST curve is extrapolated to lower temperatures, the intersection of the LCST curve with the extrapolated LLV curve occurs at approximately 150 K. The LCST curve probably merges with the UCST curve before reaching such cold temperatures. The model also predicts the merging of the UCST and the LCST curves although the shape of the curve is not quantitatively represented.

As shown in Figure 8 the fitted ν^2 values are small positive numbers except for the case without SCF additive. This negative ν^2 value is physically unrealistic and is probably a consequence of determining hexane's pure component characteristic parameters at a fixed temperature of 25^oC. Also, in these calcuations the polydispersity of the EP polymer is not incorporated

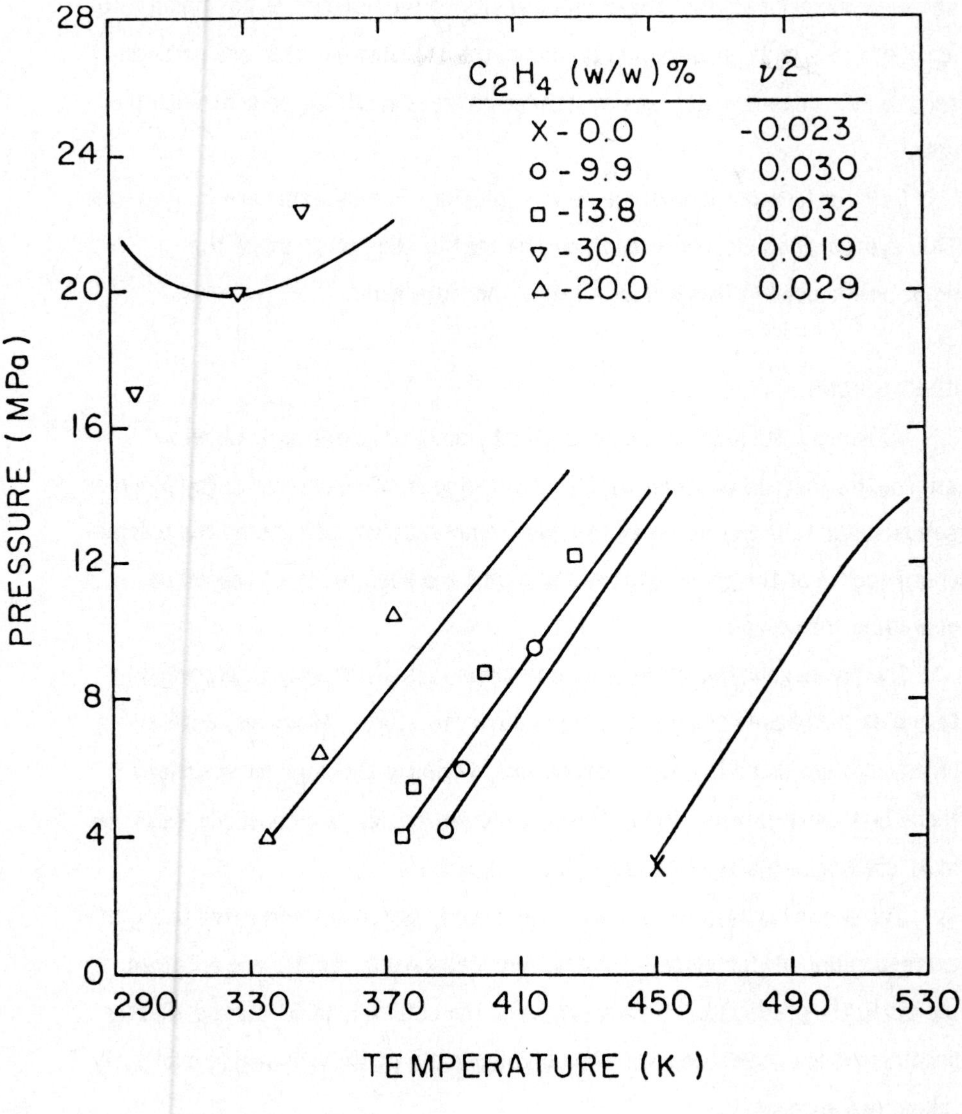

Figure 8. Comparison of the calculated and experimental LCST curves for the poly(ethylene-co-propylene)-hexane-ethylene system.

into the model. As shown by Koningsveld substantial errors in the calculations can result if the polydispersity of the polymer is not taken into account (18). It is encouraging then that the calculations that are performed for the EP - hexane - ethylene system are in reasonable agreement with the data.

Calculations for the polystyrene - toluene - SCF systems are in progress. This system should offer a less extreme test of the theory since the polystyrene used in this study is fairly 'monodisperse'.

CONCLUSIONS

Polymer solutions can be separated by an LCST phase split as an alternative method to steam stripping. Adding an SCF additive to the polymer solution shifts the LCST curve to lower temperatures, decreasing the thermal energy costs of the phase split and reducing the possibility of thermally degrading the polymer.

The phase behavior of polymer solutions is qualitatively interpreted in terms of the phase behavior of simple binary mixtures. However, with a binary system there is a three-phase line while for the polymer solutions there is a three-phase region. The three phase-region is a consequence of the multicomponent nature of the polymer solution.

The phase behavior presented in this paper are also interpreted using the corresponding states model of Patterson. Reasonable results are obtained, including the prediction of the merging of the LCST and UCST curves. Better predictions are expected when the polydispersity of the polymer is explicitly taken into account.

ACKNOWLEDGEMENT

Acknowledgement is made for the partial and technical support of the Exxon Chemical Company. This work is supported in part by the Office of Naval Research.

NOMENCLATURE

c	Prigogine parameter: a measure of the external degrees of freedom
C_p	Reduced heat capacity
L	Liquid phase region
LCEP	Lower critical end point
LCST	Lower critical solution temperature
LL	Two phase liquid-liquid region
LLV	Three phase liquid-liquid-vapor region
LV	Two phase liquid-vapor region
M_n	Number average molecular weight
M_w	Weight average molecular weight
P	Reduced pressure
r	Parameter defined by equation 2: a measure of the polymer-solvent size difference
R	Universal gas constant
SCF	Super critical fluid
T	Reduced temperature
U	Reduced configurational internal energy
UCEP	Upper critical end point
UCST	Upper critical solution temperature
V	Reduced volume
x	mole fraction
X	energy fraction
$\langle\ \rangle$	averaged property

greek symbols

π	parameter defined by equation 4: accounts for high hydrostatic pressures
τ	parameter defined by equation 3: accounts for free volume difference between solvent and polymer
χ	Flory parameter
ν	fitted parameter: a measure of the chemical difference between solvent and polymer

subscripts

i	component i
c	critical
p	polymer
s	solvent

superscripts

*	reduction parameter

REFERENCES

1. P. I. Freeman, and J. S. Rowlinson, Polymer, 1, 20 (1960).
2. J. M. Bardin, and D. Patterson, Polymer, 10, 9 (1967).
3. C. H. Baker, W. B. Brown, G. Gee, J. S. Rowlinson, D. Stubley, and R. E. Yeadon, Polymer, 3, 215 (1962).
4. C. H. Baker, C. S. Clemson, and G. Allen, Polymer, 1, 5255 (1966).
5. L. Zeman, and D. Patterson, J. Phys. Chem., 76, 1214 (1972).
6. G. Allen, and C. H. Baker, Polymer, 6, 181 (1965).
7. S. Saeki, S. Konno, N. Kuwahara, M. Nakata, and M. Kaneko, Macromolecules, 7(4), 521 (1974).
8. J. M. G. Cowie, and I. J. McEwen, Macromolecules, 7(3), 291 (1974).
9. C. D. Myrat, and J. S. Rowlinson, Polymer, 6, 645 (1965).
10. P. Ehrlich, and J. J. Kurpen, J. Polym. Sci., part A, 1, 3217 (1963.)
11. C. A. Irani, C. Cozewith, and S. S. Kasegrande, U.S. Patent 4,319,021 (1982).
12. L. Zeman, J. Biros, and D. Patterson, J. Phys. Chem., 76, 1206 (1972).
13. M. A. McHugh, and T. L. Guckes, Macromolecules, in press (1985).
14. S. Saeki, S. Konno, N. Kuwahara, M. Nakata, and M. Kaneko, Macromolecules, 7(4), 521 (1974).
15. D. Patterson, and G. Delmas, Trans. Faraday Soc., 65, 708 (1969).
16. P. J. Flory, Disc. Faraday Soc., 49, 7 (1970).
17. K. S. Siow, G. Delmas, and D. Patterson, Macromolecules, 5, 29 (1972).
18. R. Koningsveld, and A. J. Staverman, J. Polymer Sci., A-2, 6, 305 (1968).

Supercritical Fluid Technology, edited by J.M.L. Penninger, M. Radosz, M.A. McHugh and V.J. Krukonis
1985 Elsevier Science Publishers B.V., Amsterdam — Printed in The Netherlands

CONTINUOUS-FLOW APPARATUS FOR MEASURING PHASE EQUILIBRIA IN SUPERCRITICAL GAS-HEAVY HYDROCARBON SYSTEMS

M. RADOSZ
Exxon Research and Engineering Company, P.O. Box 101, Florham Park, NJ 07932

ABSTRACT

A continuous-flow apparatus is presented for measuring high-pressure phase equilibria in systems containing supercritical gases and heavy hydrocarbons. This apparatus can handle oils which have melting points higher than the ambient temperature. A windowed equilibrium cell makes it possible to observe phase separation visually and to generate large samples. The latter is a valuable feature for heavy hydrocarbon materials where relatively large samples are needed for subsequent analyses. Capabilities of this method are illustrated with phase equilibrium data obtained for supercritical gas-oil and carbon dioxide-decane systems.

INTRODUCTION

Separating heavy hydrocarbons with or from supercritical gases is an alternative to conventional subcritical separations. Heavy hydrocarbon separations, both sub- and supercritical, can be experimentally simulated with a continuous flow apparatus described below. Compared with batch units [4], this flow-cell apparatus can generate large samples and allows for short residence times which are needed for temperature sensitive heavy oils.

EQUIPMENT DESCRIPTION

The flow cell concept of phase equilibrium measurements was pioneered by Wilson [9], and Chao and co-workers [1,7]. Paulaitis proposed feeding condensable gases using a cooled metering pump [3] and measured mixture critical pressures by visual observation of critical opalescence within a view flow cell [8]. The design described below evolved around such concepts as static mixing, visual cell separator, "cold" pumping of condensable gases, gravimetric sampling for liquids and volumetric sampling for gases. Another feature of this design is the capability of handling feeds which are solids at ambient temperature, for example heavy petroleum residua, but can be liquefied at elevated temperatures.

The keystone of the apparatus is a windowed cell which serves as a high-pressure separator (Figure 1) and makes it possible to observe phase separation visually. Upstream of the cell, two high-pressure metering pumps feed the condensable components to a Kenics static mixer. This mixer provides the mixing energy needed to promote mass transfer. Downstream of the cell, two low pressure separators, one for each of the phases, separate the depressurized effluents into vapor and liquid. This way four samples are generated, vapor and liquid from each of the equilibrium phases. For the systems where GC analysis can be used, there is an option to analyze all the streams with an on-line gas chromatograph.

The cell is a windowed high-pressure Jerguson gage (17-T-40), 60cc in volume, equipped with a deentrainer. It is connected with the feeding (Figure 2) and sampling (Figure 3) sections with 1/8-inch o.d. pressure tubing. One of the feeding pumps, 2396-89 Milton Roy, is equipped with an electrical heater, as are all the parts in contact with high melting point components. The other feeding pump, Beckman 110A, is equipped with a chiller to be able to handle liquefied gases. The static mixer and cell are installed in an air bath, Bemco FTU7.0 Constant Temperature Chamber, to ensure temperature uniformity. Although the design temperature range is -100 to 800°F, the cell used for these measurements had a 600°F upper temperature limit.

EXPERIMENTAL PROCEDURE

After the whole system is cleaned and tanks filled with the feed components, the hot, and subsequently cold, feeding pumps are turned on. When the cell is filled, its pressure is adjusted using a micrometering valve mounted on the upper phase line. The liquid level is adjusted using another micrometering valve mounted on the lower phase line (Figure 3). Depending on the system, it takes up to 30 minutes to reach a steady state with constant effluent stream compositions.

During equilibration both effluent streams are directed to a knockout pot to bypass the sampling section (Figure 3). After the equilibrium is reached, the three-way valves are turned to direct the product streams to the sampling section. For the next 15 to 30 minutes the low pressure separating vessels are filled with the heavy components and the light components pass through the wet test meters to measure their volume. The amount of heavy components is determined gravimetrically. Three pairs of samples are usually collected from

Figure 1

SIMPLIFIED DIAGRAM OF CONTINUOUS FLOW APPARATUS

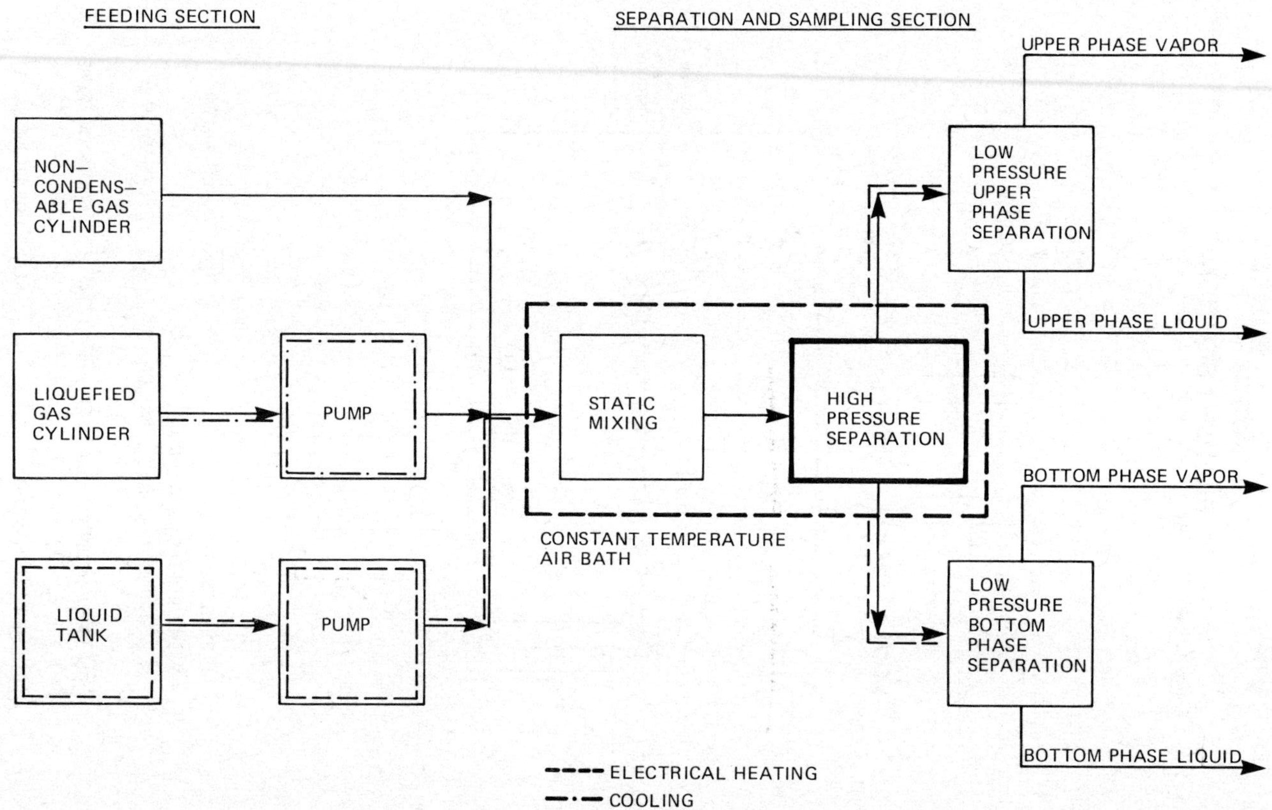

Figure 2
FEEDING SECTION

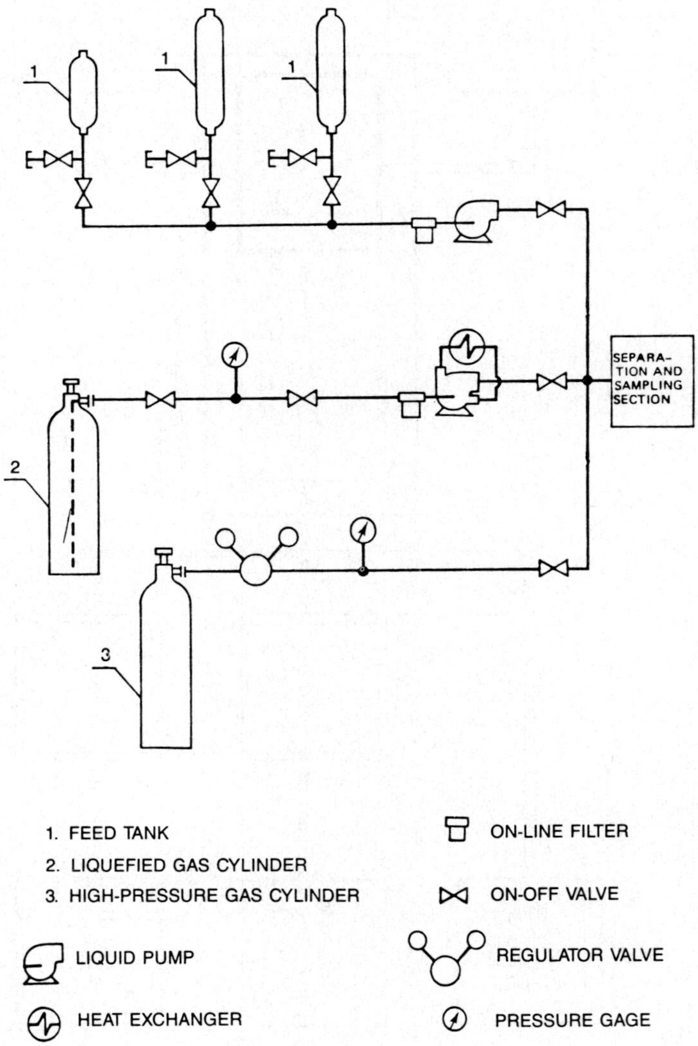

Figure 3
SEPARATION AND SAMPLING SECTION

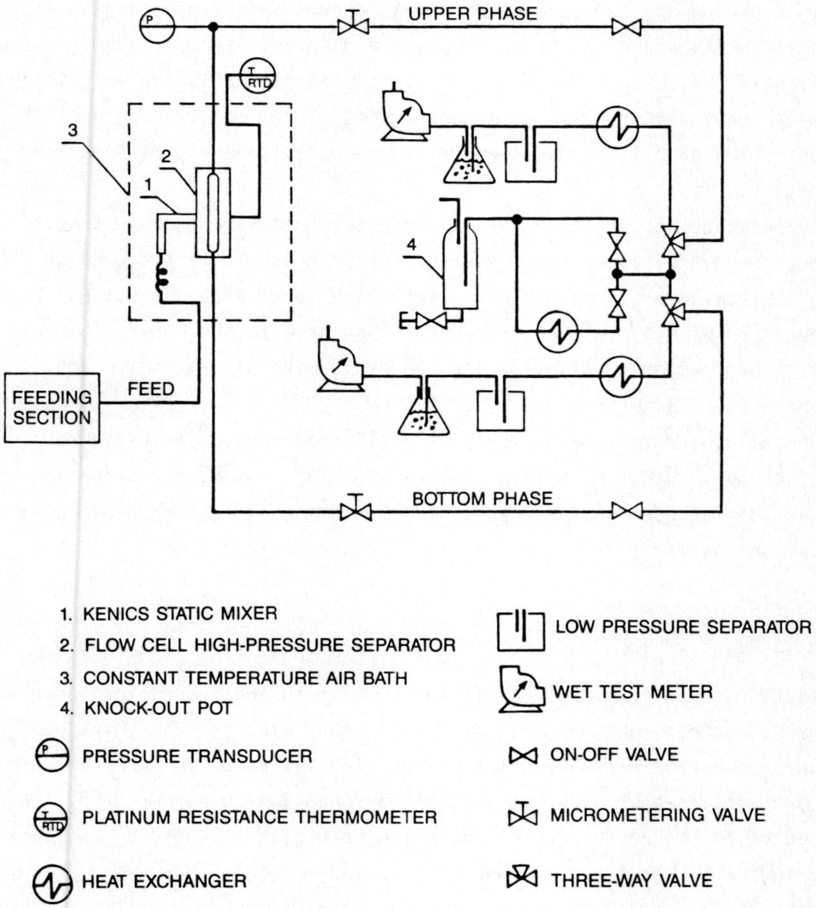

each run. Typically, the vapor samples are GC analyzed whereas the liquid samples are used for various oil tests. The equilibrium phase compositions are calculated from the material balance based on the measured vapor volume and liquid weight.

It was found that the pressure control with the micrometering valve was sufficiently smooth so that the use of a back pressure regulator became unnecessary. On the other hand, while using high pressure gases, the cell pressure is controlled by adjusting the cylinder regulator; the upper phase flow rate is maintained constant. In this mode of operation only the lower phase micrometering valve is used for the liquid level control.

For the systems of moderate relative volatility at ambient conditions, for example decane and carbon dioxide, small corrections are needed to account for solubility of decane in the vapor phase and solubility of carbon dioxide in the liquid phase. However, for the systems of high relative volatility, for example heavy oil and gas, no such corrections are required because the mutual solubilities of oil and gas are very low at the low-pressure separator conditions.

Temperature and pressure measurements are important because they directly affect the overall accuracy. Temperature is measured in the body of the cell with a platinum resistance thermometer which is accurate to 0.1°C. Pressure is measured with a pressure transducer connected to the upper phase effluent line. The pressure transducer used for this work was the Rosemount 1333G having 3.5 psi accuracy.

The oils used for measurements were filtered through a 10-micron filter prior to use. Other materials, decane (O-2128) from Fisher Scientific and gases from Matheson (hydrocarbons of C.P. purity and carbon dioxide of Coleman Instrument purity), were used without further purification.

ACCURACY AND APPLICATION

Tests were performed at conditions similar to those anticipated in the heavy hydrocarbon and gas systems. The binary used for the tests was normal decane and carbon dioxide, which was earlier investigated by Reamer and Sage [5] and confirmed at high temperatures by Chao and co-workers [6]. Results presented in Figure 4, for the 280°F isotherm also measured by Reamer and Sage [5], indicate that the agreement with Reamer's data is very good. Indeed, the largest discrepancy is 0.5 absolute percent; the reproducibility of the results from this work is better than 0.1 absolute percent.

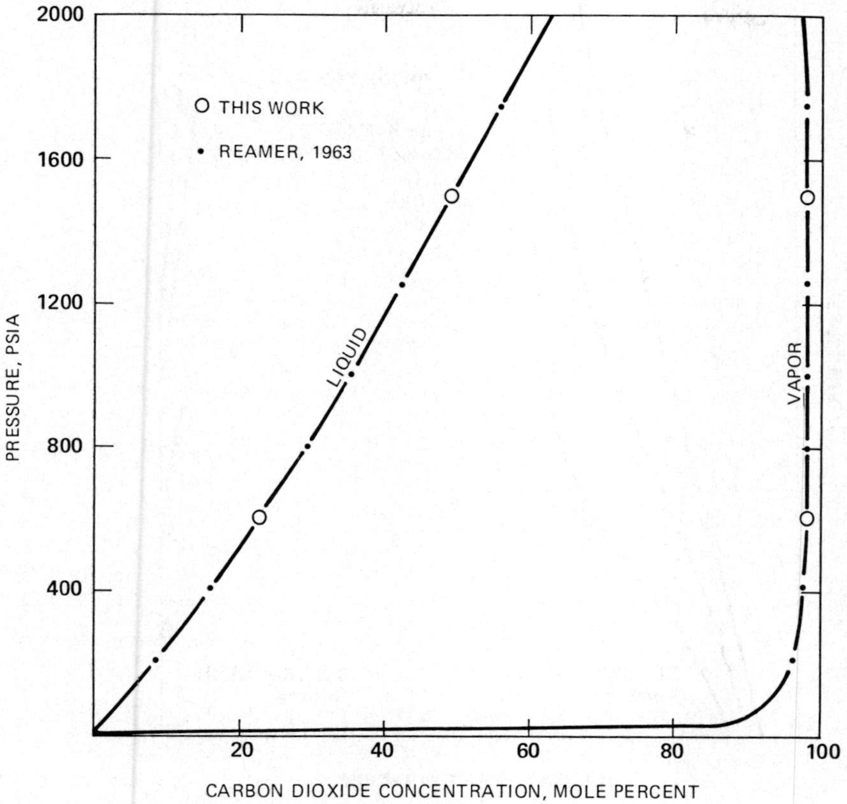

Figure 4. Vapor-liquid equilibrium for carbon dioxide and decane at 137.8°C (280°F).

Application of the apparatus described above is illustrated with an example related to separating heavy hydrocarbons (referred to as oils) with or from supercritical gases (referred to as solvents). Specifically, experimentally measured phase compositions are discussed and compared with those predicted with an equation of state.

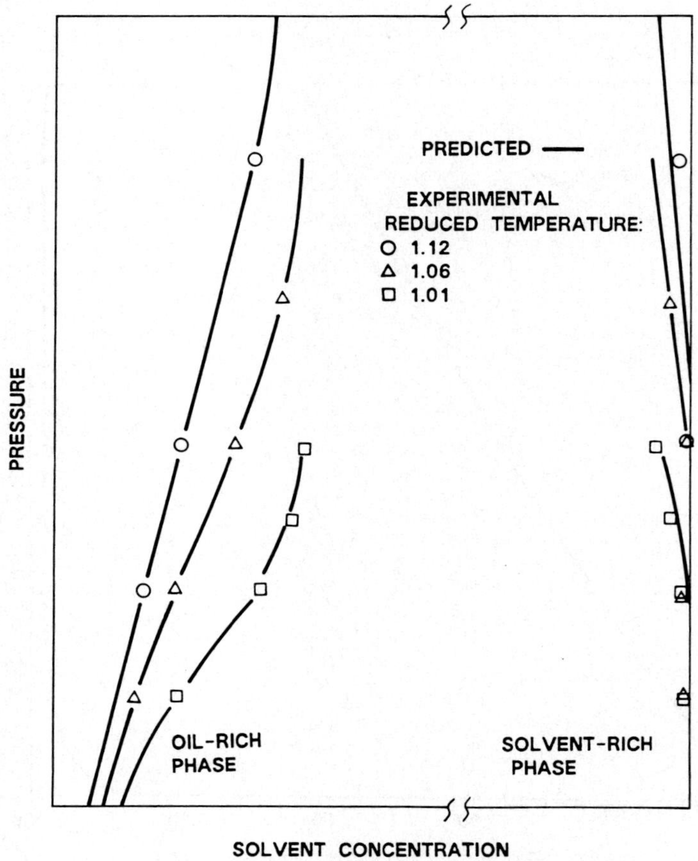

Figure 5. Equilibrium phase compositions for solvent and oil.

PHASE COMPOSITIONS

The mutual solubility of solvent and oil increases with increasing pressures, as shown in Figure 5. While the solvent solubility in oil increases with decreasing temperatures, which is also shown in Figure 5, the oil-in-solvent solubility dependence on temperature is less straightforward. It increases at high pressures, but decreases at low pressures, with decreasing temperatures.

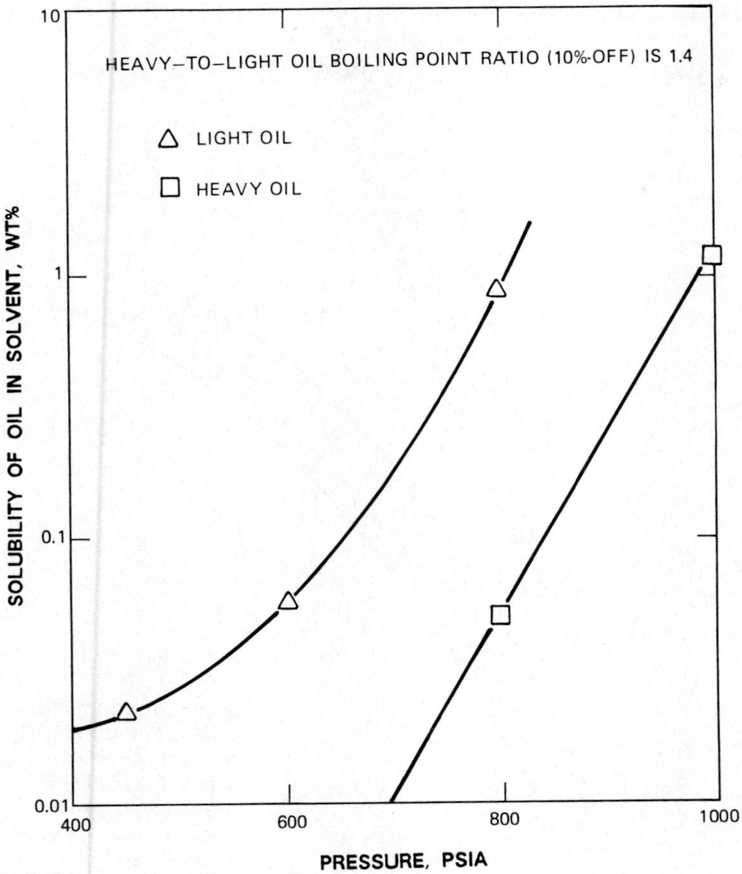

Figure 6. Solubility of oil in supercritical solvent.

Solubility of oil in solvent also depends on the molecular weight of the oil light end, which is related to the light end boiling range. This means that the lighter the oil the higher its solubility in solvent. Figure 6 shows an example for two oils where the boiling point (10%-off) for the heavier oil is 1.4 times higher than that for the lighter oil. As a result, the solubility of the lighter oil in Figure 6 is higher than that of the heavier oil by an order of magnitude.

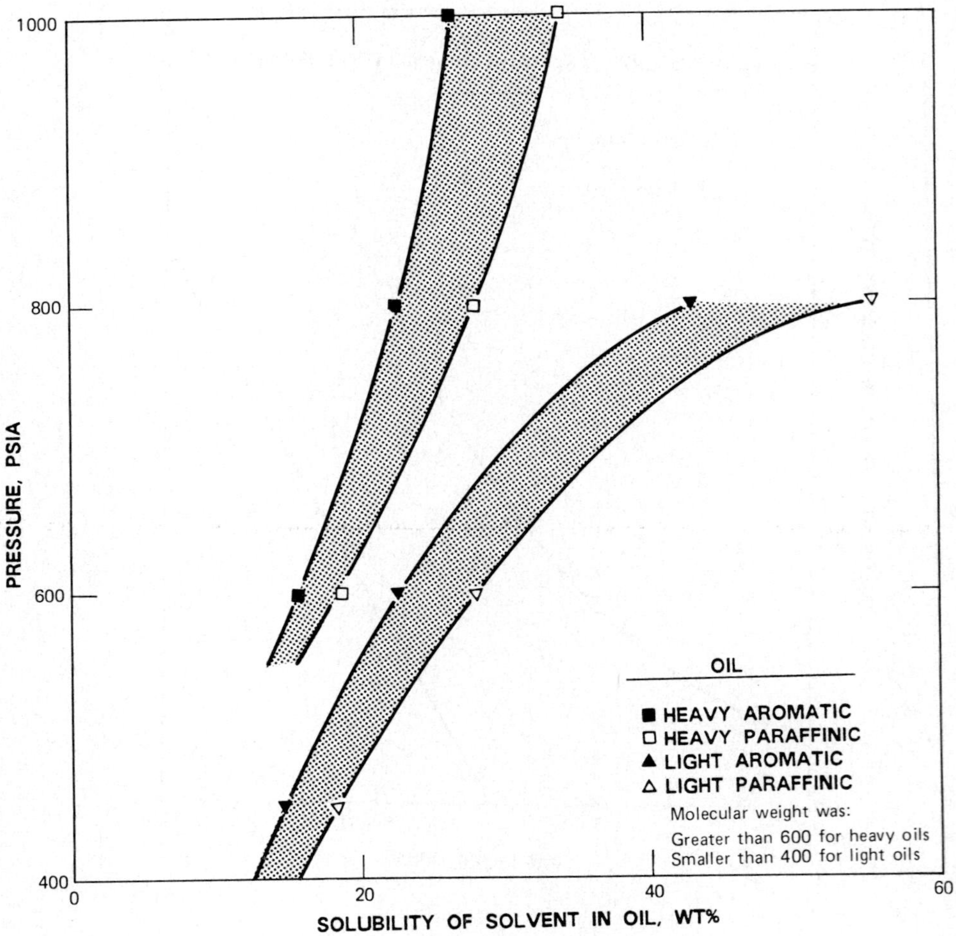

Figure 7. Solubility of supercritical solvent in oil.

On the other hand, solubility of solvent in oil depends on both oil average molecular weight and aromaticity. As shown in Figure 7, solvent solubility is higher in oils having lower molecular weights. For the oils having the same molecular weights, solvent solubility in oil decreases with increasing oil aromaticity, that is, solvent solubility in the aromatic oils is lower than that in the paraffinic oils shown in Figure 7.

COMPUTER MODEL

A computer simulaton model is normally required to reduce the need for experimental effort. In this work an Exxon version of the Redlich-Kwong equation of state [2] was used to correlate and predict phase compositions. The oils were split into two pseudocomponents, light and heavy, based on their distillation curve.

This version of the Redlich-Kwong equation of state utilizes two binary parameters, the energy interaction parameter C and the molecular size interaction parameter D [2]. Both are needed to reproduce the vapor and liquid phase compositions in the solvent and oil systems. For the pseudo-components studied, the vapor phase composition is most sensitive to D of the light pseudocomponent. The pressure-composition curve for the liquid phase, on the other hand, is sensitive to both C and D of both pseudocomponents. Changes in C shift this curve along the composition axis whereas changes in D change the slope of this curve. All the binary parameters discussed above are temperature and pressure independent, and apply to the oil-solvent interactions. For the solvents studied, the absolute values of C are in the range of 0.01 to 0.02 and the absolute values of D are in the range of 0.1 to 0.2.

Specific comparisons between predicted and experimental pressure-composition data for solvent and oil, are given in Figure 5. In addition Figure 8 shows predicted pressure and temperature dependence of the oil solubility in solvent. As expected, these predictions indicate that the oil solubility in solvent is primarily dependent on the solvent density at higher pressures but on the oil vapor pressure at lower pressures.

CONCLUSION

The continuous-flow apparatus can provide accurate vapor-liquid equilibrium data for systems of supercritical gases and heavy hydrocarbons. The examples of such data discussed above demonstrate versatility and good accuracy of this apparatus. The accuracy is sufficient to back up equation of state parameters for heavy hydrocarbon systems.

ACKNOWLEDGEMENT

The author is grateful to Exxon Research and Engineering Company for the permission to publish this paper. Mr. E. Niessen assisted in the equipment construction, start-up and measurements.

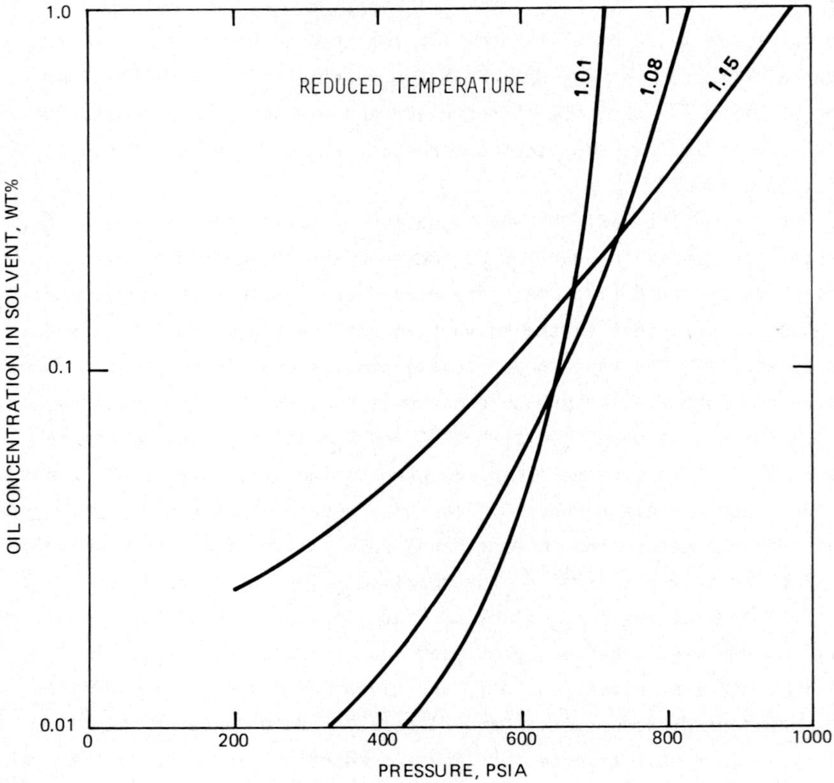

Figure 8. Solvent density at high pressures, but oil vapor pressure at low pressures, affect oil solubility in solvent.

REFERENCES

1. Chao, K. C., Lin H. M., Guo, T. M., Masuoka, H., Radosz, M., Kim, H. Y. and Leet, W. A., "Phase Equilibrium in Coal Liquefaction Processes", Electric Power Research Institute Publication AP-3450 (prepared at Purdue University), Palo Alto, August 1984.
2. Gray, R. D., Jr., Heidman, J. L., Hwang, S. C. and Tsonopoulos, C., Fluid Phase Equilibria, 13, 59 (1983).
3. Paulaitis, M. E., Gilbert, M. L. and Nash, C. A., "Separation of Ethanol-Water Mixtures With Supercritical Fluids", 2nd World Congress of Chemical Engineering, Montreal, October, 1981.
4. Radosz, M., Ber. Bunsenges. Phys. Chem., 88, 859 (1984).
5. Reamer, H. H. and Sage, B. H., J. Chem Eng. Data, 8, 508 (1963).
6. Sebastian, H. M., Simnick, J. J., Lin, H. M. and Chao, K. C., J. Chem. Eng. Data, 25, 138 (1980).
7. Simnick, J. J., Lawson, C. C., Lin, H. M. and Chao, K. C., A.I.Ch.E. Journal, 23, 469 (1977).
8. Thies, M. C. and Paulaitis, M. E., J. Chem Eng. Data, 29, 438 (1984).
9. Wilson, G. M. and Owens, R. S., "High-Temperature Vapor-Liquid Equilibrium Studies on Synthetic Fuel Systems", 70th Annual AIChE Meeting, New York, November, 1977.

Supercritical Fluid Technology, edited by J.M.L. Penninger, M. Radosz, M.A. McHugh and V.J. Krukonis
1985 Elsevier Science Publishers B.V., Amsterdam — Printed in The Netherlands

MICRO-SCALE METHODS FOR CHARACTERIZATION OF SUPERCRITICAL FLUID EXTRACTION AND FRACTIONATION PROCESSES

R. D. SMITH, H. R. UDSETH and B. W. WRIGHT
Chemical Methods and Kinetics Section, Battelle, Pacific Northwest Laboratories, P. O. Box 999, Richland, Washington 99352

ABSTRACT

The development of new supercritical fluid extraction and fractionation processes requires exploration of wide ranges of fluid temperature and pressure. As the fluid phase composition becomes more complex, conventional methods become inadequate unless augmented by increasingly sophisticated analytical instrumentation. In addition to chemical complexity, analysis is often hindered by characteristics of the analyte which make supercritical fluid extraction desirable (e.g., high molecular weight or thermal lability). Micro-scale methods are described which address these problems while providing the additional benefits of speed, reduced cost, greatly reduced sample and solvent consumption, and safety. Techniques are described in which the fluid phase from micro-scale (0.005-5 ml) high pressure cells is sampled and analyzed in either continuous or periodic modes. The methods developed in our laboratory utilize capillary supercritical fluid chromatography and direct fluid injection-mass spectrometry (or their combination), in addition to conventional analytical techniques, for the characterization of complex supercritical fluid mixtures. The application of these methods to various extraction and fractionation processes involving CO_2 and NH_3, solubility studies of simple and complex mixtures, and the study of rapid chemical processes in supercritical water is described. New methods for the automated examination of extraction processes using capillary chromatography are also described and their initial application illustrated.

INTRODUCTION

The applications of supercritical fluids continue to expand for extraction, chemical fractionation, chemical analysis (chromatography), use as reaction media, and in novel methods for the production of new materials. The primary property of a supercritical fluid which leads to such versatility is the continuously variable solvating power of the fluid. The potential applications, coupled with present limitations of theoretical methods for prediction of supercritical fluid solubilities (and other properties), has resulted in widespread laboratory investigations of these properties. Typically these studies involve an initial qualitative assessment of the application, and usually involve some determination of solubility or a simple fractionation of a mixture. If successful, further experimental studies follow aimed at optimization of conditions and, if warranted, scale-up. For supercritical carbon dioxide, the relative wealth of available experimental results can guide this development. However, it is difficult to be certain that optimum conditions have been obtained, even when using supercritical carbon dioxide. For complex systems, or those requiring higher pressures and temperatures, or more reactive fluids (e.g., ammonia, water, etc), the experimental difficulties have further hindered progress.

A major direction of research in this laboratory has been the development and application of micro-scale systems for characterization of supercritical fluid systems. The micro-scale systems utilize extremely small volumes (0.005 to 5 ml) and all incorporate on-line methods for continuous or periodic chemical analysis. The advantages of this approach include greatly reduced cost and experiment set-up time, the speed with which a wide range of experimental variables may be evaluated (e.g, pressure, temperature, solvent composition, residence time), and the almost negligible risk associated with work on this scale. The on-line analysis methods are sensitive, typically require sample flow rates of < 5 μl/min, allow characterization of even thermally labile or nonvolatile components (including those for which conventional analytical methods are often inadequate), and minimize consumption of the solvent and quantity of the material required for study. These techniques complement bench-scale evaluation and enhance productivity by rapidly defining useful process parameters.

The primary analytical methods used in our laboratory for the characterization of near and supercritical fluid solutions are capillary column supercritical fluid chromatography (SFC), direct fluid injection-mass spectrometry (DFI-MS), and their combination (SFC-MS) for extremely complex mixtures. These methods utilize the direct fluid injection (DFI) process for transfer of supercritical fluid solutes to the gas phase where a variety of detection methods may be employed. The DFI process has also provided the basis for a range of powerful characterization methods including microwave plasma emission analysis, laser spectroscopy, laser multiphoton ionization techniques, high resolution mass spectrometry and tandem mass spectrometric techniques (e.g. DFI-MS/MS) which are being explored in this laboratory. In this work, the instrumentation and methods for SFC and related mass spectrometric analytical methods used in conjunction with micro-scale methods are emphasized. Examples are presented of application of these methods to a range of extraction, fractionation, solubility measurements, and complex analytical separations. The direct study of supercritical fluid phase reaction processes is also described.

INSTRUMENTATION

Supercritical Fluid Chromatography (SFC)

A wealth of fundamental and empirical information can be obtained using chromatographic techniques with supercritical fluid mobile phases. The origin of SFC with packed columns goes back over two decades, with a period of intense activity in the late sixties (ref. 1-10). The recent interest in SFC has been due in large part to the growing general interest in supercritical fluids and the limitations in both chromatographic efficiency and detection methods with HPLC. The introduction of fused silica capillary columns with nonextractable stationary phases (ref. 11) and the compatibility with gas phase detection methods (ref. 12) have served to further increase interest.

The same properties that make supercritical fluids attractive for extraction and separations are also advantageous for SFC. The lower viscosities and higher diffusion coefficients relative to liquids result in the potential for significantly enhanced chromatographic efficiency compared to liquid chromatography. In capillary SFC the mobile phase is maintained at a temperature somewhat above its critical point, typically at reduced temperatures of 1.02 to 1.3. The density of the supercritical phase is typically several hundred times greater than that of the gas, but less than that of the liquid at pressures typical of SFC. Mobile phases with low critical temperatures allow application to thermally labile compounds. Since the solvating capability of the mobile phase is dependent upon pressure, the fluid solvating strength can be easily regulated. Separations using a variety of nonlinear or linear pressure ramps have been developed for specific applications. Additionally, the use of open tubular capillary columns results in a greatly reduced pressure drop across the column; an important consideration if the pressure programming capability of SFC is to be fully exploited. This provides the basis for very rapid pressure programming to affect high speed capillary SFC separations.

As illustrated in Figure 1, the instrumentation utilized in our laboratory for capillary SFC is quite simple. The development of small diameter (< 75 μm) fused silica capillary columns coated with crosslinked and nonextractable stationary phases now allows high resolution separations with efficiencies approaching those of conventional capillary gas chromatography to be obtained (ref. 13). A variety of polymeric bonded and crosslinked stationary phases have been used in the chromatographic columns (ref. 13). Most of the examples given later in this work utilized bonded and crosslinked 5% or 50% phenyl methylphenylpolysiloxane stationary phases.

Combined SFC-MS

The application of capillary SFC to complex mixtures clearly benefits from the use of selective detectors, such as the mass spectrometer (MS). The development of an SFC-MS interface requires that the SFC effluent undergo a pressure reduction for ionization and subsequent mass analysis. The most direct approach to this problem involves expansion of the fluid through a nozzle and production of a collimated beam using molecular beam skimmers which is then ionized and

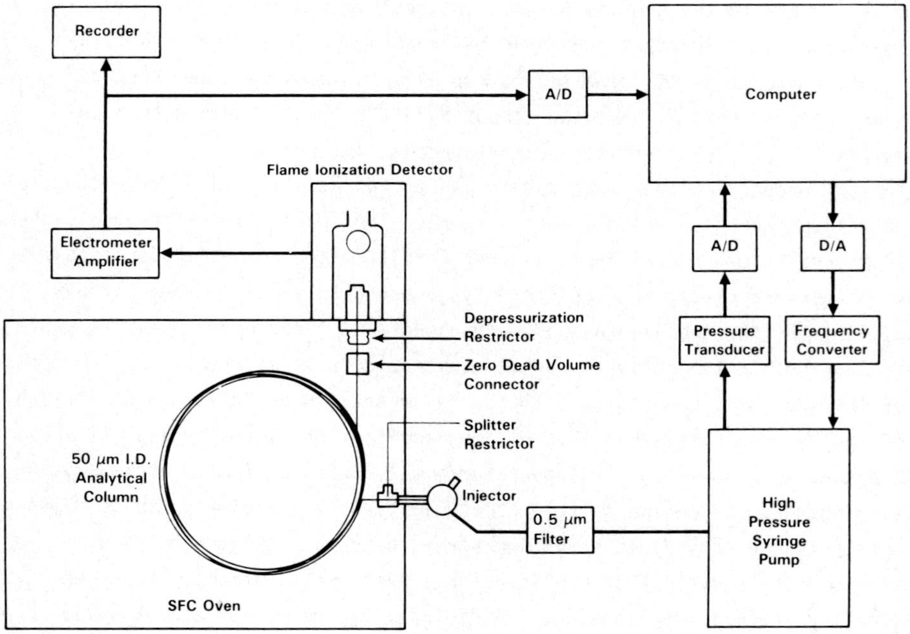

Fig. 1. Schematic illustration of instrumentation for capillary supercritical fluid chromatography (SFC).

analyzed using conventional electron impact ionization mass spectrometry. This approach was first proposed in the literature by Milne (ref. 15), and was a direct extension of work with molecular beam systems. Shortly after this initial report, Giddings, Myers, and Wahrhaftig presented a more quantitative evaluation of the molecular beam approach for "the gas phase isolation of nonvolatile molecules by high pressure jets for mass spectrometry" (ref. 16). These workers came to the conclusion that solvent clustering might not be significant and solute-solute association could be minimized if the expansion was sufficiently rapid. In addition, a solution of β-carotene in supercritical carbon dioxide was expanded through a small laser drilled orifice. This work provided the basis for the more ambitious studies of Randall and Wahrhaftig (ref. 17-19), which provided a detailed examination of the molecular beam approach to SFC-MS interfacing.

The instrumentation constructed by Randall and Wahrhaftig utilized a three-stage vacuum system which followed the classical approach to molecular beam studies (ref. 19). This work demonstrated that solutes of low volatility could be injected into a low pressure environment. The degree of solvent clustering was much less than estimated and found to be dependent on a number of experimental parameters. Randall and Wahrhaftig demonstrated that not only was solute-solute clustering negligible, as anticipated, but also that solute--solvent clustering was much less than expected. Of considerable interest was the observation that the degree of solvent-solute and solvent-solvent clustering was greatly reduced as the background pressure in the first (nozzle-skimmer) vacuum region was increased (ref. 18).

Unfortunately, the molecular beam approach does not provide a practical SFC-MS interface for routine analytical applications. The significant complexity of this instrumentation, the multi-stage vacuum system, and operational difficulties (such as beam alignment) have prevented the application to real problems.

Aside from the work of Randall and Wahrhaftig, the only other reported attempt to interface supercritical fluid chromatography was the direct introduction work of Gouw et al. (ref. 20). Although no details were provided, and additional information has not been published, this approach apparently required compound volatility for transfer to a conventional mass spectrometer electron impact ion source. Application appeared limited by compound volatility and poor sensitivity.

INSTRUMENTATION FOR CAPILLARY SFC-MS

The development of fused silica capillary columns with nonextractable stationary phases suitable for SFC presented the opportunity to eliminate the disadvantages of the molecular beam approach. The typical fluid flow rates for 50 μm I.D. columns are in the range of one μl/min. This range can be easily handled by conventional two stage mass spectrometers (as typically configured with chemical ionization capability) without the use of liquid nitrogen cryo-pumps which are required for commercial direct liquid introduction LC-MS interfaces (ref. 21).

In 1982, an effective capillary SFC-MS interface was described (ref. 22,23). In this instrument supercritical fluid conditions were maintained to a region immediately adjacent to the ion source where the fluid was rapidly decompressed by expansion through a temperature regulated orifice. The lower fluid flow rates with capillary columns avoided the multi-stage pumping required for the molecular beam approach. Rather than attempting to sample the unperturbed expanding gas by conventional electron impact ionization, this direct fluid injection (DFI) method utilized the shock fronts in the expanding jet and subsequent rapid collisional processes to disrupt cluster species and prevent droplet formation (ref. 12).

Cluster formation during expansion of a high-pressure jet through a nozzle is illustrated schematically in Figure 2. The "droplets" formed during the expansion of a dense gas result from cluster formation during adiabatic cooling in the first stages of the expansion process. Cluster size is related to the fluid pressure (P_f), the pressure in the expansion region (P_v), and other parameters involving the nature of the gas, temperature, and the orifice design. Expansion into a chamber with a region of finite background pressure (P_v), produces a shock wave system. This disrupts the highly directed jet and collisional processes subsequently result in redistribution of the directed kinetic energy of the jet among the various translational, vibrational, and rotational modes (ref. 12). The break-up of clusters is also aided by the rapid collisional processes in the chemical ionization source where the average molecule undergoes on the order of 10^4-10^5 gas phase collisions after decompression before escaping into the mass/charge separation region of the mass spectrometer. The distance from the DFI orifice to the Mach disk may be estimated from experimental work (ref. 24) as $0.67D(P_f/P_v)^{1/2}$, where D is the orifice diameter. Thus, if P_f = 400 atm, P_v = 1 torr, and D = 1 μm, the distance to the Mach disk is ∽ 0.4 mm.

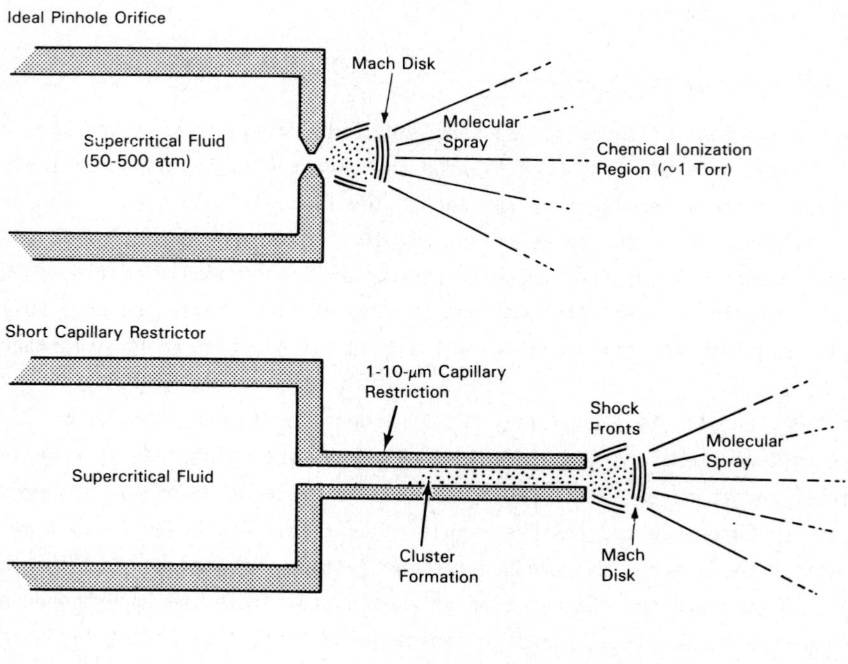

Fig. 2. Schematic illustration of the Direct Fluid Injection (DFI) process for transfer of supercritical fluid solutes to the gas phase.

The ideal pinhole type orifice shown at the top of Figure 2 is impractical for current capillary SFC-MS interfaces. In initial work with 100-200 μm I.D. capillary columns this concept was viable with laser drilled 0.5 - 1.0 μm I.D. holes (ref. 22,23). However, even in this case the minimum substrate thickness was ∽ 15 μm (and readily deformed if not precisely supported) and had a conical channel due to nature of the laser drilling process. Although, greater success was achieved with orifices drilled in 50-100 μm thick material, this typically required multiple laser "shots", and therefore the channel was less precise and reproducible. In addition, this approach presented significant mechanical difficulties associated with producing a high pressure, low dead volume seal, and the necessary precise alignment of the small capillary.

Practical restrictor designs for capillary SFC-MS may be viewed as approximating short capillary channels, schematically illustrated in the lower portion of Figure 2. The characteristics of flow through such a restrictor are complex and cannot yet be accurately predicted. However, it is clear that as the capillary restriction becomes longer, cluster size will increase. Excessively long restrictions cause precipitation of the solute in the capillary and subsequent plugging, while very short restrictions require a very small orifice which is prone to plugging. A practical compromise is the use of tapered capillary restrictions which may be produced from fused silica capillary tubing and even drawn directly on the end of the capillary column. Available results indicate that the overall result of the DFI process with a sufficiently rapid pressure drop is to produce a gas spray incorporating the nonvolatile solute molecules. Our experience with fluids as diverse as CO_2, NH_3 and supercritical water suggest that the DFI method is applicable to all fluids if saturated solutions are avoided at the point of restriction. This limitation applies only to nonvolatile compounds and is typically avoided by either dilution or adjusting the temperature of the orifice region to slightly increase the fluid solvating power.

Current Instrumentation

The basic instrumentation used for SFC and SFC-MS has been described in detail elsewhere (ref. 12,23,25) and will be only briefly considered here. As with capillary SFC, a programmable high pressure syringe pump is used to generate a pulse-free flow of a high purity fluid. Injection utilizes an HPLC valve with volumes from 0.06 to 0.2 μl and flow splitting at (typically) ambient temperatures. (Larger injection volumes are used for column coating in solubility studies, and instrumentation for micro-scale extraction or fractionation studies use micro-scale sample cells, some of which are discussed later.)

One of the three instrument configurations, currently utilized in this laboratory is shown in Figure 3. In this instrument the column is mounted in a

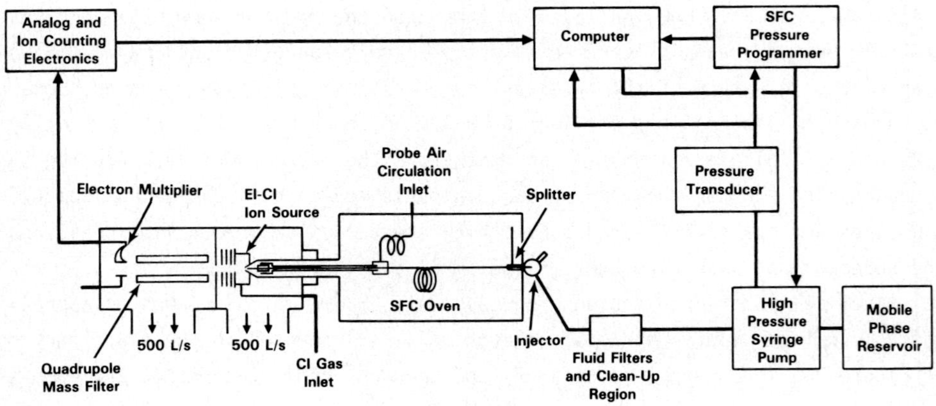

Fig. 3. Schematic illustration of instrumentation for combined capillary SFC-MS. Similar instrumentation is used for the direct study of solubilities, extraction, fractionation or reaction processes.

constant temperature gas chromatograph oven. The oven also serves to heat the air circulated through the direct fluid injection (DFI) probe. A zero dead volume union is typically used to connect the column to a short length of ∽ 4-8 μm I.D. fused silica tubing or other device to function as a flow restrictor. In more recent work, the restrictor is drawn directly on the end of the capillary column. The restrictor is heated to compensate for cooling due to decompression of the fluid. A variety of improved DFI restrictor devices are currently under investigation to provide more reproducible and controlled transfer, particularly for nonvolatile compounds. Capillary columns for SFC have been prepared in our laboratory with 10 to 200 μm I.D. fused silica tubing using techniques described previously (ref. 26,27).

The mass spectrometers utilized to date in this laboratory include conventional quadrupole mass spectrometers (one with MS/MS capability) and a high resolution double focusing instrument. Both electron impact and chemical ionization (CI) ion sources have been used for SFC detection. Methane, isobutane or ammonia are used as chemical ionization reagent gases. An advantage of SFC-MS with small diameter capillary columns is that the flow rates are small enough that reagent gas must be added to reach the operating pressure for the CI source, thus allowing any CI reagent gas to be used. Typical detection limits range from 0.1 to 10 pg depending upon the compound, analysis time, column configuration, and CI reagent gas.

APPLICATION OF MICRO-SCALE METHODS

Capillary SFC Methods

The fundamental interactions which govern the chromatographic process with supercritical fluids can be explored using a range of methods without significant instrumental modifications. In capillary SFC the supercritical fluid is used as the mobile phase and a polymeric coating of known thickness and chemical composition serves as the stationary phase. The use of relatively inert and carefully deactivated fused silica columns, which have smooth interior surfaces, eliminates or minimizes contributions due to adsorption on the silica surface. Since the pressure drop across a column is negligible due to the low fluid viscosity and flow rate, the measurement of retention allows determination of the thermodynamic partition coefficient (K) for a given solute at a defined pressure and temperature on a given column. The partition coefficient is a dimensionless number independent of column length and is equal to the concentration of the solute in the stationary phase divided by the concentration in the mobile phase. The equilibrium constant (or partition coefficient) for distribution between the two phases is related to the chromatographic retention parameter, the capacity ratio (k'), by:

$$K = k'\emptyset \qquad (1)$$

where $\emptyset$ is the phase ratio of the column (volume of the mobile phase divided by the volume of the stationary phase). Thus, the equilibrium constant for phase partitioning can be readily measured. Measurements are more straightforward than with packed columns since the volume of the stationary phase coated on open tubular capillary columns can be accurately determined. However, adsorptive processes at the fused silica interface or active sites in the stationary phase can complicate these relations. In capillary SFC an additional uncertainty related to the volume of the stationary phase can arise when the supercritical fluid has a substantial solubility in the polymer phase. For *in-situ* crosslinked polymer phases, this can be manifested as swelling of the stationary phase, effectively increasing the phase ratio. The importance of this process in capillary SFC has not yet been fully established.

A study of the effect of the polymer stationary phase film thickness is shown in Figure 4. In this study films of 5% phenyl polymethylphenyl-siloxane were coated on deactivated 100 um I.D. fused silica capillary columns and stabilized by partial *in-situ* crosslinking using free-radical techniques (ref. 26,27). The capacity ratio is given for three *n*-alkanes in a supercritical carbon dioxide mobile phase (at 40°C and 100 atm) as a function of stationary phase film thickness. Separate experiments have shown that swelling of this

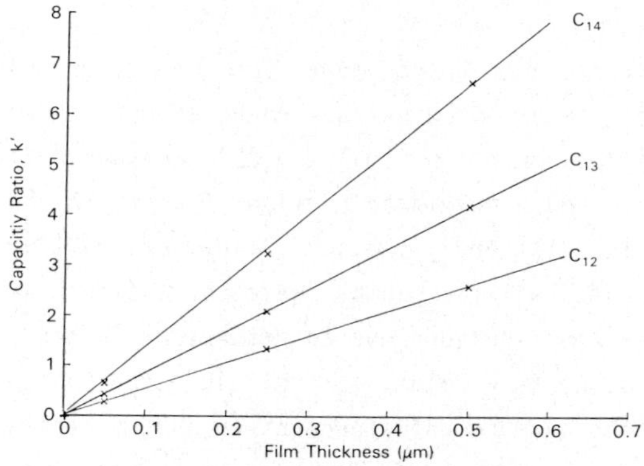

Fig. 4. Variation of the SFC capacity ratio (k') with stationary phase film thickness for three n-alkanes.

stationary phase is negligible under these conditions. The fact that the intercept for each n-alkane at zero film thickness yields a capacity ratio of 0 ± 0.1 shows that the retention mechanism for these compounds involves only partitioning between the two phases. Indeed, studies with a wide range of compounds indicate this is the operative mechanism in nearly all cases for capillary SFC using CO_2 with the exceptions apparently resulting from adsorptive contributions for highly polar compounds. These cases are relatively infrequent with supercritical carbon dioxide since fluid phase solubilities are often the limiting factor for highly polar compounds. Column deactivation techniques have been found to effectively minimize the number of active sites available for adsorption.

Similar retention studies allow the change in enthalpy (ΔH) of solute transfer from the liquid to fluid phase to be determined. The capacity ratio for a solute is determined at various temperatures while maintaining constant density conditions. The slope of a plot of capacity ratio (k') versus inverse temperature allows the change in enthalpy to be calculated since $k' \propto K$ and:

$$\ln K = - \frac{\Delta H}{RT} + \frac{\Delta S}{R} \tag{2}$$

Typical plots for two solutes using supercritical carbon dioxide and a cross-linked 5% phenyl polymethylphenylsiloxane liquid phase are shown in Figure 5. The ΔH values (-5 to -10Kcal/mole) are also indicative of partition chromatography.

Capillary SFC methods can also be utilized to determine a range of additional physicochemical data. Van Wassen et al. have previously described the

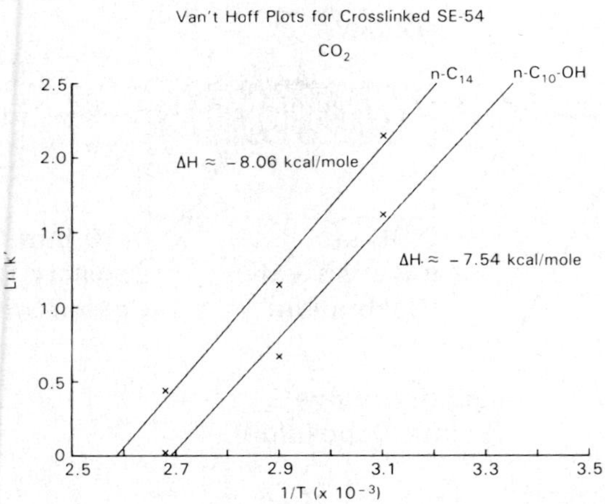

Fig. 5. Van't Hoff plots for an n-alkane and an alcohol obtained for capillary SFC with carbon dioxide at constant fluid density.

application of packed column SFC for the measurement of partial molar volumes and virial coefficients (ref. 28,29). Partial molar volumes (at infinite dilution) can be determined from the slopes of plots of log k' vs. pressure, but difficulties are encountered near the critical point (ref. 30). Diffusion coefficients can also be readily derived from the peak profile for a solute in an uncoated (but deactivated) capillary column. Although few experimental applications have yet been reported, capillary SFC techniques can be adapted for the measurement of additional thermodynamic and kinetic properties related to solvation and adsorption phenomena. These measurements utilize techniques well established in gas chromatography (ref. 31). In nearly all cases the use of deactivated open tubular columns greatly improves the ability to obtain useful fundamental data from such measurements, particularly when accurate measurement of the stationary phase volume or the absence of adsorption processes is required.

Solubility Measurements

Probably the most important measurements relevant to an application of supercritical fluids are the solubility of a solid or liquid in the fluid phase and the partition coefficient between the fluid and condensed phases. Capillary SFC methods offer a rapid and relatively reliable approach for measurement of partition coefficients. For direct measurements of equilibrium concentrations of a solid in a supercritical fluid, slightly different methods are used. Figure 6 gives a simplified schematic illustration of the instrumental arrangement used for solubility studies in this laboratory (note the similarity with

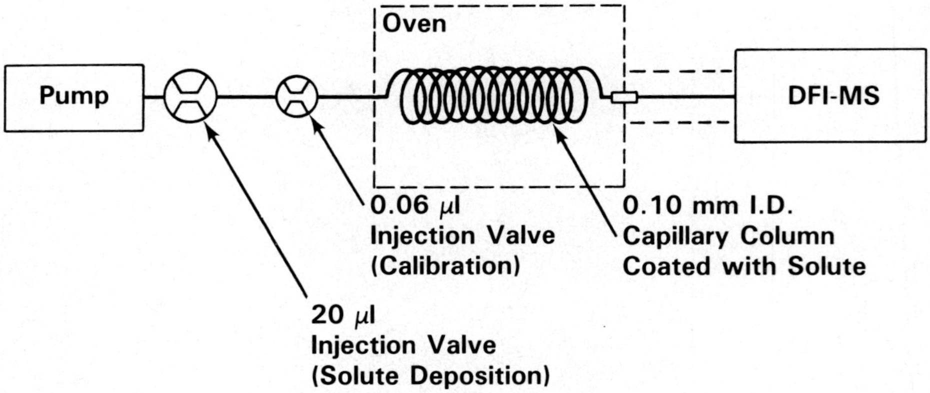

Fig. 6. Simplified diagram of instrumentation used for supercritical fluid solubility studies. The solute is coated on the capillary column surface.

the capillary SFC-MS instrumentation schematic in Figure 3). The primary requirements of this approach are that the condensed phase be coatable onto the column (or appropriate support material), that it be sufficiently stable and viscous to coat and adhere to the column and not be removed by the fluid flow. Measurements are then readily acquired for a range of pressures and temperatures.

For rapid qualitative evaluation of solubility, a solution containing the solute is injected on a short (0.5-2 m) length of uncoated capillary column using a small volume (0.06 μL) HPLC injection valve. The injection is made at the desired temperature and slightly below the critical pressure where solubility is typically negligible. The pressure is then increased rapidly and the mass spectra are observed to determine if the compound is soluble and to provide an estimate of the threshold pressure where solubility increases rapidly. The initial low pressure allows separation of the sample solvent from the compound of interest, thus allowing any solvent to be used without affecting the fluid composition. This approach is used routinely in this laboratory as the initial step in evaluation of compound solubility and has the advantages of being rapid (< 5 min), applicable to single solutes or mixtures, and having high sensitivity (typically 1-10 ng of sample is used).

Quantitative solubility measurements utilize essentially the same instrumentation as shown in Figure 6. For quantitative measurements it is necessary to assure that equilibrium is attained. This is accomplished by coating a length (1-2 m) of deactivated fused silica capillary tubing with the solute or mixture. Initial studies in this laboratory have used off-line techniques to coat the capillary column with a known amount of the sample. Any solvent can be used since it is removed by evaporation prior to the solubility measurements. These coating techniques are similar to those used for stationary phase coating methods for capillary GC or SFC (ref. 26,27). Once the column is coated, the oven temperature is selected and typically the supercritical fluid pressure is increased either continuously or in preselected steps. The mass spectrometer is used to continuously monitor the mass flux of each solute and the fluid. The mass spectrometer signal intensity for each solute divided by the solvent signal intensity, when proper mass spectrometer ion source conditions are maintained, is proportional to the solute concentration. These measurements can be made semi-quantitative by relating the signal intensity of selected ions in the mass spectra to the actual mass flux using estimated values for mass spectrometer ionization, transmission, and detection efficiencies. However, accurate quantitation requires calibration of each solubility study by introduction of known concentrations of the solute using a low volume injection valve (Fig. 6). Since the area of the resulting peak corresponds to a known weight of material, calibration of instrument response for the solute can be established in a straightforward manner. The response for the supercritical fluid can be calibrated in a similar fashion (known volume introduction), but a more effective method is to simply measure the fluid volume after decompression through the DFI orifice as a gas or liquid for a known time interval. These methods allow measurement of solubilities on a relatively rapid basis with an accuracy determined primarily by the stability of the mass spectrometer operation (typically 1-10%). Solubility measurements are practical with this approach up to approximately 0.1 weight percent; measurements at higher concentrations use equilibrium cells (described later) to obtain higher solute loadings.

Figures 7 and 8 illustrate the application of these techniques to the measurement of solubilities of complex mycotoxins of the trichothecene group in supercritical carbon dioxide and nitrous oxide. The trichothecenes are a class of polar and thermally labile fungal toxins often found as contaminants in grains and other food products. Structures for these compounds are given in Table 1. Figure 7 gives the relative solubilities determined for diacetoxyscirpenol (DAS), deoxynivalenol (DON) and T-2 toxin in carbon dioxide at 40°C and pressures up to 300 bar. These compounds show quite distinct threshold pressures (a parameter somewhat dependent upon the sensitivity of the analytical technique) above which solubility increases rapidly in the expected fashion. Figure 8 gives quantitative data for solubilities for a mixture of the trichothecenes in N_2O at two temperatures.

Both solubility studies showed surprisingly different relative solubilities for the trichothecenes than would have been anticipated on the basis of capillary SFC results using CO_2 (see Figure 9). As discussed later, elution order

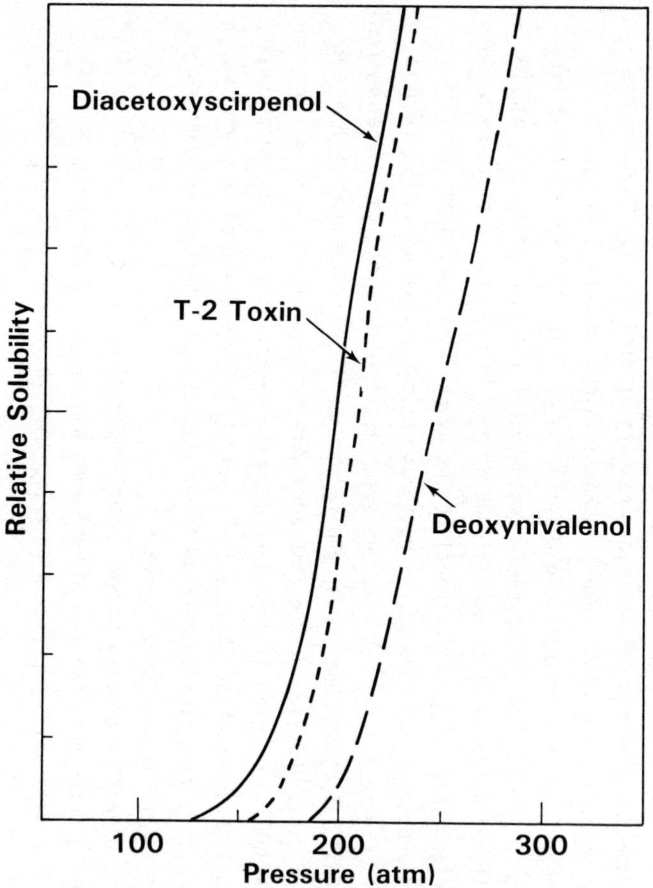

Fig. 7. Relative solubilities of a three component trichothecene mixture in CO_2 at 40°C (see text). DAS = diacetoxyscirpenol, T-2 = T-2 Toxin, DON = Deoxynivalenol.

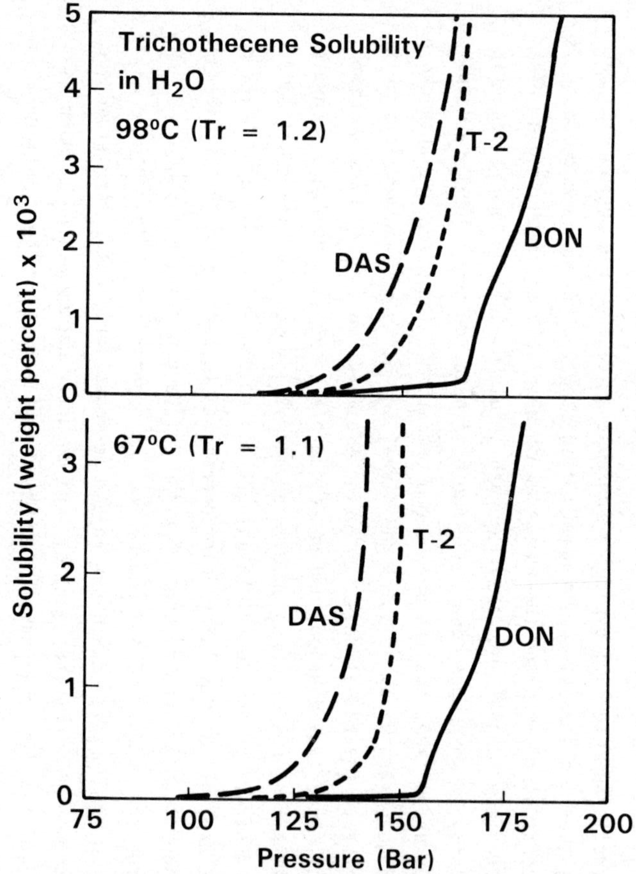

Fig. 8. Solubilities measured for a three component trichothecene mixture in N_2O at 67°C and 98°C. The solubility of DON is apparently supressed by DAS and T-2 (see text).

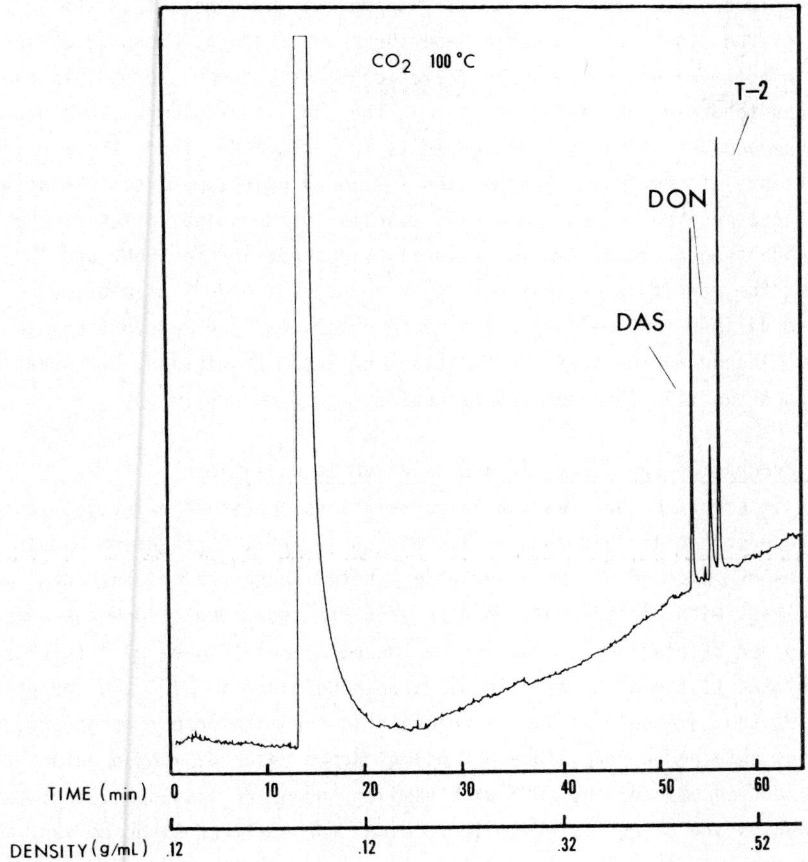

Fig. 9. Capillary SFC separation of the three component trichothecene mixture.

in pressure programmed capillary SFC usually mirrors relative solubilities of the individual components. Relative solubilities were also measured for the three-component trichothecene mixture in supercritical carbon dioxide-water mixtures (approximately 0.6 mole percent) at 108°C. The CO_2-H_2O mixture was produced by flowing CO_2 through an equilibrium cell containing water at 50°C (as described later) and using the saturated effluent for the solubility study. The threshold pressures from this study were approximately 150 bar for diacetoxyscirpenol, 175 bar for T-2 toxin, and 250 bar for deoxynivalenol. The behavior of these solubility profiles is informative. The DON showed an extremely small solubility at pressures below 250 bar, and an extremely rapid increase at this pressure. The trichothecene solubility measurements in CO_2 and N_2O show similar behavior.

The rapid increase in DON solubility appears to be a result of the initial presence of the other trichothecenes and their depletion as pressure is increased in a stepwise fashion. The dramatic increase in DON solubility closely corresponds to depletion of DAS and T-2 in the capillary column. This behavior is not observed for solubility measurements for pure DON. Thus, the equilibrium concentrations of the less polar DAS and T-2 apparently suppresses the solubility of the more polar DON. Separate studies of the solubilities of the individual trichothecenes confirms the solubility suppression for DON, and indicates that the magnitude of this effect is nearly two orders of magnitude. These observations require further study to understand the relevant physical processes, including the possible formation of a solid solution, but clearly present opportunities for improved separations in such systems.

Pressure Programmed Capillary SFC and Relative Solubilities

It is interesting that reasonable correlations appear to exist between volatility, supercritical fluid solubility, and capillary SFC retention within similar classes of compounds. The observations noted above for trichothecene mixtures contrast with measurements obtained for the individual trichothecenes and with the order of elution in capillary SFC separations (Figure 9). This is likely related to the near infinite dilution conditions in SFC and the resulting role of the threshold pressures relevant to chromato-graphic separations. Separations obtained by capillary SFC using carbon dioxide mobile phases and capillary gas chromatography (GC) with similar nonpolar liquid polymeric stationary phases are often quite similar. Even for mixtures spanning many compound classes, elution order is usually unchanged. Elution order in temperature programmed capillary GC is determined by compound volatility and the nature of the (often relatively nonspecific) interactions with the stationary phase. In pressure programmed capillary SFC using nonpolar stationary phases the elution order is related to interaction (solubility) with the mobile phase and also to the nonspecific stationary phase interactions. The similarity in elution order for capillary SFC and GC implies a rough correlation between volatility and supercritical fluid solubility or threshold pressure. This relationship is probably due to the nonspecific nature of solvation by carbon dioxide, which has no permanent dipole, and the relatively nonspecific nature of the polymeric stationary phase for most solutes. (It should be noted that for SFC on conventional packed columns there is less similarity with GC elution order, presumably due to the more specific role of the stationary phase and adsorption processes.) Another interesting observation is that the order of elution during pressure programmed operation are relatively infrequent (in the absence of adsorption processes) over a significant range of temperatures (40-140°C), although the density of the supercritical carbon dioxide required for elution of these compounds drops significantly over this range.

The advantages of capillary SFC as an analytical tool derive, in large part, from the ability to vary the fluid phase solvating power by control of pressure. Greatly improved selectivity for separation of mixtures compared to liquid chromatography can be obtained. The advantages of pressure programming are illustrated in Figure 10 which compares separations obtained for a coal tar extract using supercritical and slightly subcritical carbon dioxide (critical temperature = 31°C). Both separations used the same 10m x 100 μm fused silica column and DFI-MS detection. The capillary SFC separation was obtained at 60°C and utilized a linear pressure program from 70 to 200 bar started upon elution of the solvent peak. The liquid chromatographic separation (bottom) was conducted at 28°C and 300 bar (CO_2 density of 0.95 g/ml). Under these conditions the slightly subcritical liquid has physical properties similar to the supercritical fluid phase. The SFC separation shows rather poor efficiency by the standards of smaller diameter columns (ref. 13,32,33), but resolves many of the major components. The molecular weights labeled on the figure were determined from the mass spectra. The isobaric liquid CO_2 separation, although having a nearly identical linear velocity, shows quite different chromatographic performance and a significant loss of selectivity. In addition, the order of elution changed and several components were not detected. These characteristics are similar to packed column SFC separations under constant average pressure conditions (ref. 34).

An advantage of pressure programmed SFC compared to temperature programmed GC or gradient elution liquid chromatography is that the solvating power of the fluid can be changed on a nearly instantaneous basis. Temperature programming suffers from thermal lag and temperature gradients that limit the maximum useful rate of increase, while gradient elution requires the time for complete mixing of two liquid streams. The power of rapid programmed SFC is illustrated in Figure 11 for the separation of the same coal tar extract as shown in Figure 10. The rapid pressure programming combined with a short (0.8 m) smaller diameter (25 μm) capillary column allows much more efficient separations. The separation used supercritical CO_2 at 100°C and a pressure ramp of 100 bar/min from 75 bar at the time of injection. For carbon dioxide at 100°C this corresponds to a nearly linear density increase. Since, to a first approximation, the solubility parameter of the supercritical fluid is directly proportional to density (ref. 7), this probably provides an ideal general purpose pressure program. It is interesting to note that the higher efficiency of the separation nearly compensates for the decreased sample capacity, and we expect loss of sensitivity to be negligible for equivalent resolution. Indeed, similar resolu-

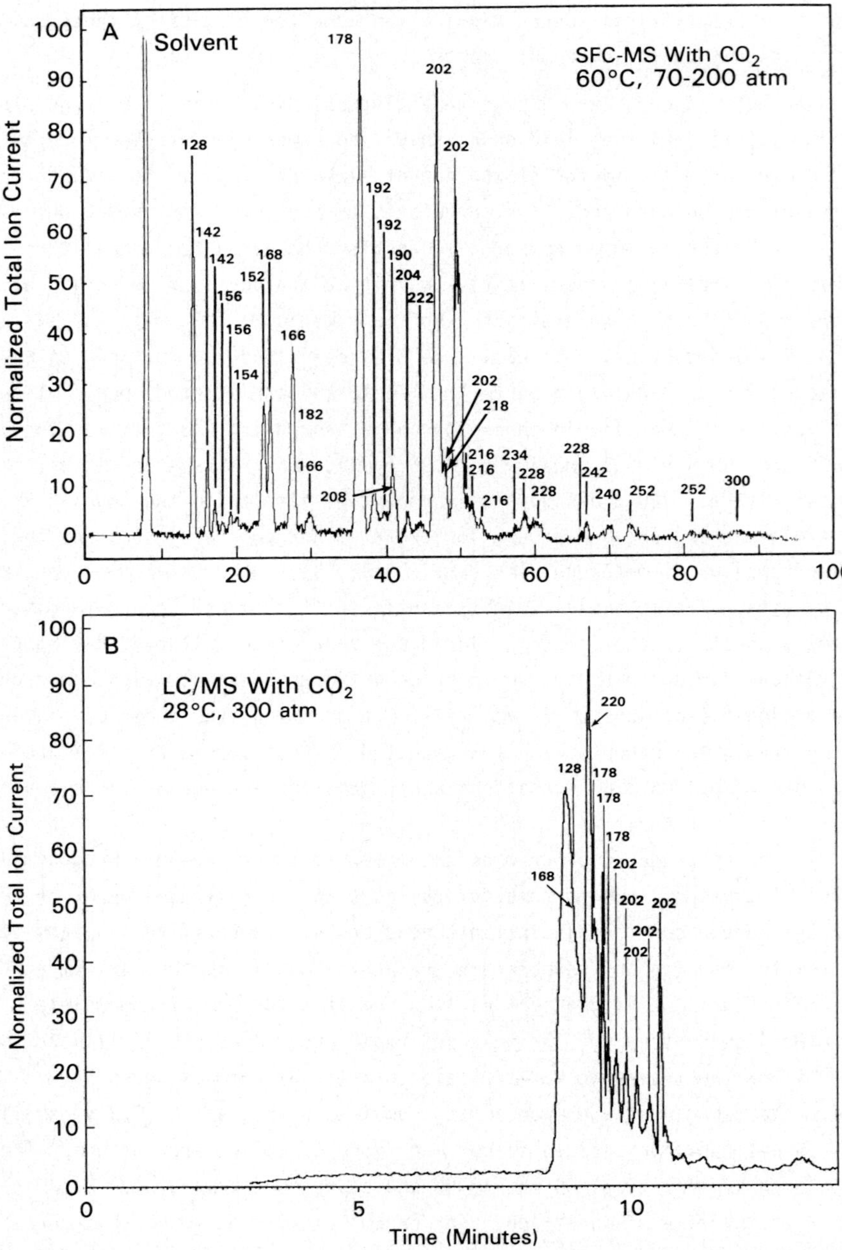

Fig. 10. Comparison of a coal tar extract separation using pressure programmed capillary SFC (top) and subcritical carbon dioxide (at 28°C) under isobaric conditions (bottom).

tions are evident in Figures 10 (top) and 11, but the rapid pressure programmed, narrow-bore column separation required less than one fiftieth the time.

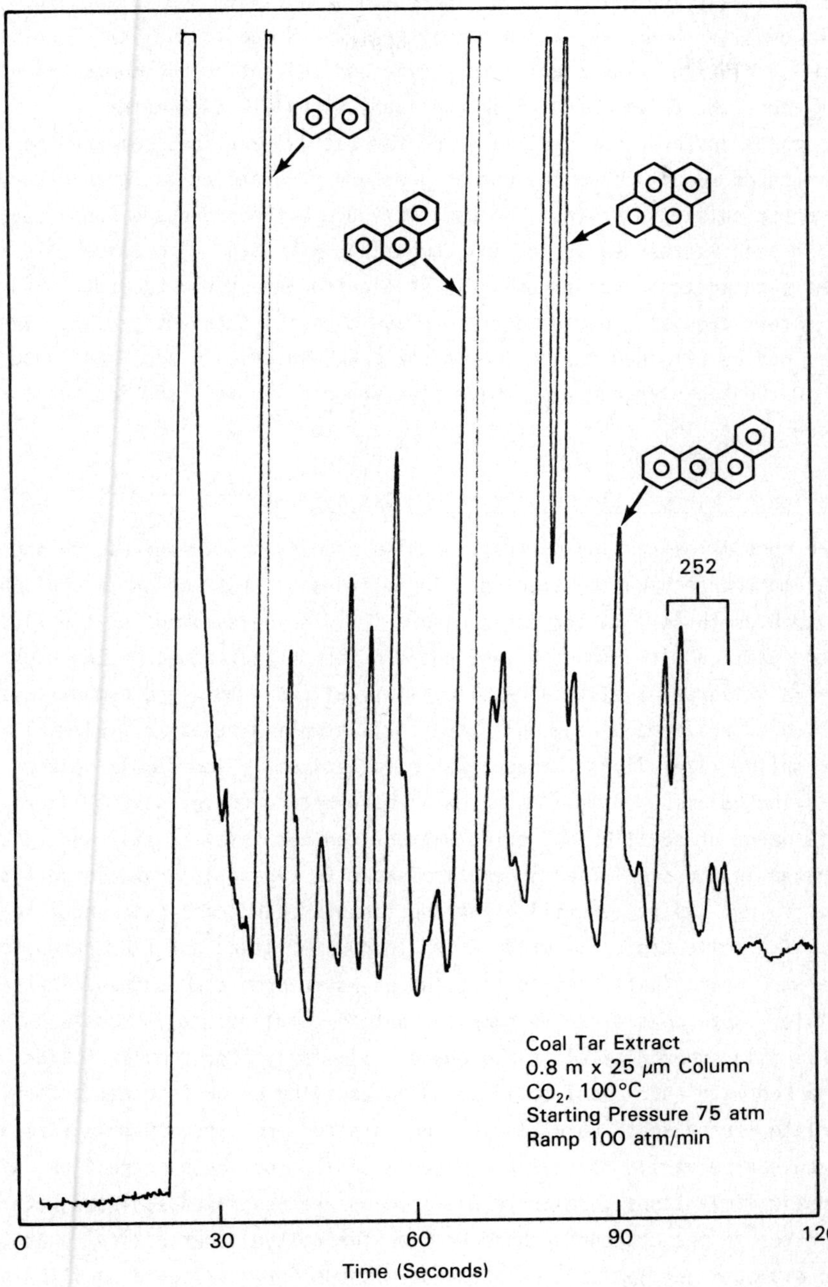

Fig. 11. Fast pressure programmed capillary SFC of a coal tar extract (See Figure 10).

Further advantages result from the reduced time for interaction of highly polar compounds with active column sites, increased sensitivity due to "band compression" as density increases, and a narrow peak width due to the low k' value at the time of elution. The small sample size and solvent requirements are also advantageous; the volume of a 25 μm capillary is only 0.5 μl/meter.

The major advantage of capillary SFC for fast separations compared to packed column methods results from the use of pressure programming to rapidly vary the mobile phase solvating power. The capillary SFC methods provide rapid separations and peak widths nearly ideally compatible with typical maximum mass spectrometer scan speeds. Quantitation is straightforward since the total SFC effluent enters the mass spectrometer. In addition, significantly higher sensitivities can be obtained due to the narrow peak widths (< 5 sec), and detection limits in the sub-picogram range have been found using selected ion monitoring techniques (ref. 35).

Application of Micro-Scale Methods to Complex Systems

More complex extraction or fractionation processes not amenable to the capillary column approach can be studied using micro-scale extraction or equilibrium cells, such as those illustrated in Figure 12. The measurement of the fluid phase composition with direct DFI-MS has a number of advantages. The high sensitivity and universal analytical applicability allows simple and complex systems to be studied while requiring only very small sampling rates (< 1 μl/min). The small sampling rate allows the application of extremely low sample volumes and cells having volumes as small as a few μl have been used to date. This can speed the attainment of equilibrium, often without the necessity of stirring. As illustrated in the simplified schematics shown in Figure 12, gas/liquid systems of small volume typically utilize introduction through the liquid phase to speed mixing. The cells typically utilize 0.5 to 10 μm filters and flow rates to the analyzer are sufficiently low so that the pressure drop across the filter is negligible. Both phases can be sampled, but the smallest cells sample only the upper (usually supercritical) fluid phase. Similarly, for complex solids, small cells packed with sufficient solid to allow equilibrium or (for extraction) appropriate steady state conditions to be obtained are used. Sample size and flow rate can be varied to evaluate kinetic limitations upon extraction rates. The kinetic limitations in micro-scale studies are usually associated with diffusion rates in the condensed phase or transfer to the supercritical phase.

To evaluate the application of these methods for gas/liquid equilibria, measurements of the solubility of water in carbon dioxide were made at 50 to 150°C using a one milliliter sample cell. The time required to reach steady state conditions varied from 2 to 5 minutes. Experimental uncertainties associ-

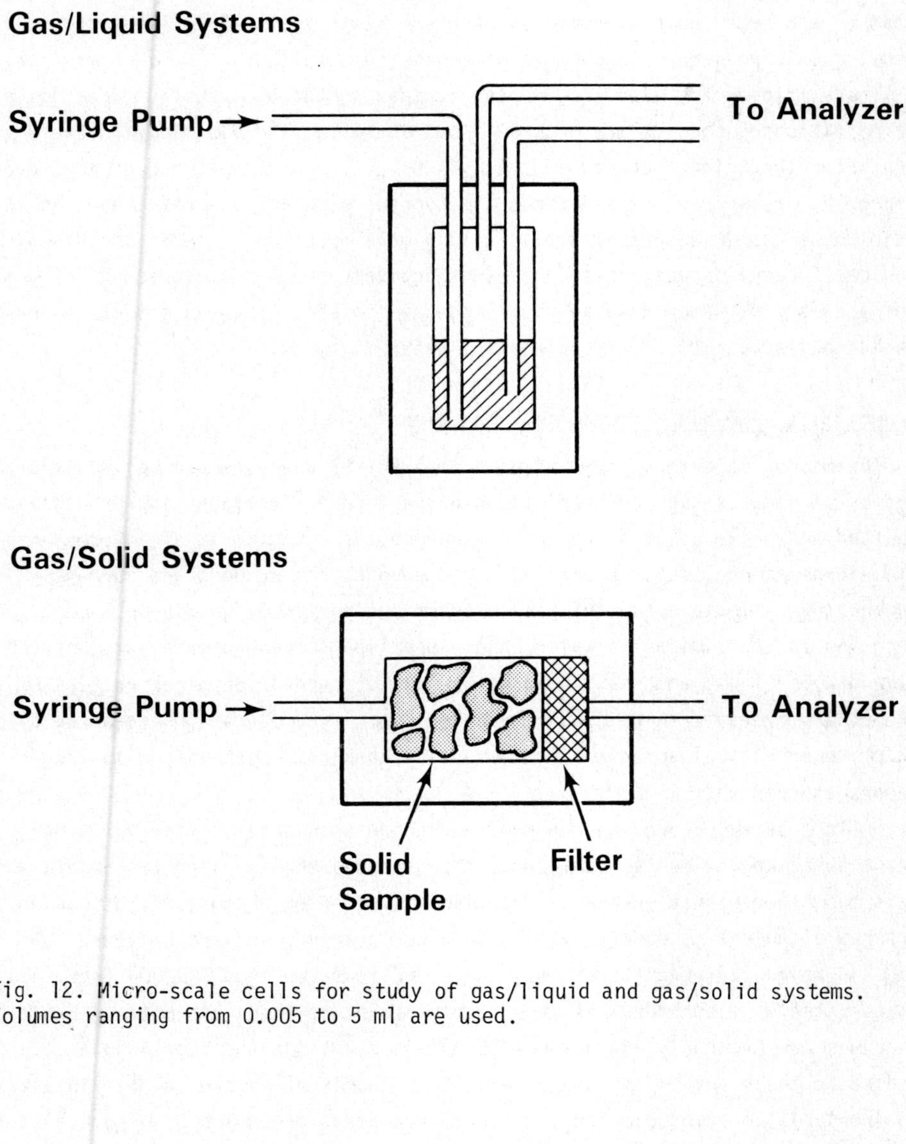

Fig. 12. Micro-scale cells for study of gas/liquid and gas/solid systems. Volumes ranging from 0.005 to 5 ml are used.

ated with these micro-scale measurements were primarily related to the effects of cell walls and the accuracy of the mass spectrometric measurements (± 5%). The results obtained at 50°C and 75°C are given in Figure 13 and compared with various experimental results from the literature (ref. 36-38) and the predictions of Won based upon a Heyden Equation-of-State (ref. 39). The micro-scale results yield reasonable agreement with previous studies.

These methods are also well suited to measure the distribution coefficients for partitioning of a solute between fluid phases. The distribution coefficients for the trichothecenes in the H_2O-CO_2 system were measured at 40°C over a range of pressures. As expected, the concentration of trichothecenes in the fluid phase was immeasurably small ($< 10^{-7}$ mole fraction). Thus, the distribution coefficient between these phases effectively precludes extraction of trichothecenes from water despite their moderate solubility in supercritical carbon dioxide or supercritical carbon dioxide-water mixtures.

Supercritical Ammonia Micro-Scale Extractions

The micro-scale techniques coupled with DFI-MS analyses are well suited for studies of complex supercritical fluid extraction or fractionation processes. The study of polar solutes with polar supercritical fluids is of interest since little experimental data is available and theoretical methods are currently inadequate. Ammonia has attracted interest due to its large dipole moment, induction forces similar to water, and dispersion forces intermediate between HCl and H_2O. Thus, ammonia should be useful for separation or extraction of more polar materials than is possible with CO_2. A complication presented by use of supercritical ammonia is it potential chemical reactivity with large classes of compounds.

Humic substances are major organic components in soils, natural waters, marine and lake sediments, peat bogs, carbonaceous shales, lignites, brown coals and other natural deposits (ref. 40,41). These substances are complex mixtures of acidic compounds, yellow to black colored, and are believed to range in molecular weight from less than 1000 to as much as 100,000 (ref. 40), although there is considerable uncertainty regarding these estimates. Humic materials are typically extracted with alkali solutions and fractionated according to their solubility characteristics. Humic acids are soluble in alkali and insoluble in acid solutions, while fulvic acids are soluble in both alkali and acid solutions. Figure 14 shows a mass spectrum obtained during the "extraction" of a fulvic acid with ammonia at 150°C and 140 bar. Approximately half of the fulvic acid could be extracted with ammonia at 400 bar and 150°C, while only about 25% of a humic acid could be extracted. The fulvic acid mass spectrum given in Figure 14 shows abundant higher molecular weight contributions, which likely extends well beyond the upper range of the mass spectrometer

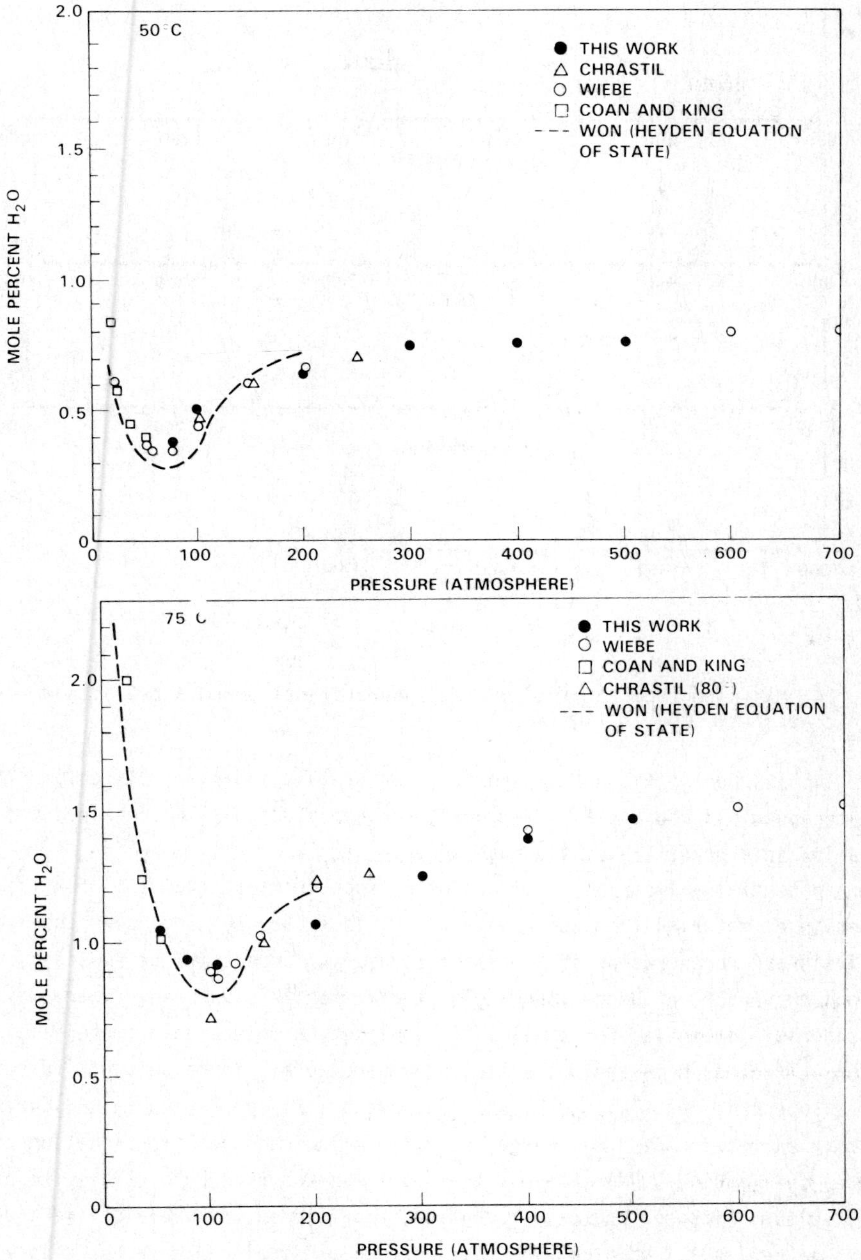

Fig. 13. Water solubility in CO_2 at 50°C and 75°C as a function of pressure. Experimental results obtained using micro-scale DFI-MS method are compared with available literature (36-39).

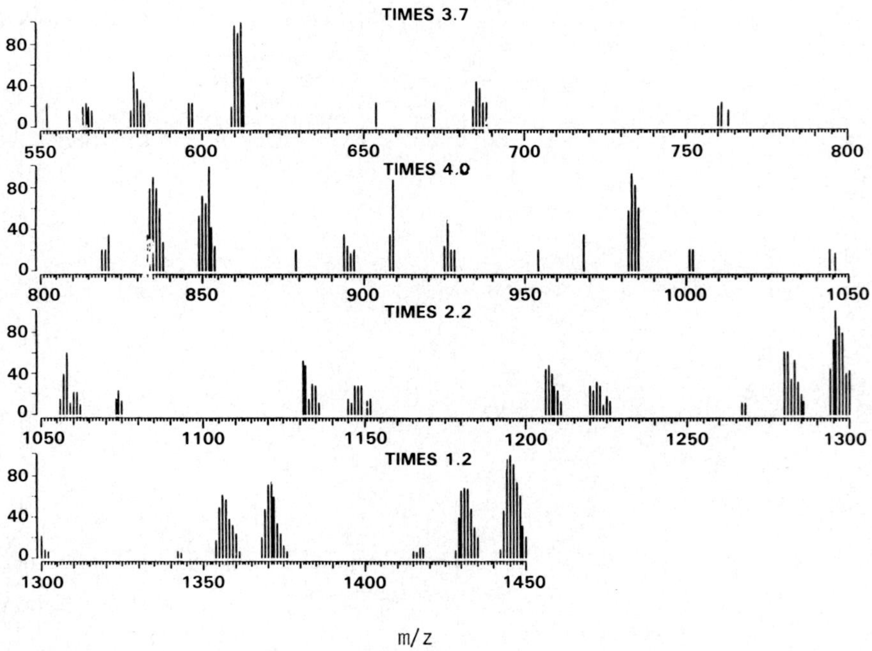

Fig. 14. Mass spectrum obtained during supercritical ammonia extraction of a Fulvic Acid at 150°C at 140 bar.

(m/z 1450). Due to the unknown composition of this material, its complexity, and the possible reactivity of ammonia, further study will be required for detailed interpretation of the mass spectra.

A potentially important application of supercritical fluids is for the extraction of material from adsorbents for various purposes including chemical analysis and regeneration of adsorbent capacity. Supercritical fluids such as ammonia have not yet been extensively studied for this purpose. There may be considerable potential for small scale applications since polar materials are often extracted which are not efficiently removed by liquid solvents or less polar supercritical fluids. Figure 15 shows a mass spectrum obtained during ammonia extraction of a silica gel column packing used for fractionation of a diesel fuel marine (DFM) sample. This fuel was known to be unstable, to have high sulfur, nitrogen and oxygen content, and form an insoluble sediment upon storage (ref. 42). Upon column fractionation a discolored layer of silica was formed at the top of the column where the DFM was introduced which was not removed during elution with conventional liquid solvents. A 300 μl cell was packed with the discolored silica and the extraction was conducted with super-

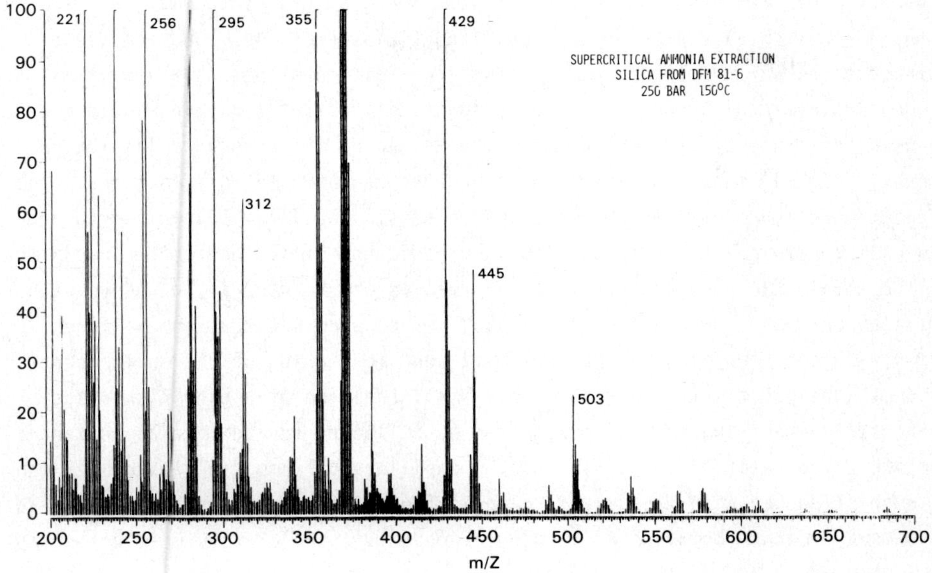

Fig. 15. Mass spectrum obtained during the ammonia extraction of an "insoluble" diesel fuel marine residue adsorbed on silica at 150°C and 250 bar.

critical ammonia at 150°C. Mass spectra were recorded while the pressure was gradually increased from 50 to 450 bar. Interestingly, although the concentration of the extract stream increased continuously between 150 and 450 bar over a three hour period at high flow rates, the relative composition of the products did not change significantly over this pressure range. It is likely that the kinetics of the desorption process make rapid fractionation directly from the silica impractical on this time scale. However, supercritical ammonia effectively removed most, if not all, of the adsorbed material from the silica. It has also been shown that a substantial fraction of the "insoluble" DFM sediments may be extracted with supercritical ammonia (ref. 42). Again, the chemical reactivity of ammonia in this process remains to be examined.

Reactions in Supercritical Fluids

Micro-scale methods are also well suited to the study of chemical reactions in supercritical fluid media. Although this constitutes a mostly unexplored area, the potential for more rapid and selective reaction processes is attractive. Opportunities include homogeneous fluid phase reactions in place of normally heterogeneous processes and elimination or more effective utilization of catalysts. Reaction products and solvents could also be efficiently separated and continuous processes are practical. Preliminary studies have described organic reactions in various organic fluids for chemical synthesis (ref. 43), heterogeneous catalysis using supercritical fluids (ref. 44), and the use of supercritical water for coal and biomass liquefaction as possible applications.

An instrumental arrangement nearly identical to that shown in Figure 6 can be used for study of systems where reactants can be introduced as liquids or gases. Slightly more complicated arrangements can be used for reactions involving introduction of solids or immiscible phases. Capillary column length and fluid linear velocity can be used to vary residence time. Heating is rapid due to the small column diameter. Injection valves can be used to introduce either small amounts of a reactant or larger volumes of premixed reaction mixtures. The DFI-MS can then be utilized for continuous monitoring of the entire reaction product stream (at low flow rates) or a small fraction at higher flow rates, where the bulk of the product stream can be collected for additional characterization. The reaction tube, typically a capillary of compatible material, can be open or packed with a catalyst. Application of these methods is limited to systems where the formation of insoluble products is relatively unimportant to avoid plugging the reaction column.

Figures 16 and 17 show mass spectra obtained during studies of reaction processes in supercritical water. These studies used 1 meter of 100 μm I.D.

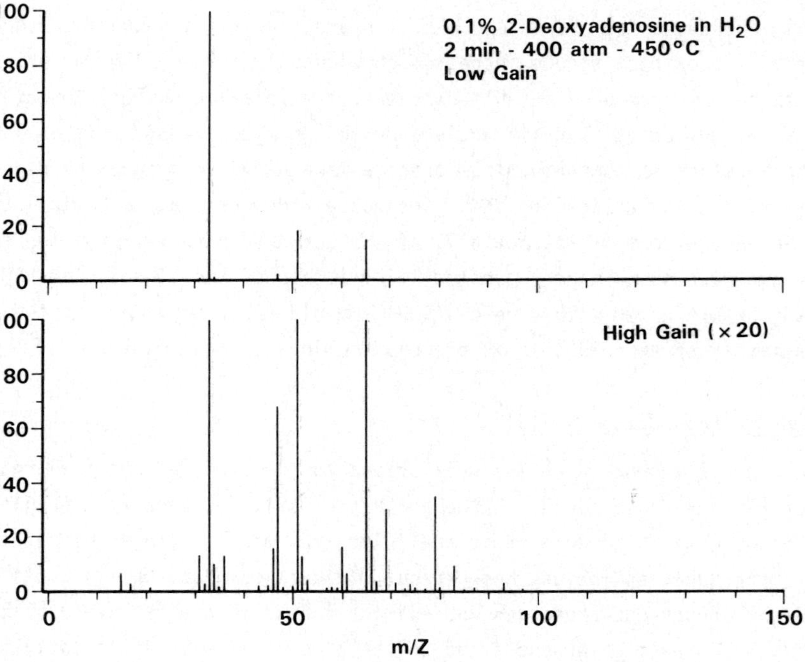

Fig. 16. Mass spectra of products from reaction of 2-deoxyadenosine in super critical water.

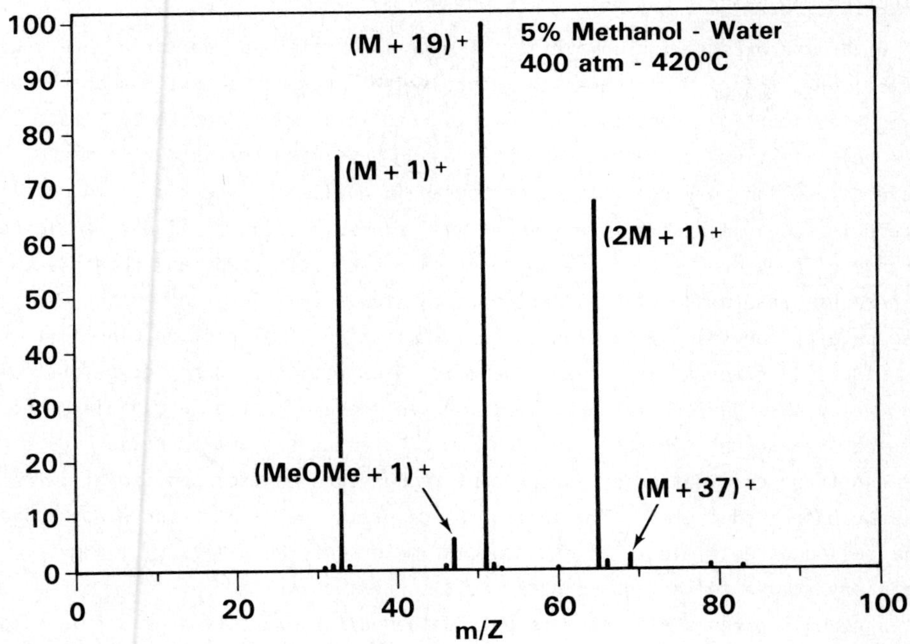

Fig. 17. Mass spectrum from reaction of 5% methanol-water solution at 40 bar 420°C. Note likely formation of dimethyl ether (m/z 47).

platinum-10% iridium tubing (fused silica is soluble in supercritical water) and residence times of 1-2 minutes. The mass spectra were obtained using water as a chemical ionization reagent; thus reaction products typically give rise to $(M+1)^+$, $(M+19)^+$, and possibly $(M+37)^+$ ions. Figure 16 shows mass spectra obtained for reaction of a 0.1% solution of labile 2-deoxyadenosine (Mol. Wt = 251) at 400 bar and 450°C. The critical pressure and temperature for water are 218 bar and 374°C, respectively. The 2-deoxyadenosine reacts to form numerous low molecular weight products which resemble those obtained by pyrolysis. There was no detectable (< 10^{-6} mole fraction) amount of the 2-deoxyadenosine observed after a 2 minute reaction period.

Figure 17 shows a mass spectrum obtained during reaction of a 5% methanol-water mixture for one minute in platinum-iridium tubing at 400 bar and 420°C. Interestingly, although most of the methanol does not react, these results indicate about 3-5% conversion to a product of molecular weight 46. The product is probably dimethylether and this reaction may involve a catalytic process involving the reactor walls. Many similar results have been obtained in these preliminary studies and we anticipate these micro-scale methods will be important tools in evaluation of chemical processes under supercritical conditions.

Automated Micro-Scale Extraction-Chromatography

Often the expense and complexity of mass spectrometric monitoring are unwarranted, or simpler chromatographic methods might be advantageous for characterizing fractionation products difficult to resolve by mass spectrometry (e.g., isomers). Instrumentation is currently being developed for automated micro-scale extraction combined with chromatographic analysis. The aim is to develop a computer controlled instrument which will expose a solid or liquid sample to a range of pressures and/or temperatures in preselected steps and obtain a chromatographic analysis of the extract at each step. This instrumentation can also be used for solubility studies for examination of extraction kinetics.

Figure 18 gives a simplified schematic illustration of a prototype instrument which uses gas chromatography; subsequent development of a capillary SFC version is also anticipated. The instrument uses an extraction cell (Figure 12) and transfer lines under independent temperature control and isolated from the chromatographic oven. The instrument deposits the extract for a controlled time period directly in the first few centimeters of the capillary column. Auxiliary sample collection methods are also available.

Figure 19 gives preliminary results obtained for extraction of a five component polycyclic aromatic hydrocarbon (PAH) mixture from a complex substrate using CO_2 at 45°C. In this example CO_2 pressure was increased in four steps from 75 to 400 bar. At each pressure a three minute extraction with on-column deposition was followed by a brief purge of CO_2 and a thirty minute chromatographic analysis. The results show that essentially no material was extracted at 75 bar and only naphthalene was removed at 150 bar. At 200 bar anthracene

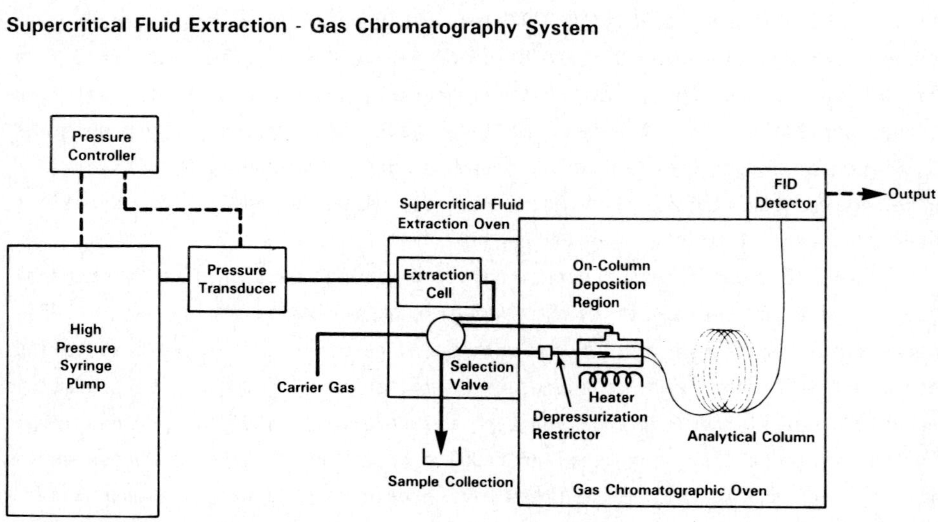

Fig. 18. Apparatus for automated micro-scale extraction with gas chromatographic analysis.

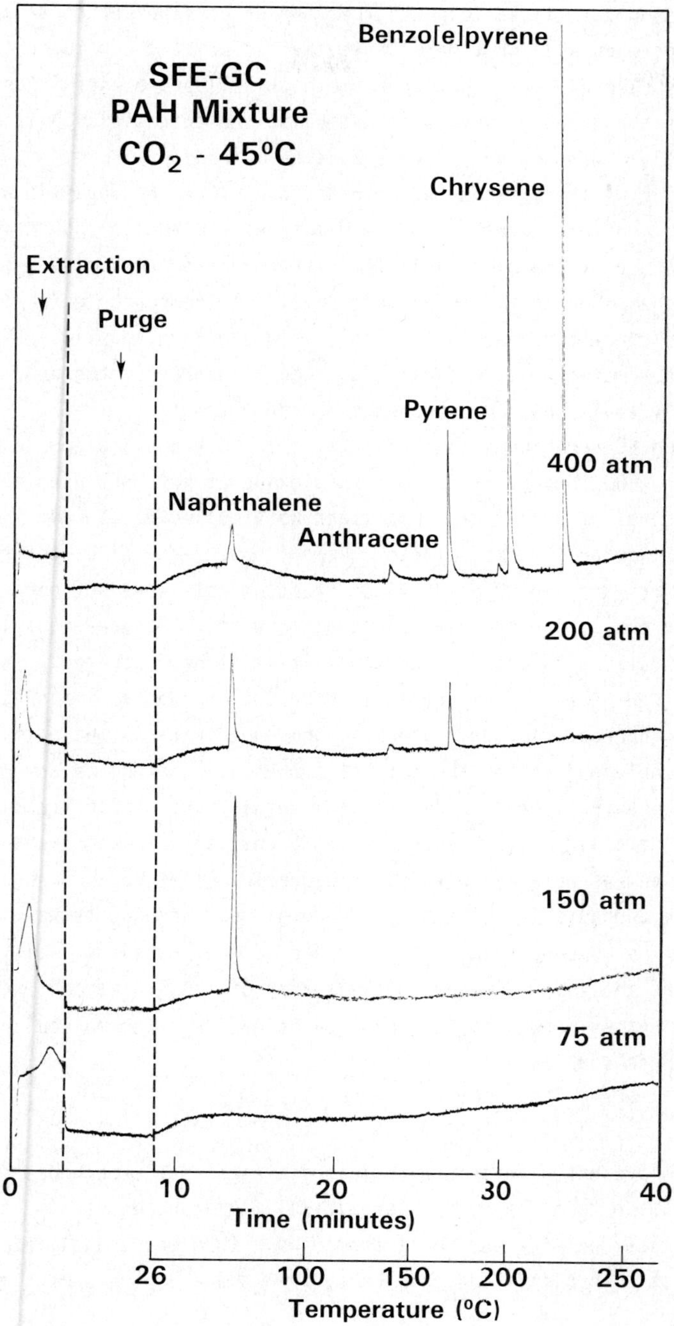

Fig. 19. Preliminary results using the automated micro-scale extraction-chromatography instrumentation (Fig. 18) for the extraction and separation of a five component PAH mixture.

and pyrene were also extracted, while all five components were observed for the 400 bar extraction. It is anticipated the further development of this automated instrumental approach will provide quantitative information on the extraction process. It is also expected that these methods will prove useful for both chemical analysis and broad surveys to establish the potential applicability of supercritical fluid extraction or fractionation methods.

The extension of these automated methods to include SFC analysis will remove any restriction due to chromatographic limitations. Potential advantages also result from the use of the same fluid for extraction and analysis. As discussed earlier, SFC is applicable to labile and nonvolatile compounds which cannot be analyzed by gas chromatography (ref. 12,13). In addition, capillary SFC provides greater chromatographic efficiencies than liquid chromatography (ref. 25) and is obviously applicable to components in the extract.

Both SFC and GC techniques have advantages in this application, as illustrated in Figure 20. Supercritical carbon dioxide at 40°C was used to extract the triacylglycerols from butter. Fractions were collected at 100, 150, 250 and 400 bar. Since only a small fluid volume (about fifty times the sample size) was used at each pressure, distinct fraction cuts were not obtained and a more integrated extraction process resulted. However, the varying solubility of the components as a function of pressure is still demonstrated. The relative fraction extracted at each pressure is labeled in the figure. This example illustrates the necessity of understanding the supercritical phase solubility and partition coefficients for the extract components as well as any kinetic limitations upon extraction in order to achieve distinct sample fractions. The figure compares capillary GC and capillary SFC analysis of each fraction. While better resolution was obtained with GC, a wider molecular weight range for the less volatile components is evident by SFC (e.g. more components were eluted by SFC than GC). The GC conditions were extended to the practical limit for heavy molecular weight analysis. Further optimization of the SFC parameters would likely lead to better resolution and elution of even heavier molecular weight materials in these samples.

CONCLUSIONS

The results presented in this work describe micro-scale methods and their initial application to a range of supercritical fluid processes. Further development and application of these tools should help to significantly expand the understanding of both simple and complex systems and aid development of practical applications.

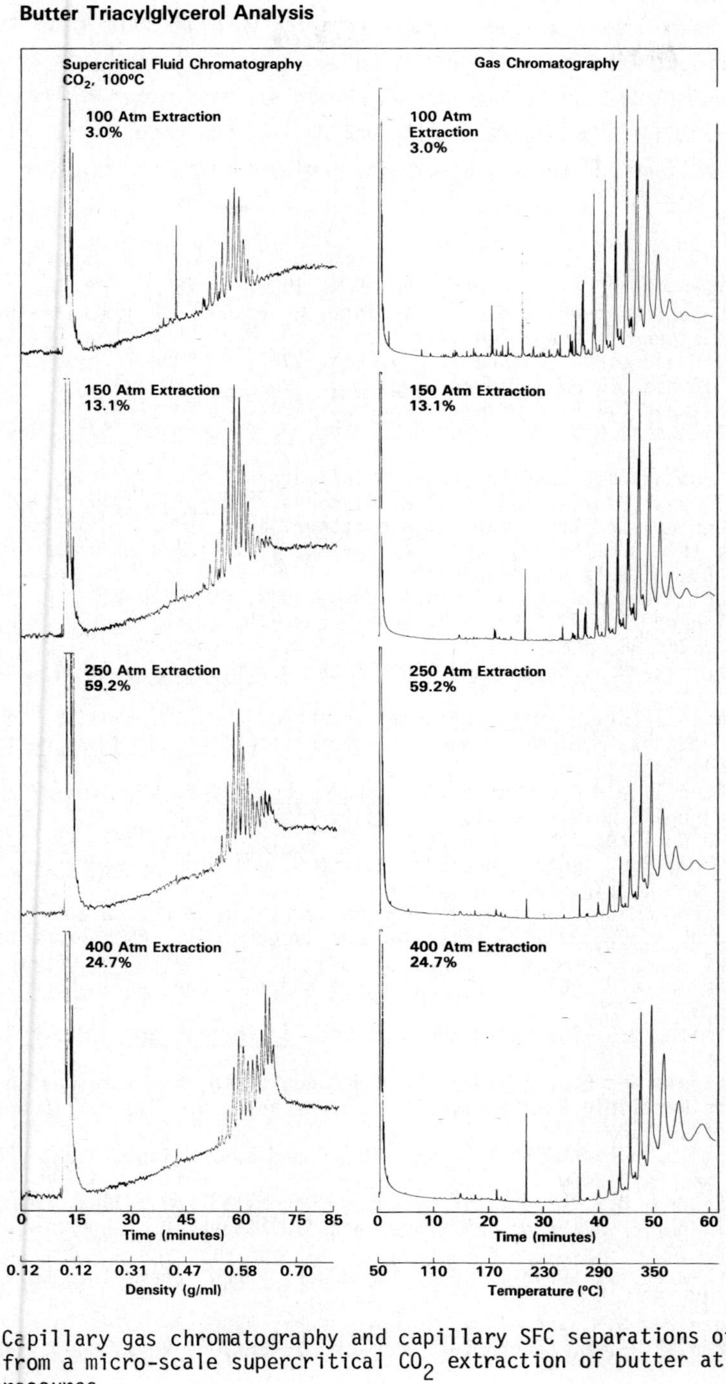

Fig. 20. Capillary gas chromatography and capillary SFC separations of products from a micro-scale supercritical CO_2 extraction of butter at various pressures.

ACKNOWLEDGEMENTS

Portions of this work have been supported by the U.S. Department of Energy, Office of Basic Energy Sciences, under Contract DE-AC06-76RLO 1830, the U.S. Army under Contract DAAK11-84-C-0007, the U.S. Army Research Office under Contract DAAG29-83-K-0172, and Battelle Memorial Institute. We also thank S. L. Frye, A. J. Kopriva, H. T. Kalinoski, R. E. Schirmer, and C. R. Yonker for contributions and helpful discussions related to this work.

REFERENCES

1. M.N. Myers and J.C. Giddings, Sep. Sci. 1966, 1, 761
2. L. McLaren, N. Myers and J.C. Giddings, Science 1968, 159, 197-199.
3. E.A. Klesper, H. Corwin and D.A. Turner, J. Org. Chem. 1962, 27, 700-701.
4. S.T. Sie and G.W.A. Fijnders, Sep. Sci. 1967, 2, 699-727.
5. S.T. Sie and G.W.A. Rijnders, Sep. Sci. 1967, 2, 729-753.
6. S.T. Sie and G.W.A. Rijnders, Sep. Sci. 1967, 2, 755-777.
7. J.C. Giddings, M.N. Myers and J.W. King, J. Chromatogr. Sci. 1969, 7, 276-283.
8. T.H. Gouw and R.E. Jentoft, Advan. Chromatogr. 1975, 13, 1-40.
9. S.T. Sie, W. Van Beersum and G.W. Rijnders, Sep. Sci. 1966, 1, 459-590.
10. R.E. Jentoft and T.H. Gouw, J. Chromatogr. Sci. 1970, 8, 138-142.
11. M. Novotny, S.R. Springston, P.A. Peaden, J.C. Fjeldsted and M. L. Lee, Anal. Chem. 1981, 53, A407-A414.
12. R.D. Smith and H.R. Udseth, Anal. Chem. 1983, 55, 2266-2272.
13. C.R. Yonker, B.W. Wright, H.R. Udseth and R.D. Smith, Ber. Bunsenges Phys. Chem. 1984, 88, 908.
14. B.W. Wright, H.R. Udseth and R.D. Smith, J. Chromatogr., 1984, 314, 253-262.
15. T.A. Milne, Int. J. Mass Spectrom. Ion Phys. 1969, 3, 153-155.
16. J.C. Giddings, M.N. Myers and A.L. Wahrhaftig, Int. J. Mass Spectrom. Ion Phys. 1970, 4, 9-20.
17. L.G. Randall and A.L. Wahrhaftig, Anal. Chem. 1978, 50, 1705-1707.
18. L.G. Randall, Ph.D. Thesis, University of Utah, 1979.
19. L.G. Randall and A.L. Wahrhaftig, Rev. Sci. Instrum. 1981, 52, 1283-1295.
20. T.H. Gouw, R.E. Jentoft and E.J. Gallegos, 6th High-Pressure Sci. Technol. AIRAPT Conf. 1979, 583-592.
21. J. Henion, "Microcolumn High-Performance Liquid Chromatography", J. of Chromatog. Lab. 1984, 28, Elsevier (P. Kucere, Ed.), Amsterdam, Chapter 8, 260-300.
22. R.D. Smith, W.D. Felix, J.C. Fjeldsted and M.L. Lee, Anal. Chem. 1982, 54, 1883.
23. R.D. Smith, J.C. Fjeldsted and M.L. Lee, J. Chromatogr. 1982, 247, 231-243.
24. J. Ashkenas and F.S. Sherman, "Proceedings of the 4th International Symposium on Rarefield Gas Dynamics", 1966; deLeevw, J. H., Ed.; Academis Press: New York, 1966.
25. R.D. Smith, H.T. Kalinoski, H.R. Udseth and B.W. Wright, Anal. Chem., 1984, 56, 2476-2480.
26. P.A. Peaden, B.W. Wright and M.L. Lee, Chromatographia 1982, 15, 335-340.
27. B.W. Wright, P.A. Peaden, M.L. Lee and T.J. Stark, J. Chromatogr. 1982, 248, 17-34.
28. V. Van Wasen, I. Swaid and G.M. Schneider, Angew. Chem. Int. Ed. Engl. 1980, 19, 575.
29. V. Wasen and G.M. Schneider, J. Phys. Chem. 1980, 84, 229.
30. M.E. Paulaitis, K.P. Johnston, and C.A. Eckert, J. Phys. Chem. 1981, 85, 1770.

31. J.R. Conder and C.L. Young "Physicochemical Measurements by Gas Chromatography", John Wiley and Sons, University Press (Belfast) Ltd., 1979.
32. B.W. Wright and R.D. Smith, Chromatographia, 1984, 18, 542.
33. B.W. Wright and R.D. Smith, J. High Res. Chromatog. Chromatog. Comm., 1985, 8, 8.
34. D.R. Gere, R. Board and D. McManigill, Anal. Chem. 1982, 54, 736-740.
35. R.D. Smith, B.W. Wright and H.R. Udseth, J. Chromatogr. Sci., in press.
36. Wiebe, R., Chem. Rev. 1941, 29, 475.
37. Coan, C.R. and A.D. King, J. Amer. Chem. Soc. 1971, 93, 1857.
38. J. Chrastil, J. Phys. Chem. 1982, 86, 3016.
39. K.W. Won, Fluid Phase. Equilib 1983, 10, 191.
40. M.M. Kononova, Soil Organic Matter, 2nd ed., Pergamon Press, Inc., 1966.
41. F.M. Swain, "Geochemistry of Humus", Organic Geochemistry (I.A. BREGER, ED.) New York, Pergamon Press, Inc. p. 81-147, 1963.
42. R.D. Smith, H. R. Udseth, and R. N. Hazlett, FUEL, in press.
43. J.O. Metzger, J. Hartmans, D. Malwitz, and P. Koll, "Chemical Engineering at Supercritical Fluid Conditions", M. E. Paulaitis, J.M. L. Penninger, R.D. Gray, and P. Davidson, Eds., Ann Arbor, 1983, Chapter 26, 515.
44. H. Tiltscher, H. Wolf, and J. Schelchshorn, Ber. Bunsenges. Phys. Chem. 1984, 88, 897.

Supercritical Fluid Technology, edited by J.M.L. Penninger, M. Radosz, M.A. McHugh and V.J. Krukonis
1985 Elsevier Science Publishers B.V., Amsterdam — Printed in The Netherlands

BINARY DIFFUSION IN SUPERCRITICAL FLUIDS

P.G. DEBENEDETTI[1,2] and R.C. REID[1]

[1]Chemical Engineering Dept., Massachusetts Institute of Technology, Cambridge, MA 02139

[2]Present address: Chemical Engineering Dept., Princeton University, Princeton, NJ 08544

ABSTRACT

The different rates of change of density and viscosity in going from the dilute gas to the dense fluid state give rise to kinematic viscosities that, in the slightly supercritical region, are exceptionally low. As a consequence, the ratio of buoyant to inertial forces is more than two orders of magnitude higher than in ordinary liquids at any given Reynolds number, and natural convection often controls the rate and mechanism of mass transfer.

Experiments were designed to measure true diffusion coefficients (in the absence of buoyancy effects) as well as apparent diffusion coefficients (under the influence of variable buoyant forces), and significant differences were observed.

Hydrodynamic behaviour is approached at high pressure, and provides a useful framework for data analysis, correlation and extrapolation.

FLUID PROPERTIES AND NATURAL CONVECTION

Consider steady laminar duct flow of an incompressible Newtonian fluid under the influence of gravitational forces and an imposed pressure gradient. If the fluid is in contact with a surface from which a solute dissolves, the resulting concentration gradient will give rise to a density gradient that will alter the velocity profile. For small concentration gradients, we can use Boussinesq's approximation, i.e., we linearize the density about its pure fluid value and neglect changes in other fluid properties. The resulting dimensionless momentum balance reads.

$$\frac{2}{Re} (\nabla^+)^2 \underline{v}^+ - \underline{\nabla}^+ \Pi^+ - \underline{g}' \left(\frac{Gr}{Re^2}\right) \frac{c^+}{2} = 0 \qquad (1)$$

where the duct radius has been used as length scale.

With $\Delta\rho$ denoting the density difference between interface and pure fluid ($\rho_i - \rho_0$) we can introduce the natural scales for buoyant, viscous and inertial forces,

Buoyant forces ~ $2Rg\Delta\rho$ (2)

Viscous forces ~ $\eta\langle v\rangle/2R$ (3)

Inertial forces ~ $\langle v\rangle^2 \rho_0$ (4)

The physical significance of the parameter $GrRe^{-2}$ follows immediately,

$$\frac{Gr}{Re^2} = \frac{(2Rg\ \Delta\rho)\cdot(\langle v\rangle^2\rho_0)}{(\eta\langle v\rangle/2R)^2} \cdot \frac{(\eta\langle v\rangle/2R)^2}{(\langle v\rangle^2\rho_0)^2} = \frac{\text{Buoyant forces}}{\text{Inertial forces}} \quad (5)$$

As shown in Equation (5), therefore, this dimensionless quantity represents the ratio of buoyant to inertial forces. Consequently, if different fluids flow inside geometrically similar ducts under diffusive mass tranfer conditions, the relative importance of natural convection is inversely proportional to the square of the kinematic viscosity of the fluid in question, at any given Reynolds number and assuming comparable density changes.

Figure 1 shows the density, viscosity and kinematic viscosity of CO_2(T_c = 304.2 K; P_c = 73.8 bar) at 310 K. Although the density of supercriti-

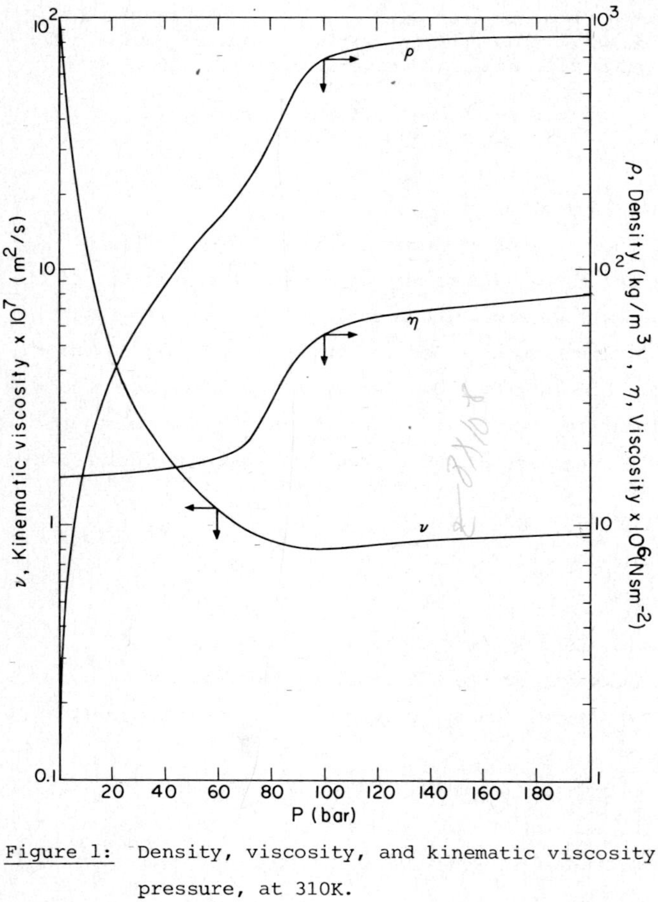

Figure 1: Density, viscosity, and kinematic viscosity of CO_2 as a function of pressure, at 310K.

cal CO_2 is liquid-like, its viscosity, being virtually pressure-independent at low pressures, is less than an order of magnitude higher than the corresponding dilute gas value. The kinematic viscosity then, is very small in this region.

In Figure 2 we compare the properties of air, water and mercury at ambient conditions with those of CO_2 at 150 bar and 310 K. We note the fact that, for ν, CO_2 has the lowest value (lower than for mercury, a liquid metal). The relative importance of buoyant forces at constant Reynolds number (scaled with the corresponding value for water) is therefore more than two orders of magnitude higher in a supercritical fluid than in normal liquids (4th column, Figure 2).

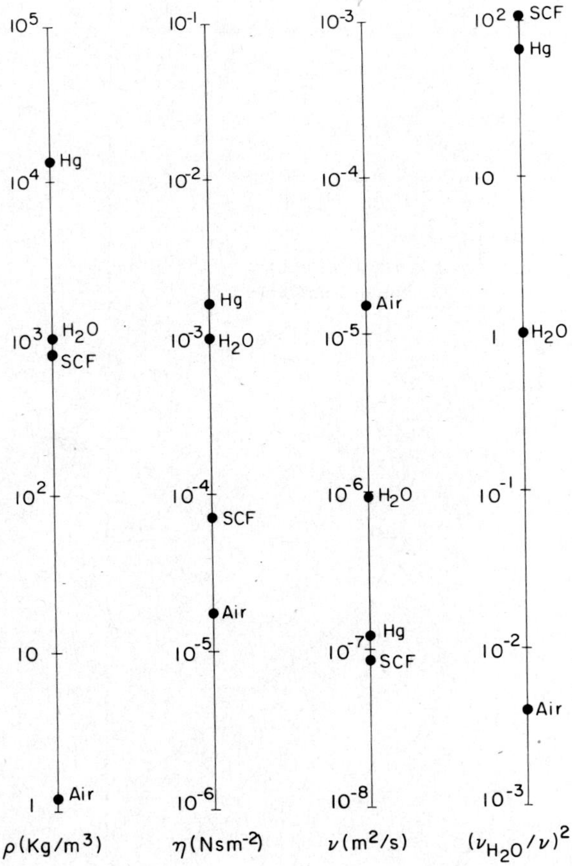

Figure 2: Comparison of physical properties of air, water, mercury (at 298K and 1 bar), and supercritical CO_2 (at 150 bar and 310K). Relative importance of natural convection at constant Reynolds numbers.

Some of the important consequences of the above discussion are summarized in Figure 3. For any value of $\Delta\rho/\rho$ (bulk-to-interface density change divided by mean density), the region lying above and to the right of the given curve represents tube geometries for which vertical, forced laminar flow is impossible due to the presence of natural convection.

For example, given a relative density change as small as 10^{-3}, any aspect ratio greater than 10^{-3} makes it impossible to attain forced laminar flow in an 8mm vertical duct.

Figure 3 is valid for $10^{-2} < Sc(2R/L) < 1$ and vertical flow. If the parameter Gr.Sc.(2R/L) (a scaled mass transfer Rayleigh number) exceeds 10^4,

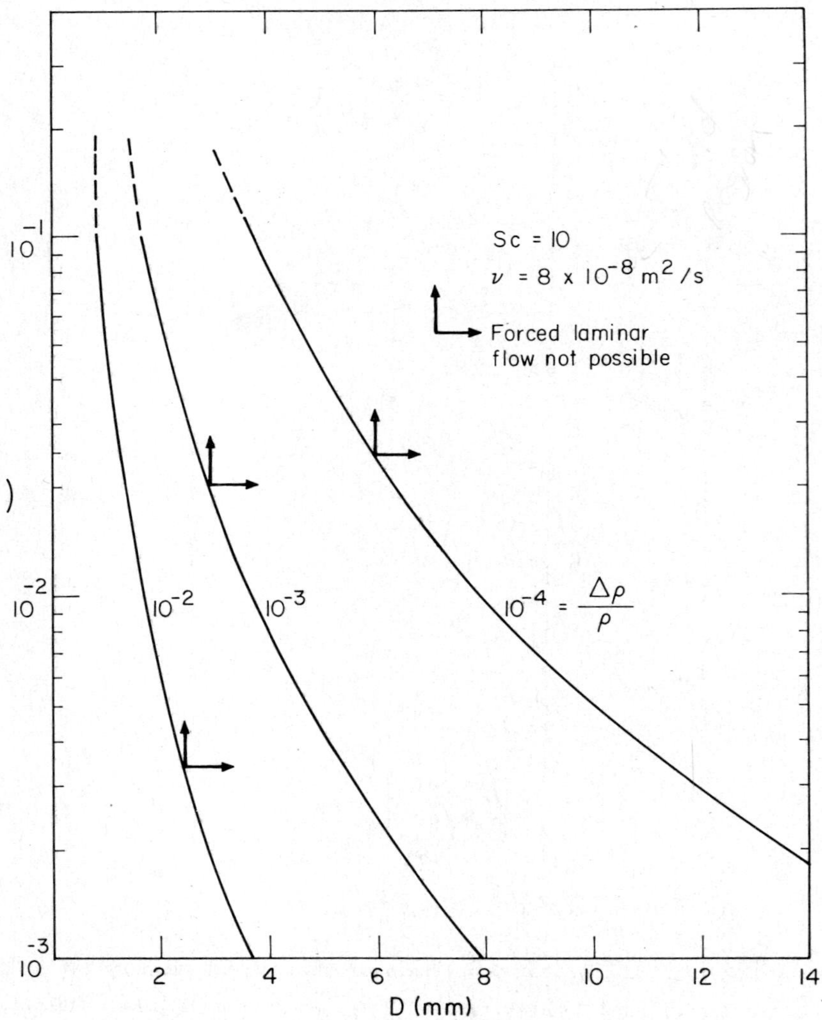

Figure 3: Cylindrical duct geometries for which vertical forced laminar flow is impossible under supercritical conditions.

forced laminar flow is impossible (Metais and Eckert, 1964); this criterion has been used to obtain Figure 3. The condition $Gr.Sc.2R/L < 10^4$ is a necessary but not sufficient criterion for forced laminar flow, so that the region lying below and to the left of each curve does not, by itself, imply laminar flow.

The kinematic viscosity is but one example of a physical property that, in the supercritical region, attains values which are not intermediate between the ideal gas and the liquid extremes. The dimensionless isothermal compressibility,

$$K'_T = \left(\frac{\partial \ln \rho}{\partial \ln P}\right)_T \tag{6}$$

is the relative density change per unit relative pressure change, at constant temperature. Figure 4 is a plot of the dimensionless isothermal compressibility versus pressure, with temperature as a parameter, for SF_6, as modelled with

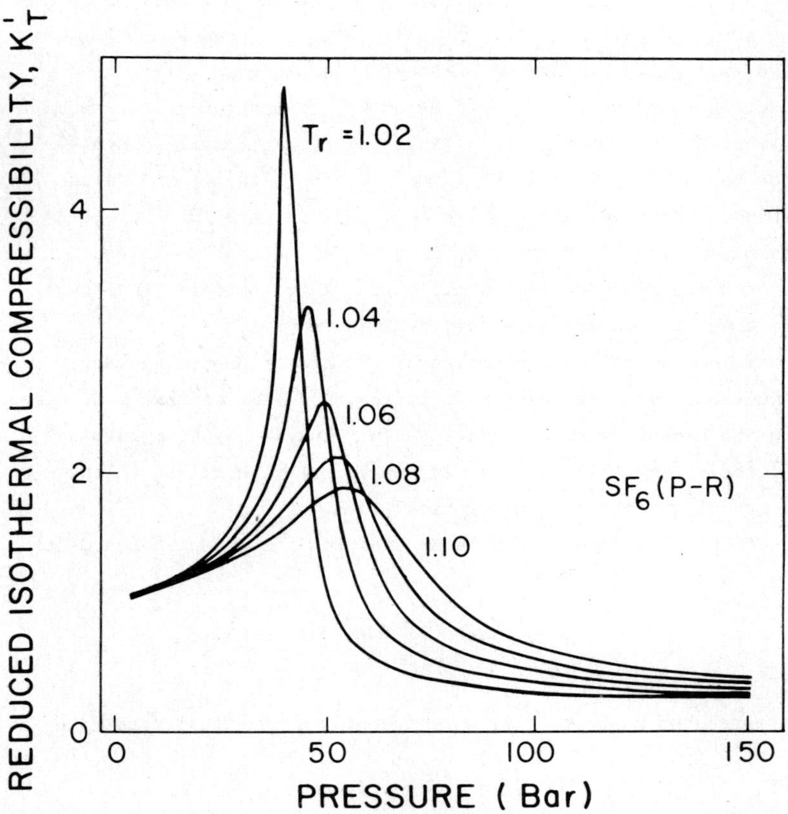

Figure 4: Dimensionless isothermal compressibility of SF_6, as modelled with the Peng-Robinson equation of state.

the Peng-Robinson equation of state (Peng and Robinson 1976). At 318 K and 100 bar (T_r = 1.05; Pr = 1.36), CO_2, for example, is ~ 280 times denser but almost three times more compressible than at 318 K and 1 bar.

These unique features of supercritical behaviour, although a consequence of the existence of the critical point, extend well beyond the region associated with non-classical, or critical, behaviour. For example, natural convection currents due to low kinematic viscosities are independent of fluid compressibility, which diverges at the critical point.

EXPERIMENTAL PROCEDURE

The experimental technique involved fully developed laminar flow of a supercritical fluid in a horizontal rectangular duct (Figure 5), the bottom surface of which was coated with the solute of interest. A brass plate (4) is tightly fitted into an enclosure made up of two aluminum hemi-cylinders (1,2); fluid flows inside the resulting channel (3).

The test section (6) is made by casting molten solute and carefully machining and polishing after solidification. The plate also contains a section (5) where laminar flow is allowed to develop, and an outlet section (7).

Fluid by-pass of the test section is prevented by a Viton gasket (8) which forces the plate against the upper surface (9'), and by the labyrinth seal (9), which results when the hemi-cylinders are forced together (10,11) and Teflon tape is placed between the upper and the lower mating surfaces (9).

The whole assembly is tightly fitted inside a 5cm (2 inches) stainless steel pipe. O-ring 13 provides sealing, while O-ring 12 is notched: the pressure in 3 is thus equal to the pressure outside 1 and 2.

When channel 3 is horizontal, and constitutes the duct's bottom surface, the flow is unaffected by gravity and binary diffusion coefficients can be measured by weighing the amount of solute that precipitates, upon decompression, from a measured quantity of solvent, at steady state. The solutions to the governing equations are given below in graphical form.

For dilute solutions of solute 1 in solvent 2, it can be shown (Debenedetti, 1984) that, if,

$$M_1/M_2 > \bar{V}_1/V_2 \qquad (7)$$

then the density decreases monotonically away from the solid-fluid interface.

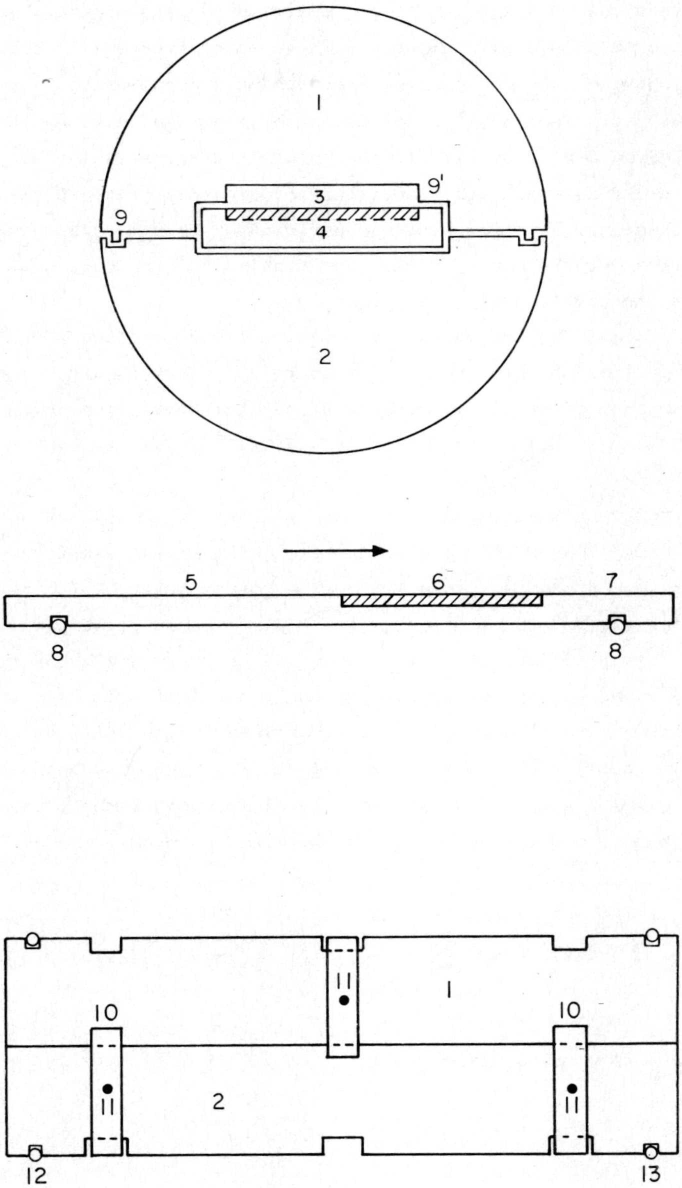

Figure 5: Schematic diagram showing rectangular duct assembly for hydrodynamic experiments.

In Equation (7), the solvent partial molar volume has been replaced by the pure (bulk) solvent molar volume, and the composition dependence of the solute partial molar volume has been neglected. These approximations are valid for dilute solutions, such as the ones we are presently considering. The above inequality was satisfied under all of the experimental conditions tested.

For any given binary system, if the solubility of the solute in the supercritical fluid decreases with temperature at constant pressure (retrograde solubility region), the dissolution process is exothermic and natural convection may arise even if Equation (7) is satisfied and the horizontal interface constitutes the duct's bottom surface.

Buoyant effects are introduced by rotating 1 and 2 inside the steel pipe. The same experiment should then give rise to different results, and information can be gathered on the importance of natural convection in supercritical fluids.

DIFFUSION IN RECTANGULAR DUCTS

Figures 6 and 7 show the solution to the differential equation governing steady diffusion of a solid solute from the bottom surface of a rectangular channel of height 2b and width 2a (where thermodynamic phase equilibrium is assumed at the solid surface) into a constant property Newtonian fluid, flowing axially under steady, laminar conditions.

This solution is valid for $b/a < 1/3$, and negligible axial or lateral diffusion. In addition, with $-2a < z < 2a$ and $-2b < y < 2b$, the quantities plotted in Figures 6 and 7 (i.e., a z^- averaged local Sherwood number, the cup-average relative saturation and a modified inverse Graetz number), are defined as follows:

$$Sh = \frac{4b}{(1+\alpha)D} \int_0^1 k\,d(z/a) = -\frac{4}{1+\alpha} \cdot \frac{1}{\langle c' \rangle} \int_0^1 \left(\frac{\partial c'}{\partial \xi^+}\right)_{\xi^+=2} d(z/a) \qquad (8)$$

$$\langle r \rangle = 1 - \frac{1}{4ab} \iint v^+ c' \, dy\, dz = 1 - \langle c' \rangle \qquad (9)$$

$$X_0 = xD/\langle v \rangle b^2 \qquad (10)$$

with

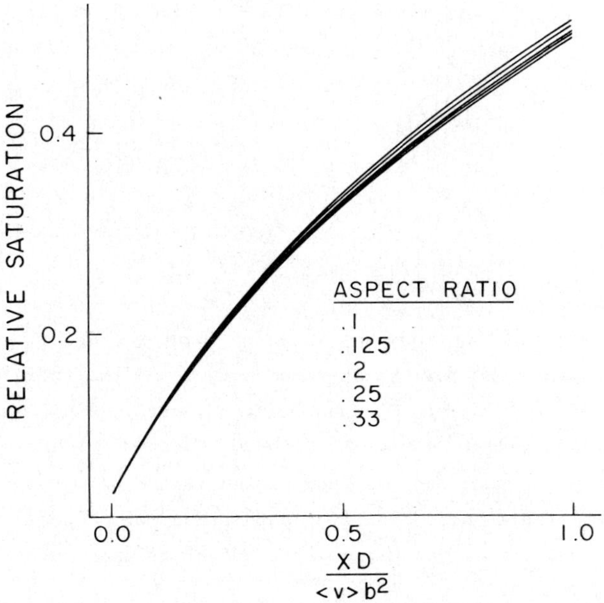

Figure 6: Relative saturation versus modified inverse Graetz number for various aspect ratios; rectangular duct flow. Solution valid for low ($\leq 1/3$) aspect ratios and high axial Peclet numbers.

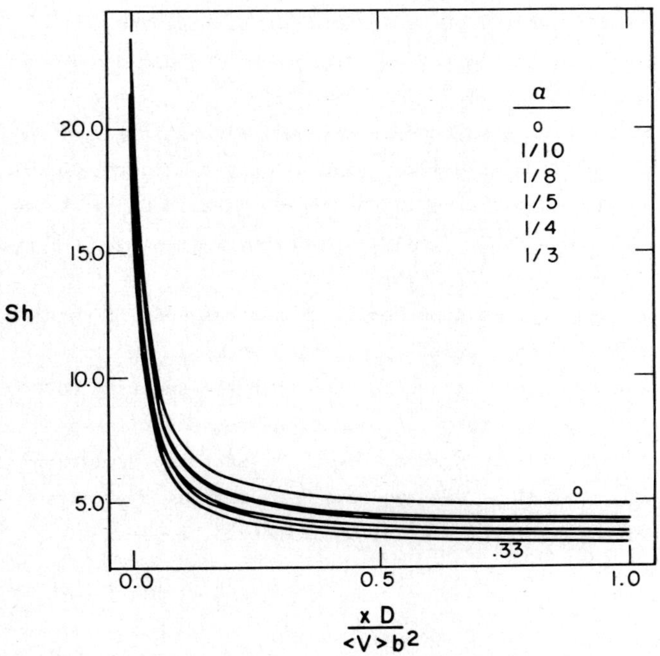

Figure 7: Z-averaged local Sherwood number, as a function of the modified inverse Graetz number for various aspect ratios; rectangular duct flow. Solution is valid for low ($\leq 1/3$) aspect ratios and high Peclet numbers.

$$\xi^+ = 1 - y/b \tag{11}$$

$$c' = 1 - c/c_i \tag{12}$$

$$v^+ = v/\langle v \rangle \tag{13}$$

$$\alpha = b/a \tag{14}$$

and x denoting the axial coordinate.

The derivation of the mathematical relationships shown in Figures 6 and 7, as well as the computational expressions for the Sherwood number and the relative saturation as a function of X_0 are given elsewhere (Debenedetti, 1984).

For a given aspect ratio, the relative saturation (and the Sherwood number) are functions of the single parameter X_0. For a given coated length (x), flow rate (<v>), and duct height (b), the measured relative saturation (which presupposes knowledge of c_i, the equilibrium solubility at T and P) is then only a function of the diffusion coefficient.

The equilibrium solubilities were measured with a flow technique similar to that used by Kurnik (1981).

RESULTS AND DISCUSSION

The measured diffusion coefficients of benzoic acid and naphthalene in supercritical sulfur hexafluoride are plotted as a function of fluid density in Figures 8 and 9 (see also Tables 1 and 2).

The measured diffusion coefficients of benzoic acid are always lower, at any given density, than the corresponding values for naphthalene, a larger molecule. This suggests a dimerization of benzoic acid in the fluid phase, a behaviour experimentally observed in CCl_4 and $CHCl_3$ (I'Haya and Shibuya, 1965), and in its vapour, C_6H_{12}, CCl_4 and C_6H_6 (Allen et al., 1966).

At any given density, the temperature dependence of the observed diffusion coefficients is higher than the $T^{\frac{1}{2}}$ hard-sphere prediction (Dymond, 1974). From Tables 3 and 4, on the other hand, it can be seen that the quantity ηDT^{-1} does not change significantly over the range of conditions tested (this is especially true for the benzoic acid-SF_6 system, where the standard deviation is only 4.2% of the mean ηDT^{-1} value). This suggests a hydrodynamic (Stokes-Einstein) description. The Stokes-Einstein equation (Einstein, 1905)

$$D = \frac{kT}{6\pi r\eta} \tag{15}$$

relates the diffusion coefficient of a spherical Brownian particle of radius r

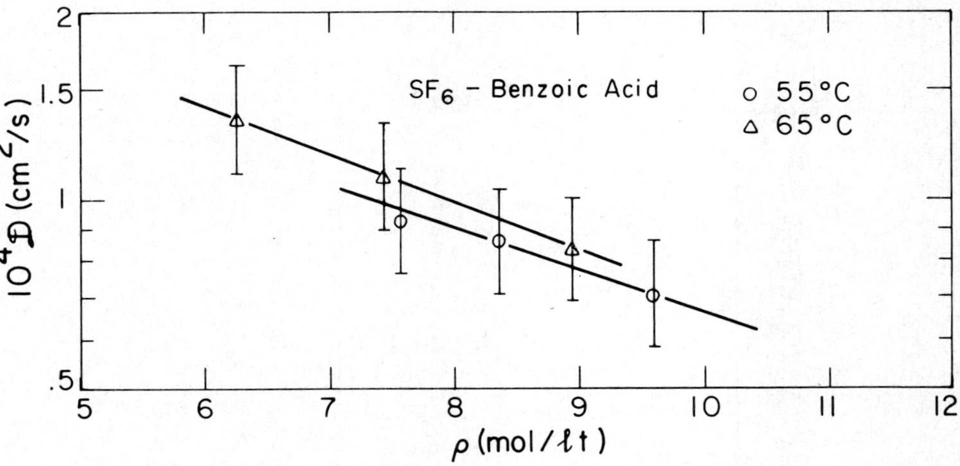

Figure 8: Diffusion coefficients of benzoic acid in supercritical SF_6, as a function of fluid density.

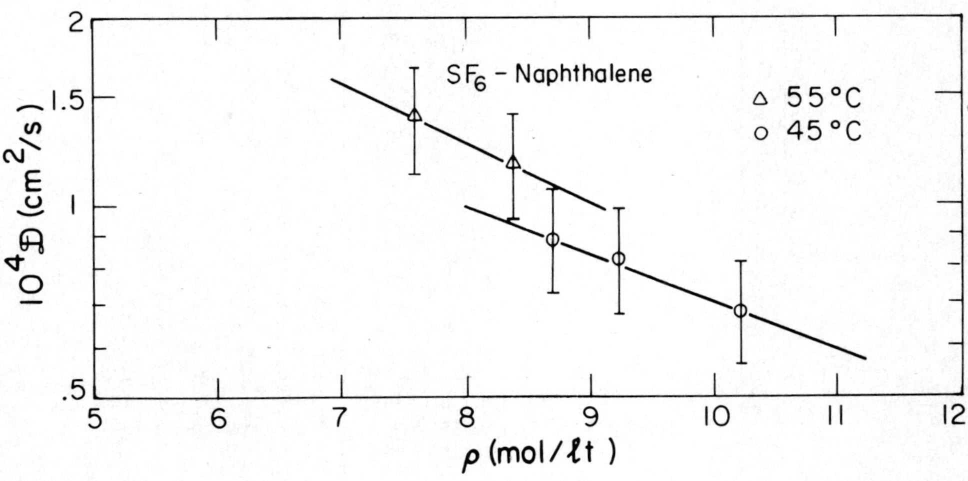

Figure 9: Diffusion coefficients of naphthalene in supercritical SF_6, as a function of fluid density.

TABLE 1 DIFFUSION COEFFICIENTS OF BENZOIC ACID IN SUPERCRITICAL SULFUR HEXAFLUORIDE

T (K)	P (bar)	T_r(b) (-)	P_r(b) (-)	ρ(a) (kg/m^3)	ρ(a) (mol/lt)	ρ_r(b) (-)	$10^5 D$ (cm^2/s)	$10^5 \rho D$ (kg/ms)
328.2	65	1.03	1.73	1106.20	7.59	1.50	9.26	1.02
328.2	80	1.03	2.13	1222.37	8.37	1.66	8.61	1.05
328.2	120	1.03	3.19	1400.42	9.59	1.90	7.02	0.98
338.2	65	1.06	1.73	917.11	6.28	1.24	13.40	1.23
338.2	80	1.06	2.13	1083.22	7.42	1.47	10.70	1.16
338.2	120	1.06	3.19	1307.25	8.95	1.77	8.15	1.06

(a) Peng-Robinson equation of state; pure SF_6

(b) Critical properties from R.C.Reid, J.M.Prausnitz, T.K.Sherwood, "The Properties of Gases and Liquids", 3rd edition, 1977, Mc Graw-Hill, New York

$T_r = T/T_c(SF_6)$

$P_r = P/P_c(SF_6)$

$\rho_r = \rho/\rho_c(SF_6)$

TABLE 2 DIFFUSION COEFFICIENTS OF NAPHTHALENE IN SUPERCRITICAL SULFUR HEXAFLUORIDE

T (K)	P (bar)	T_r(b) (-)	P_r(b) (-)	ρ(a) (kg/m^3)	ρ(a) (mol/lt)	ρ_r(b) (-)	$10^5 D$ (cm^2/s)	$10^5 \rho D$ (kg/ms)
318.2	65	1	1.73	1268.05	8.69	1.72	8.85	1.13
318.2	80	1	2.13	1349.04	9.24	1.83	8.33	1.13
318.2	120	1	3.19	1489.69	10.20	2.10	6.80	1.01
328.2	65	1.03	1.73	1106.20	7.59	1.50	13.80	1.53
328.2	80	1.03	2.13	1222.37	8.37	1.66	11.70	1.43

(a) Peng-Robinson equation of state; pure SF_6

(b) Critical properties from R.C.Reid, J.M.Prausnitz, T.K.Sherwood, "The Properties of Gases and Liquids", 3rd edition, 1977, Mc Graw-Hill, New York

$T_r = T/T_c(SF_6)$

$P_r = P/P_c(SF_6)$

$\rho_r = \rho/\rho_c(SF_6)$

to the temperature and viscosity of the surrounding fluid, when no slip exists at the solid-fluid interface.

Equation (15) implies the basic concept

$$\eta D T^{-1} = f[\text{size}] \qquad (16)$$

or, in other words, for hydrodynamic behaviour, a plot of D vs. η^{-1} should yield a straight line through the origin with a slope proportional to T. Figure 10 shows the benzoic acid-SF_6 results plotted in this manner. As can be seen, the lines have small but finite intercepts, indicating deviation from strictly hydrodynamic behaviour, a fact already noted by Feist and Schneider (1982) in connection with their studies of diffusion in supercritical CO_2.

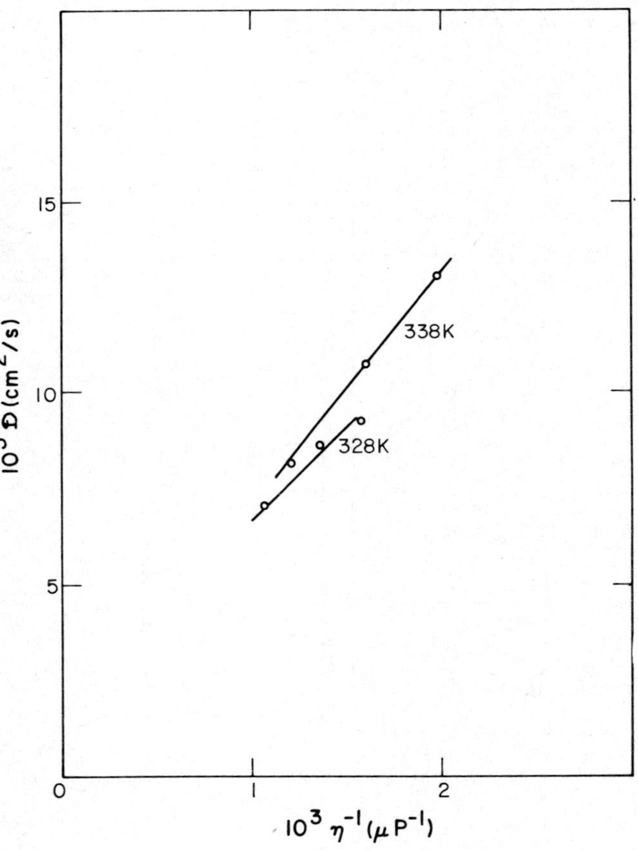

Figure 10: Diffusion coefficients of benzoic acid in supercritical SF_6, as a function of reciprocal viscosity.

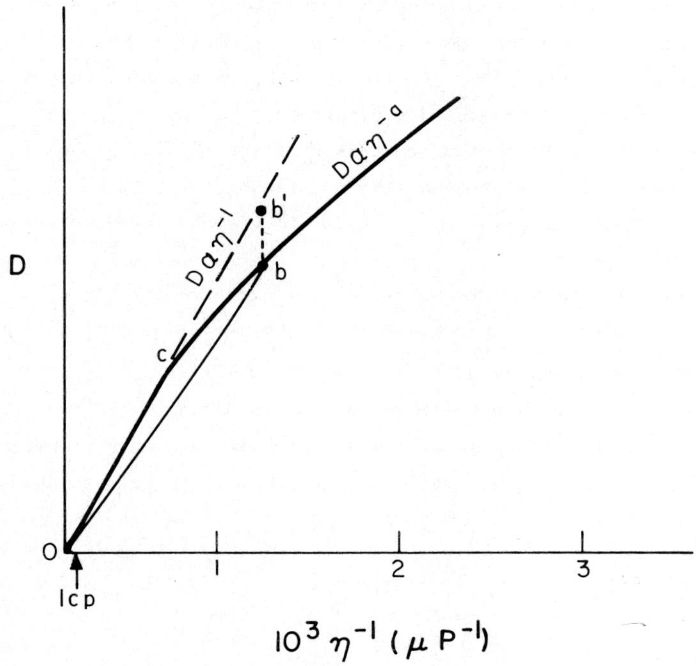

Figure 11: Isothermal viscosity dependence of the infinite dilution binary diffusion coefficient in the hydrodynamic and supercritical regimes.

The general picture that emerges is sketched in Figure 11. Hydrodynamic behaviour is approached at high viscosities; deviations from this limiting behaviour can be correlated (but not understood) by means of empirical power law relationships of the type $D \alpha \eta^{-a}$ $(a < 1)$ (Hayduck and Cheng, 1971). Supercritical viscosities fall roughly in the range $0.04 \leq \eta \leq 0.1$ cp for $1.1 \leq P_r \leq 4$ and $1 \leq T_r \leq 1.06$, which corresponds to $1 \leq 10^{-3}\ \eta^{-1} \leq 2.5$ in the units of Figure 11 (a typical liquid viscosity is shown for comparison).

The exact point at which hydrodynamic behaviour breaks down (point c) cannot, at present be predicted from first principles for any given system. However, from Figure 11 it can be concluded that predictive correlations based on the Stokes-Einstein equation (Wilke and Chang, 1955; Scheibel, 1954; Reddy-Doraiswamy, 1967; Lusis-Ratcliff, 1971) will overestimate diffusion coefficients in supercritical fluids. This was indeed found to be the case (Debenedetti, 1984), when each of the above correlations was applied to diffusion of naphthalene in supercritical CO_2 and ethylene (Iomtev and Tsekhanskaya, 1964), benzene in supercritical CO_2 (Swaid and Schneider, 1979), as well as 2-naphthol in CO_2, benzoic acid in SF_6, and naphthalene in SF_6 (Debenedetti, 1984).

In addition, at high enough viscosities (or, equivalently, at high enough pressures for any given temperature), the quantity ηDT^{-1} approaches a constant value; geometrically, this is equivalent to saying that, at small values of η^{-1}, the curve Ocb is well approximated by the line Ob. As an example, the measured diffusion coefficients of benzene in supercritical CO_2 (Swaid and Schneider, 1979) give rise to an ηDT^{-1} value that is constant to within a 4.6% standard deviation (expressed in percentage of the mean) when $\eta \geq 0.04$ cp, irrespective of the temperature and pressure (dense fluid viscosities were calculated as in Tables 3 and 4).

Thus, at high enough viscosities (or pressures, at any given temperature), hydrodynamic behaviour is approached, and this fact can be used to extrapolate experimental data by assuming constancy of ηDT^{-1}.

Buoyant effects can be introduced by performing the hydrodyanmic experiments with the two hemi-cylinders (Figure 5) rotated at an angle α with respect to the horizontal. The results of such experiments are shown, for benzoic acid diffusing into CO_2 at 160 bar and 318 K, in Figure 12.

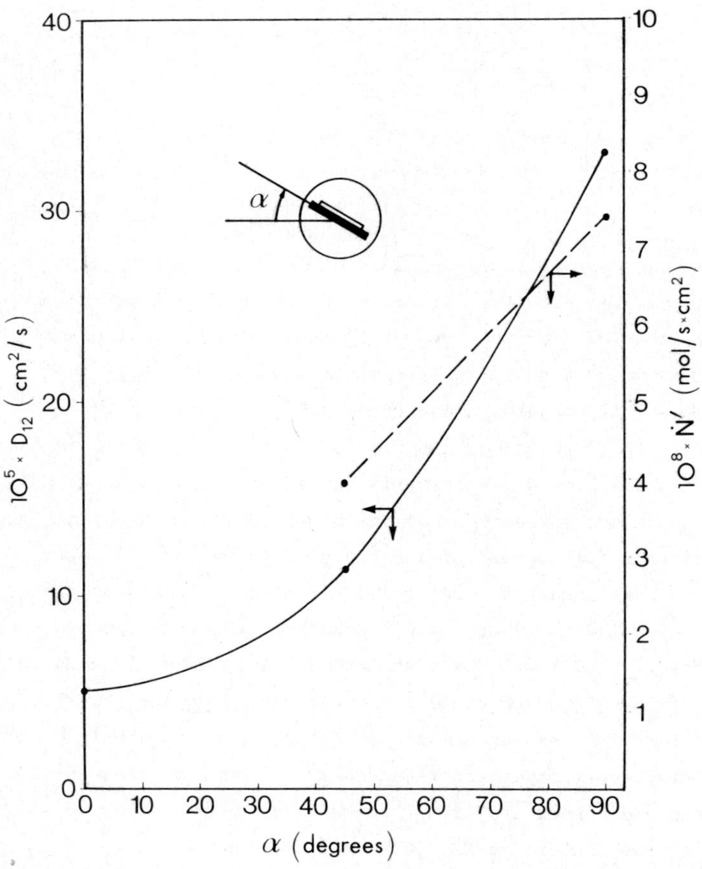

Figure 12: Effect of natural convection on mass transfer rate and apparent diffusion coefficient for the benzoic acid - CO_2 system.

TABLE 3 STOKES-EINSTEIN ANALYSIS FOR THE SF_6-BENZOIC ACID SYSTEM

T (K)	P (bar)	η^o (a) (μP)	ρ_r (-)	$10^5 D$ (cm^2/s)	Δ (b),(c) (-)	η (μP)	$10^5 D\eta T^{-1}$ (d) ($cm^2 \mu P/sK$)
328.2	65	166.91	1.50	9.26	1.787	633.65	17.88
328.2	80	166.91	1.66	8.61	1.868	737.70	19.35
328.2	120	166.91	1.90	7.02	2.001	934.63	19.99
338.2	65	171.64	1.24	13.04	1.656	504.65	19.99
338.2	80	171.64	1.47	10.7	1.772	624.12	19.75
338.2	120	171.64	1.77	8.15	1.926	823.34	19.84

(a) Low pressure viscosity from Equations (9-3.9), (9-4.3) and Appendix C; R.C.Reid, J.M.Prausnitz T.K.Sherwood, "The Properties of Gases and Liquids", 3rd edition, 1977, Mc Graw-Hill, New York

(b) $\Delta \equiv [(\eta-\eta^o)\xi + 1]^{.25}$

$\xi \equiv T_c^{1/6} / M^{1/2} P_c^{2/3}$

$[T_c]$ = K

$[P_c]$ = atmospheres

(c) Dense fluid viscosity from Equation (9-6.4); R.C.Reid, J.M.Prausnitz, T.K.Sherwood, "The Properties of Gases and Liquids", 3rd edition, 1977, Mc Graw-Hill, New York

(d) Mean = 19.467; Standard deviation = 0.812 (4.17 % of mean)

TABLE 4 STOKES-EINSTEIN ANALYSIS FOR THE SF_6 - NAPHTHALENE SYSTEM

T (K)	P (bar)	η^o (a) (μP)	ρ_r (-)	$10^5 D$ (cm^2/s)	Δ (b),(c) (-)	η (μP)	$10^5 D\eta T^{-1}$ (d) ($cm^2 \mu P/sK$)
318.2	65	162.14	1.72	8.85	1.899	775.25	21.56
318.2	80	162.14	1.83	8.33	1.959	864.11	22.62
318.2	120	162.14	2.10	6.80	2.135	1172.22	25.05
328.2	65	166.91	1.50	13.80	1.787	633.65	26.64
328.2	80	166.91	1.66	11.70	1.868	737.70	26.30

(a) Low pressure viscosity from Equations (9-3.9), (9-4.3) and Appendix C; R.C.Reid, J.M.Prausnitz T.K.Sherwood, "The Properties of Gases and Liquids", 3rd edition, 1977, Mc Graw-Hill, New York

(b) $\Delta \equiv [(\eta-\eta^o)\xi + 1]^{.25}$

$\xi \equiv T_c^{1/6} / M^{1/2} P_c^{2/3}$

$[T_c]$ = K

$[P_c]$ = atmospheres

(c) Dense fluid viscosity from Equation (9-6.4); R.C.Reid, J.M.Prausnitz, T.K.Sherwood, "The Properties of Gases and Liquids", 3rd edition, 1977, Mc Graw-Hill, New York

(d) Mean = 24.434; Standard deviation = 2.252 (9.2 % of mean)

The dotted line (mass transfer rate) in Figure 12 does not extend to 0° since the diffusion (0°) experiment was done at a different flow rate, and mass transfer rates are a function of fluid velocity. The importance of natural convection, as well as the potential for experimental error when using hydrodynamic techniques, can be seen from the fact that a 650% increase in the apparent diffusion coefficient results from a 90° rotation of the flat plate.

CONCLUSION

Significant rate enhancements due to natural convection will result in supercritical extraction processes whenever the controlling resistance to mass transfer is in the supercritical phase. This is a consequence of the unusually low kinematic viscosities of fluids in the near supercritical region.

Hydrodynamic behaviour at the molecular level is approached as a high density (or viscosity) limit, and can be used to correlate experimental data. The breakdown of hydrodynamic behaviour is currently poorly understood and deserves further attention.

REFERENCES

G. Allen, J.G. Watkinson, and K.H. Webb, Spectroch. Acta, 22, (1966), 807.
P.G. Debenedetti, PhD Thesis, Massachusetts Institute of Technology, Cambridge, MA, (1984).
J.H. Dymond, J. Chem. Phys., 60, (1974), 969.
A. Einstein, Ann. d. Phys., 17, (1905), 549.
R. Feist, and G.M. Schneider, Sep. Sci. Tech., 17(1), (1982), 261.
W. Hayduk, and S.C. Cheng, Chem. Eng. Sci., 26, (1971), 635.
M.B. Iomtev, and Y.V. Tsekhanskaya, Russ. J. Phys. Chem., 36(4), (1964), 485.
Y. I'Haya, and T. Shibuya, Bull. Chem. Soc. Japan, 38(7), (1965), 1144.
R.T. Kurnik, ScD Thesis, Massachusetts Institute of Technology, Cambridge, MA, (1981).
M.A. Lusis, and G.A. Ratcliff, Can. J. Chem. Eng., 46, (1968), 385.
B. Metais, and E.R.G. Eckert, J. Heat Transfer, 86, (1964), 295.
D.Y. Peng, and D.R. Robinson, Ind. Eng, Chem. Fundam., 15(1), (1976), 59.
K.A. Reddy, and L.K. Doraiswamy, Ind. Eng. Chem. Fundam., 6, (1967), 77.
E.G. Scheibel, Ind. Eng. Chem. Fundam., 46, (1954), 2007.
I. Swaid, and G.M. Schneider, Ber. Bunsenges. Phys. Chem., 83, (1979), 969.
C.R. Wilke, and P. Chang, AIChE J., 1, (1955), 264.

NOTATION

a = rectangular duct half-width; exponent for D vs. η^{-1} relationship

b = rectangular duct half-height

c = solute molar concentration (moles/volume)

c^+ = dimensionless molar concentration ($=c/c_i$)

c' = $1 - c^+$

D = diffusion coefficient; duct diameter in Figure 3

Gr = Grashof number ($=g(2R)^3(\Delta\rho/\rho)/\nu^2$)

g = acceleration due to gravity

$\underline{g}'$	= unit vector; parallel to the direction of gravity
K'_T	= dimensionless isothermal compressibility (defined in Equation (6))
k	= mass transfer coefficient
$\langle k \rangle$	= z-averaged mass transfer coefficient
L	= mass transfer active length of duct (coated length)
M	= molecular weight
P	= pressure
R	= duct radius
Re	= Reynolds number ($=2R\langle v\rangle/\nu$)
$\langle r \rangle$	= cup-average relative saturation
r	= radius of a Brownian sphere
Sc	= Schmidt number ($=\nu/D$)
Sh	= z-averaged Sherwood number for a rectangular duct ($=4b\langle k\rangle/(1+\alpha)D$)
T	= absolute temperature
V	= volume
v	= axial velocity in duct flow
$\langle v \rangle$	= mean axial velocity in duct flow
v^+	= dimensionless axial velocity ($= v/\langle v\rangle$)
X_0	= modified inverse Graetz number ($= xD/\langle v\rangle b^2$)
x	= axial coordinate in rectangular duct flow
y	= "height" coordinate in rectangular duct flow
z	= width coordinate in rectangular duct flow

GREEK SYMBOLS

α	= b/a; angle with respect to horizontal plane (Figure 12)
η	= viscosity
ν	= kinematic viscosity ($=\eta/\rho$)
π	= modified pressure ($=P + \rho_0 gh$)
π^+	= dimensionless modified pressure ($= \pi/\rho_0\langle v\rangle^2$)
ρ	= fluid density
ξ	= b - y
ξ^+	= ξ/b

<u>Subscripts (or underbars)</u>

1	= solute
2	= solvent
-	= vector
c	= critical property
i	= interface
o	= bulk
r	= reduced property

<u>Superscripts (or overbars)</u>

+	= dimensionless quantity
-	= partial molar quantity

Supercritical Fluid Technology, **edited by J.M.L. Penninger, M. Radosz, M.A. McHugh and V.J. Krukonis**
1985 Elsevier Science Publishers B.V., Amsterdam — Printed in The Netherlands

MASS TRANSFER IN GAS EXTRACTION

G. Brunner
Technical University Hamburg-Harburg, Eissendorfer Str. 38, 2100 Hamburg 90
W-Germany

ABSTRACT

Mass transfer rates have been observed in a solid-dense gas system and in a liquid-dense gas system.
Extraction rates of substances from solid substrates by dense supercritical gases have been determined for coffee beans and for rape oil seeds. In case of the indestructed coffee beans an unsteady extraction process prevails, which is determined by the transport of caffeine in the solid material. An approximate solution of the differential equations and a steady state approach are applicable to the long time part of the extraction. In case of the crushed seeds a steady state mass transfer prevails, which can be calculated by empirical mass transfer correlations for a fixed bed of particles.

In the overall fluid system mass transfer rates for ethanol-water mixtures are reported. Though Sherwood numbers are high, the effective mass transfer coefficients are about the same as in gas-liquid systems under normal pressure, since the diffusivity in the high pressure system is orders of magnitude lower.

INTRODUCTION

Transport coefficients in Chemical Engineering preferably are presented in form of dimensionless groups, which are used for empirical correlations. The Sherwood number in general is a function of the properties of the substances involved, expressed by the Schmidt number and of the flow regime, partly expressed by the Reynolds number. Since not all influences of the various mass exchanging devices thus can be taken into account, different dimensionless correlations are developed. In addition, many include a load factor. Another concept is that of stage efficiency, in which the mass transfer by a mass transfer device is related to the concept of ideal stage. It is determined by flow rates and phase equilibria. In order to get reliable correlations a great number of experimental data are needed, which are not available so far for the separation process of gas extraction.

On several occasions SCHNEIDER (1) has pointed out that in gas extraction favourable conditions for mass transport should occur since transport properties of the substances are similar to those of gases and density and therefore equilibrium properties are similar to those of liquids.
RANCE and CUSSLER (2) measured the rate of dissolution of iodine in supercritical carbon dioxide. According to their results, mass transfer rates are high at high temperatures and pressures. Mass transport rates increase with increasing pressure or density. Mass transport across phase boundaries is determined by the molecular transport rate (D) and the equlibrium solubility. Since at near critical states solubility strongly depends on temperature, increasing molecular transport rates by raising the temperature often are more than outweighed by the decreasing solubility. KING et al. (3) argued on base of estimating transport properties for certain examples that mass transport in gas extraction in fluid systems should be enhanced by a factor up to 20 and more as compared to conventional low pressure separation processes.

This paper is concerned with mass transport in gas extraction of light organic compounds from aqueous solutions and the extraction of compounds from solid substrates.

MASS TRANSFER AT THE SEPARATION OF ORGANIC SUBSTANCES FROM AQUEOUS SOLUTIONS: ETHANOL-WATER

The separation of ethanol from aqueous solutions is an operation connected with the production of organic chemicals with biotechnological methods. The resulting aqueous solutions have concentrations of ethanol up to about 10 wt%. The separation process is energy intensive and essential for the overall economy of the production process.

Supercritical gases at elevated pressures selectively dissolve organic compounds from aqueous solutions. Therefore gas extraction has been considered applicable for separating ethanol from water (4). In order to judge the feasibility of the process, phase equilibria and mass transfer rates are needed. Phase equilibria of mixtures of water, ethanol and carbon dioxide were measured among other by PAULAITIS et al. (5), phase equilibria of the system water-ethanol-ethane by McHUGH et al. (6). Data on mass transfer so far have not been published. In this work mainly results of experiments on mass transfer are presented. The data have been measured by K. KREIM at the Technical University of Hamburg-Harburg (7).

Phase Equilibria

Some phase equilibria data are presented first, since they are essential for understanding mass transport. In fig. 1 phase equilibria of the system water-ethanol-carbon dioxide at a pressure of 15 MPa and temperatures of 65 and 85°C are plotted. Under these conditions uniform two phase regions exist. Only one binary system (water-CO_2) exhibits a miscibility gap. Equilibrium data for the gas phases are concentrated in the lower part on the right hand side of the triangular diagram. This results in the high selectivity, mentioned above, but also in a relatively low solubility of ethanol/water in the gas as compared to liquid solvents. Fig. 2 presents experimental data on the system water-ethanol-ethane at a pressure of 15 MPa and temperatures of 65 and 80°C. Gas phase data are similar to those in the CO_2-system.

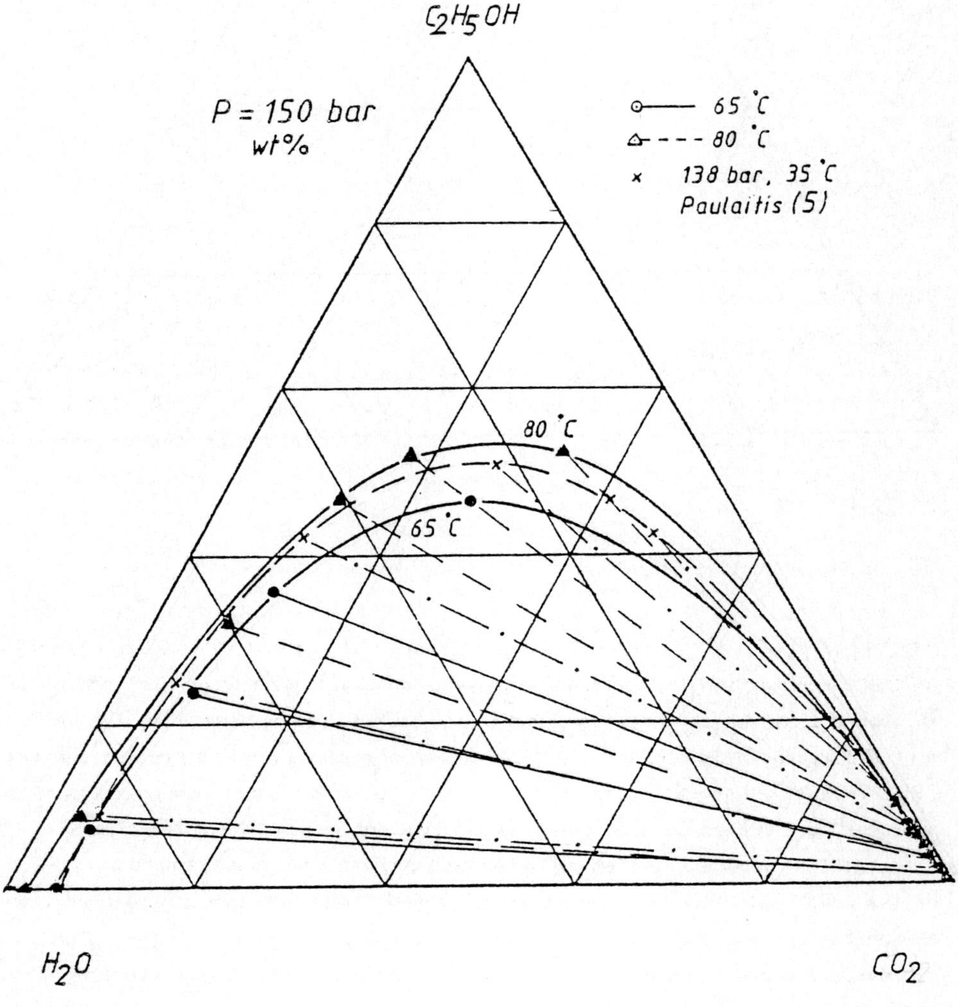

Fig. 1: Phase equilibria of water-ethanol-carbon dioxide

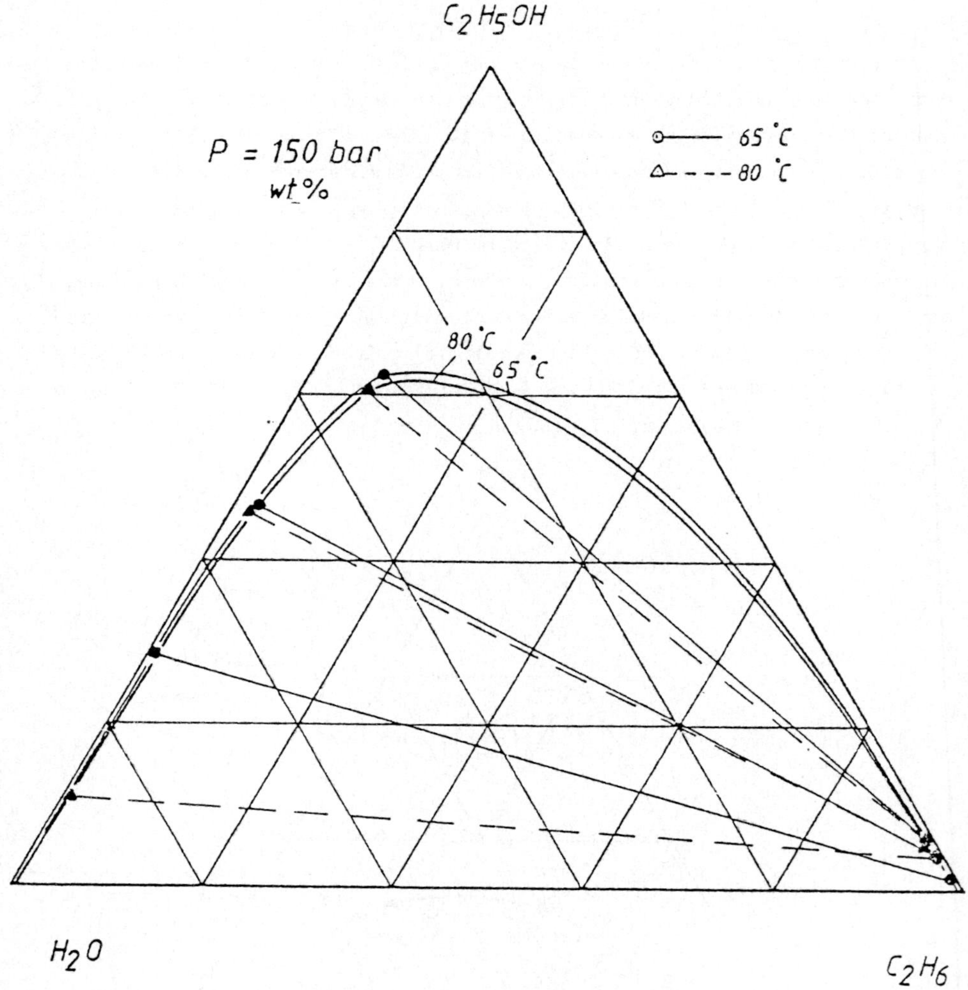

Fig. 2: Phase equilibria of water-ethanol-ethane

Mass Transfer

The investigation for separating ethanol and water have been carried out in an apparatus in which the phases can be contacted countercurrently. The set up consists essentially of two columns and the necessary pumps to provide the mass flows (fig. 3). For the investigation of mass transfer rates one column was operated as a bubble column. For this purpose the column (height 3.4 m; inner diameter 17×10^{-3}m) was filled at operating pressure up to one third of its height with the mixture to be separated. Then pure gas was bubbled from the bottom through the column. The gas was removed at the top of the column, depressurized and separated in a liquid and gaseous part. Dimensions and column internals are shown in fig. 4. The effect of pressure and temperature on

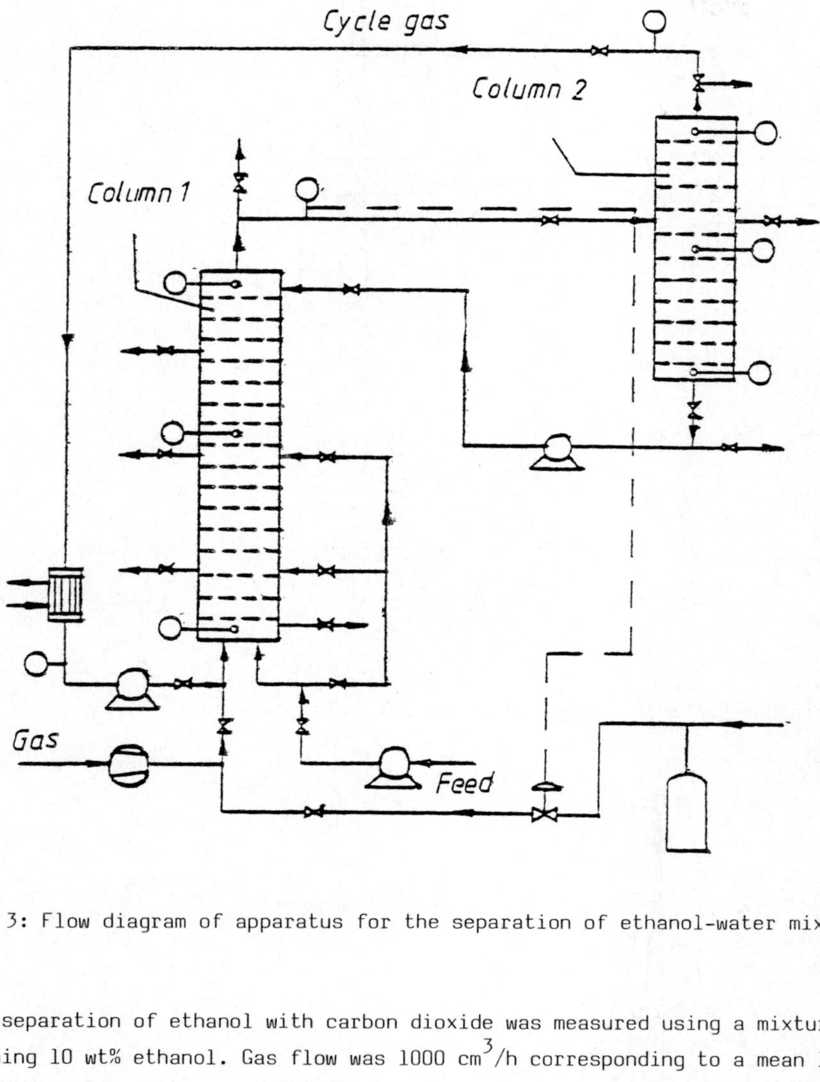

Fig. 3: Flow diagram of apparatus for the separation of ethanol-water mixtures

the separation of ethanol with carbon dioxide was measured using a mixture containing 10 wt% ethanol. Gas flow was 1000 cm^3/h corresponding to a mean linear velocity of the gas, related to the free column area, of 1.2 mm/s. In fig. 5 the depletion in ethanol of the liquid is shown for temperatures between 20^oC and 95^oC. With increasing temperature the extraction accelarates. This is due

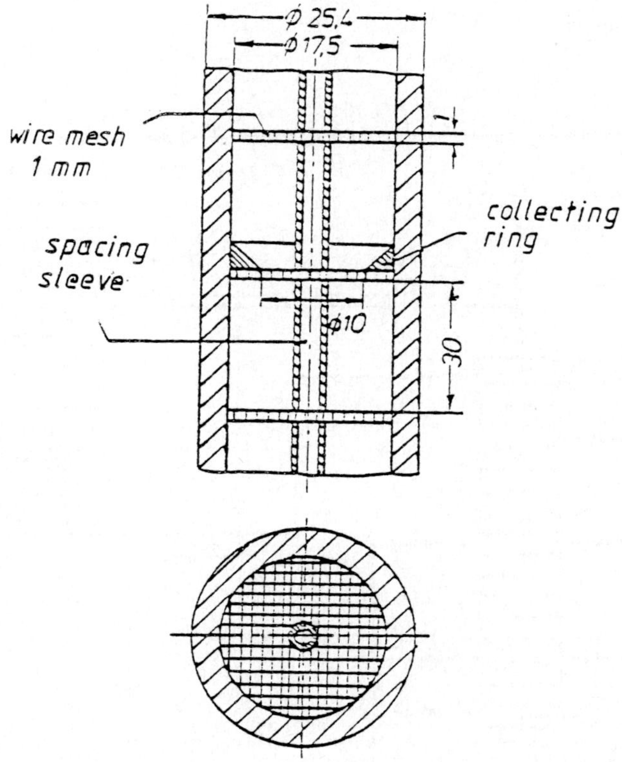

Fig. 4: Cross sections of separation column

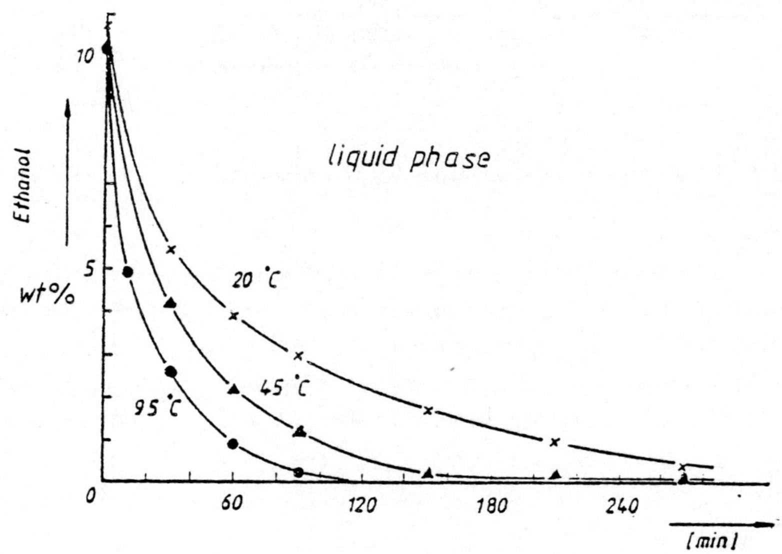

Fig. 5: Ethanol concentration of the liquid. H_2O-C_2H_5OH-CO_2; 155 bar; 1000 $cm^3/h \hat{=} 0.12$ cm/s

to the likewise effects of enhanced diffusivity and enhanced solubility of ethanol in carbon dioxide at higher temperatures (fig. 6). Ethanol concentration in the gas phase on a carbon dioxide free basis shows the selectivity of the separation. It is shown in fig. 7 as a function of extraction time and therefore also ethanol concentration in the liquid. With decreasing ethanol concentration in the liquid, the ethanol-concentration in the gas phase decreases also, the depletion being faster at higher temperatures. As can be concluded from phase equilibria data, there is an effect of pressure also. At a temperature of 65°C and pressures of 10 MPa and 15,5 MPa, higher mass transfer rates occurred at the higher pressure level. The corresponding equilibrium values for the solubility of ethanol and water in the gas phase are:

$$T = 65^{o}C: \quad P = 10\ MPa: \quad Y_{Eth} = 8.5\ wt\%;$$

$$Y_{Eth/w} = 0.7\ wt\%$$

$$P = 15.5\ Mpa: \quad X_{Eth} = 7.1\ wt\%;$$

$$Y_{Eth/w} = 1.8\ wt\%$$

The influence of ethanol concentration in the liquid on the ethanol concentration in the gas phase is shown clearly in fig. 8. The lower the initial ethanol content in the liquid the lower is also the ethanol concentration in the gas. Simultaneously the solubility in the gas phase is lowered. At a temperature of 95°C and a gas flow rate of 9100 m^3/h, corresponding to a mean linear gas velocity of 11.1 mm/s the solubility of ethanol in the gas phase at an initial concentration of 10 wt% ethanol is 1.2 wt%, which decreases to 0.15 wt% at an initial concentration of 1 wt%.

Most interesting for mass transport is the ratio of gas flow rate and quantity of liquid. At a pressure of 15.5 MPa, a temperature of 95°C an an initial concentration of 10 wt% ethanol in the liquid gas flow rate was changed resulting in a linear gas velocity from 3.3 mm/s up to 23.1 mm/s. Results of these runs are shown in fig. 9, in which the effect of gas flow rate on extraction time needed for a certain yield of ethanol have been plotted. As could be anticipated the general tendency is that with increasing gas flow rate extraction time decreases. For a yield of 90 % even the highest gas flow rates make sense, since extraction time further decreases with higher mass flow rates. But for yields up to about 60 % gas flow rates beyond 12 mm/s practically have no effect on extraction time.

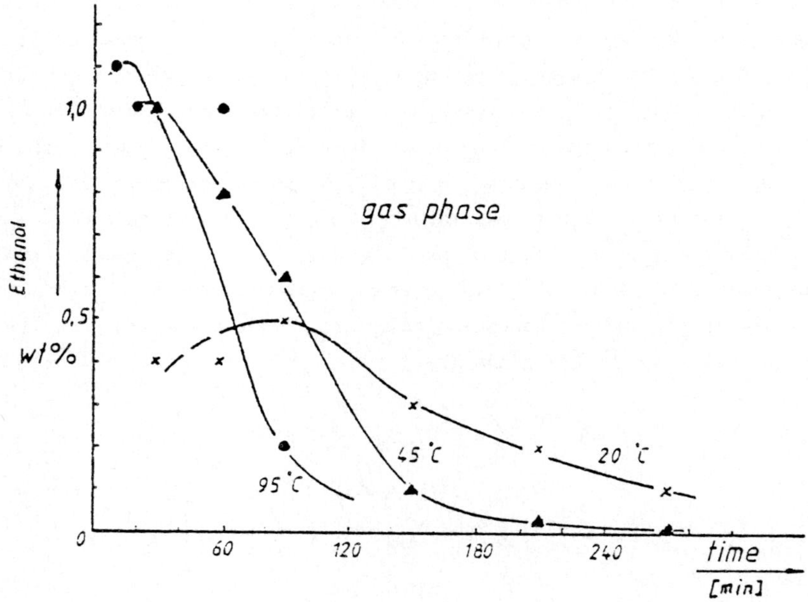

Fig. 6: Ethanol concentration of the gas phase, $H_2O-C_2H_5OH-CO_2$; 155 bar; 1000 cm^3/h ≙ 0.12 cm/s

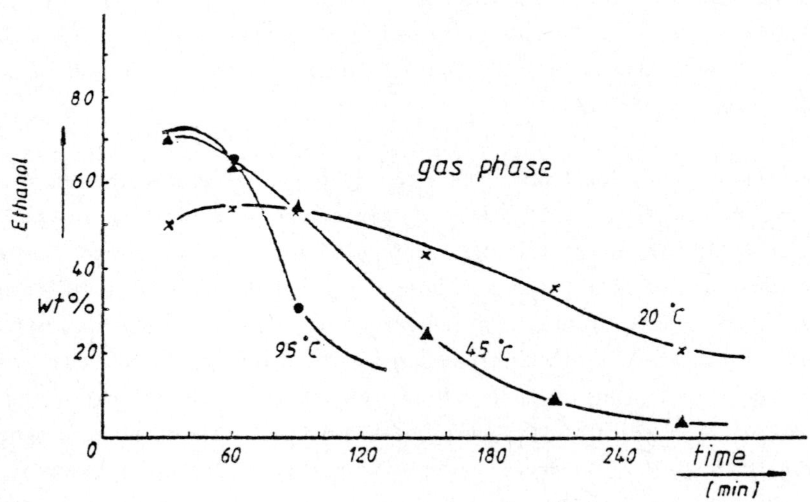

Fig. 7: Ethanol concentration in gas phase on gas free basis. $H_2O-C_2H_5OH-CO_2$; 155 bar; 1000 cm^3/h ≙ 0.12 cm/s.

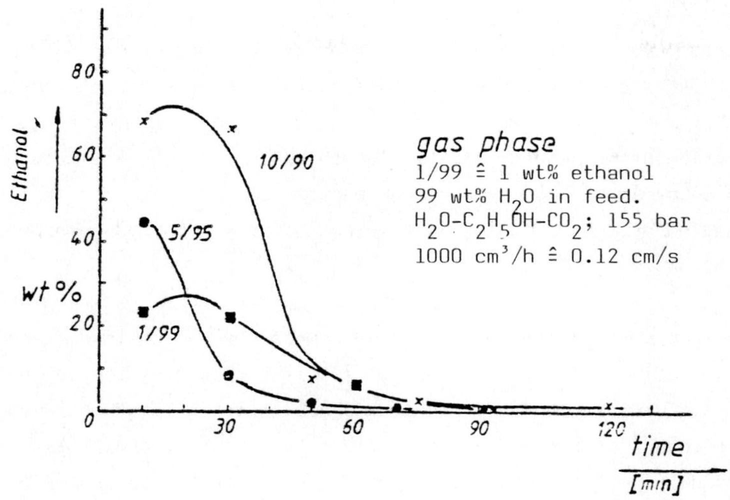

Fig. 8: Effect of feed concentration on ethanol concentration in gas phase.

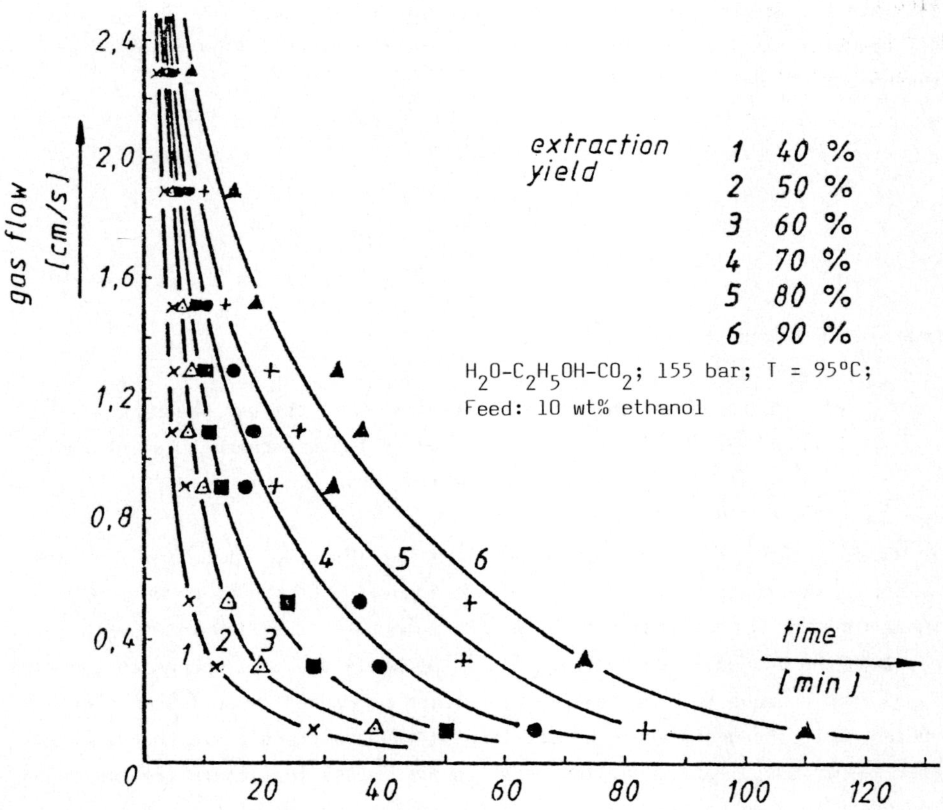

Fig. 9: Extraction yield versus gas flow and extraction time.

Dimensionless Mass Transfer Numbers

For the calculation of dimensionless mass transfer numbers from experimental data like those presented above, two difficulties and therefore uncertainties arise from

- determination of the characteristic length, which in a bubble column is reduced to determination of a mean bubble diameter, and
- determination of the diffusion coefficient of the extracted compounds in the extracting gas.

Bubbles were photographed with a movie camera through a sight glass at the bottom of the column. The mean bubble diameter at this position was determined to 3×10^{-3} m. This diameter was taken as characteristic length for mass transfer calculations. For the diffusion coefficient of the ethanol/water mixture in carbon dioxide as a first approximation the value for the diffusion of phenol in carbon dioxide was taken from a work by FEIST and SCHNEIDER (8):
$6.1 \quad 10^{-5}$ m^2/h at $P = 15.5$ MPa and $T = 95^{\circ}C$. Parameters which have been used to calculate the dimensionless numbers are summarized in table 1.
Mass transfer coefficients are calculated from the ethanol flow and the mean driving potential:

$$K = m/\Delta Cm$$

$$\Delta C_m = \frac{\Delta C_{in} - \Delta C_{out}}{\ln \frac{\Delta C_{in}}{\Delta C_{out}}}$$

wherein: in - mass flow ethanol

$\Delta C = y^* - y$

y^* = equilibrium concentration of ethanol in the gas phase

y = actual concentration of ethanol in the gas phase taken at the inlet and outlet position of the gas.

The calculated Sh-numbers are plotted in fig. 10 as a function of the Re-number. In the range of $70 < Re < 10^3$ the logarithm of the Sh-number is a linear function of the logarithm of the Re-number. At gas flow rates with $Re > 1000$ the Sh-number decreases. This is probably due to a change in the flow regime from bubbles to a froth-like dispersion or caused by coalescence of bubbles. Both phenomena would be detrimental to mass transfer in the small diameter column used. For comparison mass transfer rates for absorption und desorption processes are considered. For the desorption of ethanol from an aqueous solution containing 6.5 wt% ethanol in a packed column filled with BERL-saddles

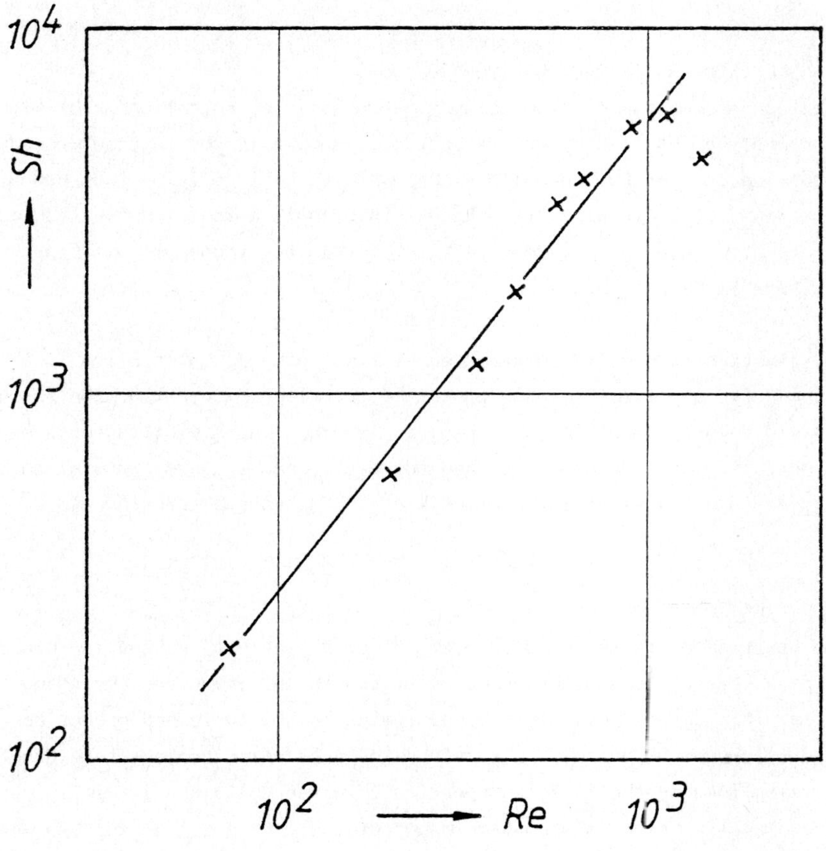

Fig. 10: Mass transfer coefficients for ethanol in the water-ethanol-carbon dioxide system; bubble column; 155 bar, 95°C; Feed concentration of ethanol 10 wt%; (Kreim 1983)

at Re = 1800 a gas side mass transfer coefficient of k = 86.4 m/h corresponding to Sh = 72.8 (30°C; 0.1 MPa) (9), during the drying process of natural gas with triethylene glycol in a tray column a mass transfer coefficient of k $\approx$ 17 m/h (30°C; 6.9 MPa) was measured (10).

MASS TRANSFER FOR GAS EXTRACTION FROM SOLIDS

Separation of substances from solid substrates is a separation problem, which can effectively be handled by gas extraction, especially for temperature sensitive substances. For two examples - the extraction of caffeine from beans and the extraction of edible oils from seeds - this application of gas extraction has been treated in details elsewhere (11). Within the scope of this paper only the results will be presented.

The extraction from solid substrates, as treated here, takes place in a fixed bed of the solid particles through which the gas flows. The extraction process can be divided into two different steps: 1. Transport of the extracted substances in the solid phase to the interface solid-gas, and 2. mass transfer through the interface and transport into the bulk of the gaseous extracting medium.

Extraction of Caffeine

The extraction of caffeine is an example for an unsteady extraction process. The solid substrate, the coffee beans, must remain undestructed, including the cell walls. This determines the transport paths in the solid and cannot be shortened by crushing the solid. Therefore the extraction process to a great extent is determined by the mass transport rate in the solid phase. Within certain limits, especially in the medium time range of the extraction process, however, the extraction is dependent on the fluid flow of the external extracting medium. Fig. 11 shows a typical extraction curve. For designing the extraction apparatus one is interested to model these extraction curves. The process can be described by a set of coupled differential equations. A general solution for this set of equations so far is not known. Approximate solutions can be obtained at least in two ways: 1. as approximate solutions of the unsteady transport process, and 2. by a quasi steady analysis. For this work both approximate methods have been adapted from solutions of unsteady heat transfer problems. This seems to be justified since in caffeine extraction low concentrations occur and for analogous solutions it must be assumed that concentration differences at the interface solid-gas are negligible. The results obtained by this method show that both approximate methods are capable of determining extraction yield and ex-

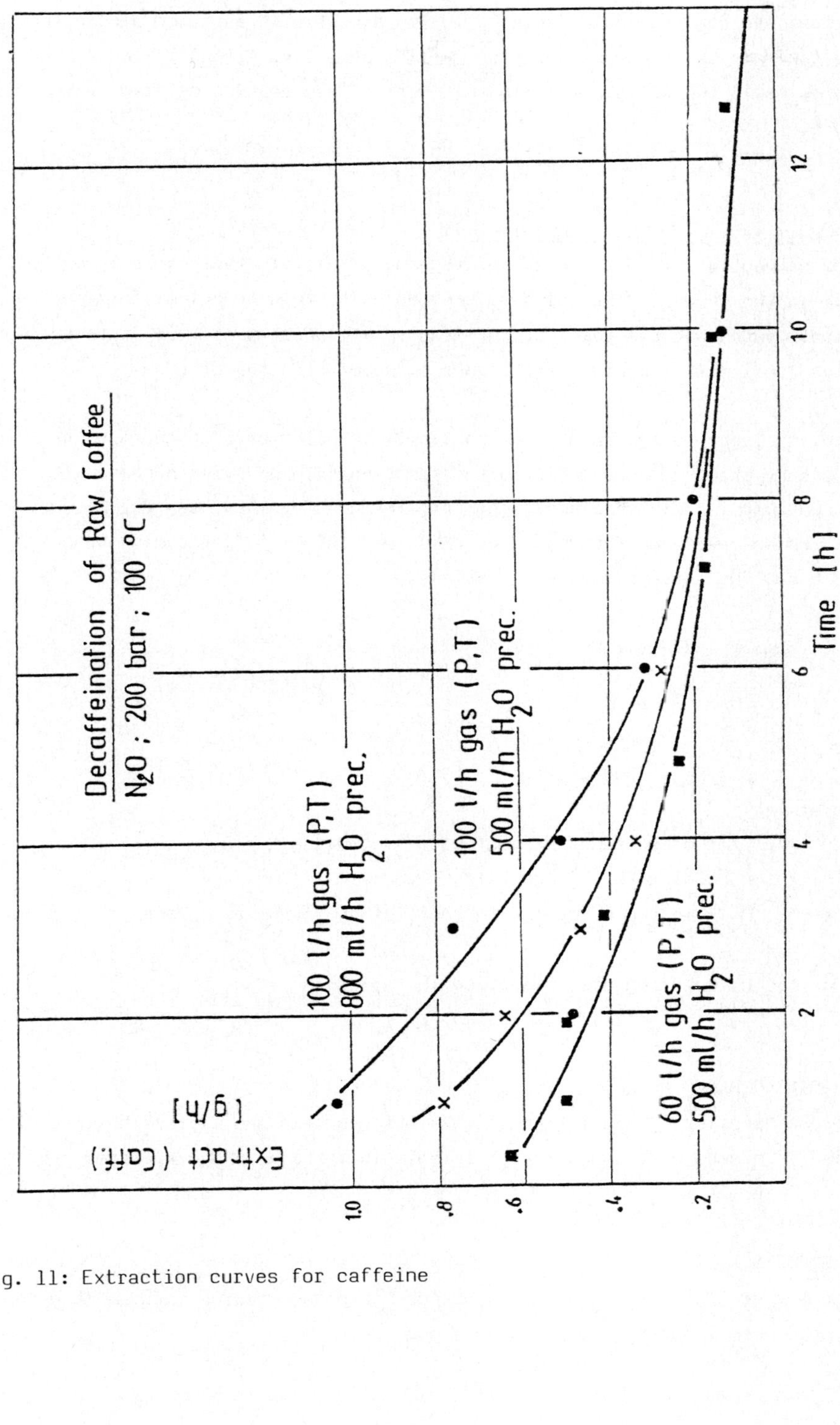

Fig. 11: Extraction curves for caffeine

traction curves for medium and long extraction times. The data used in the calculations are summarized in table 2. The most difficult part in calculating the extraction curve is the determination of the effective diffusion coefficient of caffeine in the solid phase. In this work it has been calculated from a mass balance.

Extraction of Edible Oil From Rape Seed

The extraction of edible oil from rape seeds differs from caffeine extraction mainly in two ways. 1. The solid is crushed prior to extraction. This shortens transport paths and the cells are at least partly destructed. 2. Concentrations involved during this process are higher by about a factor of 20.

Fig. 12 presents the rate of extraction. The first part of the mass transfer is a steady process, comparable to the first part in a drying process. It could be shown that in this case empirical correlations for mass transfer are applicabel, which have been determined by WAKAO and KAGUEI (12) for mass transfer in fixed beds. The correlation is as follows:

$$Sh = 2 + 1.1\ Sc^{1/3}\ Re^{0.6}$$

$$Sh = \frac{\beta_a\ d}{D_g}$$

$$3 < Re < 3000$$

wherein:

β_a = mass transfer coefficient solid-gas

d = diameter of equivalent sphere

D_g = molecular diffusion coefficient in the gas phase

Data of the calculation are summarized in table 3.

CONCLUDING REMARK

For an "a priori" calculation of mass transport in gas extraction more expertise is needed on transport coefficients in mixtures under elevated pressures.

ACKNOWLEDGEMENT

The author wishes to thank the Deutsche Forschungsgemeinschaft and the Fonds der Chemischen Industrie for financial support.

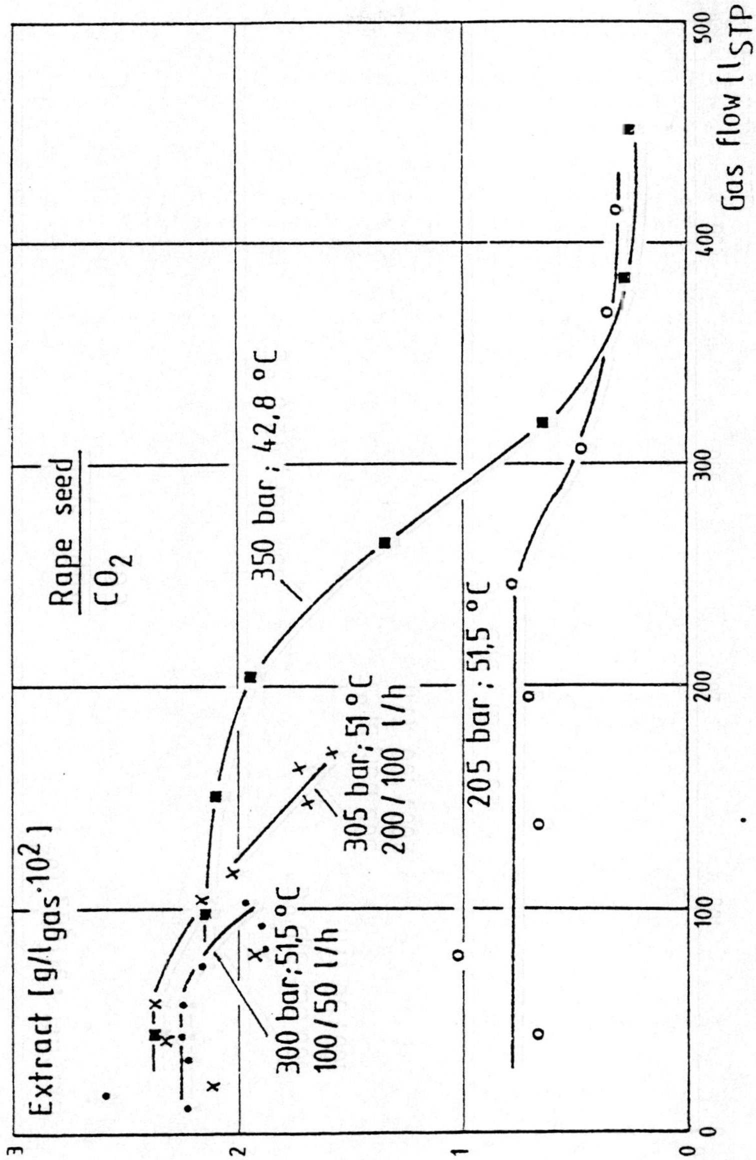

Fig. 12: Extraction of oil from rape seeds.

REFERENCES

1 Schneider, G.M., Angew. Chemie Int. Ed. 17, 716(1978), Fluid Phase Equilibria 10, 141(1983)
2 Rance, R.W. and E.L. Cussler, AIChE J. 20, 353(1974)
3 King, M.B. et al., Fluid Phase Equilibria 10, 249(1983)
4 de Filippi, R.P., Chem. and Ind. 12(1982)
5 Paulaitis, M.E., M.L. Gilbert, C.A. Nash, Proc. 2nd World Congress of Chemical Engineering, Montreal 1981
6 McHugh, M.A., M.W. Mallet, J.K. Kohn, in "Chemical Engineering at Supercritical Fluid Conditions", 1983; p. 113
7 Kreim, K., Dissertation, Technische Universität Hamburg-Harburg, 1983
8 Feist, R. and G.M. Schneider, Sep. Sci. Techn. 17, 1(1982)
9 Treybal, R.E., "Mass-Transfer Operations", McGraw-Hill, New York, 2. ed. 1968
10 Lydersen, "Mass Transfer in Engineering Practice", Wiley, New York 1983
11 Brunner, G., Ber. Bunsenges. Phys. Chem., 887(1984)
12 Wakao, W. and S. Kaguei, "Heat and Mass Transfer in Packed Beds", Gordon and Breach, New York 1982

Table 1: Parameter for the calculation of Re and Sh. Separation of H_2O and C_2H_5OH with CO_2.

P = 15.5 MPa; T = 95°C

Mean diameter of gas bubbles	3×10^{-3} m
Diffusion coefficient (C_2H_5OH in CO_2)	6.1×10^{-5} m^2/h
Density carbon dioxide	590 kg/m^3
Viscosity carbon dioxide	28.6×10^{-6} kg/ms
Free column area	2.27×10^{-4} m^2

Table 2: Parameters and mass transfer coefficients for the extraction of caffeine with nitrous oxide.

Diameter of equivalent sphere:	$d = 7 \times 10^{-3}$	
Bulk density:	$0.727\ g/cm^3$	
Effective Diffusion coefficient in the solid phase:	$D_s = 10^{-10}\ m^2/s$	
Gas:	Nitrous oxide (N_2O)	
Pressure:	$P = 20$ MPa	
Temperature:	$T = 100°C$	
Volumetric flow:	$\dot{V} = 100$ l/h	
Density:	$\rho = 0.524\ g/cm^3$	
Heat capacity:	$c_p = 1.714$ kJ/kg K	
Heat conductivity:	$\lambda = 0.0598$ W/mK	
Diffusion coefficient:	$D = 4.8 \times 10^{-4}\ cm^2/s$	
Fo	0.11	
Sh	first term solution	steady state
	4.5	4.7
β	1.2×10^{-7} m/s	1.4×10^{-7} m/s

Table 3: Parameters and mass transfer coefficients for the extraction of oil from rape seeds.

Diameter of equivalent sphere:	$d = 0.5 \times 10^{-3}$ m	
Bulk density:	$0.64\ g/cm^3$	
Gas:	Carbon dioxide (CO_2)	
Pressure:	20.5 MPa	35 MPa
Temperature:	51.5°C	42.8°C
Volumetric flow:	200 l/h STP	
Diffusion coefficient:	2.6×10^{-4}	$2.3 \times 10^{-4}\ cm^2/s$
Re	7.85	5.95
Sh	7.74	7.36
β_a	4.0×10^{-4} m/s	3.4×10^{-4} m/s
$\dot{m}_{calc.}$	4.5×10^{-4} g/s	1.0×10^{-3} g/s
$\dot{m}_{exp.}$	4.4×10^{-4} g/s	1.2×10^{-3} g/s

PART III

Process Development and Energy–Related Applications

Supercritical Fluid Technology, edited by J.M.L. Penninger, M. Radosz, M.A. McHugh and V.J. Krukonis
1985 Elsevier Science Publishers B.V., Amsterdam — Printed in The Netherlands

PROCESS DEVELOPMENT AND PLANT DESIGN FOR EXTRACTION WITH SUPERCRITICAL GASES

H. Coenen, R. Pätzold and U. Sievers

INTRODUCTION

Fundamental theoretical and experimental findings show that the extraction properties of gases are one or two orders greater in the supercritical state than under normal conditions and that supercritical gases can be used to great advantage as solvents in a wide range of materials separation operations (1-7). Possessing several favourable process characterisitis compared with distillation, rectification and extraction with liquid solvents, all of which have reached an advanced stage of technological development (8-27), extraction with supercritical gases has gained increasingly in importance as a thermal separation process with wide-ranging applications.

Of the many particular advantages of high-pressure extraction, the major ones are:

1. The solvents used are gaseous at normal pressure and ambient temperature. This means that after axtraction they can easily be expelled from both the extraction residue and the extracted products to the point of being non-detectable, and recovered.

2. Many of the gases used are physiologically safe.

3. With many gases, materials separation is carried out at a low temperature, which is extremely important when extracting natural substances.

4. The solvent properties of compressed gases can be widely varied both by appropriately adjusting pressure and temper-

ature and by introducing entraining agents. In addition, by gradually altering temperature and pressure, multi-phase extraction can be carried out and the extract fractionated into the desired products.

Extraction is usually carried out with the gas or gas mixture in the supercritical or fluid state. The shaded area in Fig. 1 is a schematic representation of the fluid state.

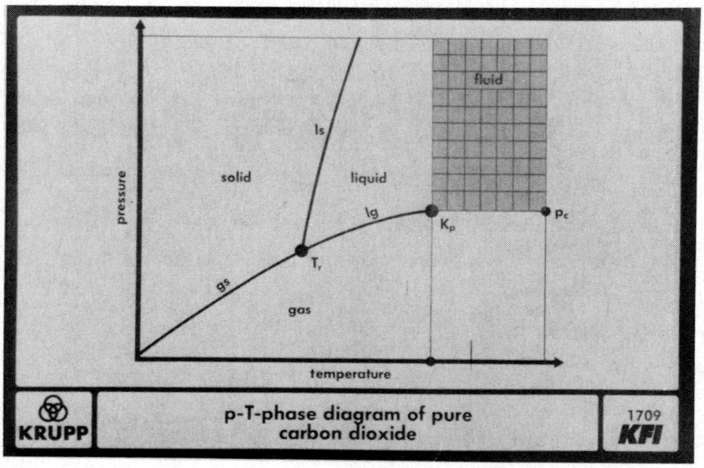

Fig. 1

Compared with the liquid or gaseous state, gases in the fluid state display several physical properties which are advantageous in terms of extraction.

Fluid gases display liquid-like density at extremely low viscosity. As well as having large diffusion coefficients, such gases are often completely miscibile with low-volatile, thermally unstable materials at certain pressures and temperatures.

The Krupp Research Institute has been working on the development of high-pressure extraction for over ten years. This work has resulted in the filing of more than 20 patents, some of them already issued, in the fields of food engineering, pharmaceuticals, petrochemicals and recycling. There have also been many technical publications (13, 18, 19, 20, 21, 24-27).

Process development and requisite pilot facilities

In the design of operational plant, recourse must be had to laboratory investigations to optimize the process parameters. The design of the equipment draws heavily on the techniques and experience gained in high-pressure technology.

The first task in a laboratory or pilot-plant test is to select the most suitable solvent for the particular separation process. The process parameters solvent ratio, pressure and temperature must be optimized in further tests, with due emphasis on minimizing the energy requirement. These tests provide the basis for designing a commercial-scale extraction plant for a specific separation operation.

Fig. 2 shows a picture of a pilot plant without heat insulation in the Krupp Research Institute.

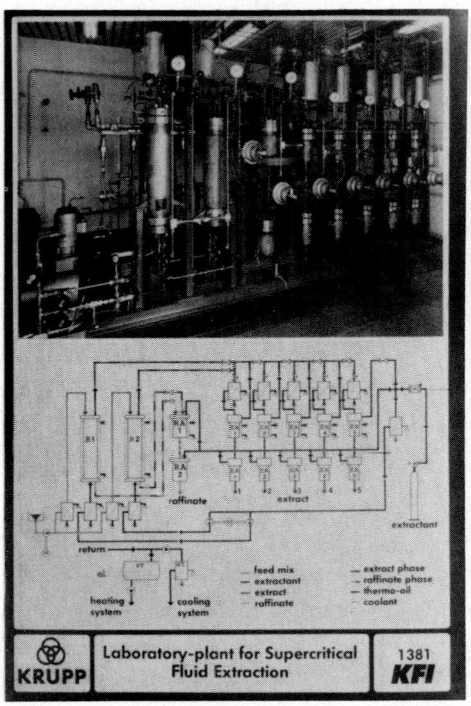

Fig. 2

The unit is designed with 2 stages on the extraction side, each with an extraction volume of 2 x 7,5 l, and five 1-l extract separators on the separator side; a further separator serves to remove the solvent from the extraction residue. Duplicate separators are used to allow buffering and solvent recovery. Owing to the fact that the unit was used for investigations involving foodstuffs, the components were made of V4A extra stainless steel (material no. 1.4571). The unit is explosion-proof, which means that in addition to carbon dioxide, other gases can be used as extractants. Hydrocarbons and chlorinated hydrocarbons have proved more effective than carbon dioxide in some applications. For this reason, the unit is also designed for high pressures and temperatures (up to 350 bar and 350 °C). The two extraction autoclaves can be operated in parallel or, for 2-stage extraction and possible product recycling, in series.

In tests using pumpable material mixtures, the raw materials are fed continuously into the top of the extractors after having been brought to extraction temperature and extraction pressure in heat exchangers and a high-pressure pump, respectively. To increase the contact surface in the autoclaves, internals such as packing or trays can be provided. The material to be extracted flows countercurrent to the supercritical gas. In the process, the supercritical gas becomes laden with predominantly one or more of the components of the raw material, depending on their properties and thermodynamic state. The laden gas phase ist removed and, in consecutive stages, expanded into the separators where the extract is fractionated and where a gradual increase in temperature can also be effected. The reduction in density this brings about results in the gas losing its capacity in the changed thermodynamic state to absorb the materials previously dissolved in the supercritical condition, and the extracted components are separated fraction-wise in the separators. The desired product may be the extract fractions or the raffinate residue in the extractor. The expanded, virtually extract-free gas from the last extract separator and the gas from the raffinate expansion tank are then brought back to extraction pressure and temperature and recycled as solvent to the extractors.

In the case of bulk materials, the extractors are operated batchwise, with the autoclaves being discontinuously filled, charged with gas, reopened and emptied. With two autoclaves, quasi-continuous operation can be achieved by alternating the autoclaves.

Apart from this pilot plant other facilities, covering the entire commercially relevant temperature range from 5 to 550 °C and a pressure range up to 350 bar, are available for developing and optimizing the supercritical extraction process for a wide variety of applications. Fig. 3 shows a schematic representation of the potential these facilities provide.

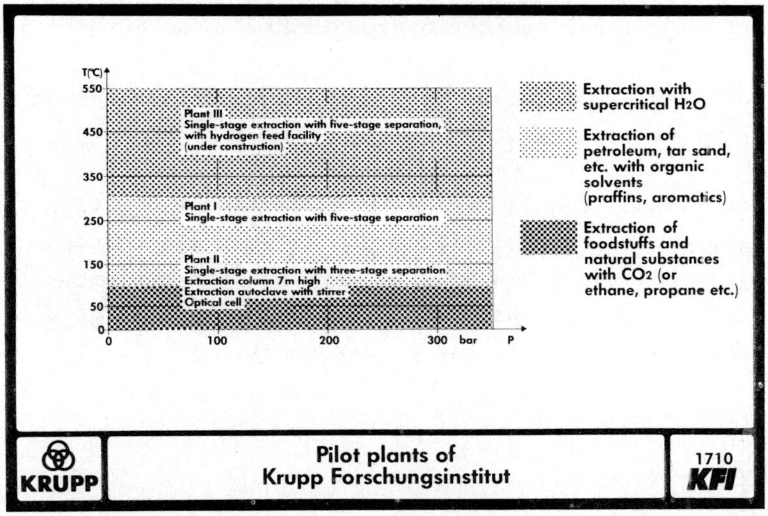

Fig. 3

Using these plants, process development has been carried out for:

the food engineering sector:

- extraction and refining of edible oils,
- extraction of colourants from paprika,
- extraction of pure lecithin from raw lecithin,
- separation of cholesterol from egg yolk,
- extraction of fat from maize starches,

for the chemical process industry:

- separation of aromatics from petroleum products,
- removal of metal from and deasphalting of petroleum stock,
- extraction of montan wax from lignite,
- extraction of liquid hydrocarbons from coal,

and for the

recycling of waste material:

- rerefining of used oil to produce lubricants
- disposal of sewage sludge and polymer waste with simultaneous extraction of liquid and gaseous hydrocarbons.

Naturally, only a few of the process developments carried out to date can be mentioned here.

The underlying assumption in designing extraction facilities is that the maximum possible solubility of low-volatile components in highly compressed gases will be in line with the pressure concentration diagram, represented schematically in Fig. 4, of a binary mixture with supercritical components.

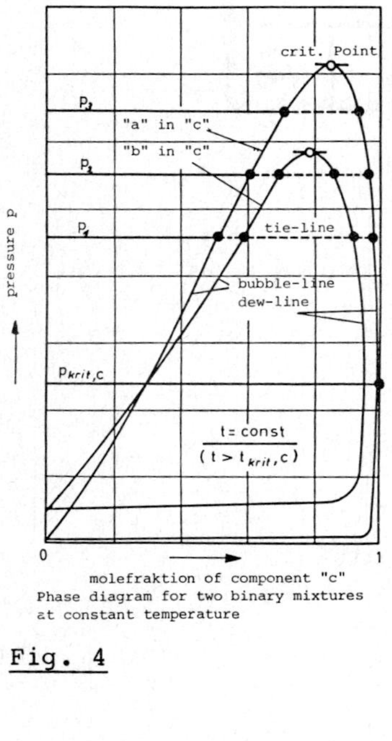

Phase diagram for two binary mixtures at constant temperature

Fig. 4

Inter alia, Hederer (28) proposed a modification of the Redlich-Kwong equation specifically for determining phase equilibria in highly compressed gases. This equation was used in conjunction with the programme package for calculating phase equilibrium developed by Sievers (29), which has proved more effective than the calculation programme published by Hederer (28).

To apply the modified Redlich-Kwong equation to multicomponent systems, 3 coefficients must be known for every pure component of the system as well as one binary interaction parameter for each binary system contained in the multicomponent one. By way of example, I should like to examine in detail computations for the seed oil/CO_2 system, which is regarded as being quasibinary. The solubility data of this system are of relevance for the highpressure extraction process employed in oil seed processing, on which Eggers and Stein have reported (30).

For the carrier gas CO_2, the coefficients of the modified Redlich-Kwong equation were taken from a paper by Brunner (31). The determination of the coefficients for seed oil and of the interaction parameters is described by Eggers, Sievers and Stein (32). The calculated solubilities for the soy oil/CO_2 system are shown in Fig. 5 together with measured values taken from Quirin (33).

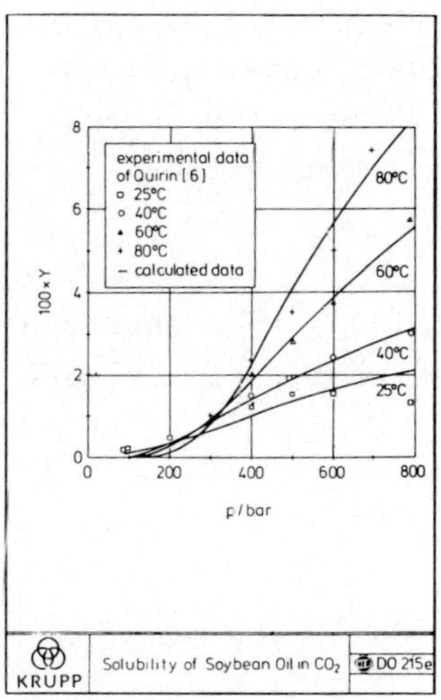

Fig. 5

The phase equilibrium data represent limiting values for high-pressure extraction. Owing to mass-transfer kinetics, the actual course of extraction will lead to the carrier gas becoming laden with smaller amounts of the low-volatile substance. The extraction curves are determined experimentally.

The modified Redlich-Kwong equation cannot predict with the required accuracy the thermodynamic properties such as density, enthalpy and entropy in the homogeneous region and in the saturated state which are also necessary for designing and optimizing high-pressure extraction facilities. For the carrier gas CO_2, which is used in particular in the high-pressure extraction of natural substances, these properties were calculated for the isochoric heat capacity in the ideal gas state using Bender's state equation (34) in conjunction with a Sievers correlation equation (35).

The thermodynamic properties of carbon dioxide are comprehensively tabulated (36). The saturation parameters vapour pressure, saturated liquid density, saturated vapour density, enthalpy change of vaporization, enthalpy and entropy of saturated liquid and of saturated vapour are tabulated between the triple point (-56.57 °C) and the critical point (31.06 °C). For the homogeneous region, density, compressibility factor, entropy, enthalpy, isobaric and isochoric thermal capacity, speed of sound, fugacity coefficient and isentropic coefficient are listed on isobars between 0.1 and 1000 bar for temperatures between -50 °C and 820 °C. For subcritical pressures (up to 70 bar) the saturation parameters are also given by way of delineation between homogeneous liquid and gas region.

As described by Eggers in an earlier paper (37), the tabulated properties serve as a basis for optimizing the energy requirement of high-pressure extraction.

Designing high-pressure extraction equipment

The extraction process makes exacting demands on the manufacturers of chemical process equipment. If the charge mixture, raffinate and extract are pumpable, the pressure vessels differ insignificantly from equipment used elsewhere in continuous operation under comparable conditions.

In the extraction of natural substances, e.g. from coffee, tea or hops, the current state of the art is such that batch processes are necessary. For filling and emptying, the vessels are brought to ambient pressure and usually have to be opened over their entire cross section.

It is this procedure that confronts extraction equipment designers with the typical problems of

- coping with pulsating loads,
- optimizing closures,
- developing semi or fully continuous charging systems.

Pulsating loads

In designing components which will be subjected to pulsating loads, a number of interacting factors have to be allowed for, namely:

a) Loading

First of all, the development over time of the two variable parameters pressure and temperature must be determined. If external forces or moments are acting on the vessel, these too must be known. The local stresses in the component are then calculated on the basis of the known design.

b) Design

The shape of components such as nozzles, transitions or welded-on parts results in stress patterns which are a major determinant of wall thicknesses. The design engineer must therefore find a solution which takes stress patterns into account while avoiding inadmissible stress peaks and obstacles to expansion.

c) Material

Great importance attaches to the choise of material because fatigue strength and notch sensitivity vary as a function of microstructure and strength.

d) Fabrication

Shop-welding is very important. Weld defects such as inclusions, undercutting, poor alignment and cracks constitute attached to a high standard of workmanship.

e) Corrosion

The corrosive effects of particular media and the resistance of the materials must be known to predict service life. It is, however, difficult to allow for corrosion in design calculations. Owing to a lack of laws for corrosion time/fatigue strength relationships, one is forced to rely extensively on experience.

In determining servic life, all the parameters mentioned have to be taken into account. Under German rules, e.g. AD-Merkblatt S2, the "decisive amplitude of stress" is determined. This is compared with experimentally established fatigue strengths dependent on the tensile strength of the materials.

Depending on the number of cycles of load stressing, dimensioning is carried out allowing a load cycle safety factor of 10 or a stress safety factor of 1.5.

Closures

The closures on vessels used in batch operation are mostly quick closures.

As the name implies, quick closures are designed to shorten, by rapid opening and closing, the interruptions which per force occur during batch operation.

Fig. 6 shows the main differences between quick closures and bolt-sealed vessel covers.

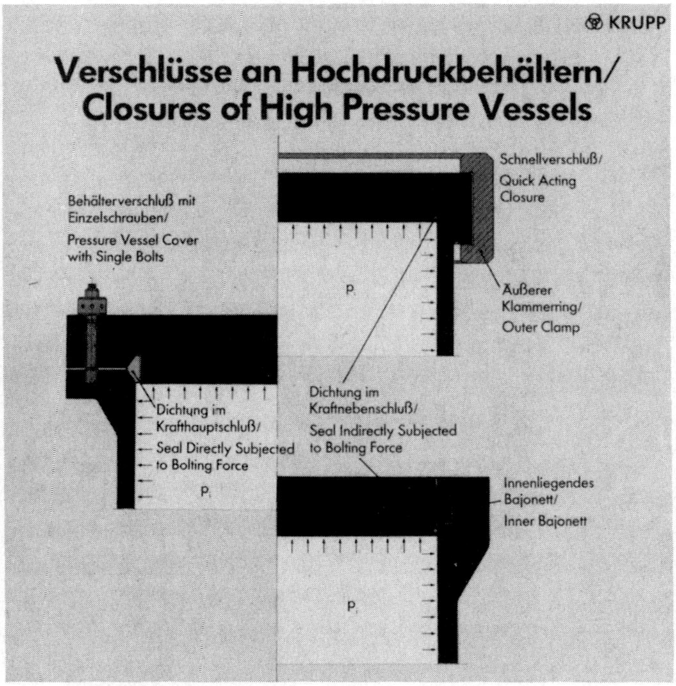

Fig. 6

As the slackening and tightening of bolts is very time-consuming, particularly on pressure vessels, quick closures usually feature either external split clamps or internal bayonet fasteners instead. The internal pressure (cover force) is transmitted into the cylinder shell by clamps in the one solution and by the bayonet projections in the other.

In both cases, for ease of operation, axial pretensioning across the cover is dispensed with, as otherwise tensioning elements would be needed. The seal cannot therefore be pretensioned to any appreciable extent; it is usually arranged so as to allow movement in the axial direction. The seal is of the soft, elastic type, offering an additional self-sealing function if suitably constructed.

The proper functioning of the seals on quick closures is always verified by means of a deformation analysis. It must be ensured that the expansion and twisting of the system, consisting of flange, cover and connections, permit reliable sealing at the sealing location.

Fig. 7 shows a quick closure of the clamp ring type for a hop extractor.

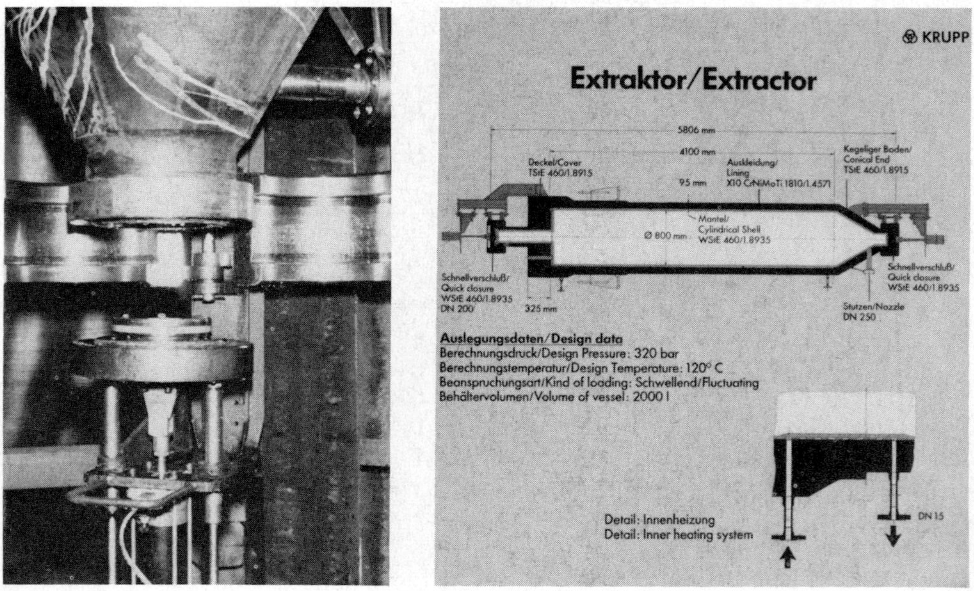

Fig. 7

Fig. 8 shows a high-pressure extractor with a vessel volume of 44 m^3. Pressure is 250 bar at 85 °C. The inlet and outlet openings have bayonet-type quick closures. The bottom cover features a discharge device with a sealing face located on the inside of the vessel. The vessel is of the Krupp multi-shell type.

It is used in the largest commercially used high-pressure extraction plant known to us.

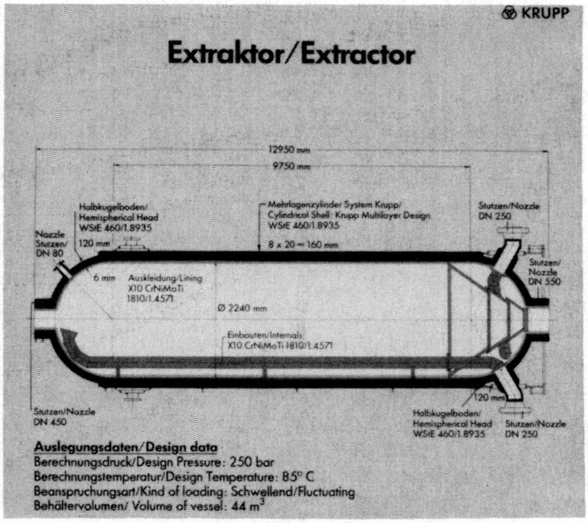

Fig. 8

Air lock systems

A discontinuous mode of operation entails regular interruption to operation. Automation is only possible to a limited extent owing to emptying and filling function which have to be performed manually.

The aim therefore is to introduce fully or semi-automatic processes, even if inferior mass transfers may result from remixing in the extractor.

The compaction that may occur in the solid and reduce considerably the mass transfer surface area can be reversed by subsequent size reduction.

If the total extraction volume as determined by plant throughput is broken down into several smaller volumes, a quasi-continuous extraction process can be achieved by setting up a cascade. In this way, the remixing in the extractor mentioned earlier can be prevented.

Piston pump systems are also conceivable which would inject and eject the solid and the gas together into and out of pressure spaces. Multiple air-lock systems with gradual pressure reduction also hold considerable potential.

In both the piston pump and multiple air-lock systems, the main problems are in overcoming wear and abrasion on the sealing faces.

Development work is being carried out at Krupp on all the systems mentioned with the aim of producing a reliable system for injecting an ejecting solids into and out of pressure spaces.

Fig. 9 shows the main differences between the described sluices for solid particles.

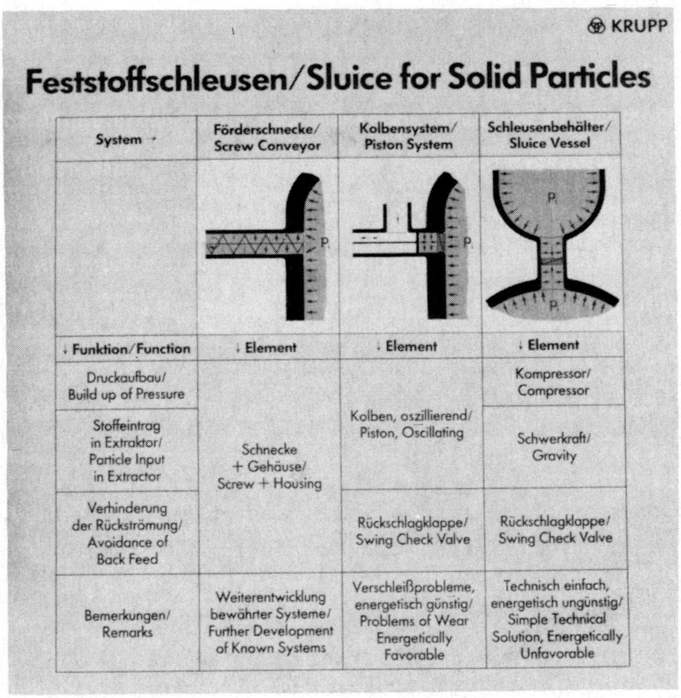
KRUPP

Feststoffschleusen/Sluice for Solid Particles

System →	Förderschnecke/ Screw Conveyor	Kolbensystem/ Piston System	Schleusenbehälter/ Sluice Vessel
	P	P	P, P
↓ Funktion/Function	↓ Element	↓ Element	↓ Element
Druckaufbau/ Build up of Pressure	Schnecke + Gehäuse/ Screw + Housing	Kolben, oszillierend/ Piston, Oscillating	Kompressor/ Compressor
Stoffeintrag in Extraktor/ Particle Input in Extractor			Schwerkraft/ Gravity
Verhinderung der Rückströmung/ Avoidance of Back Feed		Rückschlagklappe/ Swing Check Valve	Rückschlagklappe/ Swing Check Valve
Bemerkungen/ Remarks	Weiterentwicklung bewährter Systeme/ Further Development of Known Systems	Verschleißprobleme, energetisch günstig/ Problems of Wear Energetically Favorable	Technisch einfach, energetisch ungünstig/ Simple Technical Solution, Energetically Unfavorable

Fig 9

REFERENCES

1 F. Spillner, Chem. Fabrik 13 (1940) 405-424,
2 G. Brunner and S. Peter, Chem.Ing.Tech. 53 (1981) 529-542
3 M. Godlewicz, Petroleum 34 (8/1938) 1-5,
4 T.P. Zhuse, Petroleum 23 (1960) 298-300,
5 T.P. Zhuse and G.N. Yushkevich and I.E. Gekker, Maslob.-Zhirovaya Prom. 24 (6/1958) 34-37,
6 T.P. Zhuse and T.P. Safronova and G.N. Yushkevich, Izv. Akad. Nauk S.S.S.R. Otd. Tech. Nauk, Metall. Topl. 9 (11/1958) 123-125,
7 T.P.Zhuse and B.K.Sheremeta,Izv. Akad. Nauk S.S.S.R. Otd. Tech. Nauk. Metall. Topl. 10 (2/1959) 144-147,
8 O. Vitzthum and P. Hubert, D.B.Pat. (OS) 2127596 (1971/1972)-(HAG AG),
9 O. Vitzthum and P. Hubert and W. Sirtl, D.B.Pat. 2127618 (1971/1973)- (HAG AG),
10 O. Vitzthum and P. Hubert D.B.Pat. 2127611 (1974) - (Studienges. Kohle),
11 O.G. Vitzthum and P. Werkhoff and P. Hubert, J. Food Sci. 40 (1975) 911-916,
12 W. Roselius and O. Vitzthum and P. Hubert, D.B.Pat. 2043537 (1975) - (Studienges. Kohle),
W. Roselius and O. Vitzthum and P. Hubert, D.B.Pat. 2142205 (1976) - (Studienges. Kohle),
13 H. Coenen and E. Kriegel, D.B.Pat. 2846976 (1978/1981) - (Krupp),
14 J.C. Whitehead and D.F. Williams, J. Inst. Fuel 48 (397/1975) 182-184,
15 K. Zosel, Angew. Chem. 90 (1978) 748-755,
16 S.R.M. Ellis, Brit. Chem. Eng. 16 (1971) 358-361,
17 S. Peter and G. Brunner and R. Riha, Dechema-Monographien 73 (1974) 197-204,
18 H. Coenen and J. Hartwig and E. Kriegel Verfahrenstech. 14 (1980) 176-177,
19 H. Coenen and P. Rinza, Tech. Mitt. Krupp, Werks-Ber. 39 (1981) 21-31,
20 H. Coenen and E. Kriegel, D.B.Pat. (OS) 2843920 (1978) - (Krupp),
21 H. Coenen and E. Kriegel, D.B.Pat. 2850540 (1979/1982) - (Krupp),
22 K. Zosel, D.B.Pat. 1493190 (1964/1979) - (Studienges. Kohle),
23 E. Stahl and W. Schilz and E. Schütz and E. Willing, Angew. Chem. 90 (1978) 778-785,
24 H. Coenen and P. Rinza, Chem. Ing. 13, 801 (1981),
25 H. Coenen and R. Hagen, Gordian 83, 164 (1983),
26 H. Coenen and E. Kriegel, Chem. Ing. Tech. 55, 890 (1983),
27 H. Coenen and P. Rinza, vt "verfahrentechnik" 16, 628 (1982),
28 H. Hederer, Dissertation Universität Erlangen-Nürnberg 1981,
29 U. Sievers, Fortschr.-Ber. VDI-Z. Reihe 6 Nr. 110 (1982),
30 R. Eggers and W. Stein, Fette, Seifen, Anstrichmittel 86 (1984), 10-16,
31 G. Brunner, Habilitation Universität Erlangen-Nürnberg 1978
32 R. Eggers and U. Sievers and W. Stein, In preparation and International Symposium High Pressure Chemical Engineering, October 8-10, 1984, Erlangen/West-Germany,

33 K. Quirin, Fette, Seifen, Anstrichmittel 84 (1982), 460-468,
34 E. Bender, Proceedings Fifth Symposium on Thermophysical Properties. Amer. Soc. Mech. Engrs., New York 1970, 227-235,
35 U. Sievers and S. Schulz, Chemie-Ingenieur-Technik 53 (1981) 459-461,
36 U. Sievers, Fortschr.-Ber. VDI-Z. Reihe 6, Nr. 155 (1984),
37 R. Eggers, Chemie-Ingenieur-Technik 53 (1981), 551-554.

Supercritical Fluid Technology, edited by J.M.L. Penninger, M. Radosz, M.A. McHugh and V.J. Krukonis
1985 Elsevier Science Publishers B.V., Amsterdam — Printed in The Netherlands

COAL LIQUEFACTION WITH SUPERCRITICAL AMMONIA AND AMINES

WEI-YIN CHEN and FAHEEM KAZIMI

Gulf South Research Institute, P. O. Box 26518, New Orleans, LA 70186

ABSTRACT

The extraordinary high efficiency of coal liquefaction with nucleophilic nitrogen-containing solvents was recently demonstrated. Based on solubility and mass transfer considerations, liquefaction at supercritical solvent conditions has also drawn attention. The present study combines these two approaches by exploring the potential of supercritical liquefaction of coal with low molecular weight nitrogen-containing solvents.

The first year study has already indicated that these solvents are very effective in converting coal to soluble materials and this concept has potential as a significant approach to low severity coal liquefaction. It has been observed that liquefaction efficiency increases progressively with an increase in the carbon number of these primary aliphatic amines and that process optimization capitalizing on this property has resulted in 100 percent conversions at mild (275°C) conditions and low solvent to coal ratios. For example, at 275°C and with a 1:1 ratio of solvent to North Dakota lignite (by weight), the liquefaction efficiency increases from 17 percent (daf) to 100 percent when the carbon number in the primary aliphatic amines increases from 0 to 4. Wyodak coal also shows similar conversion. However, Illinois No. 6 coal provides relatively lower, but still strikingly high conversion at such mild conditions. Products were characterized by sequential Soxhlet extractions, vapor pressure osmometry, elemental analysis, ^{1}H NMR, ^{13}C NMR and GC/MS.

I. INTRODUCTION

In comparison with the other major fossil fuels (crude oil and gas), coal has a much lower hydrogen content and higher heteroatom and mineral matter concentration. To convert coal to liquid materials extensive hydrogenation, often under severe processing conditions, is required. There is accordingly an incentive for the development of lower severity technologies which could lead to significant reductions in conversion cost.

An important preliminary in the development of innovative low severity processes should include progress in the fundamental understanding of coal liquefaction chemistry, building upon the results of studies conducted over the last decade dealing with coal structure, solvent chemistry and reaction mechanisms (Bockrath, 1983; Pullen, 1983). In addition, emerging extraction technologies involving low energy input have also been considered to enhance coal liquefaction conversion at less severe conditions. The present work emphasizes a new coal liquefaction concept based on the combination of recent advancements in the chemistry of nitrogen-containing compounds as liquefaction solvents and in supercritical fluid extraction technology. Both approaches are discussed below.

A. Chemistry of Nucleophilic Solvents

Review of recent literature reveals that the initial stage of coal liquefaction could be controlled by all the following factors:

1. Cleavage of hydrogen bonds,
2. Cleavage of covalent bonds, particularly oxygen crosslinks,
3. Free radical stabilization by a hydrogen donor and hydrogen transfer agent,
4. Solubility of coal derived products in the solvent, and
5. Mass transfer limitations.

Recent reports of the Electric Power Research Institute (Atherton and Kulik, 1982, Atherton, 1984) summarized the up-to-date reaction chemistry of nitrogenous heterocyclic solvents with coals. Model compound studies indicated that basic aromatic nitrogen compounds, such as 1,2,3,4-tetrahydroquinoline (THQ), are able to catalyze not only the acid-base interaction, but also promotes tautomerization of enols to the more weakly bound keto form (which then leads to several reactions previously thought unfeasible). These secondary reactions include cleaving of ether linkages and carbonyl reductions. In tests below 300°C, nitrogen heterocycles attacked oxygen functional groups resulting in ester hydrolysis and decarboxylation. In high temperature experiments, the base-catalyzed attack of oxygen functional groups resulted in cleavage of ether linkages, deoxygenation of polycyclic phenols and dehydrogenolysis of other linkages. Furthermore, since the ring in THQ containing nitrogen is saturated with hydrogen, THQ serves as a hydrogen donor and shuttling agent in liquefaction. A combination of these reactions contributed to a complete conversion of coal to tetrahydrofuran-soluble material at 400°C and, in some cases, formation of stable covalent bonds between solvent and coal.

Several independent investigations also revealed distinctive features of nitrogen-containing solvents of even lower molecular weight such as ammonia and amines (Harris *et al.*, 1980, Tomita *et al.* 1979; Matida *et al.*, 1977, Camier *et al.*, 1981; Brooks and Sternhell, 1958; van Bodegom *et al.*, 1984). All these works concluded that carboxyl and carbonyl bonds must be broken before extensive dissolution can take place.

The hypothesis that the generation of soluble products in coal liquefaction is related to the cleavage of ether bridges or crosslinks was proposed by Fisher and Eisner (1937), and reiterated 20 years later by Takegomi *et al.*

(1963). Additional implications of the importance of ether crosslinks were recognized in the studies of steam as a reactant/solvent. In studying the reaction of steam (D_2O) with anisole at 400°C and 184 atm, Ross et al. (1982) reported that the conversion of ether-aryl linkages in anisole was about 20% in 20 minutes. Yet at this temperature the calculated half-life of the ether linkages for thermal rupture is 29 hours. This indicated that steam, which also has nucleophilic attack capability, could enhance the "depolymerization" of coal by cleaving these thermally stable linkages. This theory was recently strengthened by a study of direct coal liquefaction by steam pyrolysis (Graff and Brandes, 1984). A drastic increase in liquid yields from 23% to 50% was observed by pretreating the coal in steam at 320-360°C, 50 atm and with a residence time of less than 30 minutes.

The role of oxygen functionals in coal conversion was also discovered in an interesting correlation equation. By using a statistical method, Solomon et al. (1981) observed that volatile yields for three coal conversion processes can be correlated linearly with oxygen and aliphatic hydrogen content of the coal. The three processes are rapid pyrolysis, flash hydrogenation at 10 MP_a and liquefaction with tetralin at 400°C.

Jenkins (1980) developed a process for liquefaction by treating coal with a primary aliphatic amine. Residual amine was advantageously removed from the refined product and residue by stripping with alcohol. Jenkins' patent proposed the use of low boiling point (<100°C) amines of sufficient volatility to permit complete removal from the residue by vaporizing or stripping. Therefore, "the low molecular weight aliphatic amine employed according to the present invention is typically an amine of 1-3 carbon atoms such as methylamine." Jenkins also proposed a low temperature (50 to 200°C with 60° to 150°C preferred) in the liquefaction stage. Only partial dissolution was achieved under these conditions. Solubility and mass transfer were limited at reflux conditions of his experiments. These parameters should be enhanced by incorporating supercritical solvent operations in liquefaction.

B. Coal Liquefaction with Supercritical Solvent

To further reduce the reaction severity, the approach of supercritical fluid extraction is employed in the current liquefaction study. As early as 1879, Hannay and Hogarth postulated that the liquid condition of fluids had little to do with their solvent power beyond its indication of molecular closeness. Should this tight packing be obtained by external pressure instead of internal attraction, the result could be the same or even greater solvent power. Indeed, when compared with conventional liquid solvents, supercritical fluids offer high diffusivities, comparable densities, and low viscosities; and thus a potential for rapid penetration of solvent into coal and rapid separation of liquid products from insolubles. Subsequent phase separations are then possible with the same fluid simply by reducing the pressure at the same temperature or reducing the temperature at constant pressure. This unique separation technique is also called retrograde condensation.

Selected works on supercritical solvent extraction of coal are summarized in Table 1. The theory and applications of supercritical fluid extraction were recently reviewed by Williams (1981) who was involved in the development of coal extraction using a supercritical toluene and p-cresol/ water mixture in the early seventies (Bartle et al., 1975; Williams and Martin, 1978b). Subsequent studies involved liquefaction of coal using single ring aromatics (Blessing and Ross, 1978; Mortimer, 1980; Kershaw et al., 1980; Ceylan and Olcay, 1981; Maddocks and Gibson, 1977), aliphatic alcohols (Blessing and Ross, 1978), and pentane (Fetzer et al., 1981). In 1968, Wise filed a patent (disclosed in 1971) which involved extraction of carbonaceous materials above the critical points of a series of compounds: ethylene, propylene, methane, propane, butane, ammonia, chlorine, carbon dioxide, benzene, toluene, pyridine, dimethylamine, ethylamine, and methylamine.

Williams and Martin (1978a) disclosed a concept of extraction of oil shales and tar sands by using various supercritical solvents. Nitrogen-containing organic compounds among those suggested included aliphatic mono, di-, and tri-amines having at least four but not more than ten carbon atoms, acrylic aliphatic amines, and aromatic amine having a benzene ring. The potential of these nucleophilic amines in liquefaction was not discussed. Similar to Williams et. al., Poska (1982) and Warzel (1983) disclosed processes for the extraction of tar sand and oil shale with a variety of solvents, including amines. The extent of coal conversion in these nitrogen-containing solvents was not defined.

Quantitative analysis of coal liquefaction conversion in supercritical solvents was also examined by several investigators. Generally speaking, two approaches have been used. By using the virial equation of state, Rowlinson and Richardson (1959) found that solubility enhancement, defined as the ratio of the partial pressure of solute in the gas phase to the vapor pressure of pure solute at the same conditions, can be expressed as a linear function of the cross-virial coefficient. This model was later tested by Fong, et al., (1983) for their liquefaction data, and a maximum conversion (rather than a linear function) was observed at a particular value of the second virial coefficient of the solvent. This led to their conclusion that proper design of a multicomponent solvent system could provide optimum diffusion of solvent into and out of the coal structure during liquefaction. Optimum performance was observed with a methanol-toluene system.

TABLE 1. Selected works of supercritical fluid (SCF) extraction of coal

Authors	Institute	Reactor	Capacity	Temperature (°C)	Solvent	Highlights
Fong *et al.* (1983) Corcoran *et al.* (1983)	CIT	batch	5.5 g	250-400	alcohols, acetone BTX, pentane toluene	1. test of model 2. Optimum yield in multicomponent system
Vasilakos *et al.* (1983)	UT-Austin	batch	20 g	400	C_5-C_{12} aliphatics methanol water acetone toluene	1. Conversion vs. solubility parameter (Giddings) 2. toluene-methanol/ mixture
Squires *et al.* (1983)	Iowa Sate	tubular	0.5 g	25-400	methanol	1. SCF liquefaction vs. pyrolysis 2. Flow mode vs. batch mode
Smith *et al.* (1983)	Pacific Northwest	tubular	1 mg	280	CO_2 NH_3 1-pentane +5% 2-propanol	1. SCF/GD/MS, SCF/Ms, SCF/MS/MS
Barton (1983)	Penn State Univ.	batch	78-490 g	400	$H_2O + H_2 + CO + MoS_3$	
Blessing and Ross (1978)	SRI	autoclave	1 - 5 g	335	IPA, benzene MeOH, tolune	Product pyridine solubilities vs δ
Ross and Nguyen (1983), Ross *et al.*, (1983)	SRI	autoclave	10 g	400	water/CO	base or HCO_2 as modifiers

(Continued)

TABLE 1. Continued.

Authors	Institute	Reactor	Capacity	Temperature (°C)	Solvent	Highlights
Deshpande *et al*. (1984)	U. Pittsburgh	autoclave	10~20 g	377	water	
Finn *et al*., (1983)	NCB	autoclaves pilot plant	50 g 10 kg/hr	420-480	mixture of air aromatic and a fully hydrogenated aromatic compound	
Scarrah (1983)	Montana State University	autoclave	60 g	265,400	H_2O, H_2O/CO_2 Methanol, Hexane	
Wilhelm and Hedden (1983)	Engler-Bunte-Institute	fixed bed	140 g	100~500	aromatics, aliphatics, alcohols, H-donors	2-stages
Amestica and Wolf (1983)	Notre Dame	autoclave	-	350-450	toluene, ethanol, tetralin	
Williams and Martin (1978)	NCB, Catalytic	-	-	-	toluene P cresol/H_2O	char reactivity
Sunol (1982)	VPI	semi-batch	15 g	45-450	CO_2, H_2,O, toluene decane, pentane, xylene	
Tiffany *et al*., (1984)	W. Michigan U.	batch	2 g	400-500	water	nitrogen-containing compounds

(Continued)

TABLE 1. Continued.

Authors	Institute	Reactor	Capacity	Temperature (°C)	Solvent	Highlights
Maddocks _et al_ (1979) Maddocks and Gibbson (1977)	Catalytic, NCB	moving-bed, autoclave	-	400-440	chlorobenzene, toluene, pyridine, o-xylene, dodecane, p-cresol	
Mortimer (1980)	NCB	pilot plant	5 kg/hr	450	toluene	^{1}H NMR
Kershaw _et al_. (1980)	FRI, South Africa	hot rod autoclave	25 g 100 g	450	toluene toluene/H_2 (cat. optional)	comparative study with hydropyrolysis
Ceylan and Olcay (1981)	Karadeniz Tech. Univ.	autoclave	--	360	toluene	identification of products

TABLE 2. Properties of Coals Used in the Liquefaction Study

Sample	PSOC-1351	PSOC-1405P	PSOC-1406P
Rank	HVCB	SUBB	Lignite
State	Illinois	Wyoming	N. Dakota
Moisture	8.08	19.82	29.64
Ash	10.56	7.15	8.23
Volatile matter	37.08	38.23	31.32
Fixed carbon	44.28	34.80	30.80
C, daf, %	76.26	74.44	67.72
H, daf, %	4.85	5.46	4.97
N, daf, %	1.32	1.34	1.40
S, daf, %	5.31	2.03	1.33
O, (% by diff.)	12.26	16.73	24.58
H/C (atomic)	0.76	0.88	0.88

Based on van der Waals equation, Giddings et al., (1968, 1969) demonstrated that the Hildebrand solubility parameter can be linearly related to solvent density. This model has been used to correlate coal liquefaction efficiencies of non-polar solvents (Blessing and Ross, 1978; Vasilakos, et al., 1984). Their liquefaction yield was found to be a linear, universal function of solubility parameter.

Experimental and theoretical studies have only been partially successful in describing the complex nature of coal liquefaction. Many questions remain because the relative significance of the five factors identified in the previous discussion of various solvent systems at specific experimental conditions is usually difficult to interpret independently. As an example, chemical reactions of coal exist at the most commonly used liquefaction temperature of 400°C, and solubility alone is not sufficient to explain the complex phenomena at this temperature. The models described above assumed that solubility was the dominant factor. Recently, Squires, et al., (1983) initiated the qualitative comparison of their liquefaction data with pyrolysis data because thermal degradations cannot be overlooked in liquefaction. In fact, the interplay of C-O bond cleavages (e.g., Atherton, 1984), hydrogen donor (e.g., Bockrath, 1983) and mass transfer limitations (e.g., Hombach, 1980) make the quantitative interpretation very difficult.

It is well-known that a hydrogen donor or a hydrogen shuttling agent in a multicomponent solvent system can either enhance the liquefaction conversion or improve the product quality. In an entirely different approach, liquefaction efficiency was related to the compatibility of solubility parameters of coal and the solvent system (Hombach, 1980; Larsen, et al., 1983). A mixture of poor solvents for coal served as a better solvent when the solvent composition was chosen with respect to the solubility parameter. Similar results were also observed by Fong, et al. (1983) and Vasilakos, et al. (1983), although they took different directions in their interpretations. Fong, et al., claimed that smaller secondary components in the solvent enhanced the mass tranfer of solvent, and therefore the conversion. Vasilakos, et al. claimed that nonideal effects arose from polar/nonpolar solvent combinations. Finally, it should be noted that maximum solubility was both analytically and experimentally observed in solid-fluid equilibria at a particular pressure, even in a single component solvent (Kurnik and Reid, 1981). Therefore, the "synergistic effect" of a two component solvent must be explained with care to distinguish the effects of solubility, mass transfer, and chemical reactions.

Summarizing, all five factors noted should be considered in solvent design and data interpretation. The design of liquefaction solvents and conditions reported here is based on this rationale. In this paper, we report our study with single component solvents. The results of our multicomponent study will appear in a subsequent paper.

II. EXPERIMENTAL

Liquefaction with ammonia was performed with the system shown in Figure 1. A known amount (approximately 2g) of coal sample was placed in a 3/8" o.d. stainless steel tube about 1 ft. long. The sample was held in the tube by two unions containing mesh screen on either side of the union and some quartz wool in the middle of each union. The reactor was pressure tested for possible leaks with helium gas before the run. Then the system was evacuated for 10 min. In the mean time, a heating tape was tightly wrapped around the entire outer surface of the reactor. Four thermocouples (TC) were placed at four different spots along the tube to monitor the temperature distribution. An insulating tape was then wrapped on top of the heating tape to minimize heat loss by convection and to assure uniform temperature distribution along the tube.

The solvent density in the reactor was determined by filling the reactor with ammonia, venting a portion to a pre-evacuated cylinder. The amount vented was calculated based on the volume (717 ml) and the pressure reading of the expansion side. The difference between the amount of ammonia in the filled reactor and amount vented to the pre-evacuated cylinder provided the weight of NH_3 used in each experiment. The volumes of the reactor and expansion cylinder were determined by measuring their weight gain after filling up both compartments with liquids (methanol or water). Water and methanol gave consistent results.

When conducting liquefaction with liquid solvents, such as nbutylamine, the amount of solvent charged to the pre-evacuated reactor was measured with a buret.

The average heating rate between room and holding temperatures was about 5°C/min. Once the desired temperature was reached, the reactor section was disconnected from the main solvent transport line and from the expansion side. The mechanical shaker was then turned on and the system left at these conditions for the desired length of time.

At the end of the run the solvent and product were discharged, in most cases at reaction temperature, to a flask at atmospheric conditions. The reactor was then washed with a large amount (usually solvent:coal $\geq$ 30:1) of original liquefaction solvent. In selected cases the volatile materials produced in ammonia extraction, were condensed (by using a U tube packed with glass beads and immersed in an ice-cooled methanol-water bath). The adsorbed solvent was removed from the reactor by purging with helium for ten minutes,

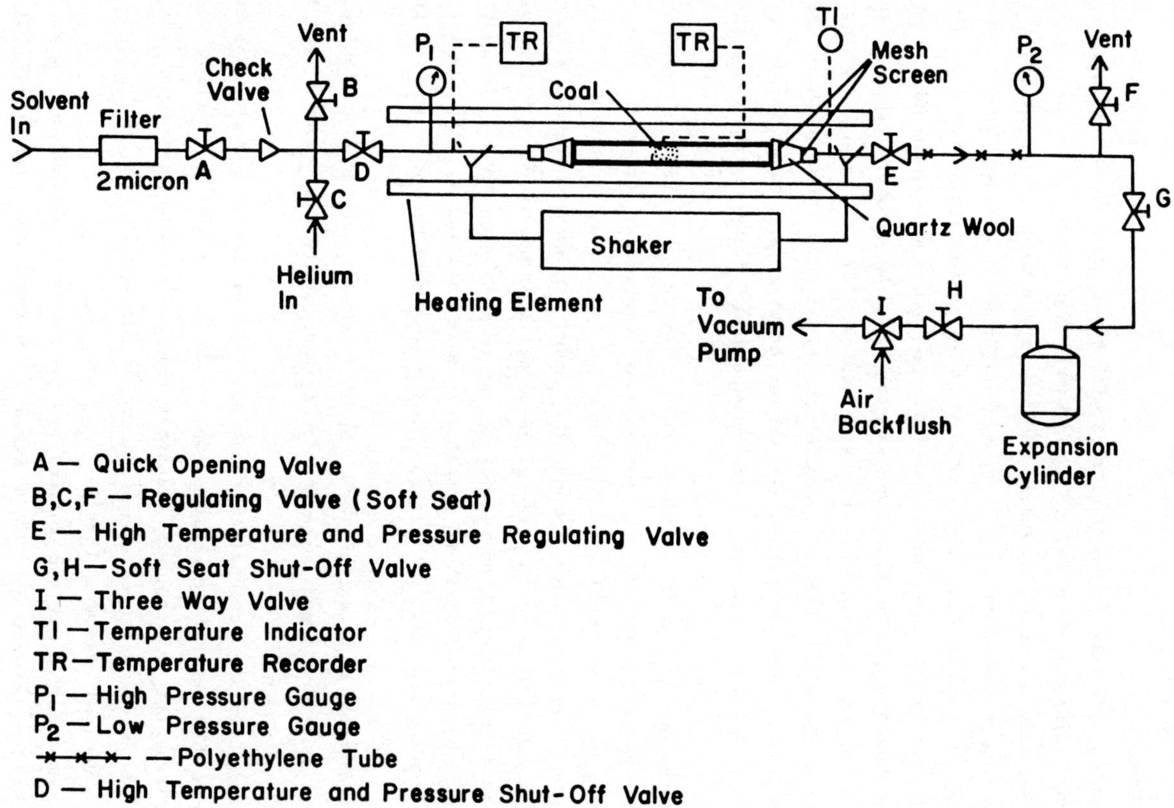

Figure 1. Experimental Arrangement

followed by vacuum stripping for five minutes. The 3/8" reactor tube was then disengaged from the system and weighed. The tube was stored in a desiccator containing $CaSO_4$ or P_2O_5 absorbent and weighed periodically until a constant weight was achieved.

Weight loss data were obtained from measurements of the whole reactor before and after the run, and from the direct measurements of samples taken out of the reactor after the experiments. The two methods were consistent within one percent which implied that products were not adsorbed on the reactor wall to any significant degree. In some cases, to retain maximum heavy products in the reactor, solvent was discharged at room temperature after the run. In these cases, condensation on the reactor wall could be significant and the procedure outlined was used to assure mass balance calculation.

It is assumed that water is completely removed by ammonia during liquefaction, i.e.,

$$\underset{\text{(daf)}}{\text{conversion to solvent extract}} = \frac{\text{as received weight loss - moisture content}}{\text{1 - moisture content - ash content}}$$

Since the primary interest is the conversion of organics in the coal to extracts, daf basis will be used in the subsequent discussion unless otherwise mentioned.

Three representative U.S. coals along with their analytical data were received from Pennsylvania State University at the beginning of the study. These coals included PSOC1351, Ill. #6 seam, high volatile C bituminous; PSOC-1405P, Upper Wyodak Seam, sub-bituminous B; and PSOC1406F, Kinnema Creek Seam, lignite. These minus 20 mesh samples were originally stored in 1-lb. cans filled with argon. After their arrival they were transferred into small vials under dry helium and stored in a helium-filled desiccator. A summary of the proximate and ultimate analyses are shown in Table 2.

III. RESULTS AND DISCUSSION

A. Coal Liquefaction with Supercritical Ammonia

The effects of temperature, reaction time, solvent density and coal type on the conversion of coal to soluble materials were examined. Similar trends were observed among the three coals we tested. North Dakota lignite data is presented here. Results of Illinois No. 6 and Wyodak coals have previously been reported (Chen, 1984a).

Temperature was the most important variable of the three investigated (Figure 2). Conversion of North Dakota lignite in ammonia increased continuously from 7 to 37 percent as reaction temperature was increased from 145 to 400°C. When compared with other liquefaction studies, ammonia gave higher

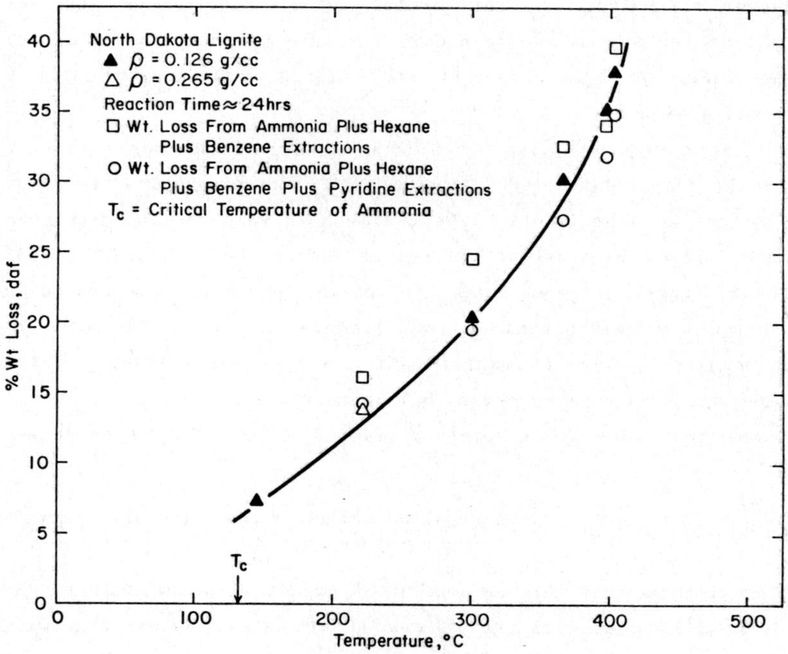

Figure 2. Effect of Temperature on the Conversion of North Dakota Lignite

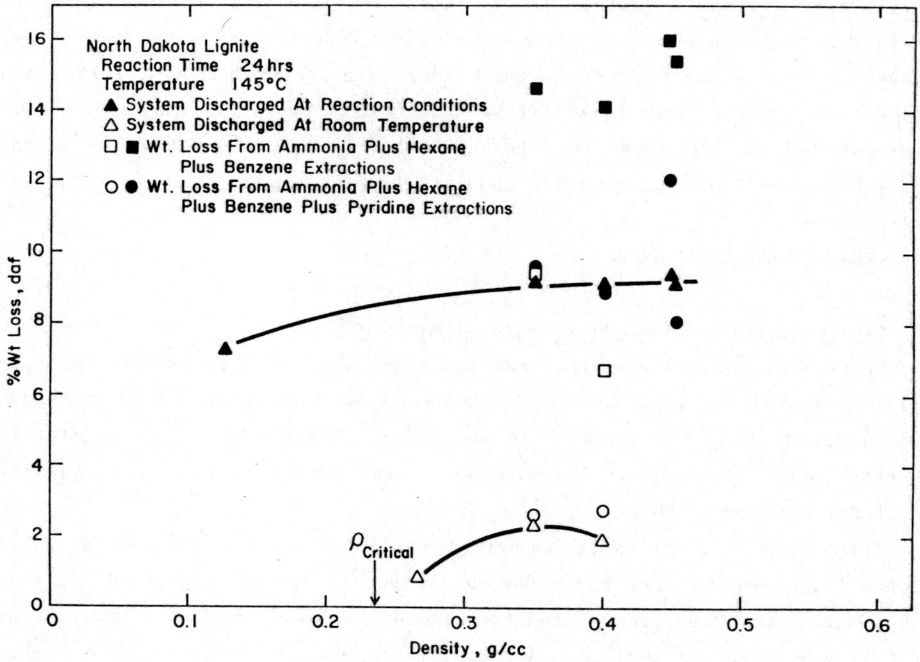

Figure 3. Effect of NH_3 Density on the Conversion of North Dakota Lignite

conversion than other non-nucleophilic supercritical fluids such as toluene (Fong, et al., 1983) and paraffins (Vasilakos, et al., 1983) but was comparable to supercritical alcohols (Blessing an Ross, 1978) and water (Barton, 1983; Deshpande, et al., 1984).

Figure 3 demonstrates that the effect of solvent density on conversion was very weak. No sharp increase in conversion was observed near the critical point. The strong temperature-dependence and weak density-dependence of yield has also been reported by other investigators of batch supercritical fluid extraction of coal such as Vasilakos, et al., (1983) and Fong, et al., (1983). Model compounds as well as theoretical calculations usually predict much higher increases in solubility around the critical point. Therefore, the weak dependence of coal conversion on solvent density could imply that factors other than solubility, i.e., chemical reactions and mass transfer limitations, are important if not dominant.

Figure 3 also illustrates that higher conversion was obtained at higher system discharge temperatures. Weight loss of coal samples was determined by measurement of the reactor system weight before and after the run. The system included coal sample, reactor tube, tube unions, quartz wool plugs, and mesh screens. Higher conversion at higher discharge temperature indicates that part of the product condensed during cooling and could not be extracted by fresh NH_3 at room temperature after discharge.

As shown in Figure 4, reaction time longer than 5 hours had little effect on conversion either at 145°or 400°C.

The effect of coal rank on liquefaction conversion is illustrated in Figure 5. A clear trend of decreasing conversion with increasing rank is observed. The atomic carbon-to-oxygen ratio, rather than carbon content, is used as an index of rank because oxygen function are generally considered to be the prime sites under nucleophilic attack. Indeed, carbon-to-oxygen ratio provides stronger correlation than carbon content or carbon-to-hydrogen ratio.

B. Comparative Studies

1. Ammonia Extraction Versus Vacuum Pyrolysis

To compare ammonia extraction with existing process concepts, vacuum pyrolysis and tetralin hydroliquefaction runs were conducted. Tetralin hydroliquefaction is described in Section B.2.

Yields of vacuum pyrolysis at 400°C are presented in Figure 6 for lignite. The yields of Illinois No. 6 coal and Wyodak coal can be found in an earlier report (Chen, 1984b). With Wyodak coal as an exception, pyrolysis of Illinois No. 6 and North Dakota coals provided only 55 percent yield from ammonia extraction. This demonstates that variables other than thermal decomposition are involved in NH_3 extraction. Pyrolysis of Wyodak coal, however,

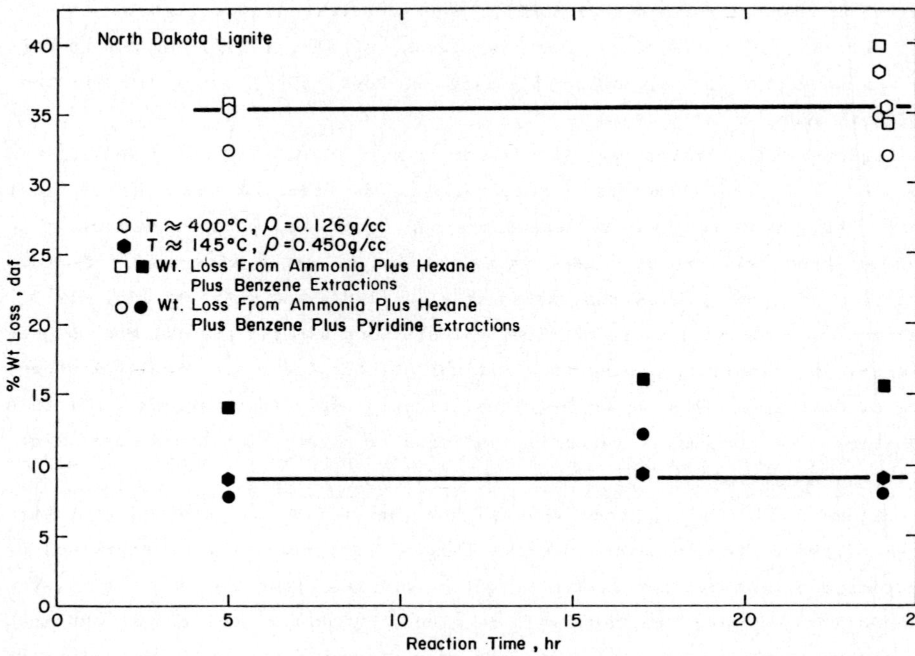

Figure 4. Effect of Reaction Time on the Conversion of North Dakota Lignite

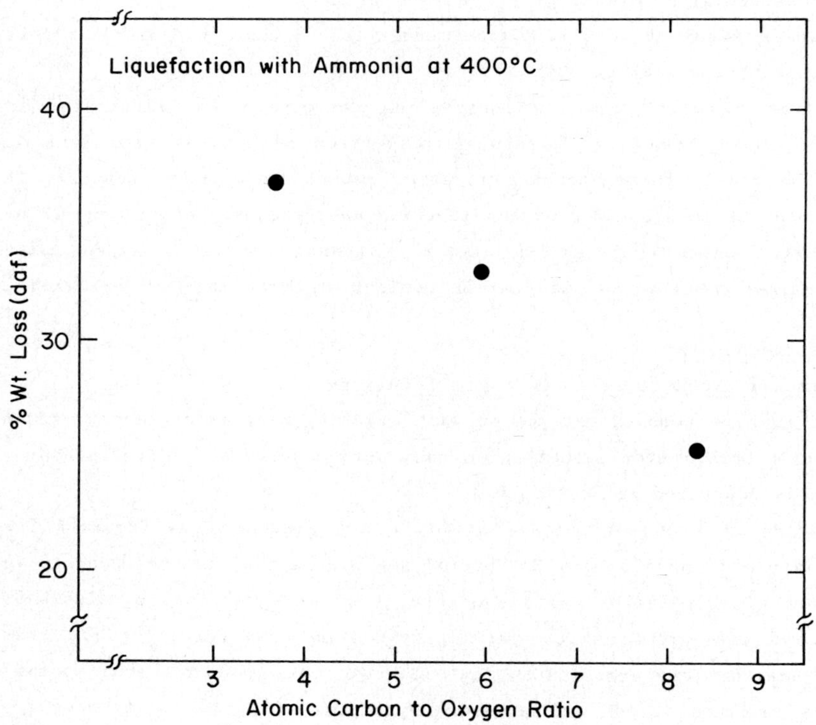

Figure 5. Effect of Coal Rank on Ammonia Extraction of Coal

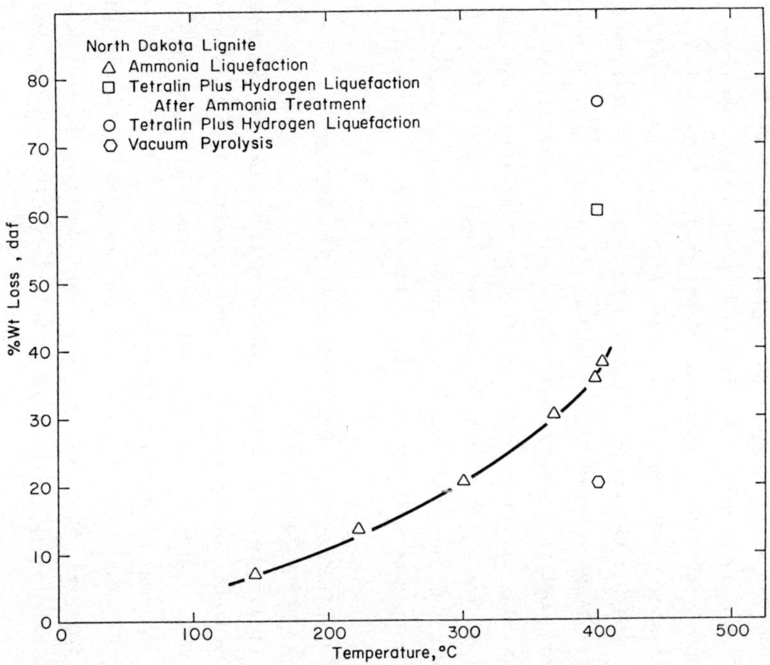

Figure 6. Comparison of NH_3 Extraction, Vacuum Pyrolysis and Tetralin Hydroliquefaction of North Dakota Lignite

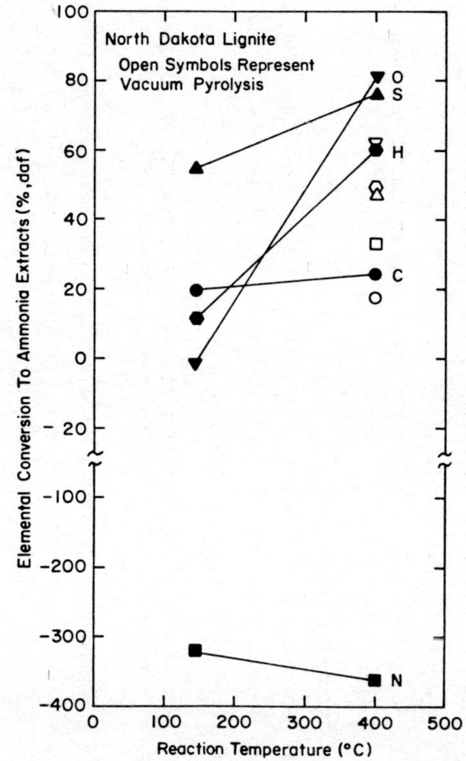

Figure 7. Elemental Conversion in NH_3 Extraction and Vacuum Pyrolysis

produced a surprisingly high yield (28%, daf,), which was only slightly lower than the extraction yield.

2. Ammonia Pretreatment Before Liquefaction with Tetralin and Hydrogen Mixture

Three liquefaction approaches are compared in this task:

--tetralin hydroliquefaction,

--tetralin hydroliquefaction with supercritical ammonia pretreatment, and

--ammonia extraction.

Combined liquefaction and Soxhlet extraction yields are presented in Figure 6. North Dakota lignite was used in all tests. The results support the following trend of conversion efficiency:

tetralin hydroliquefaction > tetralin hydroliquefaction with NH_3 pretreatment > NH_3 liquefaction

Single stage hydroliquefaction with tetralin was performed with a 3:1 solvent to coal ratio. The reactor was charged with hydrogen to 1000 psig at room temperature before heating to 400°C. Total heating time was two hours.

In the two-stage experiment, lignite was treated with supercritical NH_3 at 140°C for 17 hours. During the second stage the same tetralin hydroliquefaction reaction conditions were used. Unexpectedly, the yield from the pretreated coal (60.3% daf) was lower than the yield of single stage tetralin hydroliquefaction (76.2%). It should be mentioned, however, that the pretreatment temperature (140°C) was only slightly above the critical temperature and the oxygen functionals were basically retained in the residue after NH_3 treatment. If oxygen functionals are the primary sites subject to nucleophilic attack by solvent (Atherton, 1984), a higher temperature would be required for their removal. Our elemental analysis of the residue reinforced this concept (see Section C.2.).

C. Characterizations of Ammonia Extraction Products

1. Soxhlet Extraction

Solid residue after ammonia extraction was sequentially eluted by hexane, benzene and pyridine. Soxhlet extraction data are also shown in Figures 2-4. Note the solid lines are drawn based on NH_3 extraction data.

Several trends were observed.

--Hexane and benzene extractions provided some additional weight losses, implying that NH_3 is not able to remove the oil and asphaltene fractions completely, particularly the oil fraction. It is also possible that the incorporated nitrogen is removed in hexane and benzene extraction,

--All samples gained weight during pyridine extraction, indicating preasphaltene content of the residue is very low, if any,

--Oil and asphaltene fractions in the residue decrease with increasing temperature,

--As mentioned in the last section, weight loss after NH_3 extraction was higher at a higher discharge temperature. Figure 3 also demonstrates that Soxhlet extractions could not remove all the recondensed materials.

2. Elemental Analysis

Selected residue samples were sent to a commercial laboratory for elemental analysis. Their report for the lignite product is tabulated in Table 3, and similar data for Illinois No. 6 and Wyodak coals can be found in an earlier report (Chen, 1984b). For each coal, three sample categories were prepared. The first category was the residue produced by low temperature (145°C), but supercritical NH_3 treatment. The second category was produced by NH_3 extraction at 400°C under relatively low density. The final category of samples was vacuum pyrolysis residue.

The elemental analysis allowed us to calculate the conversion of each element to volatile species. Sulfur and nitrogen removal was of particular importance in determining the value of the char as a nonpolluting boiler fuel. Conversion calculations of each element were based on the weight loss in NH_3 extraction and elemental analyses of coal and residue. These calculated element conversions are listed in Table 3. Important observations are discussed below.

All the four major elements (C, H, O, S) show the same trend of conversion:

$$\text{NH3 } \begin{matrix}\text{extraction}\\ \text{at 400°C}\end{matrix} > \begin{matrix}\text{vacuum pyrolysis}\\ \text{at 400°C}\end{matrix} > NH_3 \begin{matrix}\text{ extraction}\\ \text{at 145°C}\end{matrix}$$

Oxygen is most sensitive to temperature change during NH_3 extraction. All three coals showed oxygen conversion above 81 percent after NH_3 extraction at 400°C. At 145°C, ammonia removed less than 19 percent oxygen. Vacuum pyrolysis resulted in the removal of no more than 61 percent of oxygen at 400°C. These results strengthened the theory that oxygen fucntionals in coal are the prime sites subject to nucleophilic attack by nitrogen-containing compounds, and there are a series of oxygen-related reactions below 300°C which enhance coal dissolution (Atherton, 1984).

At 400°C, ammonia extraction was able to remove more sulfur than vacuum pyrolysis. It is also interesting to note that about 50 percent sulfur was removed from the two low sulfur coals by NH_3 at 145°C. In fact, for North

Dakota lignite, 145°C NH_3 extraction was more effective than vacuum pyrolysis at 400°C. For Illinois No. 6 coal, high sulfur conversion was not observed at 145°C.

All the ammonia treated samples exhibited nitrogen uptake. Nitrogen content increased 2.6 times after liquefaction with ammonia. This corresponds to a nitrogen uptake of 3.4g per 100g of Illinois No. 6 coal (daf). The exact association of NH_3 with coal is not yet known. Even after the solid residue was flushed with dry helium and vacuum treated, NH_3 association still existed. Therefore, strong association is speculated. Similar amounts of solvent uptake were reported in numerous liquefaction studies in which nitrogen-containing solvents were used (Atherton and Kulik, 1982; Hanis et al., 1980; Collins et al., 1981; McNeil et al., 1983).

Among the three coals, nitrogen incorporation decreased with increasing coal rank and higher temperature also enhanced nitrogen uptake. Both observations reinforce the speculation that NH_3 forms covalent bonds with coal during extraction because low rank coal and high temperature favor higher concentrations of reactive intermediates during the reaction.

In Table 3, atomic hydrogen-to-carbon ratios of residues are reported using two methods of calculation. The first method involved direct calculation of H/C ratios based on the results of elemental analysis. In the second method, hydrogen as ammonia was excluded from the hydrogen content of residue. It is interesting to note that the residues produced at 145°C have H/C ratios close to the original coals. At 400°C, NH_3 extraction produced higher H/C residues than vacuum pyrolysis.

D. Coal Liquefaction with Supercritical Amines

Strikingly high conversion of coal to soluble material was observed in liquefaction with primary aliphatic amine at mild conditions. Figure 8 compares the data we obtained for the butylamine extraction of three coals with our vacuum pyrolysis and tetralin hydroliquefaction yields, as well as some recently reported data on liquefaction with distillate cuts and tetraline (Bockrath and Illig, 1984), THQ (Atherton and Kulik, 1982), and indole and indoline (Padrick, et al. 1984). Complete dissolution of North Dakota lignite and Wyodak coal were observed below 300°C using a solvent to coal ratio 4:1 and a contact time of five hours. This operating temperature is far below those used in advanced coal liquefaction processes. Illinois No. 6 coal produced lower extracts, but the level of dissolution at such low temperature is unusually high.

To investigate the role of supercritical solvent condition in liquefaction, several runs were performed with lower solvent to coal ratios (S:C).

TABLE 3. Elemental analysis and conversion of chemical elements to NH_3 extracts North Dakota Lignite

	Coal	Product After NH_3 Liquefaction at 145°C (ND 4)	Product After NH_3 Liquefaction at 400°C (ND 10)	Product After Vacuum Pyrolysis at 400°C (ND 21)
% C, daf	67.72	59.88	79.24	80.24
% H, daf	4.97	4.83	3.06	3.58
% N, daf	1.40	7.10	10.00	1.34
% S, daf	1.33	0.66	0.49	1.01
% O, daf, (by diff)	24.58	27.50	7.21	13.83
% Ash, dry basis	11.70	8.64	14.14	12.90
% Moisture	29.64	9.93	1.62	4.22
H/C, Atomic	0.88	0.97	0.46	0.54
H/C, Atomic (NH_3 incorporation excluded)*	--	0.77	0.23	--

Conversion to NH_3 Extracts, %

	After NH_3 Liquefaction at 145°C (ND 4)	After NH_3 Liquefaction at 400°C (ND 10)	After Vacuum Pyrolysis at 400°C (ND 21)
C	19.63	24.22	17.53
H	11.67	60.13	49.87
N	-321.34	-362.57	33.38
S	54.90	76.14	47.15
O	0.0	81.00	60.84

*This calculation is based on the assumption that NH_3 incorporates with coal at the molecular level involving no chemical reaction.

Complete conversion of lignite was still achieved at 295°C when solvent/coal was reduced to 1/1. Since the critical pressure of butylamine at 275°C corresponds to a solvent/coal ratio of 2.5/1, complete conversion at subcritical conditions could imply that the liquefaction yield is dominated by chemistry, not solubility. Further verification of this conclusion is needed.

Lignite conversion was also found to be very dependent on carbon chain length of the liquefaction amine. Four solvents were examined and the correlation between conversion and the carbon number is illustrated in Figure 9.

E. Characterization of Amine Extraction Products

To explore the governing chemistry and examine product quality, a set of characterization techniques was used after fractionation. Figure 10 summarizes the fractionation procedure for the product solution containing both the liquefaction solvent and washing solvent. Fractionation involved centrifugation, filtration and rotatary evaporation. About 8.4 percent of daf coal was separated as solid residue after centrifuging. The filter paper gained 12.1mg, which corresponded to about 2.3 percent of daf coal. Centrifuge residues contained 55% ash (dry basis). From energy dispersive x-ray analysis (EDAX 9100/40), this residue contained high concentrations of calcium and silicon. It also contained measurable amounts of Al, S, Fe, Cu, Mg and Ni.

Two distinct fractions were obtained after rotary evaporation. One was a clear liquid fraction and the other a black tarry material. It is interesting to note that the tarry fraction had a constant weight 1.84 times the original daf coal weight. This implies that extensive solvent incorporation occurred during liquefaction.

Representative chromatograms of solvent and distillate fraction after evaporation are shown in Figure 11. These two distinct chromatograms could imply that butylamine underwent chemical reactions with coal and contributed several major peaks in distillate. Since the distillate contained a large amount of butylamine (about 25 times the original coal weight) these major constituents in distillate cannot be coal-derived products. In-depth interpretation of the mass spectra of these compounds are currently being performed. It is also worth mentioning that a test liquefaction run without coal was conducted under the same conditions and the solvent chromatogram remained exactly the same as the original solvent chromatogram.

The tarry fraction were characterized by a variety of analytical tools. Some analytical results are shown in Figure 12. In sequential Soxhlet extractions, the procedure developed by Mima, et al. (1976) was employed. Conversion of this solid product to hexane, benzene and pyridine solubles were 40.2%, 38.1% and 20.0%, respectively. Only 1.7% was pyridine insoluble.

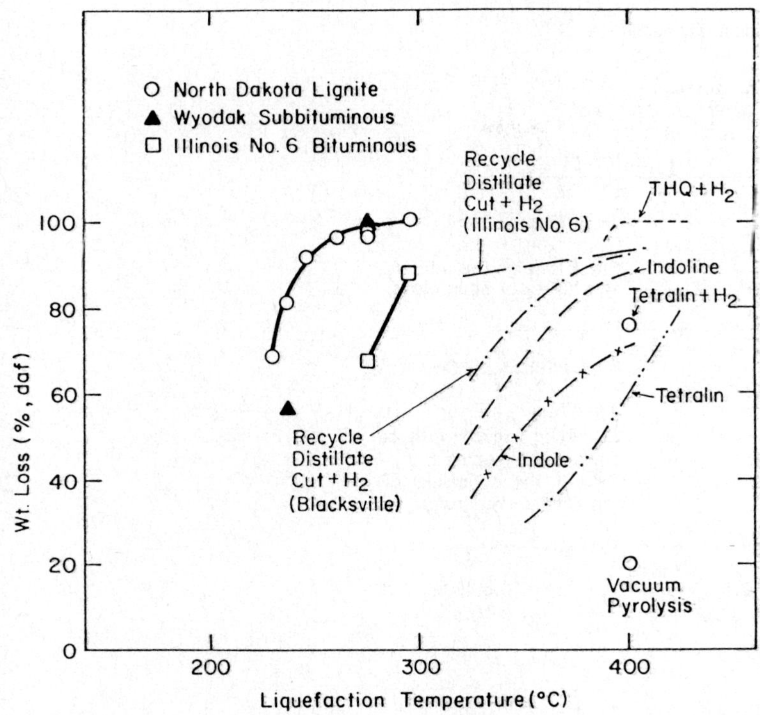

Figure 8. Effect of Temperature on Yield of Coal Liquefaction with Butylamine

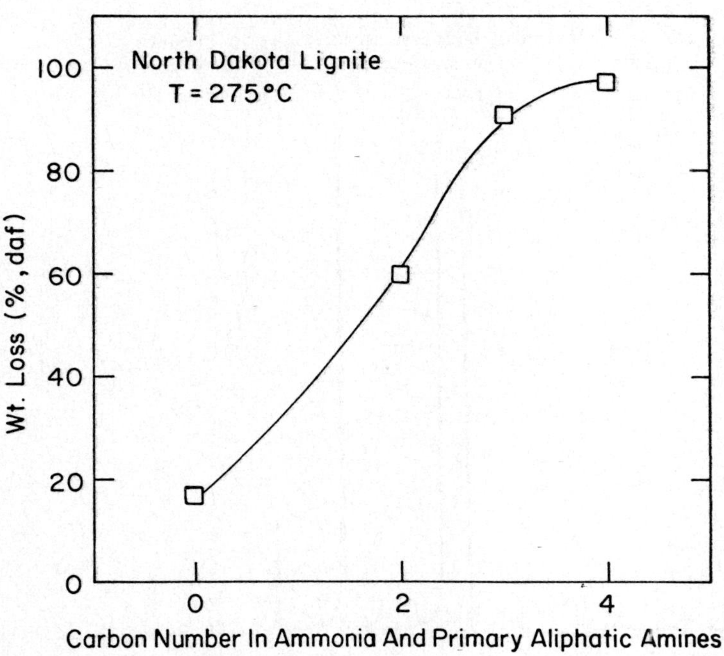

Figure 9. Carbon Number Effect of Nitrogen-Containing Solvent on Conversion

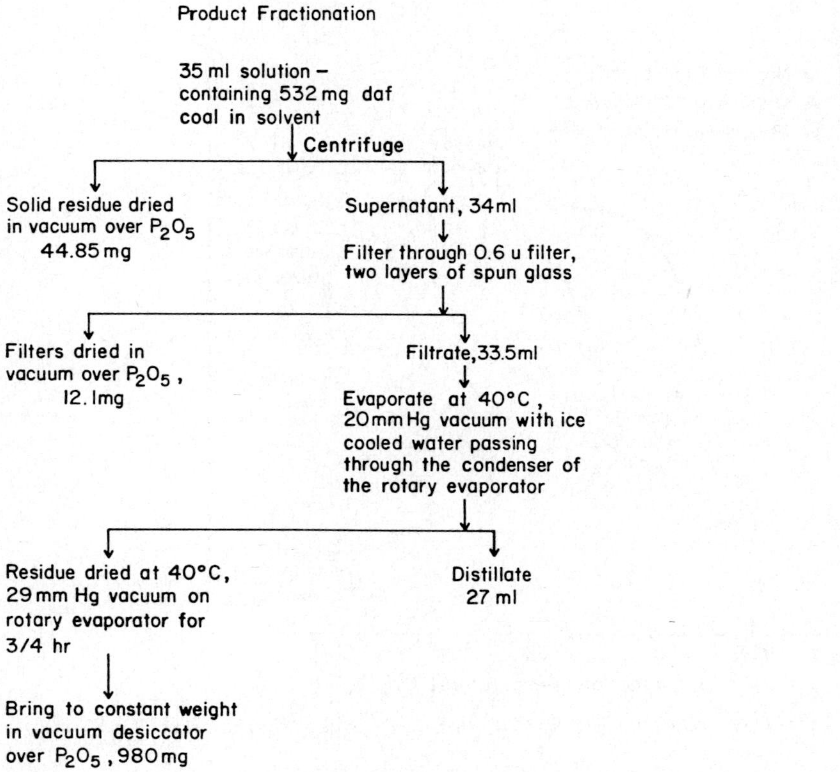

Figure 10. Fractionation Procedure for Butylamine Extraction Products

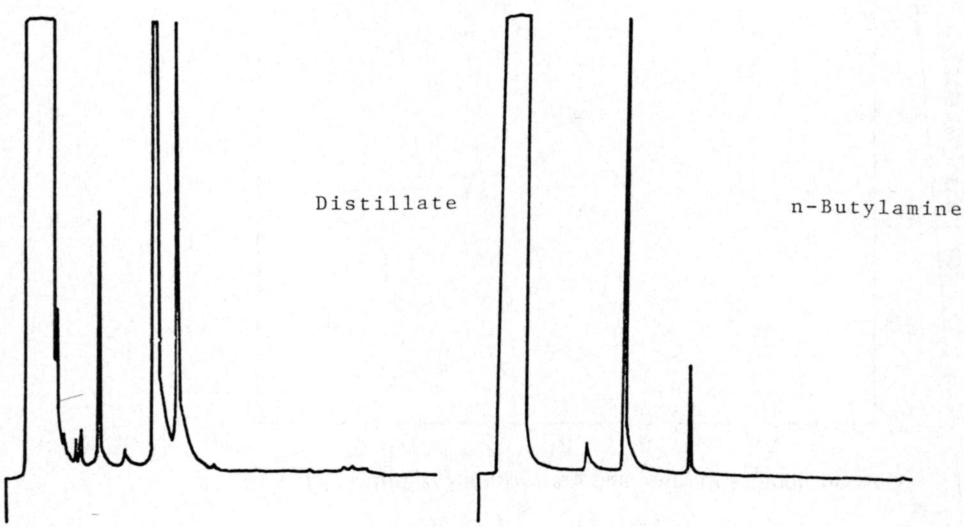

Figure 11. Chromatograms of Solvent and Distillate Fraction

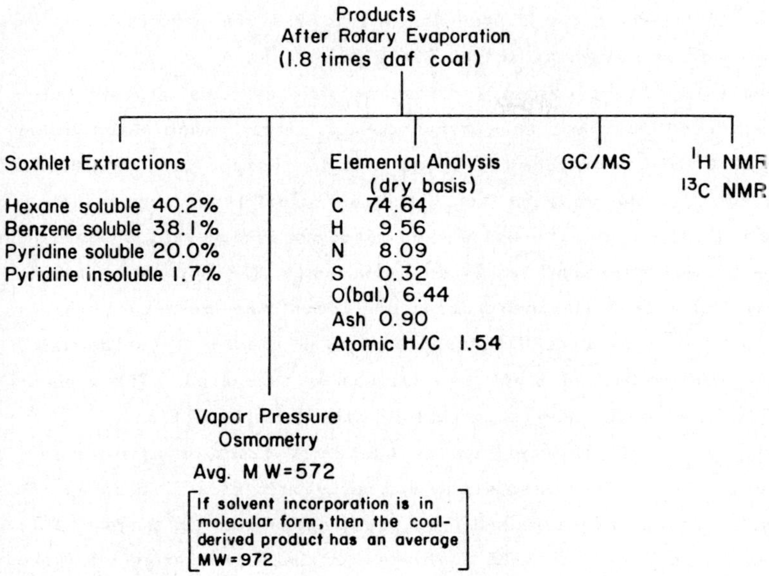

Figure 12. Analyses Performed to Tarry Product After Rotary Evaporation

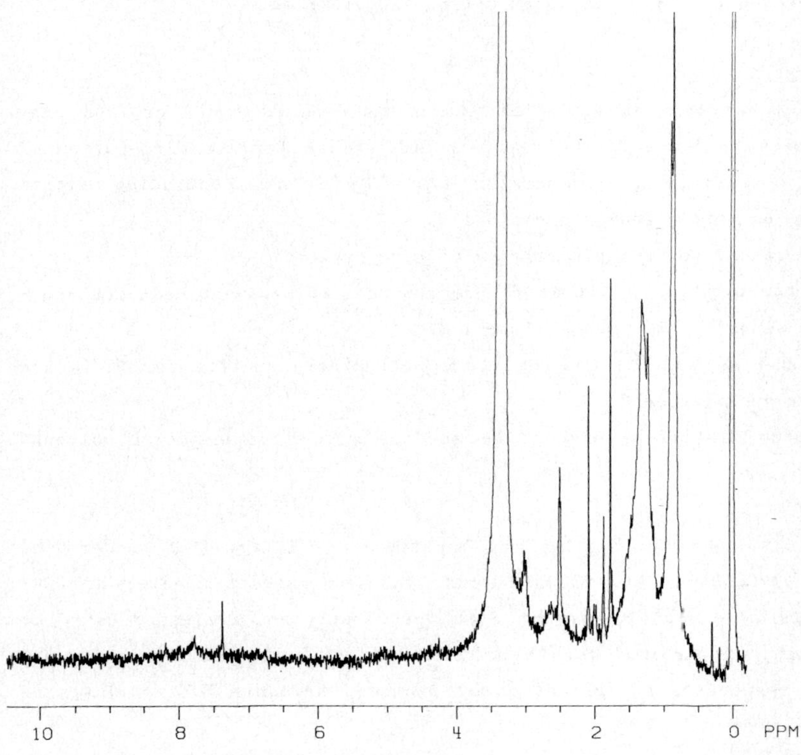

Figure 13. ^{1}H NMR Spectrum of Solid Fraction After Rotary Evaporation (dissolved in dimethylsulfoxide)

Pyridine was also used as the solvent in vapor pressure osmometry analysis. The measured average molecular weight was 572.

Elemental analysis also provided interesting insights into liquefaction chemistry. As mentioned earlier, the tarry material after evaporation had a weight 1.8 times that of the original daf coal. If the weight gain was due to incorporated solvent at the original solvent molecular level involving no chemical reaction with the coal, the calculated nitrogen incorporation would be 9.5 percent. The actual elemental analysis shows only 8.1 percent nitrogen content. This may imply that the hydrocarbon portion of the amine is incorporated into coal to a greater extent than the nitrogen element. Furthermore, elemental analysis also indicated that the coal was hydrogenated. The product oxygen content (6.4%) was much lower than that of the coal (24.6%).

Figure 13 shows the ^{1}H NMR spectrum of the tarry fraction after evaporation. The solvent used in the analysis was dimethylsulfoxide. It is interesting to note that no clear aromatic hydrogen peaks appear around 7 ppm. This observation led us to perform a ^{13}C NMR analysis. A sharp peak around 120 ppm indicates the existence of aromatic carbon. The absence of aromatic hydrogen could imply that the carbons at the outskirts of the fused aromatic clusters were fully hydrogenated or substituted during liquefaction.

CONCLUSION

This study has demonstrated that the use of amines could provide significant approach to low severity coal liquefaction. Complete dissolution of low rank coal is feasible at a temperature below 300°C. Our continuing efforts will be focused on the following questions.

1. What is the governing chemistry of this process?
2. Can the process, in terms of product quality, solvent cost and reaction severity, be further improved?
3. Can the product be upgraded to an acceptable quality for fuels and other applications?
4. Can the distillate product be used as a coal liquefaction solvent?

ACKNOWLEDGEMENT

This work is supported by the U.S. Department of Energy (DOE) under contract DE-AC22-83PC60046. Dr. Lloyd Lorenzi is the contract manager at DOE. The authors are also grateful to Dr. John Vercellotti and Professor David C. Baker (University of Alabama) for their consultation on product separation and NMR analyses, respectively. Brenda Zimny provided valuable EDXA analysis of residue.

LITERATURE REFERENCES

Amestica, L., and E.E. Wolf, "Supercritical Liquefaction of an Illinois No. 6 Coal," Proceedings of 1983 International Conference on Coal Science, International Energy Agency, p. 3, August 15-19.

Atherton, L.F., Coal Liquefacton Research Status Report, EPRI Journal, Jan/Feb, 1984, pp. 37-39.

Atherton, L.F. and C.J. Kulik, "Advanced Coal Liquefaction," Paper presented to the 1982 AIChE Annual Meeting, Los Angeles, CA, November, 1982.

Bartle, K.D., T.G. Martin, and D.F. Williams, "Chemical Nature of a Supercritical-Gas Extract of Coal at 350°C," Fuel 54, 10, 226 (1975).

Barton, P., "Supercritical Separation in Aqueous Coal Liquefaction with Impregnated Catalyst," ACS Div. Fuel Chem., Preprints, 27 (3-4), 53 (1982).

Blessing, J.E. and D.S. Ross, "Supercritical Solvents and the Dissolution of Coal and Lignite," in "Organic Chemistry of Coal," ed. by J.W. Larsen, ACS Symposium Series, 71, 171 (1978).

Bockrath, B.C., "Chemistry of Hydrogen Donor Solvents," in "Coal Science, Volume II," Ed. by M.L. Gorbaty, J.W. Larsen, and I. Wender, Academic Press, New York, 1983.

Bockrath, B.C., and E.G. Illig, "Coal Liquefaction Utilization at Low Temperatures,"Preprint of ACD Div. of Fuel Chemistry, 29(5), 76 (1984).

Brooks, J.D., and S. Sternhell, "The Action of Alkalies on Low-Rank Coals," Fuel, 37, 124 (1958).

Camier, R.J., S.R. Siemon, H.A.J. Battaerd and B.R. Stanmore, "The Nature and Possible Significance of Coal," in "Coal Structure," Ed. by M.L. Gorbaty and K. Duchi, ACS Advances in Chemistry Series, 192, 311 (1981).

Ceylan, R., and A. Olcay, "Supercritical-Gas Extraction of Turkish Coking Coal," Fuel, 60, 3, 197 (1981).

Chen, W.Y., "Enhanced Coal Liquefaction with Ammonia and Amines in Single and Binary Solvent Systems," Quarterly Report to U.S. Dept. of Energy, DE-PC/60046-2, 1984a.

Chen, W.Y., "Enhanced Coal Liquefaction with Ammonia and Amines in Single and Binary Solvent Systems," Quarterly Report to U.S. Dept. of Energy, DE-PC/60046-3, 1984b.

Corcoran, W., W. Fong, P. Pichaichanarong, P. Chan, and D. Lawson, "Supercritical-Multiple-Solvent Extraction from Coal," JPL Invention Report, NOP-15767/5224, 1983.

Deshpande, G.V., G.D. Holder, A.A. Bishop, J. Gopal, and I. Wender, "Extraction of Coal Using Supercritical Water," Fuel, 63 (7), 956 (1984).

Fetzer, J.C., J.A. Graham, R.F. Arrendale, M.S. Klee, and L.B. Rogers, "Characterization of Carbonaceous Materials Using Extraction with Supercritical Pentane," Separation Science and Technology, 16, 1, 97 (1981).

Finn, M.J., C.J. Bower, and R.D. Hughes, "The Extraction of Coal with Supercritictal Hydrocarbon Mixtures," Fluid Phase Equilibria, 10, 327 (1983).

Fisher, C. H., and A. Eisner, "Primary Liquefaction by Destructive Hydrogeneration of the Oxygen, Nitrogen and Sulfur Linkages," Ind. Eng. Chem., 29(12), 1371 (1937).

Fong, W.S., P.C.F. Chan, P. Pichaichanarong, W.H. Corcoran, and D.D. Lawson, "Experimental Observations on a Systematic Approach to Supercritical Extraction of Coal," in "Chemical Engineering at Supercritical Fluid Conditions," ed. by M.E. Paulaitis, J.M.L. Penninger, R.D. Gray, Jr., and P. Davidson, p. 377, Ann Arbor Science, 1983.

Giddings, J.C., M.N. Myers, L. McLaren, and R.D. Keller, "High Pressure Gas Chromatography of Nonvolative Species," Science, 162, (10), 67 (1968).

Giddings, J.C., M.N. Myers, and J.W. King, "Dense Gas Chromatography at Pressures to 2000 Atmospheres," J. Chromatographic Sci., 7 (5), 276 (1969).

Graff, R. A., and S.D. Brandes, "Coal Liquefaction by Steam Pyrolysis," ACS Preprints, Div. of Fuel Chem., 29 (2), 104 (1984).

Hannay, J.B., and J. Hogarth, "On the Solubility of Solids in Gases," Proc. Roy. Soc. (London), Ser. A, 29, 324 (1879).

Harris, R.L., L.H. Simons, and J.J. Lagowski, "The Dissolution of Lignite in Anhydrous Liquid Ammonia," ACS Preprints, Div. of Fuel Chem., 25 (1), 264 (1980).

Homback, H.P., "General Aspects of Coal Solubility," Fuel, 59 (7), 465 (1980).

Jenkins, R., "Method for Treating Coal to Obtain Refined Carbonaceous Materials," U.S. Pat. 4,319,980; March, 1982.

Kershaw, J.R., G. Barrass, D. Gray, and J. Zezko, "Process Effects on the Nature of Coal Liquefaction Products," Fuel, 59, 6, 413 (1980).

Kurnik, R.T., and R.C. Reid, "Solubility Extrema in Solid-Fluid Equilibria," AIChE Journal, 27 (5), 861 (1981).

Larsen, J.W., T.K. Green, and I. Chiri, "Contributions of Non-Covalent Interactions to the Macromolecular Structure of Bituminous Coals," Proceedings of 1983 International Conference on Coal Science, p. 277.

Maddocks, R.R., and J. Gibson, "Supercritical Extraction of Coal," Chem. Eng. Progr., p. 59, June 1977.

Maddocks, R.R., J. Gibson, and D.F. Williams, "Supercritical Extraction of Coal," Chem. Eng. Progr., June 1979, p. 49.

Matida, M., Y. Nishiyama and Y. Tamai, "Gasification of Coals Treated with Non-Aqueous Solvents 1. Liquid Ammonia Treatment of Bituminous Coal," Fuel, 56 (4), 177 (1977).

Mima, M.J., H. Schultz, and W.E. McKinstry, "Method for the Determination of Benzene Insolubles, Asphaltenes, and Oils in Coal Derived Liquids," PERC-RI-7616.

Mortimer, J., "Thermal Behavior of a Supercritical Gas Extract of Coal," Fuel, 59 (11), 782 (1980).

Padrick, T.D. and S.J. Lockwood, "Liquefation of Wodak Coal in Indoline," Preprint of ACS Div. of Fuel Chem., 29(5), 92 (1984).

Poska, F.L., "Supercritical Tar Sand Extraction," U.S. Pat. 4,341,619; July, 1982.

Pullen, J.R., "Solvent Extraction of Coal," in "Coal Science, Volume II," Ed. by M.L. Gorbaty, J.W. Larsen, and I. Wender, Academic Press, New York, 1983.

Ross, D.S. and Q. Nguyen, "Coal Conversion in Aqueous Systems," Fluid Phase Equilibria, 10, 319 (1983).

Ross, D.S., T.K. Green, and R. Mansani, "Coal Conversion in Supercritical Water Systems," Proceeding of 1983 International Conference on Coal Science, International Engergy Agency, p. 10, August 15-19.

Ross, D.S., D.F. McMillen, W.C. Ogier, R.H. Fleming, and G.M. Hum, "Exploratory Study of Coal Conversion Chemistry," Quarterly Report No. 4, DOE/PC/40785-4, 1982.

Rowlinson, J.S. and M.J. Richardson "The Solubility of Solids in Compressed Gases" Advan. Chem. Phys. 2, 85 (1959).

Smith, R.D., B.W. Wright, and H.R. Udseth, "Supercritical Fluid Methods for Coal Extraction, Separation, and Analysis," Preprints of ACS Div. of Fueld Chem., 28 (4), 235 (1983).

Solomon, P.R., R.H. Hobbs, D.G. Hamblen, W.Y. Chen, A.I. LaCava, and R.A. Graff, "Correlation of Coal Volatile Yield with Oxygen and Aliphatic Hydrogen," Fuel, 60(4), 342 (1981).

Squires, T.G., T. Aida, Y.Y. Chen, and B.F. Smith, "The Role of Thermal Chemical Processes in Supercritical Gas Extraction of Coal," Preprints of ACS, Div. of Fuel Chem., 28 (4), 228 (1983).

Sunol, A.K., "Supercritical Extraction of Coal," Ph.D. Dissertation, Virginia Polytechnic Institute and State University, 1982.

Takegami, Y., S. Kajiyama and C. Tokokawa, "Studies on the Chemical Structure of Coal II - Hydrogenolysis of Bituminous Coal and Humic Acid in the Presence of the Adkins Copper-Chromite Catalyst," Fuel, 42, 291 (1963).

Tiffany, D.M., T.J. Houser, M.E. McCarville, and M.E. Houghton, "Reactivity of Some Nitrogen-Containing Compounds at Supercritical Water Conditions," Preprint of ACS Div. of Fuel Chem., 29 (5), 56 (1984).

Tomita, A., T. Tano, Y. Oikawa and Y. Tamai, "Gasification of Coals Treated with Non-Aqueous Solvent. 5. Properties of Coals Treated with Ammonia, "Fuel, 58(8), 609 (1979).

van Bodegom, B., J.A.B. van Veen, G.M.M. van Kessel, M.W.A. Sinnige-Nijssen and H.C.M. Stuiver, "Action of Solvents on Coal at Low Temperature 1. Low-rank Coals," Fuel, 63 (3), 346 (1984).

Vasilakos, N.P., J.M. Dobbs, and A.S. Parisi, "Solvent Effects in Supercritical Extraction of Coal," Preprints of ACS, Div. of Fuel Chem., 28 (4), 212 (1983).

Warzel, F.M., "Solid Liquid Extraction," U.S. Pat. 4,376,693; March, 1983.

Wilhelm, A. and K. Hedden, "Nonisothermal Extraction of Coal with Solvents in Liquid and Supercritical State," Proceedings of 1983 International Conference on Coal Science, International Energy Agency, p. 6, August 15-19.

Williams, D.F., "Extraction with Supercritical Gases," Chem. Eng. Sci., 36 11, 1769 (1981).

Williams, D.F. and T.G. Martin, "Extraction of Oil Shales and Tar Sands, "U.S. Patent 4,108,760, August, 1978a.

Williams, D.F., and T.G. Martin, "Gas Extraction of a Western Sub-Bituminous Coal," Report prepared for National Coal Board, England, for Electric Power Research Institute under Project 779-1, Report No. 699, February 1978b.

Wise, W.S.D., "Method of Extracting Materials," U.S. Patent No. 3,558,468, Jan. 26, 1971.

Supercritical Fluid Technology, edited by J.M.L. Penninger, M. Radosz, M.A. McHugh and V.J. Krukonis
1985 Elsevier Science Publishers B.V., Amsterdam — Printed in The Netherlands

EXTRACTION OF OIL FROM WYOMING COAL WITH AQUEOUS SOLVENTS AT ELEVATED PRESSURE

J.M.L. PENNINGER*

Occidental Research Corporation,

2100 Main Street, Irvine California 92714 (USA)

ABSTRACT

Wyoming coal is extracted in a flow of steam admixed with other components, at temperatures and pressures ranging from 350 to 450 C and 3.6 to 10.9 MPa.

Relative to Fischer Assay analysis the oil yield is almost tripled when carbonmonoxide is added to steam at a pressure of 10.9 MPa. Hydrogen is far less effective as co-component: nitrogen-steam doesnot improve oil yield,while carbondioxide-steam results in lower than Fischer assay oil yields. Upon cooling to ambient of the hot extractor effluent, oils separate as the top layer of the aqueous condensate.

The oils consists for 85 % or more of hexane soluble material with the balance made up of asphalthenes. Preasphalthenes are selectively retained by the char.

INTRODUCTION

Extraction with solvents at (super)critical conditions is one of the technologies pioneered to transform coal into liquid hydrocarbons. The use of critical fluids, as compared to liquids which are practized in every H-donor coal liquefaction scheme, has the advantage of facile separation of extracted oil from residues, such as inorganic coal constituents. There still remains a separation problem between solvent and extracted oil. By manipulating temperature and pressure the oil may separate but full recovery of the solvent may need destillation.

One has learned during the last decade of renewed coal liquefaction research that the organic solvents also undergo chemical reactions in the extraction process. These reactions decrease the H-donor capability of the solvent (e.g. isomerization and other structural rearrangements) and chemically bind solvent molecules to the extracted oil molecules or the residual char. Organic solvents, which operate at near-critical conditions, also exhibit this undesired chemical behaviour.

---

*present address: Boerhaavelaan 18, 7550 BC Hengelo (O) (The Netherlands)

A preferred solvent for coal would therefore have following properties:

- a critical temperature in the range where coal pyrolysis reactions occur
- no adduct formation with either coal oil or char
- no loss of solvent power by structural rearrangements
- high solubility for oils at the extraction temperature and pressure, zero solubility at some lower temperature level so that oil is separated from solvent by temperature swing.

Supprizingly, water comes very close:

- it has a critical temperature (374 C) where coal pyrolysis starts
- it is completely stable at (super) critical conditions
- hydrocarbons show strong miscibility with water at elevated temperature and pressure but separate spontaneously and complete upon cooling to ambient
- adducts of water with either hydrocarbons or char donot form.

However, in order to choose for critical fluid extraction conditions water should also show enhanced oil yields over simple atmospheric pyrolysis; it should be kept in mind that this enhancement criterium was originally behind the idea of (super)critical coal liquefaction. But it is now more and more recognized that many of the published high oil yields obtained with organic solvents are the result of adduct formation with the solvent.

Early work describes the conversion of coal with water and carbonmonoxide (refs. 1 and 2).

The COsteam process, as it was named later, became the subject for many investigations; a review is given by Ross (ref. 3). Typical for this work is that the liquefactions are done with coal slurried in high boiling organic solvent, with steam and CO as the reducing agent.

Only in the recent past the solvent capability of water near its critical point has been appreciated in coal liquefaction. The batch operations in autoclaves, where coal and water remain in contact also after cooling of the reactor, make it difficult to determine whether or not a single oil phase was formed and separated from the char at the extraction conditions. The oil becomes entrapped in the char upon cooling, so that a solid product is usually observed directly when the autoclave is opened.

Only in a few investigations the presence of a separate oil phase over coal has been unambiguously established by blow-down of the hot autoclave in a cooled separator (ref. 4) or by positioning the coal charge in the autoclave so that the condensate,formed by cooling of the reactor, doesnot contact the residual char (refs. 3 and 5).

All in all water appears of sufficient potential to explore the technical feasibility of an extraction process for coal with compressed steam as continously flowing solvent. In the following paragraphs we report results obtained with a bench scale extraction plant. More in particular we studied the effects of mixtures of steam and co-components, such as CO_2, N_2, H_2, CO, alkali, on coal conversion and oil yield.

EXPERIMENTAL PROCEDURES

Figure 1 illustrates schematically the bench scale extractcr. It consists of an extraction container which is heated electrically from the outside, an ambient temperature gas-liquid separator and facilities to supply a continuous flow of water and other compressed gases to the extractor.

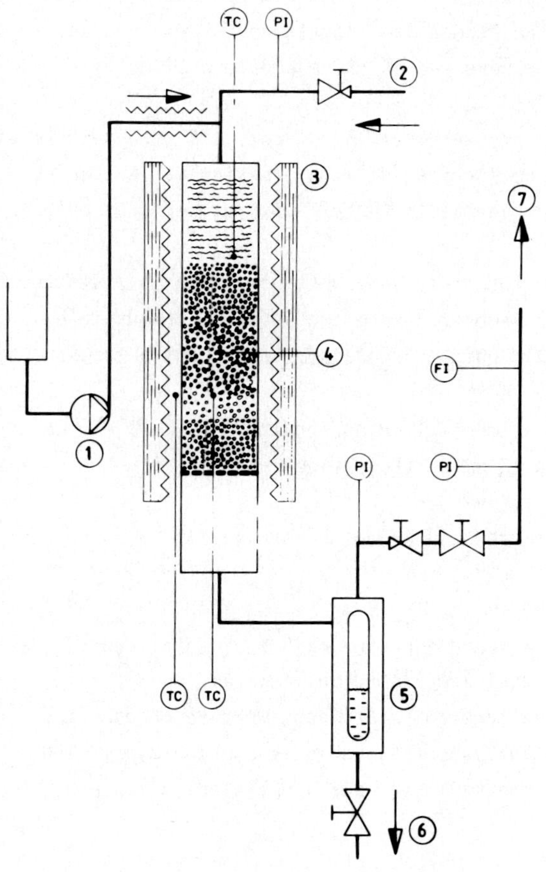

Fig. 1. Flow diagram of extraction unit. 1. Feedwater pump, 2. co-component. 3. Electrical furnace. 4. Coal bed. 5. HP separator, 6. Aqueous condensate. 7. Off gases.

The AISI 316 extraction container has a diameter of 34.2 mm and holds on a support plate a bed of 200 grams of coal, crushed and sized between 3.3 and 10 mm. The bed height is approximately 280 mm.

Temperatures are measured of the outside extractor wall at half coal bed height, in the center of the coal bed and at the top of the coal bed. The extractor is heated in about 20 to 30 minutes to a wall temperature of 400 to 450 C and held at that level for several hours. The temperature in the center of the bed is lower by 50-100 C but this difference tends to decrease with extraction time.

Water through a metering high pressure pump, and compressed gas from a high pressure tank, are supplied to the top of the extractor and flow downward. In the upper section of the extractor water is made to evaporize in a bed of porcelain chips, prior to contacting the coal bed.

The hot extractor effluent leaves at the bottom end and condenses in the separator at ambient temperature. The condensate is retained, the non-condensable gases are vented through a pressure relief valve and controlled by a mass flow meter.

The aqueous condensate is taken out periodically. It consists of a milky emulsion of oil in water. For complete recovery of oil the emulsion was extracted with diethylether and, after drying over $CaCl_2$ the oil was recovered by evaporization of the ether.

Oil, in our experiments, is therefore only the substance which is extracted with compressed steam from coal and recovered from the aqueous condensate by ether extraction. Any other material, which is retained in the extractor is classified as char.

Extractions were generally done at the following conditions:

- 200 gram of coal : 3.3 to 10 mm particle size
- feedrate of steam : 180 grams/h
- feedrate of co-component : 180 l/h (20 C, barometric pressure)
- extractor inlet pressure : 3.6-11 MPa
- extractor temperature : 350-440 C

Detailed information on extraction conditions of each individual run is listed in Table 1. Properties of feed coal are listed in Table 2.

As a base of reference for the extraction experiments we have choosen the yields obtained from Fischer Assay Analysis (FA); this is a batch type, slow heat pyrolysis in autogeneous atmosphere. These data are listed in Table 3.

TABLE 1

Extraction conditions of individual runs[a]

#	Solvent composition (molar ratio)	feedrate (grammole-min^{-1})	Pressure (MPa)	Extraction time (minutes)	Temperature (C)	Other characteristics
1	Steam-N_2; (1.33:1)	0.29	7.1	0-219	~ 400	-
				219-300	400-500	
2	Steam-CO_2;(1.33:1)	0.28	5.6	240	~ 420	-
3	CO;	0.125	7.0	30	~ 400	reactor effluent lines were plugged after 30 minutes
4	Steam-CO; (1.33:1)	0.29	3.6	0-110	~300-360	
				110-200	~ 450	
				200-260	~500-600	
5	Steam-CO; (1.33:1)	0.29	7.3	388	380-450	
6	Steam-CO; (1.33:1)	0.29	10.9	270	400-440	
7	Steam-CO; (1.33:1)	0.29	7.1	275	350-430	5% wt Na_2CO_3 in water
8	Steam-H_2-CO (1.33:1)	0.29	6.9	0-275	~ 350	
				275-375	~ 420	
9	Steam-CO; (1.33:1)	0.29	10.9	0-140	~ 350	coal sized between 0.3 and 0.85 mm
				140-240	~ 510	and impregnated with 1% wt Na_2CO_3
10	Steam-CO; (0.44:1)	0.16	6.9	659	~ 400	coal charged 355 grams
11	Steam-CO; (0.5:1 to 1:1)	non flow	10-14	3230	370-410	batch extraction[b]

[a]coal charged 200 grams except runs 10 and 11.
coal sized from 3.3 to 10 mm, except run 9.

[b]Batch extraction; coal charge of 409 grams. Extractor is charged with steam-CO untill listed pressure and left for several hours. Extractor is blown-down and effluent is condensed. After recharging with steam-CO the extraction is continued untill second blow-down. This procedure of charging and blow-down was repeated seventimes during the listed extraction time, with the same coal charge.

TABLE 2

Analysis of feed coal

Rank: Subbituminous, Wyoming

Elemental analysis (wt % as received)

C	60.89
H	5.56
S	0.64
N	1.15
Ash	6.05
Moisture	10.25
O	15.46 (bal)

TABLE 3

Fischer Assay Analysis of Wyoming coal ("as received" basis)

Product	Yield wt %
char	: 57.4
water	: 20.0
non-condensable gases	: 12.0
TAR	: 6.5

RESULTS

Oil yield

In all experiments we have observed that coal generates oil constituents which are volatile with the aqueous solvent and are continously removed from the coal bed. The volatily in steam is illustrated by Fig. 2 which shows that approximately half of the amount of volatile oil has a boiling point higher than the temperature of extraction. Upon cooling of the extractor effluent an opaque aqueous condensate is obtained which separates spontaneously at 50-70 C in a oil layer floating on top of water. No oil remained in transfer lines between extractor and separator; hence steam is an efficient carrier for oils.

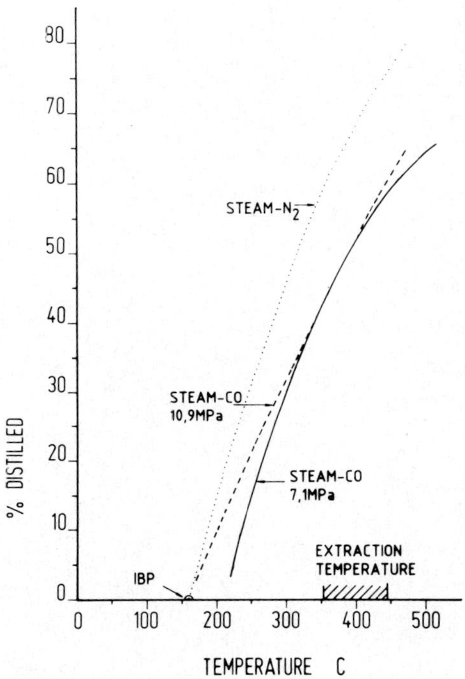

Fig. 2. Distillation curves of extracted oils; full line: vacuum distillation, broken/dotted line: GC simulated distillation.

The oils have HHVs from 37.6 MJ/kg to 38.3 MJ/kg (Fuel oil # 6 has 42.3 MJ/kg by comparison), pour points between 27 anc 35 C and viscosities of approximately 0.1 Pa:s (38 C).

The amount of oil produced is a distinct function of the composition of the solvent. By admixing different permanent gases with the steam such as nitrogen, carbondioxide, carbonmonoxide, hydrogen, it follows that the oil yield from Wyoming coal can be strongly effected. This is illustrated in Fig. 3. With the Fischer Assay yield of Wyoming coal as a reference level (FA line) we conclude that steam-N_2 doesnot enhance oil yield; CO_2 has even a negative effect on oil yield.

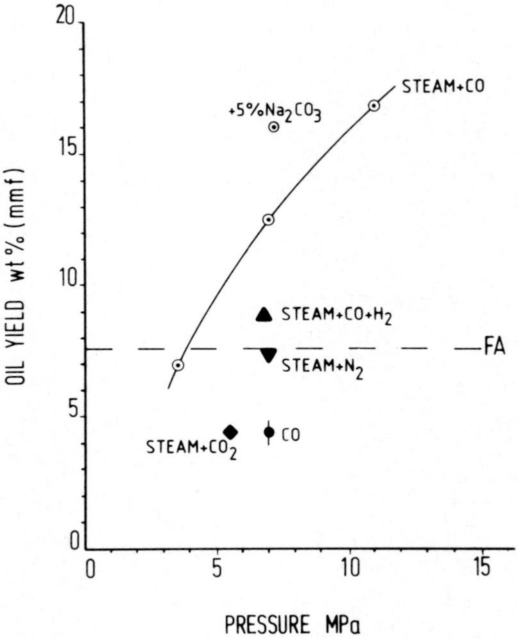

Fig. 3. Oil yields after 250 minutes of extraction

A promoting effect has steam-CO: this is even more pronounced by an increase in pressure to 10.9 MPa. This enhances the oil yield by a factor of 2.5-3 over FA level. On the contrary, carbonmonoxide by itself is a poor solvent; not only was the oil yield very low but the transfer lines between extractor and condensor were plugged by solid tar already after 30 minutes of operation. This in contrast to any other aqueous solvent, which could be kept in operation for any length of time. Substituting part of carbonmonoxide by hydrogen has a reverse effect on oil yield (Fig. 7). This is supprizing because hydrogen is in fact an essential reactant when coal is liquefied with donor solvents. Hydrogen for some reason is not active in the prevailing aqueous environment; indeed a similar observation made Ross et all (ref. 3). Addition of Na_2CO_3 to steam enhances the oil yield which is especially noticable at longer extraction times (Fig. 8).

The differentiation in oil yields as observed can have physical as well as chemical reasons. Physical in the sense that solvent properties of steam are inversily effected by CO_2 or promoted by CO. Chemical in the sense that certain combinations of steam and co-components activate specific chemistry which favour or retard the formation of oil. The rate of oil production over a extended time span and the composition of the oil might shed some light on this question. The oil production with the extraction time is illustrated in Fig. 4 through 8.

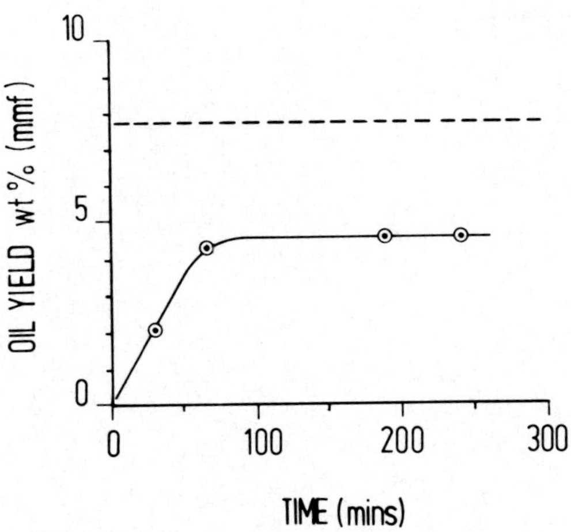

Fig. 4. Steam-CO_2 extraction (# 2)

Mixtures of steam- N_2 and steam-CO_2 both shown an initial oil production but this levels off very soon: the steam-N_2 mixture approaches the FA level while the steam-CO_2 mixture reaches only approximately 60 % of FA level. Mixtures of steam-CO are most productive especially at the highest pressure of 10.9 MPa. The curves suggest that more oil can still be recovered e.g. by extended extraction. Fig. 7 shows very clearly how oil production is retarted when H_2 is substituted for CO.

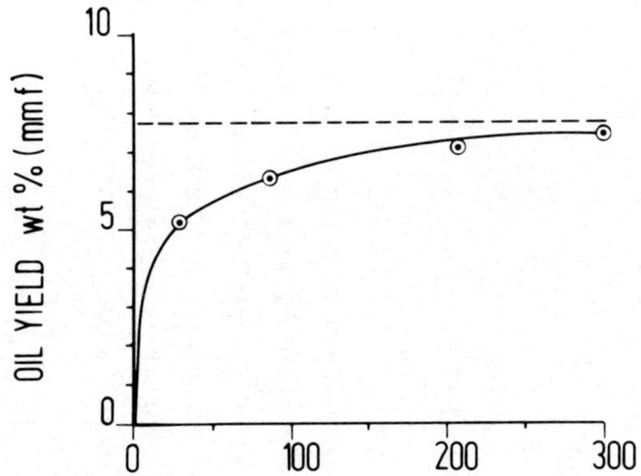

Fig. 5. Steam-N_2 extraction (# 1)

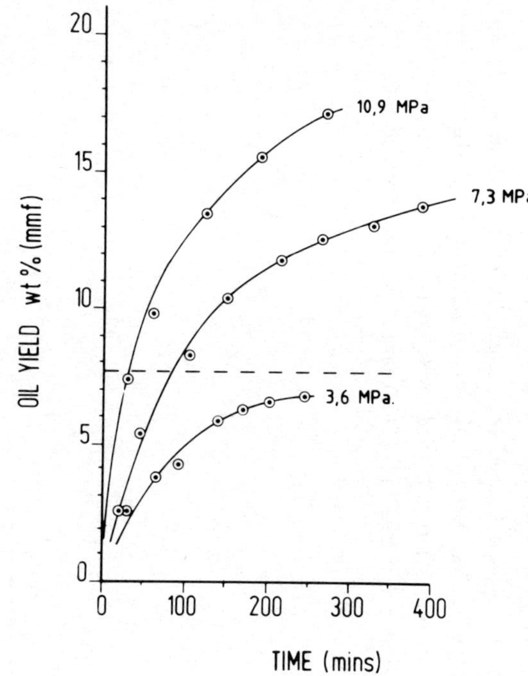

Fig. 6. Steam-CO extractions (# 4, 5, 6).

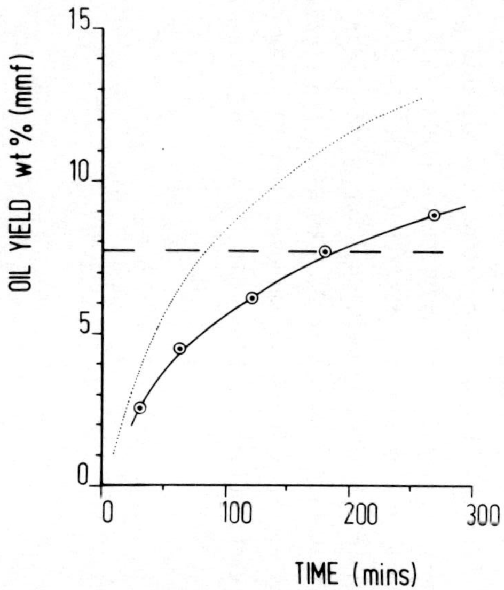

Fig. 7. Steam-H_2-CO extraction (# 8)
Dotted line is oil yield of run # 5.

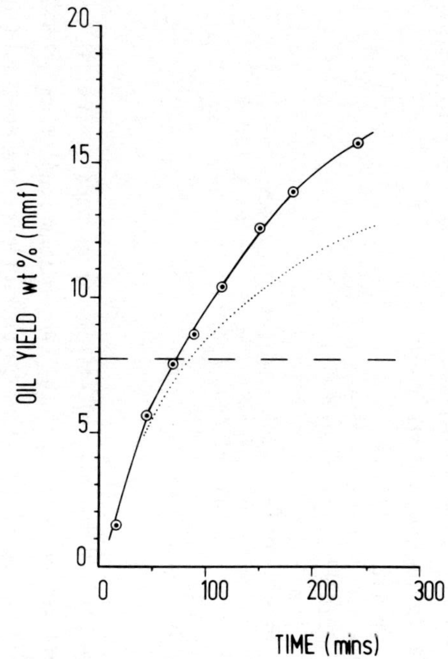

Fig. 8. Steam-CO extraction, 5 % wt Na_2CO_3 in feedwater (# 7)
Dotted line is oil yield of run # 5.

All in all these kinetic curves show that oil extraction is not simply a matter of thermal coal pyrolysis with dissolution and extraction of the pyrolysate; the date indicate special chemical interactions between the coal and the constituents of the solvent as the key to oil yield.

Solvent selectivity

Apart from oils, the extractions produce gases, water and char. These gases include CO_2, CH_4 and small amounts of C_2 hydrocarbons. The total amounts of gases and water (called volatiles) follow from a mass balance by substracting the amounts of oil and char from the initial mass of coal, all on dry basis. We introduce now the following parameters:

$$\text{Selectivity to Oil: } \frac{\text{Weight of Oil}}{\text{Weight of Oil + Volatiles}} * 100\ \% \qquad (1)$$

$$\text{Conversion of Coal: } \frac{\text{Weight of feed coal - weight of char}}{\text{Weight of feed coal}} * 100\ \% \qquad (2)$$

(on moisture and mineral free basis).

In Fig. 9 we have plotted the oil selectivity as a function of coal conversion for the different extraction runs. The coal conversions range from a low 33 % in FA to a high 58 % in one of the steam-CO extractions. One must hold in mind however that extraction times differ among runs sothat restrain should be excercised in relating conversion level to conversion rate.

A second conclusion is that, on a mass basis the major product from extraction is the combined mass of gases and water, with oil only making up the balance. However, the composition and pressure of the solvent exercises a strong influence on this balance. Favorable for oil generation are solvents consisting of steam and CO; increase of pressure and addition of Na_2CO_3 (5 % wt in feed water) improve the oil selectivity by almost a factor of 2 over FA , at conversion levels which are 1.5 to 1.8 times higher. Coal impregnated with Na_2CO_3 (1.5 wt) and with a size distribution from 0.3 to 0.85 mm gives the highest conversion level but the oil selectivity is hardly improved over FA. The same is true for the steam-H_2-CO solvent; relatively high coal conversion but no improvement in oil selectivity over FA. The poor performance of steam-N_2 and especially steam-CO_2 shows up very clearly; although the conversion level is slightly higher than FA the oil selectivity is remarkably worse, steam-CO_2 even by a factor of 2!

As a conclusion we can report the statement made at the end of the previous section. Treatment of coal with aqueous solvents can indeed result in higher coal conversion levels and higher oil selectivities than FA but the succes of this depends highly on the composition and pressure of the aqueous solvent. It appears that specifically steam-CO activates chemical reactions in/with the coal which favor the generation of oil constituents. Addition of Na_2CO_3 to the steam enhances this chemistry; molecular hydrogen is not a favored constituent for oil formation. In the presence of steam-N_2 and steam-CO_2 conversions are slower and for a very large degree directed to formation of non-oil volatile materials.

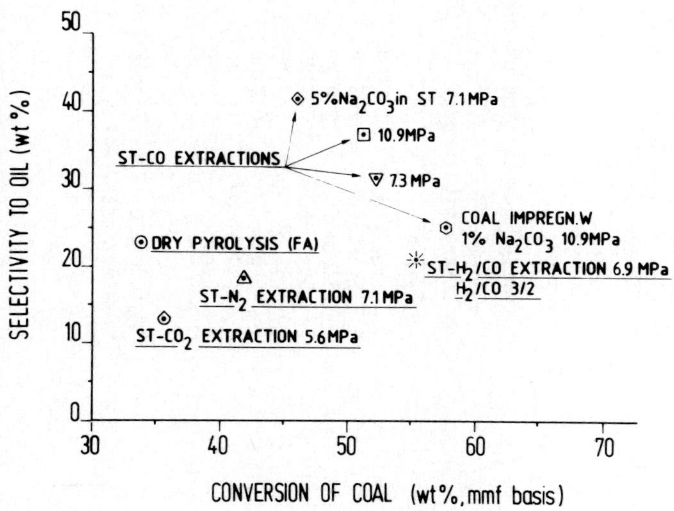

Fig. 9. Effect of extraction conditions on oil selectivity in relation of total coal conversion.

TABLE 4

Elemental composition of extracted oils (wt % as received)

#	Extraction process		C	H	S	N	Ash	Moist.	O (bal)	H/C molar
	FA[b]		58.43	9.92	0.44	0.54	0		25.8	2.04
1	Steam-N_2	7.1 MPa	80.94	8.98	0.52	0.39	0.1	0.50	8.57	1.33
5	Steam-CO	7.3 MPa	82.41	8.40	0.33	0.70	0.1	0.13	7.91	1.22
6	Steam-CO	10.9 MPa	81.97	8.25	0.36	0.86	0.1	0.06	8.40	1.21
7	Steam-CO (5% wt Na_2CO_3 in steam)	7.1 MPa	81.09	8.82	0.39	0.65	0.16	0.32	8.57	1.30
9	Steam-CO (1% wt Na_2CO_3 in coal)	10.9 MPa	83.41	8.52	0.49	0.83	0.13	0.29	6.33	1.22
8	Steam-CO-H_2		83.90	8.18	0.69	0.95	0.49	0	5.79	1.17

TABLE 5

Elemental composition of chars (wt % as received)

#	Extraction proces		C	H	S	N	Ash	Moist.[a]	O (bal)	H/C molar
	FA[b]		76.56	3.43	0.60	1.75	9.39	0.98	7.29	0.54
1	Steam-N_2	7.1 MPa	79.70	3.07	0.27	1.32	11.44	0.11	4.09	0.46
5	Steam-CO	7.3 MPa	79.22	3.73	0.24	0.56	11.76	0.35	4.14	0.56
6	Steam-CO	10.9 MPa	78.76	3.76	0.28	1.22	12.36	0.10	3.52	0.57
7	Steam-CO (5% wt Na_2CO_3 in steam)	7.1 MPa	73.4	3.63	0.41	1.37	14.19	0.87	6.07	0.59
9	Steam-CO (1% wt Na_2CO_3 in coal)	10.9 MPa	79.13	2.98	0.20	1.05	14.36	0.77	1.51	0.45
8	Steam-CO-H_2	6.9 MPa	80.16	3.38	0.14	1.40	11.74	0.65	2.53	0.50

[a]Moisture by Karl-Fischer titration.
[b]Fischer assay analysis

Elemental composition of extraction products

Elemental compositions of Wyoming coal, oils and chars are listed in Tables 2, 4 and 5.

A general conclusion drawn from these tables is that oils are richer but chars are poorer in hydrogen than feed coal. This is also true for the Fischer assay oils. Formation of oils appears to be associated with a redistribution of hydrogen atoms, resulting in hydrogen-rich species, such as oil and gases and a hydrogen-poor char. Alternatively, oils result primarily from hydrogen-rich coal entities and extraction leave the hydrogen-poorer entities as char (precursors). Typical is that all aqueous runs, contrary to FA, result in oils and chars with a relatively low oxygen content. The nature of the co-component appears of secundary importance. Noticable is also that Fischer assay oils have much larger H/C ratios than oils obtained by aqueous extractions. This is even more remarkable when one realizes that Fischer assay is carried out only in autogeneous atmosphere, while in the extraction with steam-CO some hydrogen is formed as a result of shift reaction, which potentially could increase H/C ratio. Nevertheless, oils obtained by steam-CO extractions have a relatively low H/C ratio.

We have calculated in which ratio C and H atoms of feed coal are distributed over the extraction products viz oil, char and volatiles. The results are illustrated in Figs. 10 and 11. The X-axis represents the percentage of coal-C (resp. coal-H) which is recovered as oil plus volatiles; the Y-axis gives the relative distribution of extracted C (resp. H) over oil and volatiles, the total being always 100 %. It follows that aqueous solvents enhance the total conversion of coal C (resp. H); steam-CO gives the best performance of all. A higher fraction of the extracted C (resp. H) moreover is recovered as oil constituents by increase of the pressure of steam-CO and by addition of Na_2CO_3 to steam-CO.

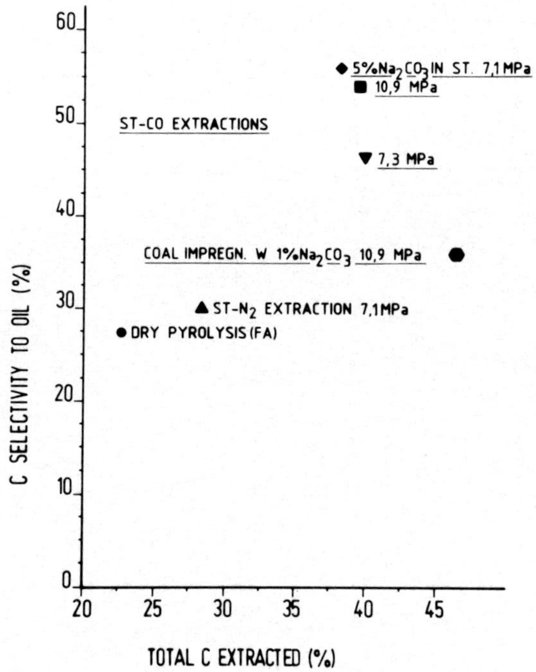

Fig. 10. C in oil as a fraction of total extracted C.

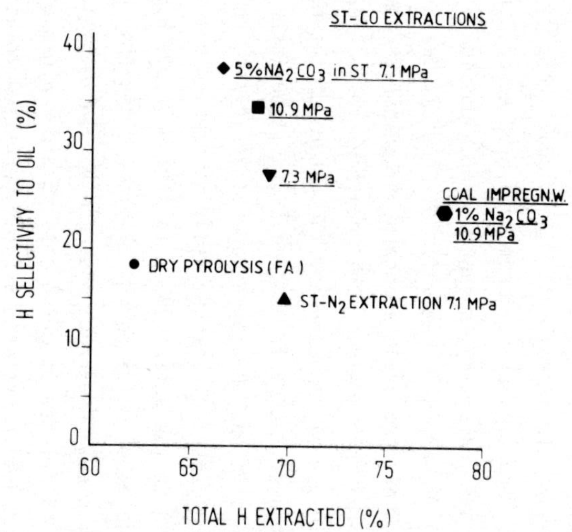

Fig. 11. H in oil as a fraction of total extracted H.

Organic solvent fractionation of extraction products

By convention, coal liquefaction products are characterized by its solubility in series of organic solvents with increasing polarity. We have determined solubility of oils and chars in n-hexane, toluene and pyridin (Tables 6 and 7).

TABLE 6

Solvent fractionation of oils (wt %)

#	Extraction process		Solubility in Hexane	Toluene	Pyridin	Pyr. Insoluble
6	Steam-CO	10.9 MPa	87.0	12.8	0.5	1.2
9	Steam-CO (1% wt Na_2CO_3 on coal)	10.9 MPa	94.4	5.2	0.1	0.3
5	Steam-CO	7.3 MPa	86.5	12.9	0.4	0.2
8	Steam-CO-H_2	6.9 MPa	85.4	12.9	0.5	1.2

TABLE 7

Solvent fractionation of chars (wt %)

#	Extraction process		Solubility in Hexane	Toluene	Pyridin	Pyr. Insoluble
6	Steam-CO	10.9 MPa	0.34	0.53	6.62	balance
9	Steam-CO (1% wt Na_2CO_3 on coal)	10.9 MPa	n.d.[a]	n.d.	<0.1	"
5	Steam-CO	7.3 MPa	n.d.	n.d.	2.7	"
8	Steam-CO-H_2	6.9 MPa	0.72	0.53	0.01	"

[a]n.d.: not detectable

All oils are very soluble in the non-polar n-hexane (85 % wt or more); the balance dissolves nearly entirely in toluene. Less than 1 % is soluble only in pyridin. Chars on the other hand are virtually insoluble in pyridin. Hence, the aqueous extractant efficiently sweeps the coal bed and removes extractable material as soon as same is formed by chemical transformation in the coal. This chemistry must proceed first through the stages from insoluble material to preasphalthenes. The preasphalthenes subsequently transform into asphalthenes and hexane solubles. The selective solubility of each of the four classes of material in the aqueous solvent at extraction conditions is clearly demonstrated with Table 8.

TABLE 8

Relative distribution over oil and char of Hexane solubles, Asphalthenes, Preasphalthenes and Pyridin Insolubles.

#	Extraction conditions	Hexane in oil	Solubles in char	Asphalthenes Oil	Char	Preasphalt. Oil	Char	Pyr.Insol. Oil	Char
6	Steam-CO 10.9 MPa	98.8 %	1.2 %	89.0%	11.0%	1.11%	98.9%	0	100%
5	Steam-CO 7.3 MPa	99.7 %	0.3 %	100 %	0	3.9 %	96.1%	0	100%
10	Steam-CO 6.9 MPa	99.3 %	0.7 %	48 %	52 %	0.3 %	99.7%	0	100%

Hexane solubles appear to have a high solubility in the aqueous solvent under extraction conditions and are virtually completely removed from the char.

Asphalthenes are also preferentially extracted with the aqueous agent but its solubility is noticebly effected by the molar fraction of steam and the feed rate of the aqueous solvent. A feed rate of 30.51 grmole/kg coal-hr with a steam content of 30.7 % molar (run # 10) leaves 52 % of the formed asphalthenes back in the char; on the other hand asphalthenes are extracted for almost 90 % at a solvent feedrate of 87.5 grmole/kg coal-hr and a steam content of 57.1 % molar (run # 5). Preasphalthenes are nearly completely retained in the char and are therefore virtually insoluble in the aqueous solvent.

Rates of extraction and coal conversion

The first step in the sequence from coal to oil is the transformation of Pyridin Insolubles into preasphalthenes. The rate equation for this transformation is not known, but in order to compare the effects of solvent composition and extraction conditions on the rate of this first coal conversion step we assume a first-order rate relation with respect to Pyridin Insolubles. From the amount of Insolubles in coal at the start of extraction, the amount of Insolubles left in the char after extraction and the extraction time an apparent first order rate constant is calculated. The values found at different extraction conditions are listed in Table 9.

It follows that the conversion rate is higher in a continuously flowing solvent system by almost one order of magnitude over non-flow extraction conditions. CO, as a co-component with steam is more active than N_2. Higher pressure favors also the conversion rate. It is intriguing to observe that addition of Na_2CO_3 to the system, either as a co-component of the solvent or as an impregnate of the coal doesnot effect the conversion rate of Pyridin Insolubles into preasphalthenes.

TABLE 9

Apparent first-order rate constant k for conversion of Pyridin Insolubles

#	Solvent composition	Pressure (MPa)	Temperature range (C)	10^6k (sec^{-1})
1	Steam-N_2	7.1	370-480	18.0
5	Steam-CO	7.3	370-450	24.1
7	St-CO + 5% wt Na_2CO_3 in steam	7.1	370-430	27.8
6	Steam-CO	10.9	370-440	36.7
9	St-CO + 1% wt Na_2CO_3 in coal	10.9	370-525	35.7
10	Steam-CO lower solvent rate	6.9	370-400	15.4
11	Steam-CO non-flow extraction	10-14	370-410	3.1

How do aqueous solvents compare with organic critical fluids in coal extraction? Table 10 gives a comparison with supercritical toluene and mesitylene extraction of a German coal, but we realize that an honest comparison of solvent effects can be made only with the same coal. The rate data listed are maximum values measured during the extraction runs. We conclude that steam-CO at comparable pressure and temperature is as effective as supercritical toluene or mesitylene.

With regards to coal conversion and oil yield we have made a comparison in Table 11. Also here we conclude that steam-CO scores favorable as an alternative to organic critical fluids. One should note the data of Scarrah on steam-CO_2 extraction of lignite; a high conversion of coal combined with a low oil yield. We found that steam-CO_2 is an inefficient solvent at much lower pressure for Wyoming coal. Scarrahs data show that increase of pressure apparently doesnot improve the performance of steam-CO_2 as an extractive coal liquefaction agent.

TABLE 10

Comparison of oil formation rates

Extraction process	Rate of oil formation (kg oil/kg coal-sec)
Steam-CO, Wyoming coal 10.9 MPa 400 C	$10.5 * 10^{-5}$
Toluene , Gasflammkohle[a] 10 MPa 400 C	$2 * 10^{-5}$
Mesitylene, Gasflammkohle[a] 10 MPa 400 C	$13 * 10^{-5}$

[a]ref. 6.

TABLE 11

Coal conversion and oil yield obtained with different solvents and coals

Extraction conditions	Coal conversion (%)	Oil yield (%)
Wyoming coal		
Steam-CO, 10.9 MPa, 400 C	51.3	19.0
Steam-CO, 7.2 MPa, 400 C	46.0	19.9
(5% wt Na_2CO_3 in steam)		
Gasflammkohle Ensdorf[a]		
Toluene, 10 MPa, up to 550 C	36	20
Methanol, " "	?	6
Toluene-Tetralin (25%) "	50	30
Lignite[b]		
Steam-CO_2, 27.9 MPa, 400 C	40.3	10.2
Illinois # 6[c]		
Toluene, 11 MPa, 400 C	20.8	?

[a]ref. 6.

[b]ref. 7.

[c]ref. 8.

CONCLUSIONS

The results of this exploratory research teach that high pressure steam in combination with co-components, such as CO and Na_2CO_3 is an effective solvent for extracting oil from Wyoming coal under continuous flow conditions. The extracted oils condense with the steam as a milky emulsion which subsequently separate by gravity. The aqueous solvent extracts only hexane soluble matter and asphalthenes but leaves preasphalthenes back in the char. The solvent influences the pyrolysis chemistry of Wyoming coal profoundly; oils have substantially lower oxygen content and a higher carbon content than oils obtained in autogeneous pyrolysis. The rate of coal conversion into preasphalthenes is enhanced by increase of extraction pressure and addition of CO to steam. The selectivity of oil formation increases with the extraction pressure and with addition of CO and Na_2CO_3 to steam. Na_2CO_3 as a co-component of steam-CO doesnot effect the first conversion step of coal into preasphalthenes; it does effect the preasphalthene conversion so that the oil selectivity is increased. CO as co-component is preferred over H_2, CO_2 and N_2, when maximum oil yields are sought after. We expect that oil yields and coal conversion rates can be increased over the levels reported here by increasing the extraction pressure over 11 MPa. Indeed water holds promise to become an efficient oil extraction agent for coal without the drawbacks of more conventional organic solvents.

REFERENCES

1 F. Fischer and H. Schrader, Brennst.Chem., 2 (1921), 161-172.
2 H.R. Appell, I. Wender and R.D. Miller, Prepr.ACS Div.Fuel Chem., 12 (1968), 220-224.
3 D.S. Ross, in: Coal Science, 3 (1984), 301-338, Academic Press.
4 US 4113446, (1978)
5 N.P. Vasilakos, J.M. Dobbs and A.S. Parisi, Ind.Eng.Chem.Prccess Des.Dev., 24 (1985), 121-128.
6 A. Wilhelm and K. Hedden, Forschungsbericht T82-177 Bundesministerium für Forschung und Technologie (1982).
7 W.P. Scarrah, in "Chemical Engineering at Supercritical Fluid Conditions", M. Paulaitis, J. Penninger, R. Gray and Ph. Davidson (eds) Ann Arbor Science Publishers (1983), p. 395.
8 W.S. Fong, P.C.F. Chan, P. Pichaichanarong, W.H. Corcoran and D.D. Lawson, ibid, p. 377.

Supercritical Fluid Technology, edited by J.M.L. Penninger, M. Radosz, M.A. McHugh and V.J. Krukonis
1985 Elsevier Science Publishers B.V., Amsterdam — Printed in The Netherlands

CATALYTIC LIQUEFACTION OF COAL IN SUPERCRITICAL TETRALIN

L. A. AMESTICA and E. E. WOLF
Chemical Engineering Department, University of Notre Dame, Notre Dame, IN 46556

ABSTRACT

A cobalt molybdenum catalyst used during coal liquefaction in the presence of a supercritical tetralin solvent, increases the rate of hydrotreating reactions leading to lighter products, but has a small effect in coal conversion. The extent of hydrotreating is favored by supercritical conditions and exhibit a slight dependence on the catalyst particle size and pore size distribution. Catalysts with bimodal pore size distribution exhibit higher hydrocracking activity leading to lighter products.

INTRODUCTION

In catalytic coal liquefaction, the catalyst provides the dual function of cracking and hydrogenation of the coal derived fragments. Cracking reactions consist of breaking carbon-carbon bonds and heteroatom-carbon bonds. Hydrogenation of the cracked fragments occurs directly from hydrogen donation from the solvent, or via hydrogen supplied by the catalyst. The catalyst can also hydrogenate the solvent in situ thus increasing its hydrogenation capacity. Supported cobalt molybdenum (CoMo) has been found to activate both coal dissolution and heteroatom removal (1-5). One of the problems that can arise when using supported catalysts in coal liquefaction, are pore diffusion limitations. Ruckenstein et al. (6) calculated that for molecules of about 50 A size, the optimun pore diameter is about 100 A and above. Yen et al. (7) found that the rate of conversion of asphaltenes to oil in a catalyst with 60 Å average pore diameter, is strongly limited by pore diffusion, and reported that catalysts with particle sizes in the range 270μ to 720μ affected both coal dissolution and oil yield. However, Ruether (8) found no effect when using catalysts with particle sizes between 100μ to 1/8 in.

Diffusional limitations that can occur in catalytic coal liquefaction could in principle be offset by the use of supercritical solvents. Supercritical solvents exhibit and increase diffusivity and solubility power. In addition, mass transfer

resistances due to gas absorption into the fluid and interphase transport, are minimized since the reactions occurs in a single fluid phase. Thus, in principle, catalytic liquefaction in the presence of supercritical hydrogen donor solvents can provide an enhancement of the rate of coal dissolution. The objective of this study is to investigate such process by using a tetralin hydrogen mixture at supercritical conditions in the presence of a commercial cobalt molybdenum catalyst. Kinetic data of catalytic coal dissolution were obtained at subcritical and supercritical conditions. In addition, the effect of different pore size distributions, particle sizes, and catalyst loading on conversion and product distribution, were studied at supercritical conditions.

EXPERIMENTAL

Apparatus and experimental procedure. The experiments were conducted in a 300 cc autoclave described elsewhere (9). In order to define the initial time for the kinetic experiments, a slurry of coal and half the solvent was injected into the reactor from an auxiliary vessel. The reactor was equipped with a wet gas flowmeter to measure the gas evolution; traps were used to analyze the gas sulfur content, and a sample loop was available to collect gases for gas chromatographic analysis. An agitator rotating at 1200 rpm provided mixing, as well as support for a basket attached to its shaft containing the catalyst pellets.

All experiments were carried out at 450°C and with a tetralin to coal ratio of 7/1(cc/g). The dry weight of coal charged to the reactor was 20 g. for supercritical conditions and 10 g. for subcritical conditions. The operating pressures were 1100 psia for subcritical conditions and 2200 psia for supercritical conditions. These conditions were selected on the basis of calculations of model hydrogen-tetralin-naphthalene-anthracene mixtures reported elsewhere (9). The catalyst loading is based on the percentage of coal feed.

The reactor was charged with half of the tetralin charge, and with the catalyst. It was then closed, leak tested, flushed with nitrogen and heated to the reaction temperature of 450°C. The remaining of the tetralin mixed with coal and pressurized with hydrogen, was then injected as a slurry into the reactor from the auxiliary injection vessel. The reactor temperature decreased when the coal was injected but it recovered to 450°C in approximately two minutes. The reaction was then conducted for a predetermined elapsed time and then it was quenched by flowing water through the cooling coils. The temperature decreased to 350°C in about three minutes at which point the reactor pressure was let down to atmospheric pressure. The reactor was opened when reaching room temperature, and the liquids and solids were removed for further fractionation and analysis. The ultimate analysis of the coal is presented in Table 1.

Catalysts. Two Cobalt Molybdenum (CoMo) catalysts were used: an Amocat 1A catalyst, developed by Amoco, and a commercial CoMo Harshaw HT-400 catalyst. The characteristics of these catalysts are summarized in Table 2. The Amocat catalyst has a bimodal pore size distribution and it was used in two forms, i as 1/16 in. pellets, and u as a powder with particle sizes below 270μ. The Harshaw catalyst has a unimodal pore size distribution and it was used as 1/16 and 1/8 in. pellets. Prior to the liquefaction runs, the catalysts were sulfided for 90 minutes at 300°C in a stream containing 10% H_2S and 90% H_2. Surface area measurements of these catalysts were conducted in a Quantasorb N_2 adsorption apparatus.

RESULTS

The kinetic data obtained during liquefaction of an Illinois N#6 coal at supercritical and subcritical conditions were correlated by the following kinetic model:

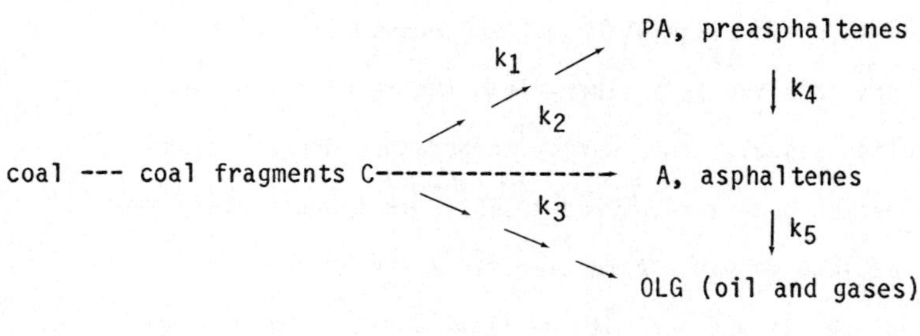

The coal dissolution reaction to PA, A, and OLG was assumed to be second order in (C-Ceq), i.e. the unconverted coal C, minus Ceq, its unreactive fraction. The values of Ceq were 0.077 and 0.129 corresponding to 92.3% and 87.1% conversion, obtained after 2 hours of reaction time at super and subcritical conditions respectively.

TABLE 1

Ultimate Analysis, Maceral Composition and Particle Size Distribution of Illinois No. 6 Coal.[a]

A. Ultimate Analysis, wt%

Moisture	2.6
C	66.4
H	4.4
N	1.3
S	2.9
O	10.7
Oxide Ash	12.3

TABLE 2 Catalyst Characterization

Composition	AMOCAT 1A	Harshaw HT-400
% MoO_3	15.9	15.0
% CoO	2.9	3.0
Surface Area, m^2/g	172	220
Pore Volume, g/ml	0.792	0.5
Aver. Pore Diameter, Å	120	91
Pore Size Distribution	Bimodal	Jnimodal

The cracking series reactions of PA to A and to OLG were assumed to be first order. Mass balance equations for each component in a batch reactor, were solved including the temperature profile for the heating and cooling periods fitted to a polynomial expression. The rate constants k_i, i=1,5 were assumed to follow Arrhenius behavior. The rate constants were obtained by using the non-linear parameter estimation method of Levengerd and Marquardt (10).

Experimental conversion versus time values obtained using the Amocat 1A catalyst at sub and supercritical conditions, were used to obtain the values of the rate constants and activation energies summarized in Table 3. Also listed in table 3 are results corresponding to non-catalytic liquefaction. Supercritical conditions gives larger rate constants than subcritical whereas the catalyst tend to give lower activation energies and higher cracking rates.

The experimental and calculated conversion and yields for catalytic liquefaction are presented in figs. 1 and 2 for subcritical and supercritical conditions respectively. The predicted concentration-time profiles for the unreacted coal, PA, A, and OLG are shown in fig. 3 for both sub and supercritical conditions. The presence of the catalyst increases conversion by 2.5 and 5% at subcritical and supercritical conditions respectively. Supercritical conditions gave an initially higher coal conversion, higher PA and OLG yields, and lower A yield. As reaction time increases, the yield of heavy fractions decreases in favor of the yield of the lighter fractions. This trend is more pronounced at supercritical conditions than at subcritical conditions.

TABLE 3

Rate Parameters for the Coal Liquefaction Kinetic Network at Supercritical and Subcritical Conditions

	Supercritical		Subcritical	
A. Non Catalytic				
Reaction	k_o^{sp} (min^{-1})	E^{sp} (Kcal/mol)	k_o^{sb} (min^{-1})	E^{sb} (Kcal/mol)
Coal $\xrightarrow{k_1}$ preasphaltenes	90.1×10^{6}	25.8	4.8×10^{6}	22.1
Coal $\xrightarrow{k_2}$ asphaltenes	123.3×10^{6}	27.3	$930. \times 10^{9}$	39.1
Coal $\xrightarrow{k_3}$ OLG	8.6×10^{16}	55.4	1.5×10^{15}	49.8
Preasphaltenes $\xrightarrow{k_4}$ asphaltenes	10.9×10^{6}	28.3	3.9×10^{3}	17.1
Asphaltenes $\xrightarrow{k_5}$ OLG	7.5×10^{6}	30.4	121.5×10^{3}	24.2
Equilibrium Conversion (%)	88.1		85.1	
B. Catalytic				
Coal $\xrightarrow{k_1}$ preasphaltenes	4.54×10^{6}	22.5	7.15×10^{6}	23.9
Coal $\xrightarrow{k_2}$ asphaltenes	1.32×10^{6}	21.8	1.43×10^{6}	21.0
Coal $\xrightarrow{k_3}$ OLG	861.3×10^{6}	30.3	199.29×10^{6}	28.1
preasphaltenes $\xrightarrow{k_4}$ asphaltenes	496.9×10^{6}	33.4	0.91×10^{6}	25.9
asphaltenes $\xrightarrow{k_5}$ OLG	2731.2×10^{6}	37.6	205.91×10^{6}	34.9

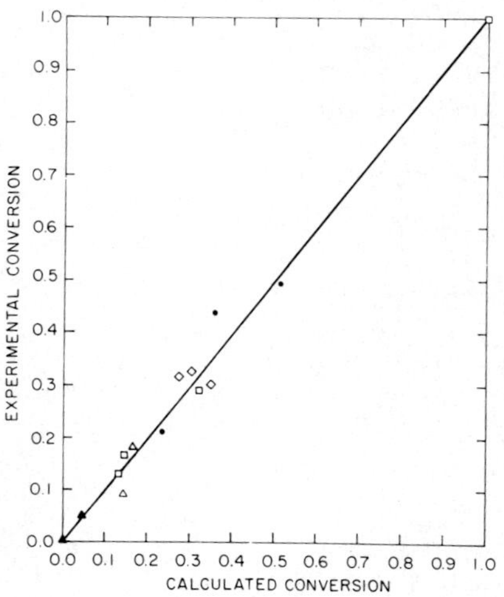

Fig. 1 Experimental versus calculated conversion, subcritical; □ unreacted coal, △ preasphaltenes, ◇ asphaltenes, ● OLG.

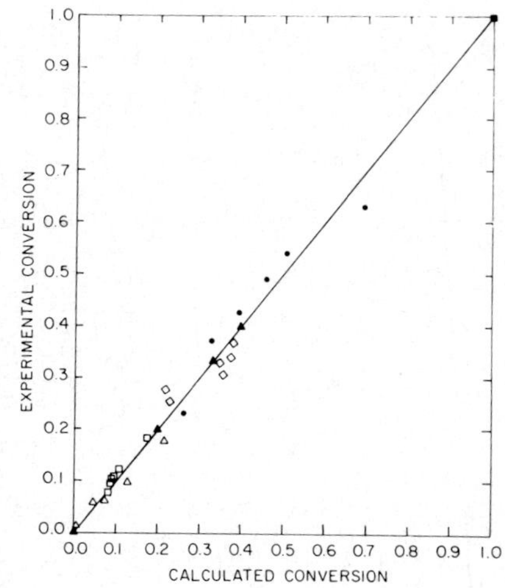

Fig. 2 Experimental versus calculated conversion supercritical; □ unreacted coal, △ preasphaltenes, ◇ asphaltenes, ● OLG.

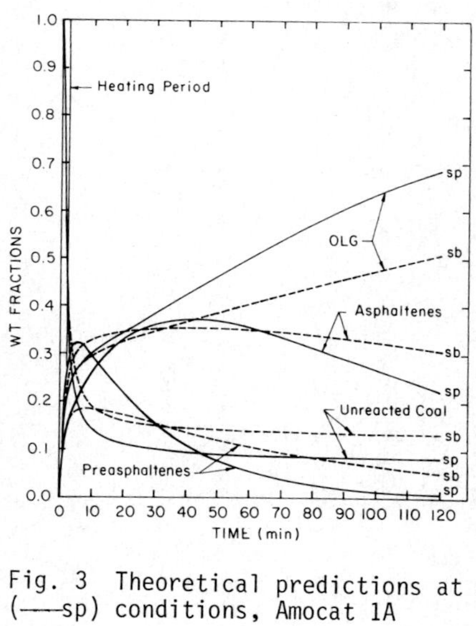

Fig. 3 Theoretical predictions at subcritical (---sb) and supercritical (——sp) conditions, Amocat 1A

The conversion and product distribution obtained using different catalyst loadings, catalyst particle sizes, and type of catalysts are summarized in table 4 and figs. 4 to 7. Table 4A and 4B shows that increasing the loading of the Amocat 1A catalyst have a small effect in total coal conversion, however as displayed in fig. 4, as catalyst loading increases, the product distribution shift significantly from the heavy PA products to the lighter OLG. This effect becomes more pronounced as reaction time increases. The presence of catalysts however, has a small effect in the product distribution at subcritical conditions (fig. 5).

Decreasing the size of the Amocat 1A catalyst from 1/16 to 270μ increases coal conversion by about 4%, see tables 4A and 4C. The effect of particle size in product distribution depends on the reaction time, the small particles gave lighter products at the long residence time of 2 hrs (fig. 6). However for the Harshaw HT-400 catalyst decreasing the particle size from 1/8 to 1/16 in. does not affect significantly conversion and tends to increase the OLG yield only at long contact times.

Comparison of the coal conversion obtained with the Amocat 1A and Harshaw catalysts at the same loadings and particle sizes shows that both gives similar conversions (see tables 4C and 4D), however, the Amocat catalyst yielded less PA and OLG than the Harshaw catalyst (fig. 7). However the highest conversion was attained with the 270μ Amocat 1A catalyst particles.

The BET surface area of the Amocat 1A catalyst was measured after sulfidation and after the various elapsed reaction times. It was found that after sulfidation, the surface area decreased by 11%, from 175 to 156 m^2/g. There was a further smaller decrease of surface area to 141 m^2/g after two hours of reaction time.

TABLE 4 Coal Conversion, Product Distribution and Desulfurization at Supercritical Condition. Effect of Catalyst Type, Catalyst Loading and Catalyst Particle Size.

Reaction time, t_f (min)	Conversion (%)	Product Distribution, % Preasphaltenes	Asphaltenes	OLG[a]	Sulfur eliminated (%)
A: Amocat 1A catalyst, 1/16" particle size, loading = 10 wt%					
30	89.1	11.0	41.2	47.8 (2.5)	24.2
120	92.2	1.6	30.0	68.4 (5.6)	20.2
B: Amocat 1A catalyst, 1/16" particle size, loading = 20 wt%					
30	91.2	5.5	39.1	55.4 (2.1)	29.4
120	92.5	0.9	18.8	80.3 (6.5)	-
C: Amocat 1A catalyst, powder < 270μ particle size, loading = 10 wt%					
30	93.0	9.6	50.7	39.7 (3.1)	13.6
120	96.2	1.4	24.3	74.3 (-)	24.7
D: Harshaw HT-400 catalyst, 1/16" particle size, loading = 10 wt%					
30	89.3	14.5	41.9	43.6 (4.6)	23.3
120	89.1	4.2	22.4	73.4 (6.4)	25.5
E: Harshaw HT-400 catalyst, 1/8" particle size, loading = 10 wt%					
30	90.0	13.9	43.9	42.2 (3.1)	22.2
120	89.0	3.6	28.4	68.0 (7.1)	-

a: OLG represents oil and gas. The values in parentheses indicate the amount of gas.

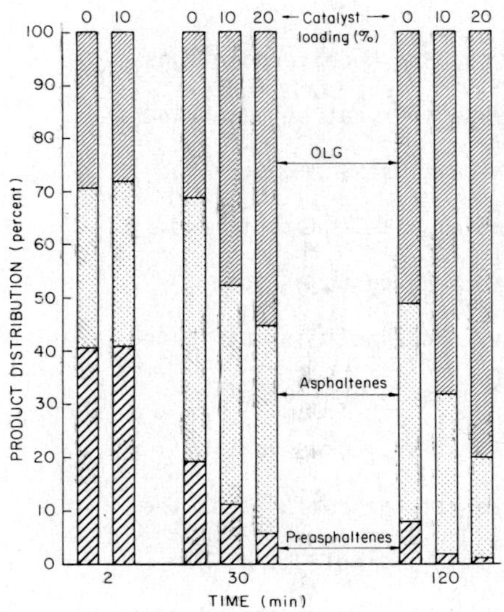

Fig. 4 Effect of Amocat 1A catalyst loading on product distribution at supercritical conditions.

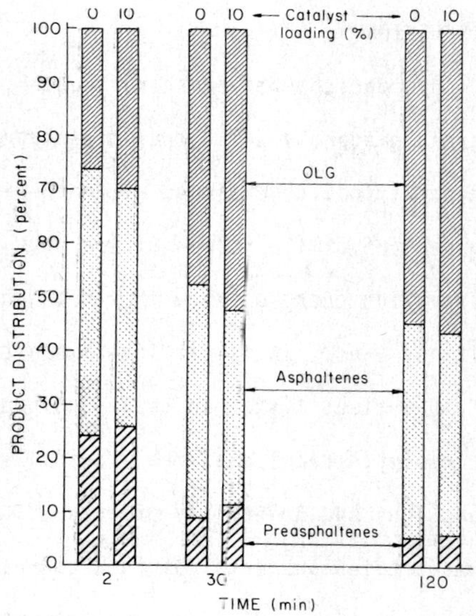

Fig. 5 Effect of Amocat 1A catalyst loading on product distribution at subcritical cordtions.

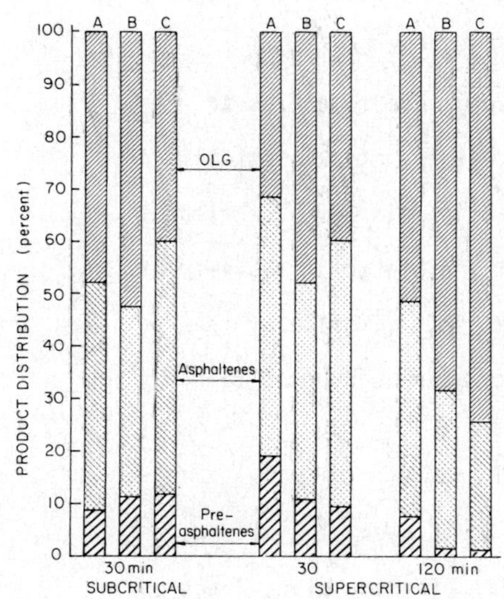

Fig. 6 Effect of catalyst particle size. A: no-catalyst, B: 1/16 Amocat 1A, C: 270μ Amocat 1A

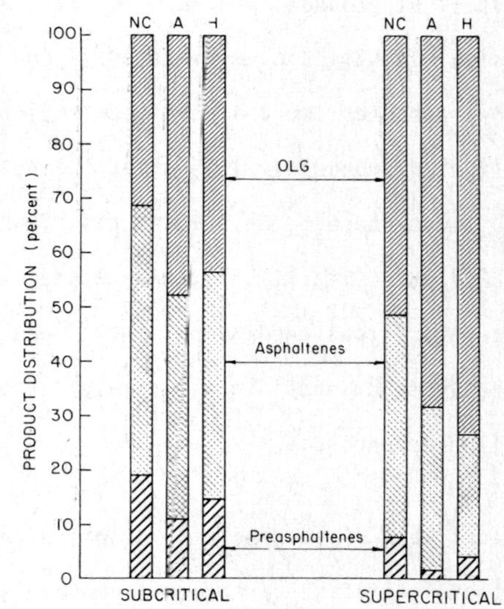

Fig. 7 Effect of catalyst pore size; NC: no catalyst, A: Amocat 1A, H: Harshaw HT-400

DISCUSSION

Comparison between experimental data and the model predictions agree reasonably well, coal dissolved very rapidily at both sub and supercritical conditions, with and without catalysts. However, inspection of the values of the preexponential rate constants and activation energies show that both change depending on the conditions used. It is difficult to ascertain the kinetic significance of the values listed in Table 3 since coal dissolution comprise thermolysis reactions, free radical stabilization, condensation reactions and solubility effects. Thus, is not unexpected that the lumped rate constants obtained cannot be given a simple kinetic interpretation. However, once the constants for a given set of conditions were established, the model predicted fairly well the conversions versus time profiles.

The main effect of the catalysts used was to increase the yield of light products as shown in figs. 3 and 4, without affecting very significantly coal conversion. Thus the catalyst basically hydrocracked the coal derived heavy preasphaltene molecules to lighter products. This effect is not so evident at subcritical conditions wherein the product distribution is similar with or without catalysts (fig. 5). It thus appears that the higher diffusivity at supercritical conditions, increase the rate of diffusion of the heavy fractions into the catalyst where they are converted to lighter products.

When discussing diffusional effects in this complex mixture, the significant variation in molecular size of the various fractions involved, must be accounted for. Thus while the OLG and light asphaltene fractions are not diffusion limited, the heavy preasphaltene fractions with average molecular weights of 400

(11-13) and diameters larger than the pore size, are going to be strongly affected by diffusional effects and thus by particle size and pore size distribution.

Decreasing the size of the Amocat 1A catalysts from 1/16 in to 270μ increases coal conversion by about 4%, yielding more asphaltenes, which after 2 hours of reaction time are converted further cracked to OLG products (fig. 6). Thus it appears that the small catalysts particles are more effective in initially converting the PA---> A fraction but at the expense of some catalysts fouling which prevents further hydrocracking of the asphaltenes to OLG. As reaction time increases the molecular weight shifts to lighter products which in turn can be hydrocracked to the lighter OLG products. Particle size effects with the Harshaw catalysts were not significant probably due to diffusional effects arising from its unimodal pore structure which excludes molecules larger than 70Å from entering into the catalysts internal area. This effect is demonstrated when comparing the product distribution of the Harshaw versus the Amocat 1A catalysts (fig. 7). It can be clearly seen that the Harshaw catalyst is effective in catalyzing the conversion of the A------> OLG fraction but it is not as active to convert PA-----> A as the Amocat catalyst.

In summary, catalytic coal liquefaction at supercritical conditions increases the rate of the hydrotreating reactions leading to lighter products but has a small effect in coal conversion. The extent of hydrotreating is favored by supercritical conditions and exhibit some dependence on the catalyst particle size and pore size distribution. Catalysts with bimodal pore size distribution enables hydrocracking of the large preasphaltenes molecules to lighter products.

REFERENCES

1. Mahoney, J. A., J. J. Helstrom and R. J. Pellet, "Coal Liquefaction Catalyst Development", Final Report, DOE, Contract No. DE-AC22-79ET-14803, 1982.

2. Mitchell, P. C. H., "The Chemistry of Some HDS Catalysts Containing Molybdenum", Climax Molydbenum Company, London, 1967.

3. Kranich, W. L., K. Guruz, A. H. Weiss, and A. Foutsitzis, "Catalytic Hydroliquefaction of North Dakota Lignite, Part II. Kinetics of Dissolution", Fuel Processing Technology, 6, 203-213, 1982.

4. Brooks, J. A., R. J. Bertolocini, L. C. Gutberlet, and D. K. Kim, "Catalyst Development for Coal Liquefaction", Annual Report, Amoco Oil Co., Project Research 408-1, 1976.

5. Bertolocini, R. J. , L. C. Gutberlet, D. K. Kim and K. K . Robinson, "Catalyst Development for Coal Liquefaction", Annual Report, Amoco Oil Co., Research Project 408-1, 1977.

6. Ruckeinstein, E., and M.C. Tsai, "Optimum Pore Size for the Catalytic Conversion of Large Molecules", AIChE Journal, 27(4), 1981.

7. Yen, Y. K., D. E. Furloni, and S. W. Walker, "Batch Autoclave Studies Catalytic Hydrodosulfierization of Coal", Ind. Eng. Chem. Proc. Res. and Dev., 15(1), 24, 1976.

8. Ruether, J. A., "Kinetics of Heterogeneously Catalyzed Coal Hydroliquefaction", Ind. Eng. Chem. Process Des. and Dev., 16(2), 1977.

9. Amestica, L. A., E. E. Wolf, "Kinetics and FTIR Studies of Coal Liquefaction with a Supercritical Tetralin Solvent", Submitted for publication.

10. Marquardt, D. W., J. Appl. Math., 11(2), 431, (1963).

11. Fercasiu, M., "Fractionation and Structural Characterization of Coal Liquids", Fuel, 56, 9, 1977.

12. Fercasiu, M., T. O. Mitchell, and D. D. Whitehurst, "Asphaltols-Keys to Coal Liquefaction", Chemtech, 680 November, 1977.

13. Fercasiu, M., "Short Time Reaction Products of Coal Liquefaction and their Relevance to the Structure of Coal", Am. Chem. Soc. Division Fuel Chemistry, Preprints, 24(1), 121, 1979.

Supercritical Fluid Technology, edited by J.M.L. Penninger, M. Radosz, M.A. McHugh and V.J. Krukonis
1985 Elsevier Science Publishers B.V., Amsterdam — Printed in The Netherlands

APPLICATION OF SUPERCRITICAL FLUID EXTRACTION IN THE HYDROLIQUEFACTION OF LIGNITE

P.J. Göttsch[1], U. Lenz[1], W.O. Eisenbach[2], W. Dolkemeyer[3], F. Gajewski[3]
[1]Rheinische Braunkohlenwerke AG, P.O.Box 410840, 5000 Köln 51 (West Germany)
[2]Max-Planck-Institut, Kaiser-Wilhelm-Platz 1, 4330 Mülheim/Ruhr (West Germany)
[3]Union Rheinische Braunkohlen Kraftstoff AG, P.O.Box 8, 5047 Wesseling (West Germany)

ABSTRACT

The use of SCF Extraction instead of vacuum distillation in the recovery of recycle solvent from the hot separator bottom product in hydroliquefaction of lignite is discussed. Experiments were performed in a continuous flow process developing unit. Preliminary economics show advantage over vacuum distillation as a result of higher recovery of product; solvent loss, however, remain critical for profitibility.

INTRODUCTION

From 1941 to 1944 hydroliquefaction of Rhenish lignite according to the Bergius-Pier process was carried out on a technical scale by the Union Rheinishce Braunkohlen Kraftstoff AG (UK Wesseling).

The annual production amounted to 250,000 tonnes of liquid products.

Since 1978 the Rheinische Braunkohlenwerke AG (Rheinbraun) has been operating a continuous process development unit for hydroliquefaction of lignite (HVB) on the same site. In this plant about 0.5 tonnes of lignite per day are subjected to hydroliquefaction. The aim is to further develop this process with modern technology to plants of commercial size.

Fig. 1 shows a flow sheet of the HVB process. At the bottom of the hot separator a slurry is produced containing ash, catalyst, unconverted coal, and heavy oil which is to be separated and subsequently used as a recycle solvent. In the HVB process steps, used in the past such as centrifuging and low-temperature carbonization are now replaced by vacuum distillation to produce the recycle oil.

This modification has the advantage that a recycle oil free of asphaltenes is produced. As a consequence it was possible to lower the process pressure and increase the coal throughput as shown in Table 1. A disadvantage of vacuum distillation is, however, that a considerable amount of heavy oil remains in the residue.

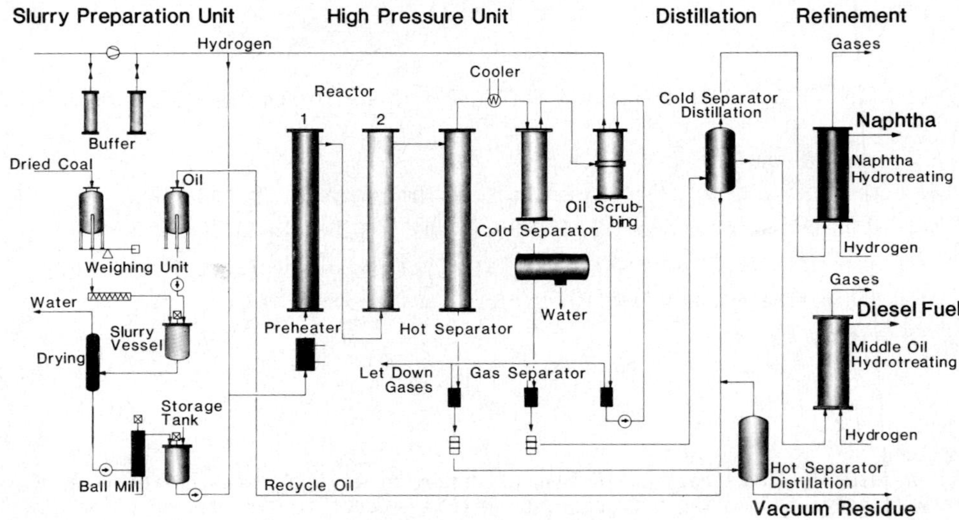

Fig. 1. Flow Sheet of the HVB - Test Plant

Table 1. Brown Coal Liquefaction
Operating Conditions and Results

	old (1943)	new (1983)
Operating Pressure (bars)	710	200 - 300
Partial H_2 Pressure (bars)	510	180 - 270
Reaction Temperature (°C)	478	450 - 470
Processing of Residue	Centrifuging Carbonizing	Distilling / SCF–Extraction Gasifying
Specific Coal Throughput (t/m^3h)	0,4	0,6
H2 Consumption (wt %) (in terms of m.a.f. coal)	6,5	6,0

APPLICATION OF SCF EXTRACTION

Another very effective process for solids separation is the SCF extraction. Since basic investigations at the Mülheim Max-Planck-Institut (MPI) für Kohlenforschung had shown that this process can be used to produce asphaltene-free extracts from various kinds of bituminous materials, Rheinbraun and the Mülheim MPI started a joint program on the application of SCF extraction to residues from lignite hydroliquefaction. Two different processing routes are shown in

Fig. 2.

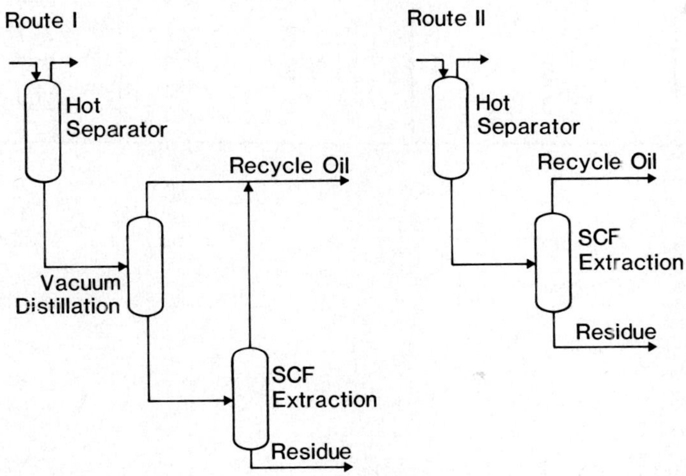

Fig. 2. Alternatives for SCF - Extraction in HVB-Process

With the first route the residue from vacuum distillation is subjected to SCF extraction in order to produce an additional oil fraction.

With the second route the entire hot-separator bottom product is subjected to SCF extraction.

SEMI-BATCHWISE OPERATED UNIT

First of all at the Mülheim MPI both routes were investigated in a semi-batchwise operated unit.

In the course of these investigations low-boiling hydrocarbons were used as solvents, the critical data of which are shown in Table 2.

These investigations were performed with several batches of hot-separator bottom product and vacuum residue. The average analytical data of both products are shown in Table 3.

Table 2. Critical Data of Solvents Investigated

	Pressure bar	Temperature °C	Density g/l
propane	42	97	224
i-butane	36	136	222
n-butane	37,5	152	225
n-pentane	33,3	197	244

Table 3. Analytical Data of Feed Materials

		Hot Separator Bottom (wt.-%)	Vacuum Residue (wt.-%)
Elementary Analyses	C	80 - 81	70 - 73
	H	8 - 9	5 - 7
	O	2	3
	N	0,3	0,7
	S	1	3
Solid Matter		12 - 15	25 - 30
Ash		9 - 10	20 - 22
Asphaltenes		15 - 20	35 - 40
CCT		9	35 - 40
Softening Point ring and ball		20 - 30 °C	120 °C

As can be seen from the ash and solids contents it is possible to remove about 50% of the hot-separator bottom product as vacuum distillate.

From the solvents under investigation pentane proved to be best suited as the highest extraction yields in the shortest period of time were obtained.

Table 4 compares the results of the pentane-extraction of hot-separator bottom product and vacuum residue. As can be seen from the solids contents of the residues, the best oil recovery is achieved with direct extraction of the hot-separator bottom product. A quite remarkable finding was that the SCF extracts contained a rather high amount of asphaltenes while in extraction of petroleum-derived residues all asphaltenes are rejected.

Molecular weight determination of the coal-derived residues showed rather low values ranging from 300 to 400 g/mol for the asphaltenes extracted. This is only one tenth of the values normally found with petroleum-derived asphaltenes. This led to the question how these asphaltenes would behave as a recycle solvent component. First autoclave experiments showed that when only a small amount of these asphaltenes is added to the recycle solvent their influence is even positive. But these findings had still to be checked in the continuous process.

Table 4. Analytical Data: Products of SCF Extraction

		Hot Separator Bottom		Vacuum Residue	
		SCF Extract (wt.-%)	Residue (wt.-%)	SCF Extract (wt.-%)	Residue (wt.-%)
Yield		80 - 82	18 - 20	48 - 50	50 - 52
Elementary Analyses	C	88	40 - 45	89	55
	H	8,5	3,5	7,7	4,0
	O	1	3	2,5	13
	N	0,3	1	0,3	1
	S	1	4,5	1	4,5
Solid Matter		0	55 - 60	0	50
Ash		0	54	0	40
Asphaltenes		13	--	35	--
CCT		6	--	12	--
Softening Point ring and ball		25 °C	--	65 - 70 °C	--

At this point the question arose whether higher-boiling hydrocarbons can be used favourably as a solvent. There are three reasons why this question was not followed up. They are:

1) When the hot-separator bottom product is extracted with pentane the solids content of the residue is increased up to 60%. A higher solids content resulting from the use of even higher-boiling solvents will lead to residue

discharge problems.

2) When higher boiling solvents are used the critical temperature of the solvent will interfere with the initial boiling point of the hot-separator bottom product resulting in problems with the product/solvent separation.

3) A higher extraction yield is obtained mainly by an increased recovery of asphaltenes and pre-asphaltenes which are thought to cause trouble in the hydrogenation step when a larger amount is recycled.

The diagram in Fig. 3 compares the product distributions of the two processing modes described as Route I and Route II. Since by following Route II more oil at a higher space velocity had been obtained it was decided to carry out all further investigations with undistilled hot-separator bottom product.

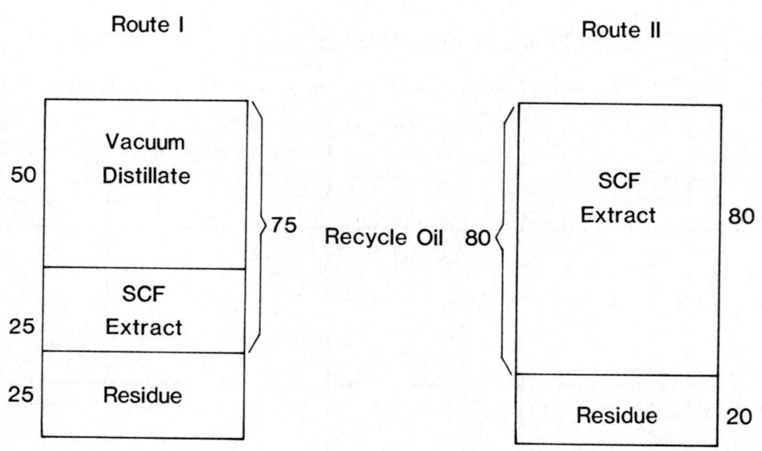

Fig. 3. Mass balance of SCF - Alternatives

CONTINUOUS FLOW UNIT

For these investigations a continuous flow unit for extracting the hot-separator bottom product was designed and built at the Mülheim MPI. The aim of this operating mode was to demonstrate this kind of extraction on a technical scale. The second reason for choosing the continuous flow unit was to obtain a sufficient quantity of SCF extract for use as a recycle solvent component for the HVB process development unit. Fig. 4 shows a flow sheet of the continuous extraction unit.

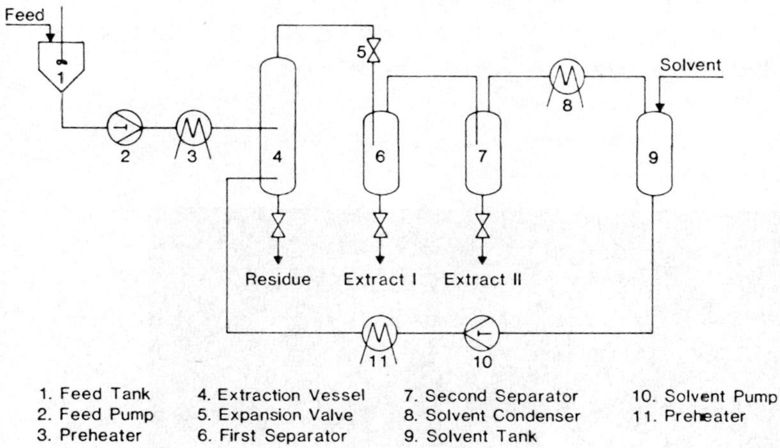

Fig. 4. Continuous Flow SCF - Extraction Unit

The hot-separator bottom product is stored in the heated tank 1 which is equipped with a stirrer. With feed pump 2 the product is pressurized and fed to preheater 3 from where it enters extraction vessel 4. The solvent leaves solvent tank 9 and is pressurized by solvent pump 10. While passing preheater 11 the solvent reaches supercritical state. The supercritical fluid subsequently enters extraction vessel 4 from the bottom. While streaming upwards the supercritical fluid contacts the product and is laden with the extract. The laden supercritical phase exits the extraction vessel overhead. At expansion valve 5 the supercritical phase is depressurized and enters the first separator 6. During this stage the supercritical state disintegrates. The liquid product obtained settles at the separator bottom. The vaporized solvent leaves the separator overhead and enters the second separator 7 which is kept at a lower temperature than the first one. This second separation step is necessary because some low-boiling components of the hot-separator bottom product escape from the first separator with the solvent stream. In the second separator these low-boiling components are condensed together with a small amount of the solvent; otherwise they would bind too much of the solvent thus lowering its quality. The pure solvent is then liquefied in condenser 8 and returned to solvent tank 9.

The extraction residue settled at the bottom of extraction vessel 4 is led down to a pressure drum where it solidifies. It may then be used in a gasification step for hydrogen production. SCF extract I from the first separator is returned to the HVB process as a recycle solvent component. SCF extract II from the second separator which amounts only to a few percent of the feed is recirculated to a distillation step in the HVB process.

Fig. 5 shows the entire extraction unit at the Mülheim MPI.

Fig. 5. Extraction unit at the Mülheim MPI

In December 1982 the first test runs with the continuous flow unit were started. After overcoming the usual start-up problems the unit operated continuously for more than 40 hours.

In 1983 an overall output of more than 1,000 kg of hot-separator bottom product was reached in a series of test runs. The throughput of the extraction unit amounted to almost 5 kg/h. In most cases the solvent-to-product ratio chosen was 4 to 1.

In mid-1983 the HVB process development unit was run twice with recycling SCF extract in place of vacuum distillate. For producing the initial amount of SCF extract several charges of hot-separator bottom product from different runs of the HVB unit were extracted and mixed. With this initial quantity of the SCF extract it was possible to operate the HVB process development unit for one week. In order to check whether the process was in balance the hot-separator bottom product from this run was returned to the Mülheim MPI for re-extraction. With the SCF extract from this product a second run of the HVB process development unit was made.

Three mass balances were made for each of the two HVB process development unit runs. For each balance the hot-separator bottom product, the SCF extract and the SCF residue were analysed. As can be taken from Table 5, there are no variations in the yield or the product quality between the first and the second run proving the process to be in balance and enabling us to calculate the oil yield of the overall HVB process.

Table 5. Yields and Analytical Data of Materials received from integrated mode operation

		Feed (wt.-%)		SCF Extract (wt.-%)		Residue (wt.-%)	
		I	II	I	II	I	II
Yield		100	100	79,5	79,6	20,5	20,4
Elementary Analyses	C	78,0	79,3	88,1	87,8	44,3	42,1
	H	7,8	7,9	8,9	8,9	3,0	3,1
	O	2,9	1,7	1,3	1,4	14,5	8,4
	N	0,7	0,7	0,8	0,9	0,5	0,5
	S	3,4	2,7	0,9	0,8	12,5	9,5
Asphaltenes		10,3	8,9	10,8	9,3	22,5	18,9
Preashaltenes		3,6	1,9	1,0	2,4	13,8	13,3
Ash		9,8	7,9	0,0	0,0	53,8	56,0

I = first recycle
II= second recycle

Fig. 6 shows the product distribution of both operating routes, viz. the HVB unit with vacuum distillation and SCF extraction, in relation to 100 parts of lignite (maf) and 6 parts of hydrogen. The output obtained comprises 15 parts of water, 8 parts of carbon oxides, 13 parts of hydrocarbon gases, and in addition, 49 parts of oil and 21 parts of residue with vacuum distillation, and 55 parts of oil, and 15 parts of residue with SCF extraction.

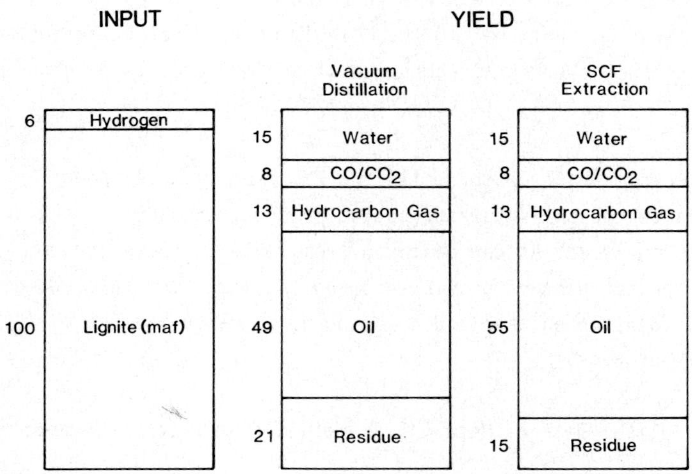

Fig. 6. Comparison of Product Distribution

All the above experiments were performed with pentane as a solvent. Since pentane, however, is too expensive to be used in a commercial-size plant some further experiments were carried out using a low-boiling gasoline fraction from the HVB process. In accordance with the higher boiling range compared with pentane the temperature of the extraction vessel was slightly increased. These tests proved that the SCF extraction process works with a process-derived solvent, too.

The solvent composition remained unchanged over the experiment as was verified by GC analysis.

Both the extraction yield and product quality corresponded more or less with those obtained with pentane.

ECONOMIC CONSIDERATIONS

As has been shown by our experiments, the advantages of SCF extraction are evident when considering the extraction yield. In order to check whether SCF extraction is an attractive alternative to vacuum distillation in a commercial-size plant as well a cost comparison was made.

For this purpose basic engineering data were evaluated for both a vacuum distillation and SCF extraction as part of a HVB pilot plant having a daily throughput of 200 tonnes of lignite (maf). On the basis of these data for both processing modes capital requirements and operating costs were estimated. While the capital requirements differ only to a small extent favouring SCF extraction, the operating costs of SCF extraction, mainly due to solvent losses, exceed by far those of vacuum distillation. These cost estimates, however, had to be compared to the product proceeds. Both the vacuum distillate and the SCF extract were taken as vacuum gas oil while the residues were evaluated on the basis of their individual heating values.

As can be seen from Fig. 7, the higher operating costs of SCF extraction are more than compensated by its higher proceeds resulting from a better oil yield.

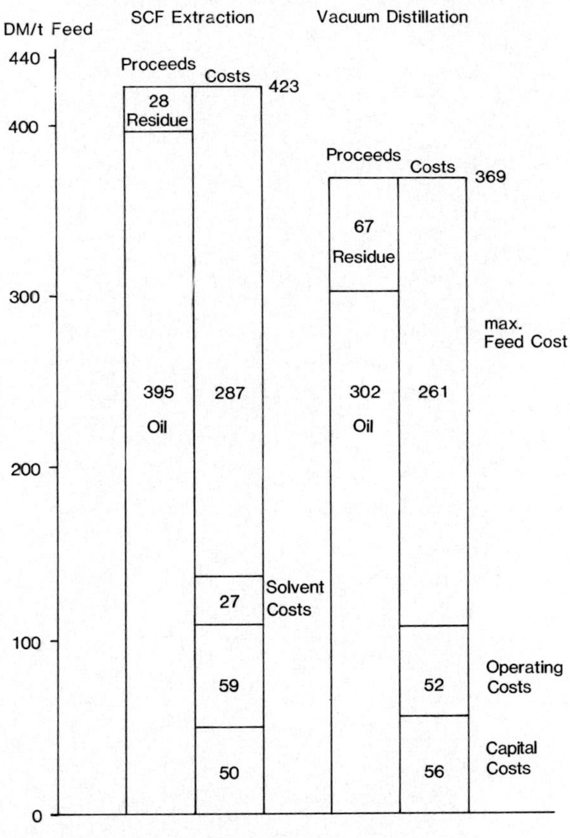

Fig. 7. Profitability Comparison

CONCLUSIONS

SCF extraction was successfully used for residues from lignite hydroliquefaction. The operation of a continuous SCF extraction unit proved that, in principle, the process can be scaled-up to commercial plant size. An economic evaluation on a pilot plant basis showed the significant advantage of SCF extraction over vacuum distillation. A major cost factor of SCF extraction are solvent losses. Their minimization will have to be the objective of further investigation.

It should be noted that with various modes of residual-oil conversion (where, however, cycle oil availability is no problem) SCF extraction might become an interesting alternative to vacuum distillation or to vacuum residue processing.

Supercritical Fluid Technology, edited by J.M.L. Penninger, M. Radosz, M.A. McHugh and V.J Krukonis
1985 Elsevier Science Publishers B.V., Amsterdam — Printed in The Netherlands

SUPERCRITICAL FLUID EXTRACTION OF COAL: DEVELOPMENT OF A 'SECOND GENERATION' PROCESS CONCEPT

Achim Wilhelm* and Kurt Hedden
Engler-Bunte-Institut der Universität Karlsruhe, W-Germany
*present address: LURGI GmbH, 6000 Frankfurt/M., W-Germany

ABSTRACT

For a number of years beginning 1976 work has been proceeding at the Engler-Bunte Institute, Karlsruhe, directed at the development of a process for supercritical coal liquefaction. Early experiments, initially set up for optimizing supercritical coal extraction with toluene in a moving-bed extractor as originally proposed by the NCB, Great-Britain, did not lead to the desired increase of the low coal conversion degrees observed earlier. However, these experiments provided new insights into elementary processes employed in coal extraction. The better understanding of the complex superposition of chemical reactions, phase equilibria and mass transport steps during coal extraction lead to a new process concept for supercritical coal extraction. This paper briefly summarises some of our experimental findings, whose interpretation provided ideas to the new process. This 'second generation' process for supercritical coal extraction employs two stages: subcritical solubilizing of coal in the presence of a donor-solvent followed by supercritical extraction with a non-donor solvent. It is called the Donor-Solvent-Supercritical-Extraction Process (DSSE-Process). It combines the advantages of hydrogenating coal extraction with donor solvents (high degree of coal conversion) with the advantages of supercritical coal extraction (feasible solid separation) and, therefore, has a potential to be used for large scale coal liquefaction with supercritical fluids.

TECHNICAL BACKGROUND AND SCOPE

Most of today's coal liquefaction processes follow the two principal routes proposed early this century by Bergius and Pott/Broche. Bergius-type liquefaction processes convert coal in one stage into liquid distillates using a finely dispersed one-way catalyst and hydrogen under high pressure. The Pott/Broche-type coal liquefaction sequence employs two stages: non-catalytical conversion of coal with hydrogenating solvents (H-donor solvents) into an ash free extract followed by catalytical hydrocracking of the coal extract into distillates.

The economy of coal conversion strongly depends on hydrogen utilization. In the Bergius-type processes a large proportion of hydrogen unselectively reacts to undesired hydrocarbon gases. Pott and Broche's two-stage approach, however, promises a more economic use of hydrogen, since key conversion steps are more efficiently optimized in two stages (e.g., premium catalysts with high selectivity can be used for converting the ash-free coal extract to liquid products in the second stage). However, all efforts to scale up the Pott/Broche process failed, mainly because no feasible method was found for separating finely dispersed minerals and undissolved coal from the viscous extract solution of the first stage.

More than 50 years later the National Coal Board (NCB) in England pioneered the use of supercritical toluene for coal extraction (ref. 1). The unique combination of high, liquid-like density with low, gas-like viscosity as is typically encountered in supercritical fluids makes it possible to overcome the classical solid-separation problem. However, coal extraction with supercritical solvents like toluene gives low coal conversion degrees and extract yields (ca. 30 % for the extraction of hard coal). The low extract yields are a major disadvantage for a large scale technical use of supercritical extraction in coal liquefaction since the desired product (extract) is produced with much lower yield than the undesired byproduct (residue char).

The above described dilemma was the starting situation for our investigations. The aim of our work (ref. 2,3,4) was to find a way how to process coal with supercritical solvents to high conversion degrees and to develop a technically and economically feasible route for large scale supercritical coal extraction as a first stage of a Pott/Broche type coal liquefaction process.

NON-ISOTHERMAL FLUID EXTRACTION TECHNIQUE

To improve our understanding of the complex coal extraction mechanism we first studied the elementary processes employed in coal extraction in the entire temperature range that coal undergoes in a technical extraction reactor. Fig. 1 shows the NCB-designed moving-bed extractor of the SCE-Process (ref. 5) and a scheme of the nonisothermal experimental technique we used in our investigations. In order to be close to the moving-bed technology of the SCE process (Fig. 1a) we chose a similar reaction system for our investigations. Our laboratory reactor (Fig. 1b) contains the coal in a fixed bed, which can be subjected to a defined temperature-time history. Both the reactor system and the pressurized solvent before entering the coal bed are slowly heated according to a temperature program from room temperature to 550 °C at linear heating rates of 2 to 8 °C/min. The extract content of the solvent and the volume of the gaseous coal decomposition products are measured in short time intervals at increasing temperature and are used to calculate formation rates and integral yields of extract and gas. The liquid solvent flowrate and the pressure are held constant during the experiment. Details of the experimental and analytical procedures are given in ref. 2.

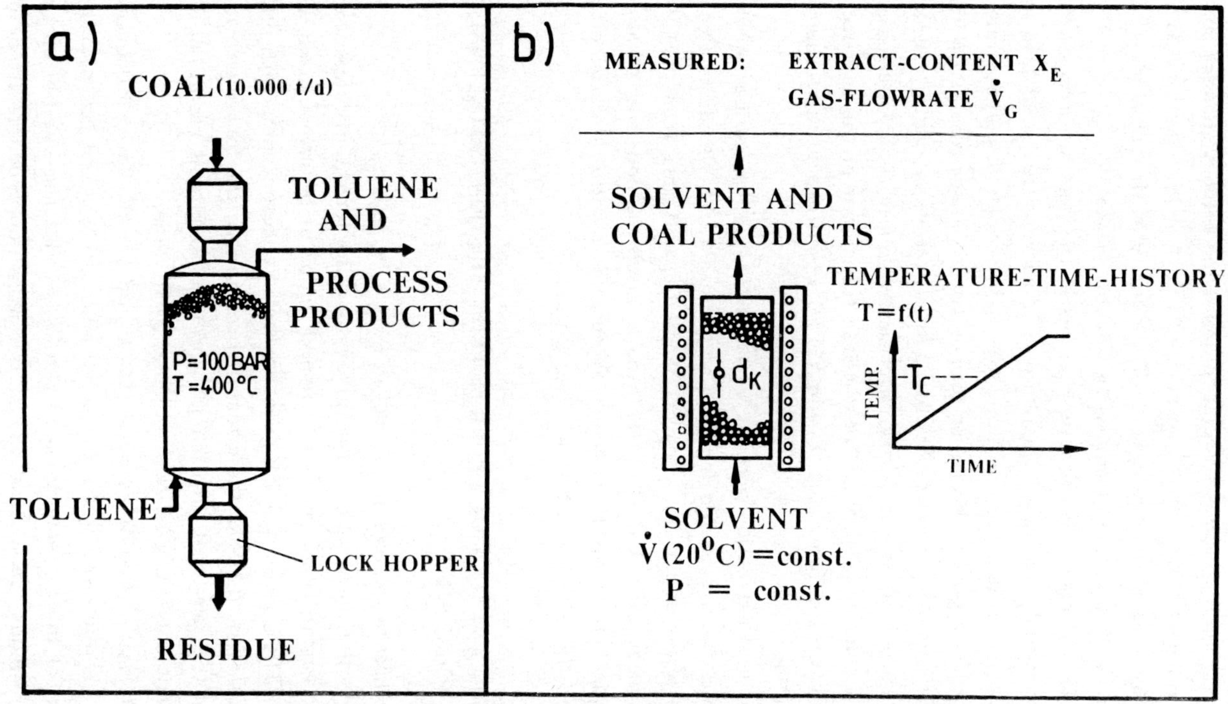

FIG. 1. TECHNICAL MOVING-BED EXTRACTOR OF THE SCE-PROCESS (a) AND NONISOTHERMAL FLUID EXTRACTION TECHNIQUE (b)

CHANGES IN SOLVENT AND COAL DURING HEATING

Fig. 2 shows schematically how solvent and coal change with rising temperature. At constant pressure above the critical point the solvent changes continously without a phase transition from a pressurized liquid at the start to a supercritical fluid near the end of the experiment. The density of the solvent decreases thereby by a factor 2 to 3, the viscosity by a factor 10^2 to 10^3.

A coal particle undergoes also drastic changes in the heating process. Bituminous coal as is commonly used for liquefaction is a microporous solid with thermoplastic behaviour. When heated, it softens like a thermoplastic polymer and formes an intermediate plastic phase. At higher temperatures gas evolves as the result of cracking reactions and generates bubbles, which can expand the soft particle. At even higher temperatures the plastic intermediate phase repolymerizes again to a macroporous solid or semicoke. These phase transitions of the coal occur at moderate heating rates between 300 °C and 450 °C which is also the temperature range for extraction.

In a technical coal extraction process as well as in our non-isothermal experiments solvent and coal are heated together and interact. This leads to a complex superposition of phase equilibria, chemical reactions, and mass transport during the extraction procedure. At first, therefore, we used the non-isothermal experimental technique to separate these overlapping effects and to determine the role of single steps in the complex extraction mechanism.

CHARACTERISTIC EXTRACTION PROFILES FROM THE NON-ISOTHERMAL FLUID EXTRACTION TECHNIQUE

Typical experimental results are shown in Fig. 3, 4, and 5.

Fig. 3 exhibits the temperature-time history and the formation rates of extract and gas with increasing temperature for the extraction of bituminous coal with toluene at 10 MPa. The curves with different symbols represent runs with different coal particle sizes. Fig. 4 shows the effect of pressure on the formation rates for the extraction of the same coal with toluene and Fig. 5 contains results obtained with other aromatic solvents (benzene, xylene, mesitylene).

The extract formation begins between 150 °C and 200 °C, levels off between 250 °C and 300 °C, and rises again to reach a maximum at about 400 °C. This double-step rise of the extract formation rate is encountered in all of our experiments with bituminous coals. Gaseous coal decomposition products are only formed at temperatures above 400 °C, as the circular data points (right ordinate) indicate.

A comparison of the results in Fig. 3, 4, and 5 shows, that a variation of extraction conditions leads to formation rate profiles of different shape. The inter-

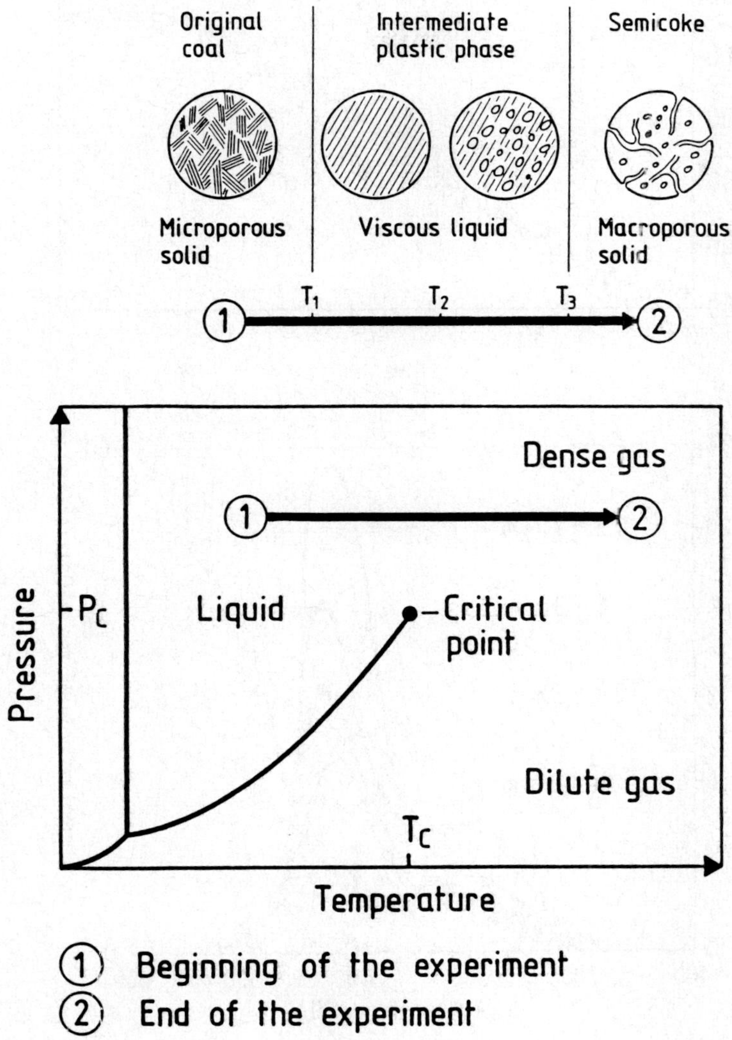

FIG. 2. CHANGES IN COAL AND SOLVENT PROPERTIES DURING HEATING

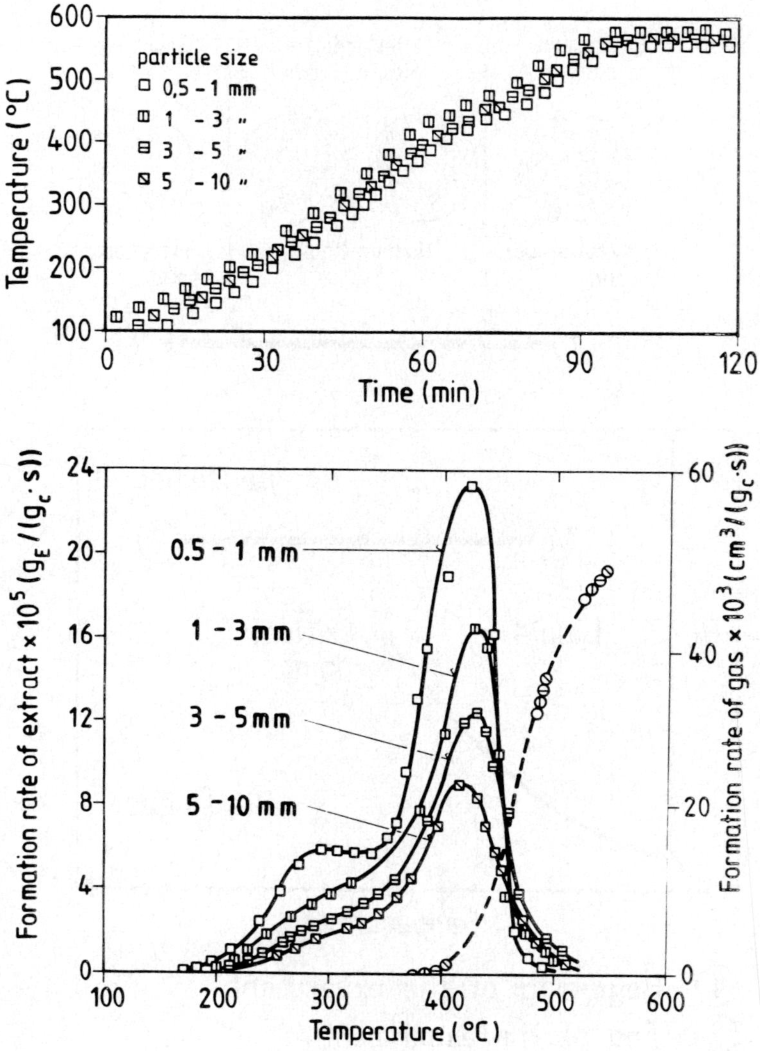

FIG. 3. EXTRACTION OF BITUMINOUS COAL WITH TOLUENE
PARAMETER VARIED: COAL PARTICLE SIZE

CONSTANT: PRESSURE 10 MPa, HEATING RATE 5 °C/min
WEIGHT OF COAL SAMPLE 150 g
LIQUID SOLVENT FLOWRATE 2 dm^3/h

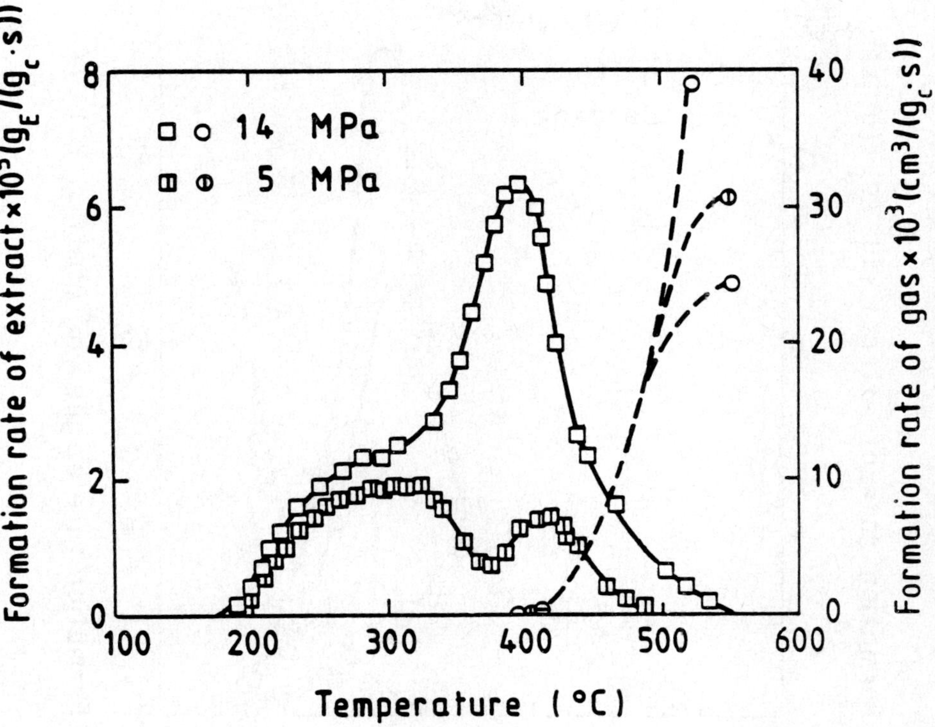

FIG. 4. EXTRACTION OF BITUMINOUS COAL WITH TOLUENE
PARAMETER VARIED: PRESSURE

CONSTANT: HEATING RATE 2.5 °C/min
WEIGHT OF COAL SAMPLE 150 g
PARTICLE SIZE 0.5-1 mm
LIQUID SOLVENT FLOWRATE 2 dm^3/h

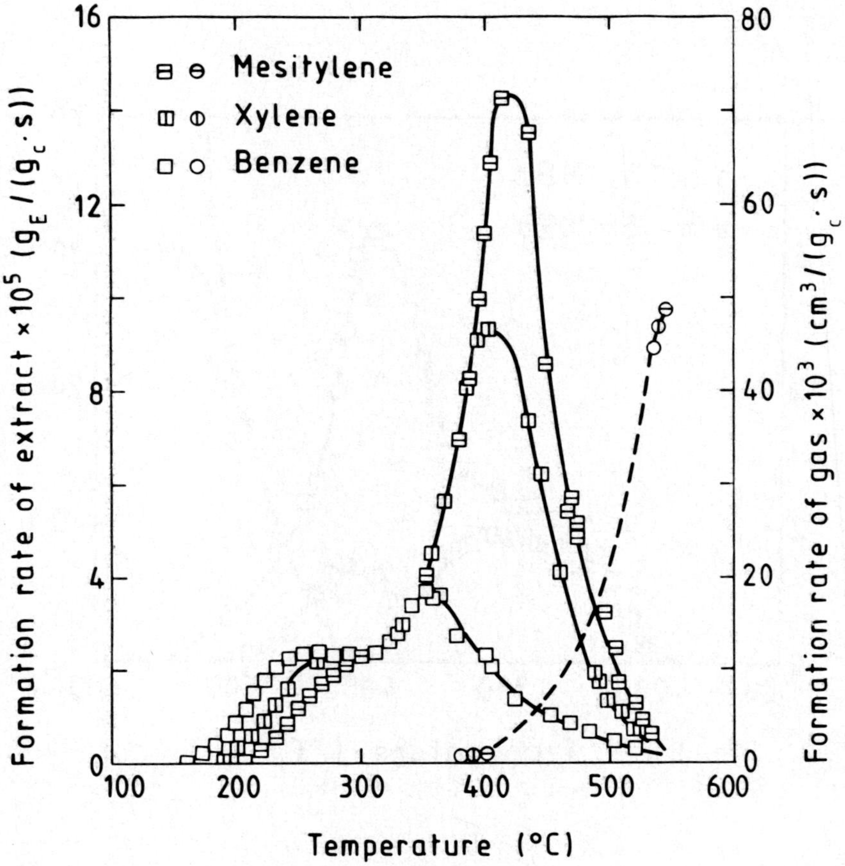

FIG. 5. EXTRACTION OF BITUMINOUS COAL WITH DIFFERENT AROMATIC SOLVENTS

CONSTANT: PRESSURE 10 MPa, HEATING RATE 2.5 °C/min, WEIGHT OF COAL SAMPLE 150 g

PARTICLE SIZE 1-3 mm

LIQUID SOLVENT FLOWRATE 2 dm^3/h

pretation of the characteristic shape differences makes it possible to separate overlapping effects of chemical reactions, solubilization and mass transport during the extraction procedure. The experimental program performed with the non-isothermal technique, therefore, includes a systematic variation of many process parameters which are listed in Table 1.

Table 1
Extraction parameters varied in the experimental program.

Process Parameters:	- Pressure - Heating Rate - Final Temperature - Coal Particle Size
Coal Types:	- Fettkohle - Gaskohle - Gasflammkohle - Flammkohle - Braunkohle
Solvents:	- Benzene - Toluene - Xylene - Mesitylene - 1-Methylnaphthalene - Hexane - i-Octane - Methanol - i-Propanol
Hydrogenation:	- Toluene + Tetralin - Toluene + Hydrogen - Toluene + Tetralin + Hydrogen - Toluene + Hydrogen + Catalyst - Toluene + Tetralin + Hydrogen + Catalyst

THE ROLE OF CHEMICAL REACTIONS, PHASE EQUILIBRIA, AND MASS TRANSPORT DURING SUPERCRITICAL COAL EXTRACTION

For the interpretation of the experimental results, in particular the characteristic shape differences of the rate profiles, we formulated an extraction mechanism (Fig. 6), which describes the interaction of chemical reaction, phase equilibria and mass transport during supercritical coal extraction.

Coal is considered to be composed of polymerized material (m_p), nonpolymerized material (m_{NP}) and material inert with respect to extraction (m_c), which can only decompose into gas and coke. Only the nonpolymerized material (m_{NP}) can be recovered by extraction without chemical reactions (Fig. 6: route A, dissolving and transport). The polymerized material (m_p) can be extracted only after thermal decomposition, which leads first to the formation of a plastic coal phase. Part of the plastic phase is dissolved and recovered as extract (Fig. 6: route B,

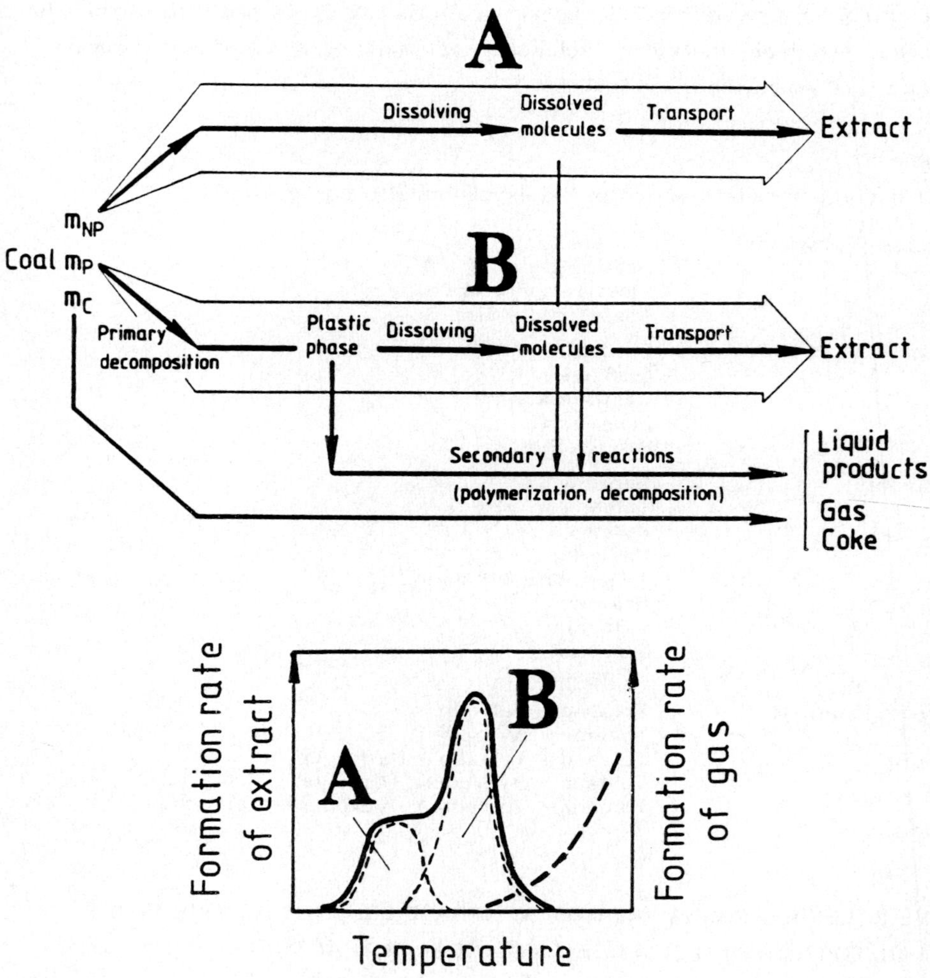

FIG. 6. INTERACTION OF CHEMICAL REACTIONS, PHASE EQUILIBRIA, AND MASS TRANSPORT DURING SUPERCRITICAL COAL EXTRACTION

primary decomposition, dissolving and transport). The remaining reacts mainly to coke, gas, and liquid products via secondary reactions.

The measured extraction profiles and their characteristic changes caused by variation of parameters can be interpreted consistently with the mechanism in Fig. 6 as demonstrated in the following on the examples given in Fig. 3, 4, and 5.

Extraction according to route A leads to the first increase in the extract formation rate profiles while extraction coupled with chemical reactions (route B) causes the socond rise to the maximum (compare Fig. 3, 4, and 5 with the scheme in Fig. 6, bottom). The gas formed at temperatures above 400 °C results mainly from secondary reactions.

The differently shaped rate profiles obtained in experiments with different coal particle sizes (Fig. 3) are caused by changes in mass transport conditions. Mass transport steps are superimposed to both route A and B and, therefore, a particle size effect is found in the entire temperature range, at sub- and supercritical temperatures.

A variation of pressure (Fig. 4) or of solvent (Fig. 5) leads to large differences only in the second extraction peak at supercritical temperatures of the solvent, whereas the formation rate of extract remains unchanged for the solvent's subcritical (liquid) temperatur region. The model indicates, that these differences originate from a phase equilibrium. The shape differences in the extraction rate profiles of Fig. 4 and 5 at temperatures above 350 °C are caused by a density effect on the solubility of the plastic coal phase in the supercritical solvent in the dissolving step of route B (comp. Fig. 6).

These selected examples show how we used the mechanism in Fig. 6 to interpret the experimental information and to find out, which role single steps in the extraction mechanism play.

IMPROVING THE YIELD BY OPTIMIZING EXTRACTION CONDITIONS

The interpretation of the extraction rate profiles from non-isothermal experiments using the mechanistic model in Fig. 6 indicate, that mass transport and solubility effects are among the limiting factors for the extract yields. These findings suggest, that higher yields should be obtainable by improving mass transport and dissolving steps during extraction. Fig. 7, however, shows, that an optimization of process parameters with respect to these effects (e.g., coal particle size (Fig. 7a), solvent flowrate (Fig. 7b), heating rate (Fig. 7c)), pressure (Fig. 7d)) is not sufficient to increase coal conversion and extract yield substantially. For toluene and other non-donor solvents coal conversion remains always below 35 %.

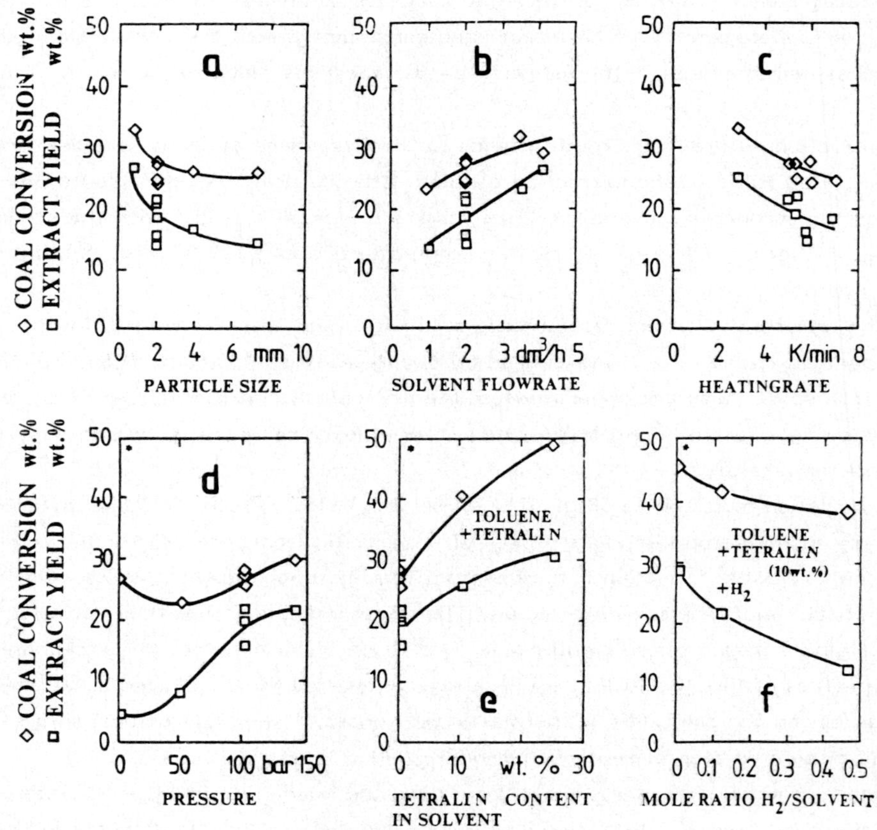

FIG. 7. EFFECT OF EXTRACTION PARAMETERS ON COAL CONVERSION AND EXTRACT YIELD
EXTRACTION CONDITIONS (IF NOT SPECIFIED OTHERWISE):
COAL SAMPLE: BITUMINOUS COAL, INITIAL WEIGHT 150 g, PARTICLE SIZE 1-3 mm (*0.2-0.5 mm),
SOLVENT: TOLUENE, LIQUID SOLVENT FLOWRATE 2 dm^3/h,
PRESSURE 10 MPa, HEATING RATE 5 °C/min (*2.5 °C/min)

Only the additional use of hydrogenating substances enhances the yield drastically as required to meet our initial technical goal. Coal conversion and extract yield strongly increase with increasing concentration of a hydrogen-donor added to the solvent (Fig. 7e, hydrogen-donor: tetralin). An addition of molecular hydrogen to the supercritical solvent can increase or reduce the yields with increasing molar ratio of H_2/solvent. However, molecular hydrogen reduces the yields in supercritical coal extraction, when an H-donor-solvent is used simultaneously (Fig. 7f).

PHYSICAL AND CHEMICAL EFFECTS OF MOLECULAR HYDROGEN AND A HYDROGEN DONOR ADDED TO THE SUPERCRITICAL SOLVENT

The experiments, where hydrogenating substances were added to the supercritical solvent, and its interpretation with the model provided also new insights.

The extraction mechanism (Fig. 8) indicates, that the processing of the intermediate plastic coal phase is most important for improving the yields in supercritical coal extraction. Coal conversion and extract yields are mainly determined by the two competing reaction pathways of the plastic phase:

1. dissolving and transport, which lead to the extract, and
2. secondary reactions, which lead mainly to a repolymerization of the plastic coal into coke.

If toluene and other non-donor solvents are used, a large proportion of the plastic phase repolymerizes to coke. The addition of a hydrogen-donor, however, strongly inhibits these reactions and, hence, the plastic coal mainly reacts along the extraction pathway; consequently coal conversion and extract yield are enhanced. The addition of hydrogen also causes an inhibition of the repolymerization reactions, but at the same time, it lowers the density of the supercritical fluid and thereby strongly decreases the solubility of the plastic coal in the dissolving step of route B. Hence, the repolymerization pathway as well as the extraction pathway are inhibited at the same time. The addition of hydrogen to a supercritical solvent, therefore, can improve or decrease the yields in supercritical coal extraction depending on which of two is stronger affected: the solubility in the dissolving step or the chemistry of secondary reactions.

Besides, the presence of an H-donor in the supercritical solvent effects strongly the physical properties of the plastic coal phase during extraction. Fig. 9 compares bituminous coal particles after extraction with toluene and a mixture of toluene and tetralin (10 wt.%), respectively.

If toluene is used as solvent, the coal particles retain their original size and shape during extraction (Fig. 9a). However, if an H-donor is present in the solvent, the plasticity of the coal particles is so much enhanced that they melt and form a highly porous plastic mass (Fig. 9b).

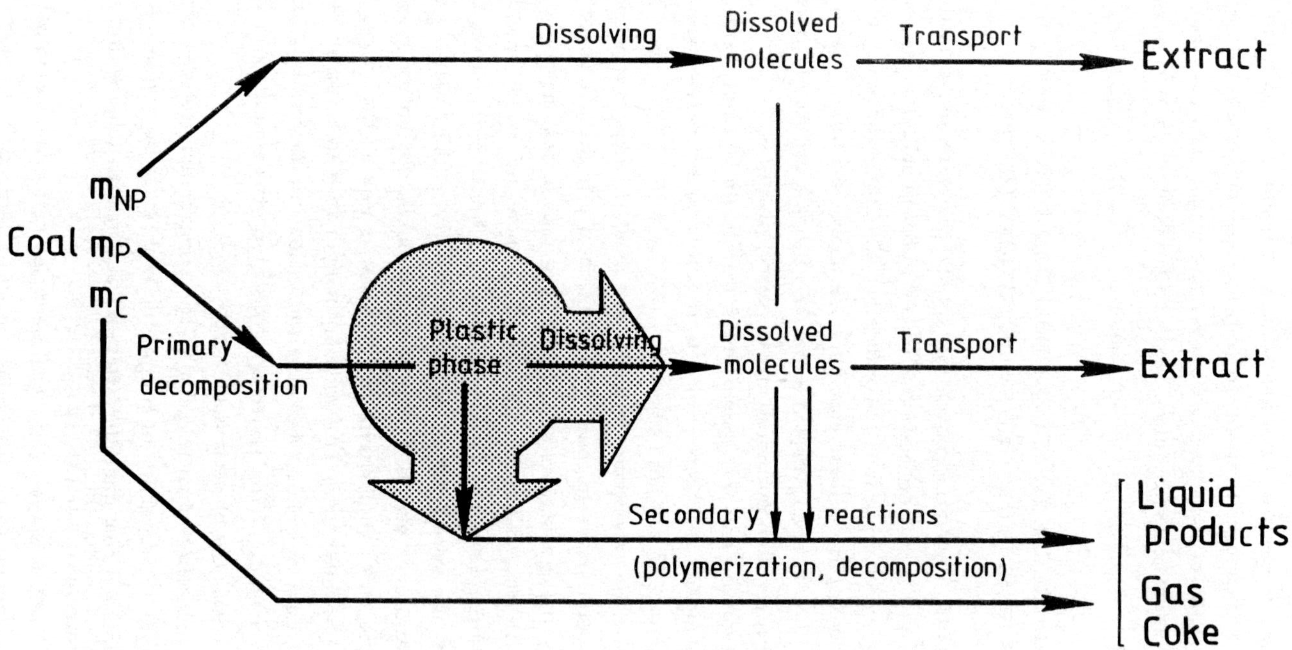

FIG. 8. COMPETING REACTION PATHWAYS OF PLASTIC COAL

a) SOLVENT : TOLUENE

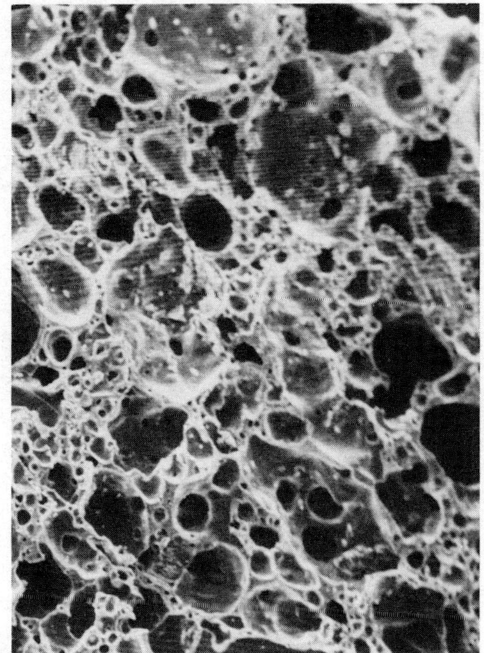

b) SOLVENT : 70 wt.% TOLUENE
30 wt.% TETRALIN

FIG. 9. EFFECT OF A DONOR-SOLVENT ON COAL PLASTICITY

This observation focussed our interest on the physical properties of plastic coal and finally lead to the principle idea of a new concept for processing coal with a supercritical solvent in order to obtain high extract yields.

THE NEW PROCESS CONCEPT FOR SUPERCRITICAL COAL EXTRACTION

Based on the better understanding of the complex interaction of chemical reactions, phase equilibria, and mass transport during the extraction procedure we propose a 'second generation' process for supercritical coal extraction which is entirely different from the original moving bed concept.

The main feature of the new process concept is a functional separation of the depolymerization of the coal and the supercritical extraction into two stages. This two-stage approach makes it possible to optimize independently each stage for its specific function.

In Reactor I (see Fig. 10) a slurry of fine ground coal and a donor solvent is heated up and plasticized to a subcritical viscous solution. The high concentration of an H-donor used in this stage prevents repolymerization reactions, enhances the coal plasticity, and leads to a nearly complete solubilizing of the coal particles.

The fact that the coal has been transformed in stage one into a viscous liquid makes it possible to use a liquid-liquid extractor type for the supercritical extraction in stage two. The jet-pipe-nozzle extractor used as second stage (Reactor II) gives very high contact surface areas between the subcritical coal solution and the supercritical solvent and thus enhances mass transport rates.

The solution emitted by the jet-pipe-nozzle is supercritical. The fluid's viscosity is reduced and undissolved solid material is separated more easily. Both the solid-free solution and the residual slurry are flashed and give as bottom products extract and residue, respectively. The vapors containing the supercritical solvent and the dehydrogenated H-donor are recycled, cooled down, and split into 2- and 3-ring aromatics on the one hand and into one-ring aromatics on the other hand. After rehydrogenation the 2- and 3-ring aromatics are used as H-donors for the coal slurry preparation and the depolymerization in Reactor I, whereas the one-ring aromatics are degassed, dewatered, and recycled as supercritical solvents to Reactor II. Both solvents are generated in the process from the coal itself. Hydrogen, which is consumed in the overall process, is not used in the two stages in direct contact with the coal. Its indirect introduction via rehydrogenation of the spent donor-solvent avoids detrimental effects on the solubiltiy of the plasticized coal in the supercritical solvent in the extraction stage.

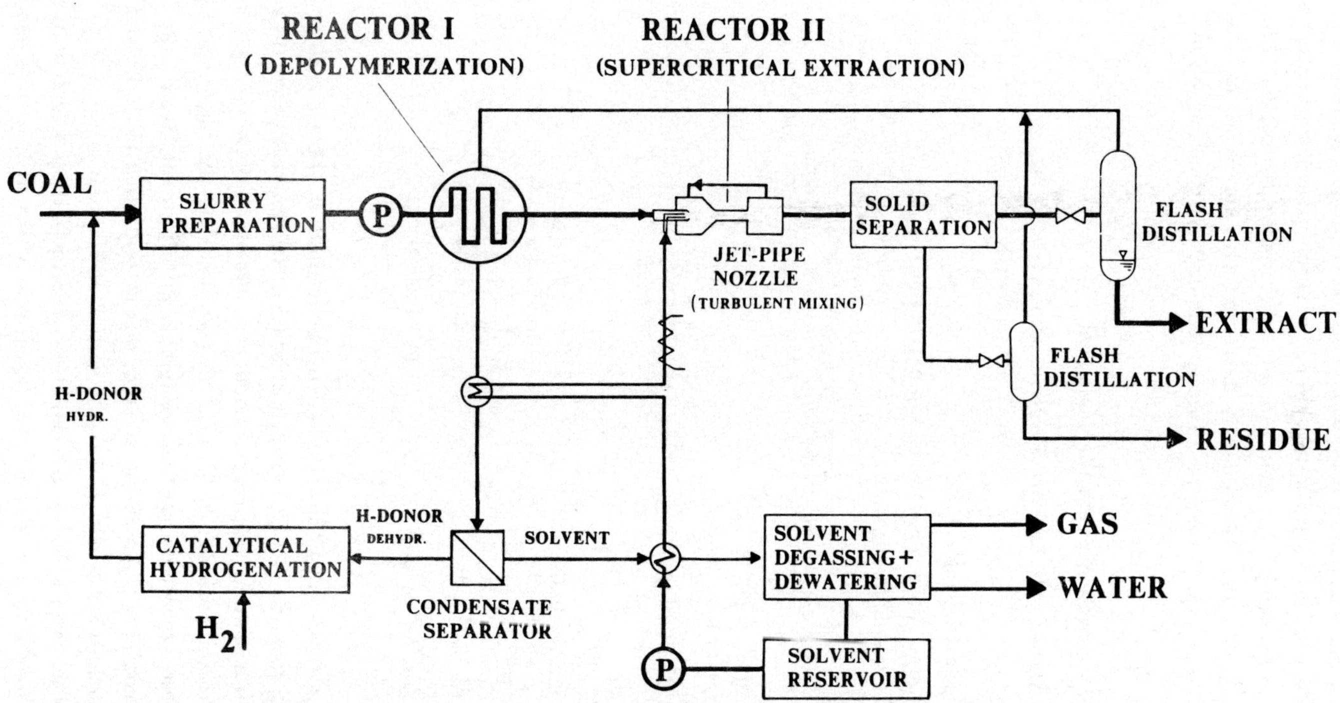

FIG. 10. A 'SECOND GENERATION' APPROACH FOR SUPERCRITICAL COAL EXTRACTION: THE DSSE-PROCESS

FIRST TEST OF THE NEW PROCESS CONCEPT

The proposed two-stage approach for coal extraction was tested by simulating the coal processing through the two stages in batch experiments. Coal was first plasticized in presence of a donor-solvent in a shaking autoclave. The resulting viscous solution was subsequently extracted with supercritical toluene. Extraction conditions and results for a bituminious coal are summarized in Table 2.

Table 2
Results for DSSE-extraction of a bituminous coal.

Coal sample: Gasflammkohle Ensdorf (particle size 0.2-0.5 mm)

water	: 1.0 wt.%
ash (m.f.)	: 6.5 wt.%
volatile matter (m.a.f.)	: 38.5 wt.%

Extraction conditions (batchwise two-stage extraction):

stage 1 (depolymerization)	: 10 minutes at T = 410 °C
	H-donor : tetralin
	coal/tetralin = 1:1
stage 2 (supercritical extraction)	: solvent : toluene
	P = 100 bar, T = 380 °C

Coal conversion and yields (m.a.f.):

coal conversion degree	67 wt.%
extract	44 wt.%
liquid distillates	22 wt.%
gases	1 wt.%

Properties of the extract:

elemental analysis	: C	82.4	wt.%
	H	6.4	wt.%
	O	7.2	wt.%
	N	1.6	wt.%
	S	0.64	wt.%
ash	:	< 0.1	wt.%
softening temperature	:	120-140 °C	

The two-stage treatment of bituminous coal presented in Table 2 results in a degree of coal conversion of 67 %. An optimization of the extraction conditions of the two stages in a continuous process mode most likely will enhance the coal conversion degree to values above 70 %. The low gas yield and the high liquid yield show, that the two-stage treatment converts coal with high selectivity into liquid and soluble hydrocarbons avoiding a consumption of hydrogen in undesired reactions to hydrocarbon gases.

The coal extract is practically ashfree and characterized by a higher hydrogen to carbon ratio than extracts obtained by liquid extraction with anthracene-oil-type solvents. This indicates, that the DSSE-extracts might be more easily converted to the desired hydrocarbon distillates in a subsequent catalytic hydrocracking stage.

INCENTIVES FOR DEVELOPING THE DSSE-PROCESS

The Donor-Solvent-Supercritical-Extraction Process combines several desirable features making it suitable as first stage of a Pott/Broche-type coal liquefaction process, e.g.:

--- high coal conversion (∿70 %) compared to "first generation" supercritical extraction processes (∿30 %)

--- high yield to liquid distillates already in the extraction stage

--- low production of hydrocarbon gases and, therefore, a high hydrogen utilization efficiency

--- feasible large-scale solid separation

--- no need of large reaction vessels.

The DSSE-process combines the advantages of both hydrogenating coal extraction with liquid H-donor solvents (high coal conversion) and coal extraction with supercritical solvents (practicable solid separation) and, therefore, might open a technically feasible route for large scale coal processing with supercritical fluids. In addition, the high selectivity to liquid hydrocarbons and the low production of hydrocarbon gases promises an efficient use of hydrogen, which is of paramount importance for coal liquefaction process economics.

REFERENCES

1 J.C. Whitehead, D.F. Williams: Journal of the Institute of Fuel, 48, 182-184 (1975).

2 A. Wilhelm, Dissertation, University of Karlsruhe (TH), 1981.

3 A. Wilhelm, K. Hedden: "Extraktion von Kohle mit Lösungsmitteln in unterkritischer und überkritischer Phase unter hydrierenden und nichthydrierenden Bedingungen", Forschungsbericht FB-T-82-177, Bundesministerium für Forschung und Technologie, 1982.

4 A. Wilhelm, K. Hedden: Erdöl und Kohle - Erdgas - Petrochemie, 33, 269-274 (1983).

5 R.R. Maddocks, D.F. Williams: Chemical Engineering Progress, 75 (6), 49-55 (1979).

PART IV

Special Applications

Supercritical Fluid Technology, **edited by J.M.L. Penninger, M. Radosz, M.A. McHugh and V.J. Krukonis**
1985 Elsevier Science Publishers B.V., Amsterdam — Printed in The Netherlands

SUPERCRITICAL EXTRACTION (SCE) OF DIXYLENOL SULFONE (DXS)

Jennie R. Patrick, Research Center, Philip Morris U.S.A., Richmond, Virginia
Fred Palmer, R&D Center, General Electric Co., Schenectady, N.Y.

ABSTRACT

This study employed supercritical carbon dioxide, ethylene, chlorotrifluoromethane, and mixtures of carbon dioxide/ethylene and acetophenone/carbon dioxide to extract 2,6-xylenol, xylenol sulfonic acid (XSA) and quinone type compounds from dixylenol sulfone (DXS). Dixylenol sulfone purification is significant because DXS is the monomer used to synthesize polycarbonate polymers. The three-part study to purify DXS included: 1) Removal of color-causing quinone type compounds, 2) one-stage, and 3) two-stage extraction of other impurities in DXS. The results indicated close solubility of supercritical carbon dioxide and ethylene for the DXS system. The major impurities, 2,6-xylenol and XSA, decreased as much as 100% after the two-stage extraction.

INTRODUCTION

This study employs supercritical fluids to purify dixylenol sulfone (DXS). Dixylenol sulfone is a monomer used in the synthesis of various polycarbonate polymers (Van Hoorn, 1968; Borman, 1972). Dixylenol sulfone is produced by the following reaction (Fox, 1968):

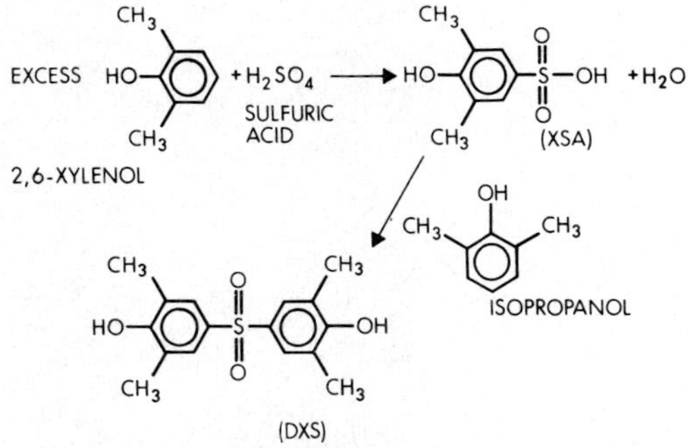

It is desirable to remove the reactants in addition to other secondary impurities, including 2,6-xylenol, xylenol sulfonic acid, and tetramethyldiquinone in the purification of DXS. These impurities usually produce polymers with poor mechanical properties and low light transmittance.

Reducing the impurities level in the DXS monomer before polymerization lessens the purification requirements for the aforementioned polymers. Also, prior purification can considerably reduce the number of process steps in the polymerization scheme. Thus, the desire to produce a more efficient and less complicated polymerization process scheme serves as the impetus for this study, which is Part I of a proposed two-part series. In Part I the focus is to produce quantitative data indicating the most appropriate solvent system as well as entrainers. The objectives in this study are as follows: 1. Remove primary impurities of 2,6-xylenol and xylenol sulfonic acid (XSA) from DXS. 2. Remove "quinones" type impurities believed responsible for color in the DXS.

EXPERIMENTAL APPARATUS/PROCEDURE

Figure 1 is a schematic diagram of the experimental apparatus. The system consists of a one-liter high pressure reactor made from 316 stainless steel, two product collectors, a pressurizer (surge tank), and a horizontal opposed double-ended air driven two-head compressor which transports

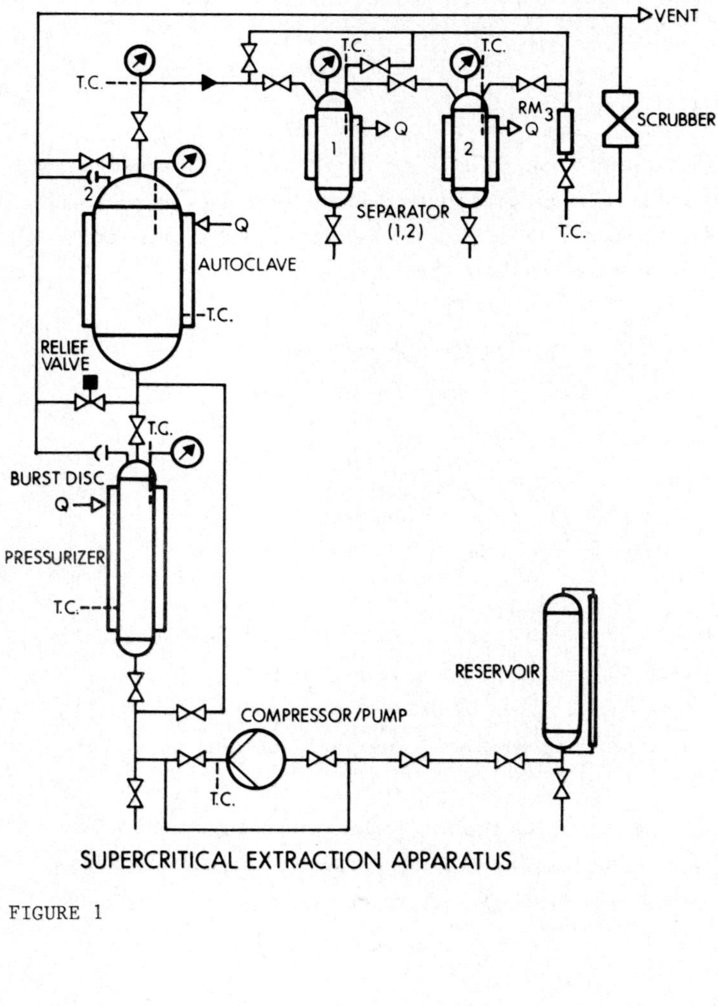

FIGURE 1

solvents from the gas cylinders into the system.

The experiments are initiated by charging the monomer into the autoclave. The gas solvent is then compressed and sent to the surge tank/pressurizer. The valve between the autoclave and the surge tank is kept closed temporarily until temperature and pressure are at the desired supercritical conditions in the surge tank. The valve is then opened and the conditions in the autoclave and surge tank are allowed to come to equilibrium. The DXS monomer and the supercritical fluid are vigorously mixed with a mechanical stirrer for a specified time. The supercritical fluid is slowly expanded and throttled off. As the fluid is purged, it is gradually replaced with nitrogen at the same condition until the system is fluid-free. The nitrogen is then purged and the system allowed to cool. A sample is then taken from the reactor to be analyzed by high pressure liquid chromatography (HPLC).

Following the initial HPLC analysis, the sample is placed back into the autoclave and the above experimental procedures repeated. The sample is then resubmitted for HPLC analysis.

RESULTS AND DISCUSSION

One-stage Extraction

Shown in Table 1 are results from one-stage batch extractions of DXS samples A and B. Table 1 exhibits the extraction temperature, pressure, and solvent. Also shown are the remaining amounts of XSA and 2,6-xylenol after supercritical extraction.

TABLE 1

One-stage supercritical extraction of DXS

EXPERIMENT #	PRESSURE psia	TEMPERATURE °C	SOLVENT	XSA PPM	2,6-XYLENOL PPM
Control A	(untreated)	(untreated)	(untreated)	1647	289
	1315	71	CO_2	1632	292
	1415	66	Acet./CO_2	905	-
	1515	177	CO_2	415	22
	765	177	Freon-13	477	61
	2015	204	CO_2	214	21
Control B	(untreated)	(untreated)	(untreated)	980	275
	2515	204	CO_2	289	62
	2015	204	C_2H_4	174	12
	2015	204	CO_2/C_2H_4	77	<1
	2015	232	CO_2	39	19
	2015	232	CO_2	58	<1
	2015	232	CO_2/C_2H_4	53	<1
	2015	232	C_2H_4	38	1
	2015	260	CO_2	54	6

Quantitatively, supercritical carbon dioxide at 2015 psia and 204°C (400 F) proved more effective at removing impurities from sample A than supercritical Freon(R) 13 and supercritical carbon dioxide at less stringent conditions (See Table 1).

Impurities removal appeared maximum circa 2015 psia and 232°C (450 F). Xylenol sulfonic acid is less soluble than 2,6-xylenol in both supercritical carbon dioxide and ethylene. Relatively small variations in both temperature and pressure caused considerable change in solubility. Additional experiments are required for accurate characterization of temperature and pressure effects. Pure supercritical carbon dioxide and supercritical ethylene exhibit close degrees of solubility for this system.

Two-stage Extraction

Results for two-stage extraction of dixylenol sulfone are presented in Table 2. The two-stage extraction process removed essentially all of 2,6-xylenol from DXS (see Table 2). The two-stage extraction process reduced levels of XSA by unobservable amounts up to 74%. The two-stage process decreased the levels of 2,6-xylenol by 71% to 100%.

TABLE 2

Two-stage supercritical extraction of DXS.

Pressure psia	Temp. °C	Solvent	XSA PPM	2,6-xylenol PPM	XSA PPM	2,6-xylenol PPM
2015	149	CO_2	-	--	523	130
2015	204	CO_2	214	21	55	3
2015	232	CO_2	39	19	39	6
2015	204	C_2H_4	174	12	128	0
2015	204	*CO_2/C_2H_4	77	<1	71	0
2015	232	*CO_2/C_2H_4	53	<1	21	0

* CO_2/C_2H_4 - 50%/50% composition

Color Removal

Potential color removing solvents tested in this study are listed in Appendix A. These solvents were screened by visual observation as potential entrainers. In addition to the visual screening of potential entrainers, visible (VIS) spectra measurements were also made. An example of the visible spectra measurements made on DXS samples is shown in Appendix B.

Of the potential color removing solvent listed in Appendix A, only acetophenone substantially removed color from DXS. Thus, acetophenone was used as an entrainer in the extractions. Unfortunately, acetophenone proved not feasible for the system because:

(a) It interfered with available analytical test procedure for determination of impurities levels in DXS.

(b) Energy intensive drying would be required to remove acetophenone from the DXS.

Physical observation of DXS specimens immediately after extraction suggest possible color removal. But, the analytical results do not support this physical observation. The analytical results indicated that even two-stage extractions did not enhance color removal from DXS. Whether or not a chemical change causing more color occurs with the DXS from the time of extraction to the time of color analysis remains uncertain.

Nevertheless, as shown by the VIS spectra, a decrease in color occurred using acetophenone as an entrainer with supercritical carbon dioxide. Extraction using pure supercritical carbon dioxide, Freon$^{(R)}$ 13, and ethylene resulted in an increase in color as indicated by the VIS spectra.

CONCLUSIONS/RECOMMENDATIONS

Both supercritical carbon dioxide and supercritical ethylene reduced xylenol sulfonic acid and 2,6-xylenol levels in dixylenol sulfone to acceptable specifications. A mixture of supercritical carbon dioxide and ethylene demonstrated similar purification capabilities. However, optimization of process variables and solubility would be required to enhance the effectiveness and efficiency of this DXS purification process.

Satisfactory levels of color removal from DXS were not achieved in this study. Closer examination of both structure and properties of acetophenone may offer insight into the identification of an acceptable solvent for this purpose.

ACKNOWLEDGEMENTS

I wish to thank Stan P. Urbaetis (General Electric Co.) for his help with the experimental apparatus. Also, many thanks to Carol Cowsert (General Electric Co.) and Anne Donathan (Philip Morris) who typed this manuscript.

REFERENCES

1 W. F. H. Borman, Solvent Resistant Polyesters, U. S. Pat. 3,652,499 (1972).
2 D. W. Fox and P. Shenian, U. S. Pat. 3,383,421 (1968).
3 H. Van Hoorn, Dynamic Mechanical Study of the Effect of Chemical Variations on the Internal Mobility of Linear Epoxy Resins, J. Appl. Polym. Sci., 12, 871-888 (1968).

APPENDIX A

Tests for color removal.

SOLVENTS	OBSERVABLE COLOR CHANGE	SOLUBILITY WITH DXS
Acetone	Yes	Soluble
Acetophenone	Yes	No
Benzyl Alcohol	No	No
Cyclopentanone	Yes	Soluble
Diacetone	Yes	Slightly
Isophorone	Yes	Soluble
Methylene Chloride	No	No
Octyl Alcohol	Yes	Soluble
Pyridene	No	No
Toluene	No	Slightly
Water	No	No

APPENDIX B

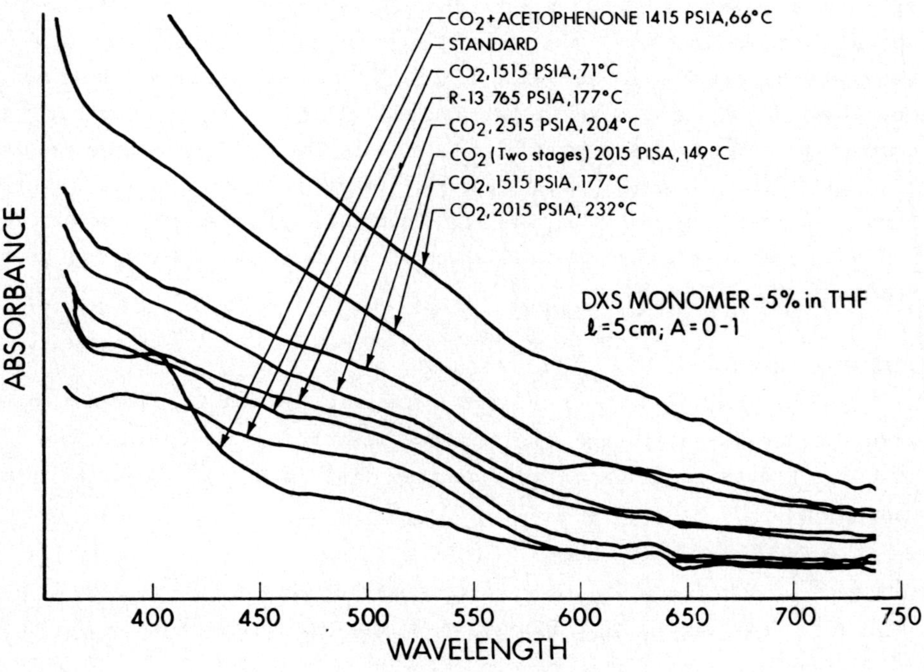

Supercritical Fluid Technology, edited by J.M.L. Penninger, M. Radosz, M.A. McHugh and V.J. Krukonis
1985 Elsevier Science Publishers B.V., Amsterdam — Printed in The Netherlands

HIGH PRESSURE EXTRACTION OF LIGNIN FROM WOOD

R. BEER and S. PETER
Lehrstuhl für Technische Chemie, Universität Erlangen-Nürnberg, Egerlandstr. 3
D-8520 Erlangen (West Germany)

ABSTRACT

The digestion of wood from spruce by mixtures of water and methylamine was investigated at conditions where the binary system water/methylamine is supercritical. Within the temperature range from 150 to 180 °C a product is extracted which is very soluble in the supercritical extracting agent. Thereby the extracted product is obtained as a solution with a concentration of about 60 wt.%. The digestion reaction is of first order, similar to the sulphite and sulphate process. The phase equilibria of the binary systems methylamine/water and methylamine/extracted product and the ternary system methylamine/water/extracted product were measured at temperatures from 150 to 180 °C and pressures up to 10 MPa.

INTRODUCTION

In wood the cellulose is a part of a sophisticated assembly, in which the long and strong cellulosic chains are twisted to fibers and serve as a frame substance. The hydrophobic lignin acts as a protective coating (ref.1). For the production of alpha cellulose and other valuable products from ligno-cellulose materials such as wood, the compound material must be disintegrated.

Of the various methods for the disintegration of ligno-cellulosic materials the soda, sulphate and sulphite processes are the best known and most widely used. The soda and the sulphate processes are alkaline cooks (ref.2). The soda cook is used primarily in disintegrating deciduous woods as aspen, birch, white maple, basswood, beech etc., while the sulphate process is well adapted for digestion of highly resinous coniferous woods.

In the soda process the dissolving agent strongly acts on the ligno-cellulosic structure and yields a purified cellulosic fiber which is weak and possesses in general inferior chemical properties. In the sulphate process a milder action on the cellulose fibers occurs but the produced pulp still contains a considerable amount of lignin and other contaminating products.

The sulphite process is an acid cook and produces a higher grade pulp than the above mentioned alkaline digestions (ref.3,4). The application thereof, however, is limited to the processing of the less resinous woods, such as spruce, fir and hemlock.

All processes yield a waste liquor from which residual digestant and valuable by-products, such as lignin, are difficult to recover, in poor quality only. At present the waste liquor is largely, concentrated by evaporation and incinerated. Some of the waste materials are used in the preparation of fertilizers, spreading or dispersing agents, ethanol etc.

The products which are extracted by the commonly used sulphate and sulphite processes contain sulphur compounds in rather considerable amounts, hence their value as a base material for further chemical processing is limited. In the past proposals were repeatedly made for processes which yield a sulphur-free extract. Peterson and Wise discussed the treatment of ligno-cellulosic materials with solutions of amines in water at low pressures (ref.5). Kleinert (ref.6) proposes the use of a mixture of 42.2 wt.% ethanol and 57.8 wt.% water as an extracting agent at temperatures of about 185 °C. In a process developed at the Batelle-Geneva laboratories (ref.7) aqueous phenolic solutions are applied for upgrading biomass under acidic conditions.

In this work near-critical extraction was investigated, as a new way to recover lignin from wood. The main objective of this approach is to produce a sulphur-free lignin extract and concurrently avoid the excess of waste liquor of conventional processes.

PHASE EQUILIBRIA

Preliminary experiments showed that among suitable aliphatic amines methylamine is the most promising. Hence the phase equilibria of the binary system methylamine/water, methylamine/extracted product and the ternary system methylamine/water/extracted product was investigated.

For the investigations the static method was applied. The components were placed in a cylindrical thermostated pressure vessel and the pressure was raised to the required level. The mixture was stirred until equilibrium was reached. Subsequently samples were taken from the coexisting phases and analysed. The volume of the samples was small in comparison to the volume of the phases so that the pressure was not affected by the sampling procedure .

The results obtained with the binary system methylamine/water at 150, 160 and 180 °C are presented in Fig. 1. At 150 °C the two phase region covers the whole range of composition. Above the critical temperature of methylamine (156.9 °C) the isotherms as a function of pressure and composition form a closed loop. The critical pressures are 7.0 MPa at 160 °C and 8.4 MPa at 180 °C. The concentrations at the critical pressures are 87 wt.% and 84 wt.% methylamine respectively (according to linear bisector).

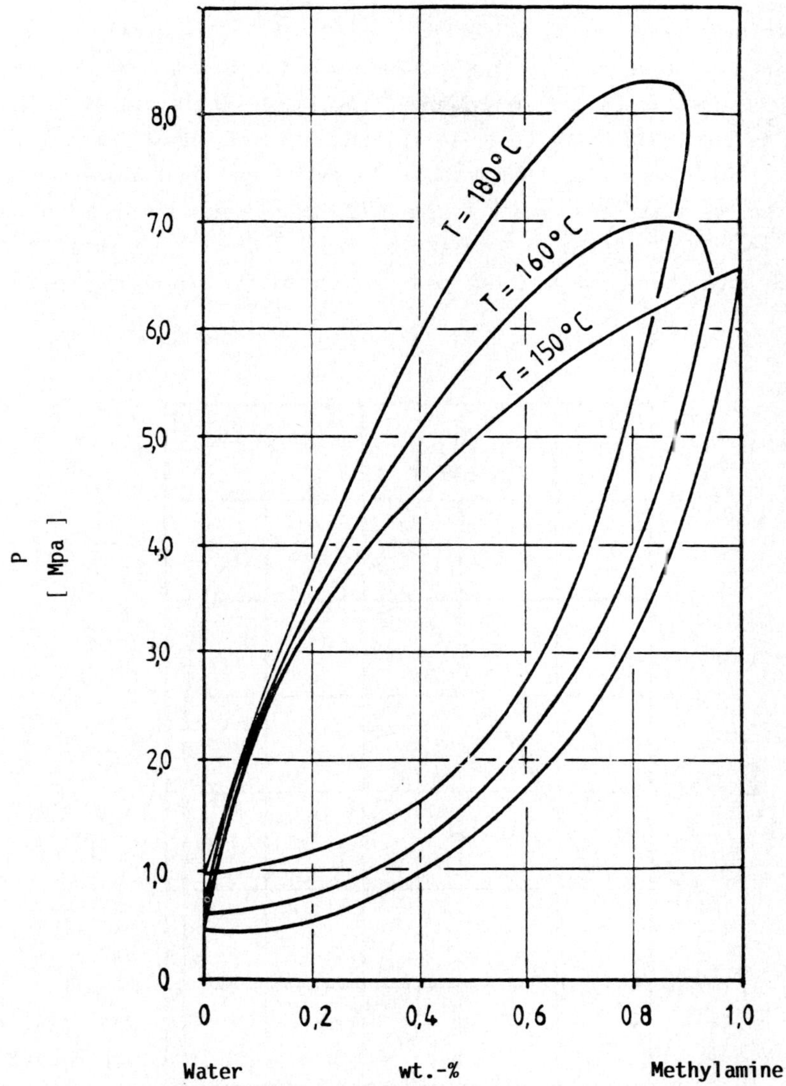

Fig. 1. P-x-diagram of the binary system methylamine/water

Extraction of spruce wood with a mixture of 83 wt.% methylamine and 17 wt.% water at 180 °C and 11.0 MPa gives a solid product which is miscible with water. The loading of the extracting agent with lignin amounted to about 10 wt.%. The extract is probably an adduct with methylamine because methylamine can be set free by digesting the concentrated aqueous extract with sodiumhydroxide. In the following the extracted product is named methylaminelignin although its chemical constitution is not known. This extract was used to measure phase equilibria.

The system methylamine/methylaminelignin at 180 °C is presented in Fig. 2. As methylamine is above its critical temperature the binodal curve forms a loop which does not touch the methylamine coordinate. The critical pressure is found to be about 9.8 MPa, the critical composition in 78 wt.% methylamine (according to the linear bisection method). These results in combination with Fig. 1 do conclude that the ternary system methylamine/water/methylaminelignin at 11.0 MPa and 180 °C is single-phase i.e. supercritical. The three binary systems which form the ternary system have no miscibility gap under these conditions.

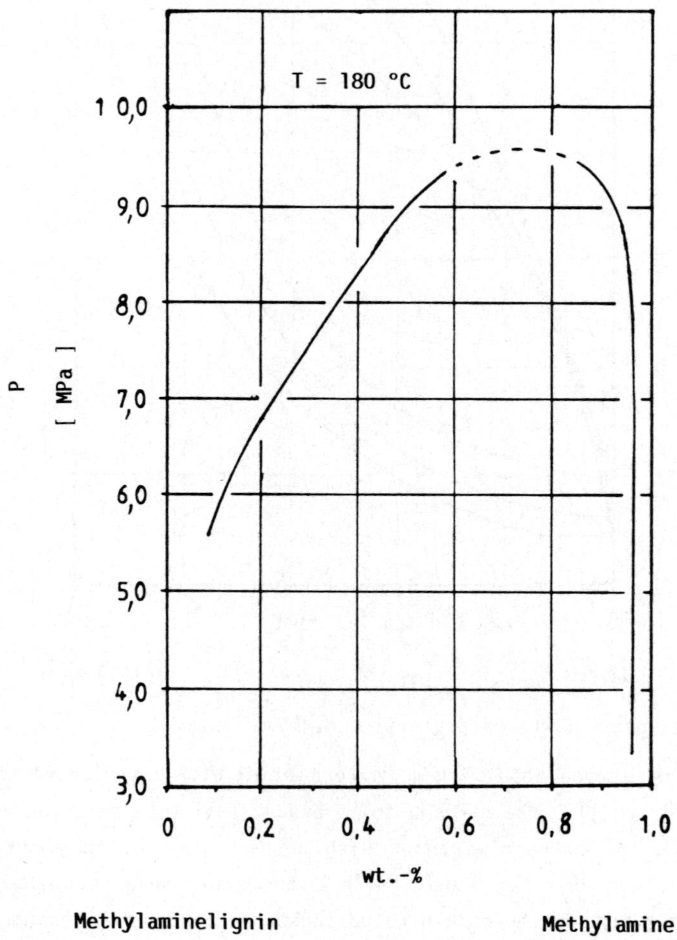

Fig. 2. P-x-diagram of the binary system methylaminelignin/methylamine

In order to study the possibility for regenerating the supercritical extracting agent composed of methylamine and water the phase equilibria in the ternary system methylamine/water/methylaminelignin were investigated at 180 °C and 5.2 MPa. The results are presented in Fig. 3. A system of type II with two miscibility gaps exists. The gas phase contains only small amounts of lignate. Thus an expansion of the solution from 11.0 to about 5.0 MPa is sufficient for the regeneration of the extracting agent.

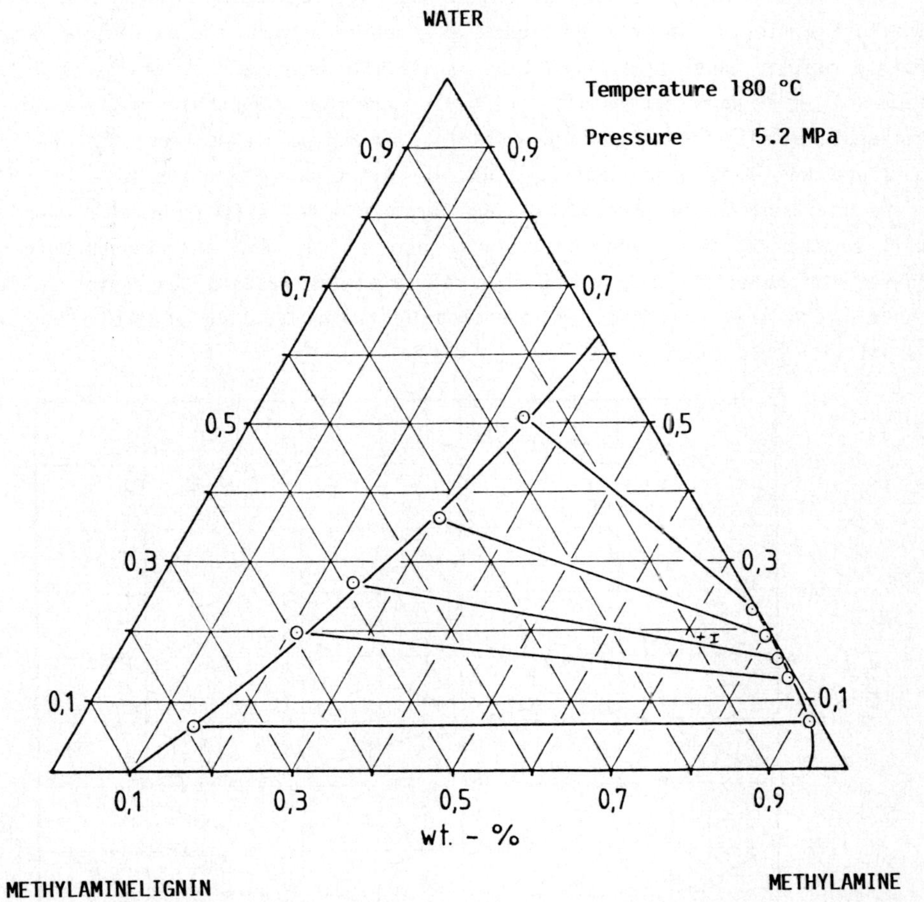

Fig. 3 Phase equilibria in the ternary system of methylaminelignin/methylamine/water.

EXTRACTION OF WOOD UNDER NEAR-CRITICAL CONDITIONS

Extraction of spruce wood at 180 °C and 11.0 MPa with a supercritical mixture of 83 wt.% methylamine and 17 wt.% water produces a solid product which is miscible with water. After an extraction time of about 3 hours the remaining α-cellulose contains about 7 wt.% of lignin.

Since the experiments proved that amines in general and methylamine in particular are efficient solvents for lignin the influence of water upon the dissolution of lignin was investigated. From the extraction of vegetable materials (e.g. hops, coffee beans, wood etc.) it is known that the rate of mass transfer is affected by the water content. The extraction rate is low for dry vegetable materials (ref.8,9) but increases with the content of swelling moisture. In general, there exists an optimal content of swelling water.

The extraction was studied at 180 °C and 11.0 MPa using wood with different moisture contents. These conditions were chosen because the extraction occurs fairly rapidly under these conditions while the decomposition of the cellulose is slow. At higher temperatures cellulose decomposes faster while at lower temperatures dissolution of lignin is slow. During each experiment the swelling moisture was kept approximately constant. After an extraction time of three hours the content of α-cellulose and lignin of the residue was determined as well as the degree of wood disintegration (see Fig. 4). Ten minutes were required for heating the extraction vessel to the extraction temperature. After three hours the extraction is quenched by rapidly cooling of the extraction vessel with cold water.

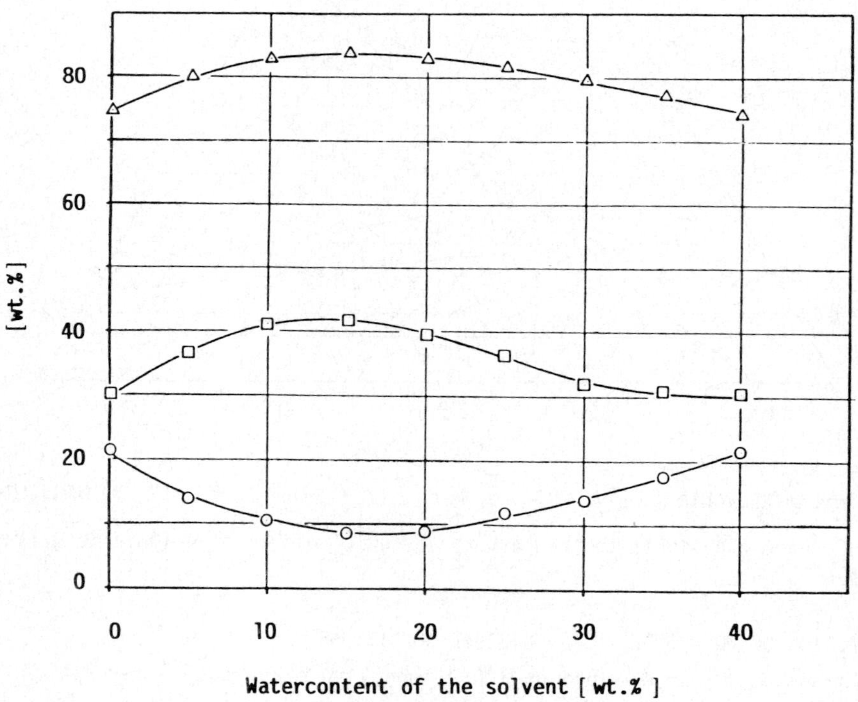

Fig. 4 Residual lignin [⊙], wood disintegration [⊡] and α-cellulose content of the residue [△] as a function of the water content of the extraction agent. Extraction time: 3 hours; 180 °C, 10 MPa

The results of the experiments are presented in Fig. 4. A minimum lignin content in the residual cellulose is reached when the extraction agent contains about 17 wt.% water. With this composition of the extracting agent the diminution of the wood and the content of α-cellulose in the undissolved material are at a maximum.

Moreover the extraction rate depends on the swelling moisture of the wood. The swelling moisture of the wood can only be altered slowly. The humidity content of the wood must be adjusted before the extraction process. The water content of the extracting agent should be chosen so that during the extraction no substantial change of the swelling moisture occurs. For a humidity content of the spruce wood chips of 44 wt.% the optimal water content of the extracting agent is about 20 wt.% at 180 °C and 11.0 MPa. Under these conditions the selectivity Q (ratio of dissolved lignin to dissolved carbohydrates) of the digestion is at a maximum as can be seen from Fig. 5.

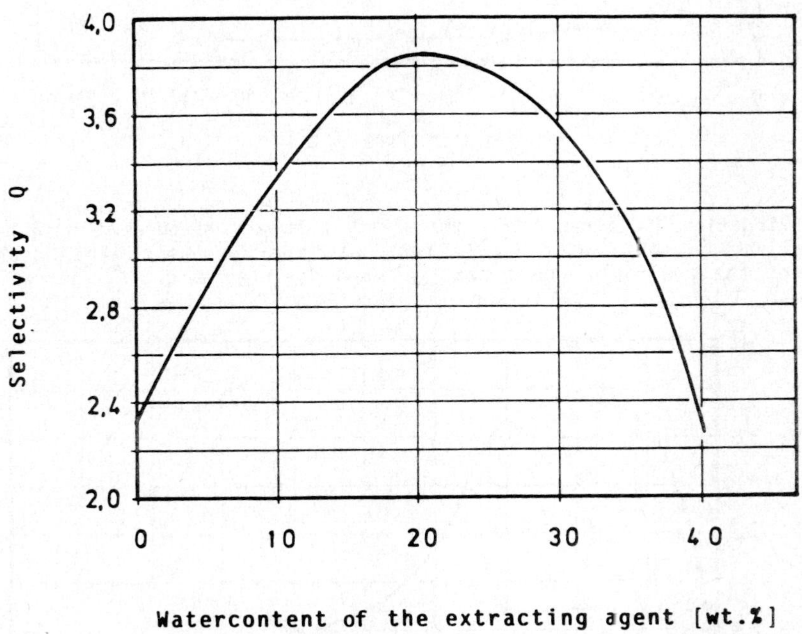

Fig. 5 Selectivity Q vs. water content of methylamine.
Extraction time: 3 hours; T = 180 °C; p = 11 MPa; wood moisture 44 wt.%

The extraction of lignin from spruce as a function of time is presented in Fig. 6. Rapid dissolution of the lignin occurs within the first three hours; thereafter the rate of dissolution decreases markedly. After three hours the

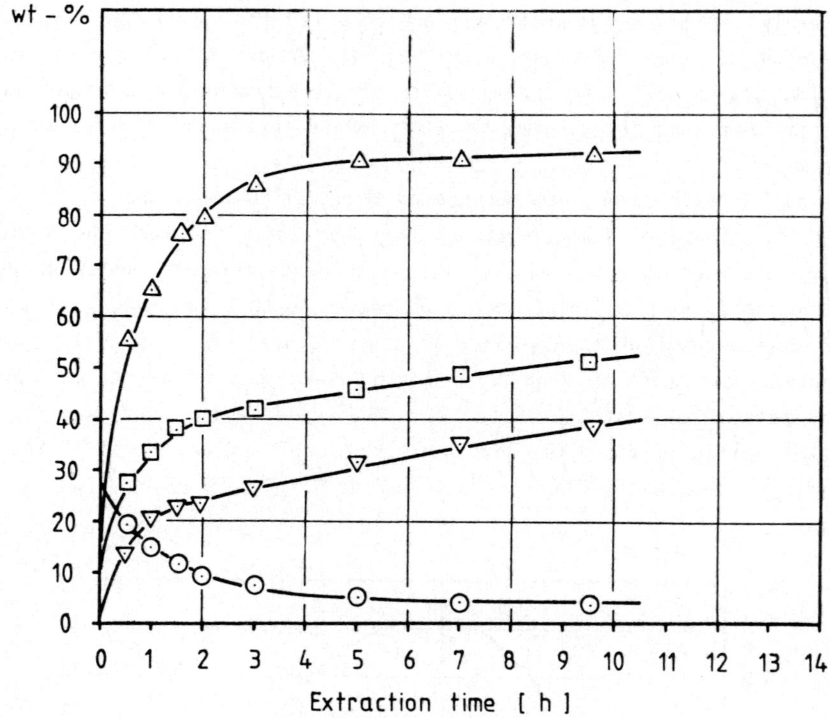

Fig. 6. Extraction of lignin from spruce with a binary mixture of 83 wt.% methylamine and 17 wt.% water (⊙ residual lignin content after cooking, ▽ decrease of the nonlignin substances, ⊡ wood disintegration, △ decrease of the lignin). T 180 °C, p 11 MPa, wood moisture 44 wt.%

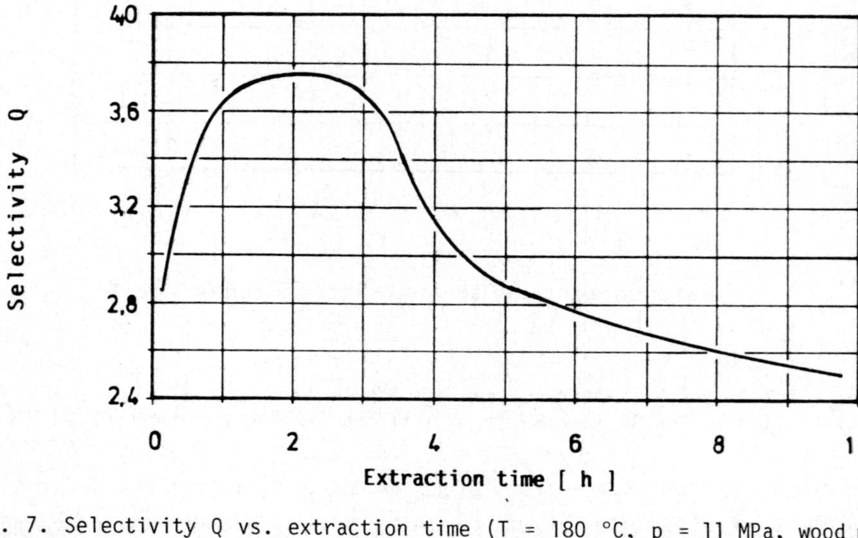

Fig. 7. Selectivity Q vs. extraction time (T = 180 °C, p = 11 MPa, wood moisture 44 wt.%, extracting agent 17 wt.% water 83 wt.% methylamine).

lignin content of the residual cellulosic is reduced to about 7 wt.%. The rate of dissolution of the other compounds also decreases in the course of three hours but remains at a much higher level for the rest of the experiment. Therefore the ratio of dissolved lignin to dissolved carbohydrates reaches a maximum at about 2 hours extraction time as shown in Fig. 7.

The logarithmus of the residual lignin content is plotted in Fig. 8 as a function of the extraction time at temperatures of 165 and 180 °C and a pressure of 11.0 MPa. The digestion reaction has first order kinetics. It consists of two sections of which the first has a higher reaction rate than the second. This behaviour for which a satisfactory explanation is not yet fully understood is also observed in the Kraft and the organosolv processes. From the temperature dependence of the reaction rate of both sections activation enery of about 130 kJoule/mole is obtained.

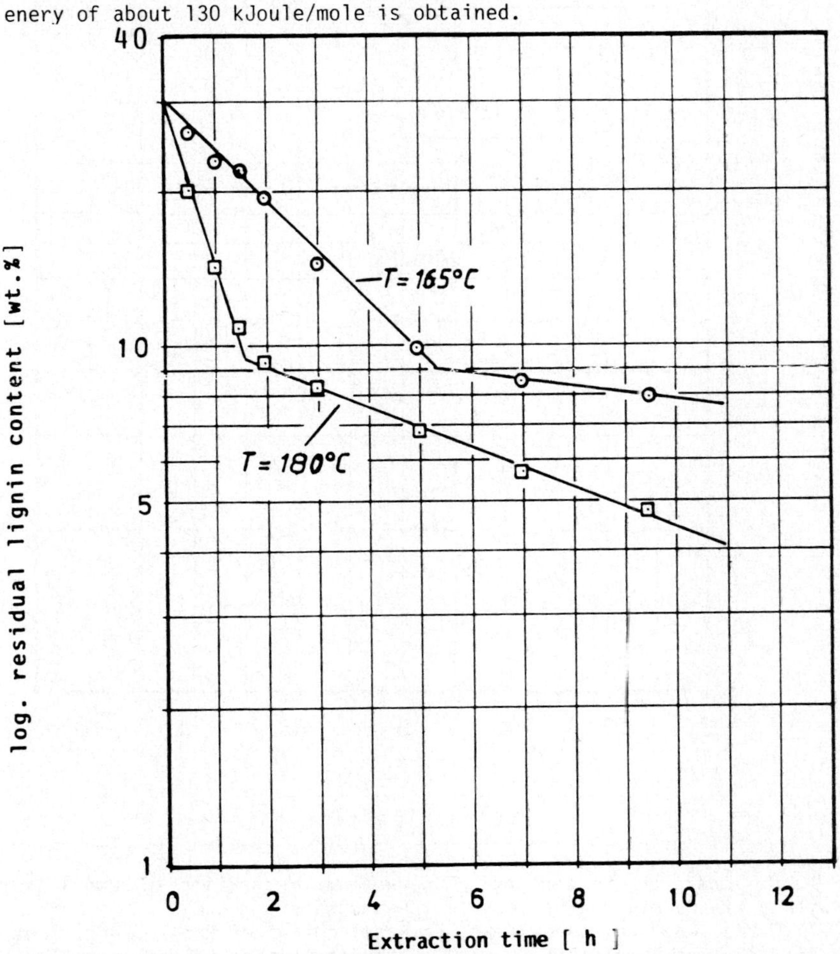

Fig. 8. Residual lignin content of spruce as a function of extraction time (wood-moisture 44 wt.%, extracting agent 83 wt.% methylamine and 17 wt.% water, 11.0 MPa).

If the swelling moisture of the wood and the corresponding optimal water content of the extracting agent are raised, the transformation from fast into slow rate occurs at a lower content of residual lignin, as is shown in Fig. 9. With a wood humidity of about 80 wt.% the residual lignin of spruce chips could be reduced to about 4 wt.% within 3 hours. In a flow reactor where the extracting agent is circulated we found the point of inversion to be shifted to an even lower lignin content.

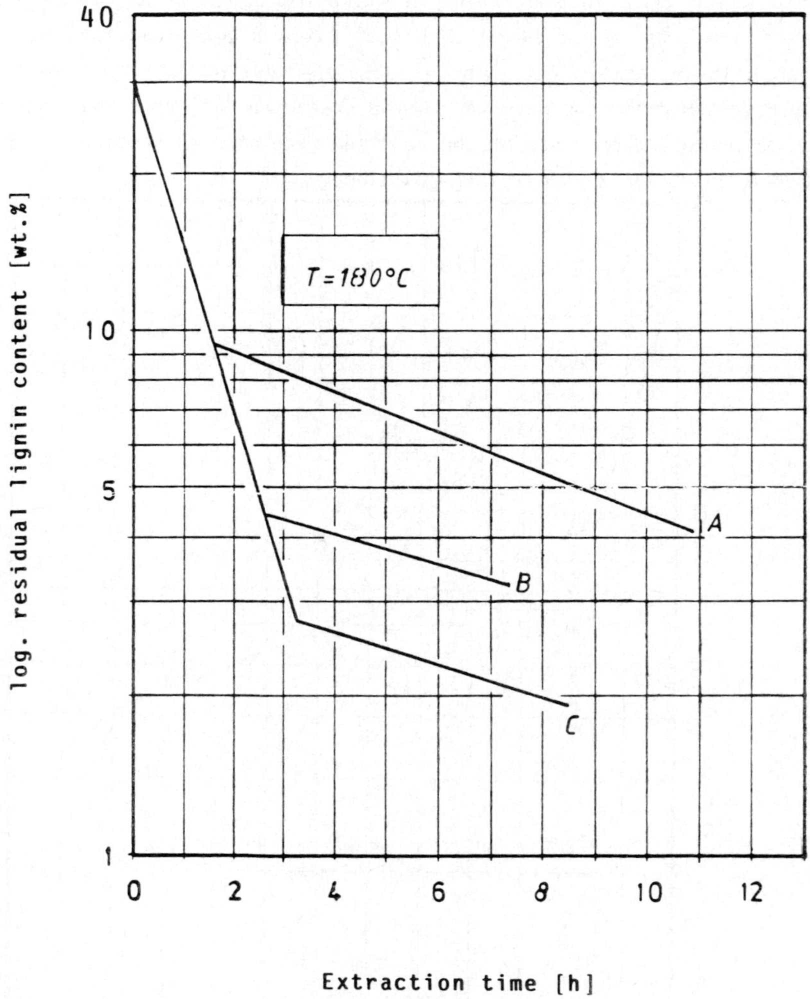

Fig. 9. Residual lignin content of spruce as a function of the extraction time with a binary mixture of methylamine and water.
a. p= 11.0 MPa, wood moisture 44 wt.%, water content of the solvent 17 wt.%(*)
b. p= 11.0 MPa, wood moisture 80 wt.%, water content of the solvent 32 wt.%(*)
c. p= 11.0 MPa, wood moisture 80 wt.%, water content of the solvent 32 wt.%(**)
(*supercritical conditions, ** supercritical conditions in a flow reactor)

DISCUSSION

The results described above reveal a favourable design for a process to produce α-cellulose from wood by supercritical fluid extraction. In Fig. 10, a schematic diagram shows how such an extraction could be realized. The supercritical extracting agent flows through the extraction vessel 1 which is filled with the wood chips to be extracted. The extraction takes place at about 180 °C and a pressure between about 8.0 to 11.0 MPa. The loaded extracting agent is expanded into the regenerator 2 by means of the expansion valve 5. If necessary the heat exchanger 4 compensates the Joule-Thomson effect.

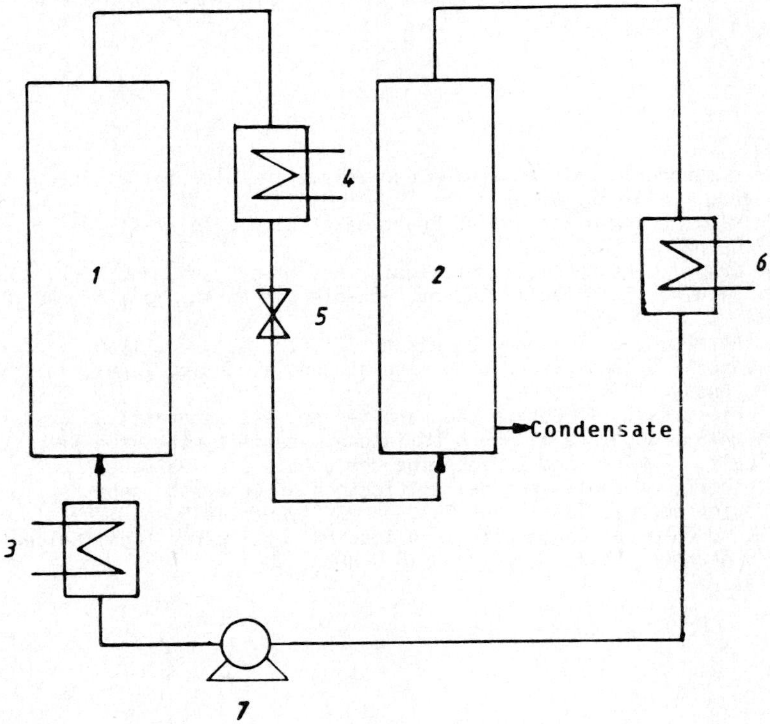

1	extraction vessel	5	expansion valve
2	regeneration vessel	7	pump
3, 4, 6	heat exchanger		

Fig. 10. Schematic diagram of a plant for the extraction of ligno-cellulosic materials with near-critical fluids.

For the example of an extracting agent which consists of 17 wt.% of water and 83 wt.% of methylamine the loading of the solvent at 11.0 MPa and 180 °C amounts to about 10 wt.%. The composition of the loaded solvent is marked in Fig.3 by the point I. By decreasing the pressure to 5.2MPa at 180 °C the solution decomposes along the tie-line which passes the point I in a gas phase poor in lignin and a liquid phase which contains the extracted components. Thereby a solution of about 43 wt.% methylaminelignin, 30 wt.% water and 27 wt.% amine is formed. The precipitated solution is withdrawn from the regenerator and simultaneously the pressure is reduced to atmospheric pressure. As a result, the amine and water are vaporized so that a solid extract is obtained. The gaseous mixture of amine and steam is cooled, condensed and recirculated. Since the extracted product contains less than 500 ppm of sulphur it may be of interest as a base product for further chemical processing.

By comparison, use of a subcritical liquid mixture of amine and water or ethanol and water for the extraction of lignin results in solutions of lignin of 7 to 10 wt.%.

REFERENCES

(1) K. Freudenberg, Lignin im Rahmen der polymeren Naturstoffe, Angew. Chem. 68. Jahrg. 1956, no. 3, p. 84.
(2) T. Enkvist, New Aspects on the Reaction of Lignin in Kraft Pulping, Paperi ja. Puu, no. 8, 1962, p. 395.
(3) F. E. Brauns, The Chemistry of Lignin, Academic Press, New York, 1952.
(4) R. M. Pichler, Sulfitaufschluß mit organischen Basen, Holzforschung no. 23 (1969) H. 6, p. 192.
(5) F. C. Peterson, L. E. Wise, US Patent 2218478, Oct. 21, 1930.
(6) T. Kleinert, Organosolv pulping with aqueous alcohols, Tappi, Vol. 57, no. 8, August 1974, p. 99.
(7) J. Krieger, Battelle-Geneva Laboratories process separates, upgrades all three main components of ligno-cellulose materials without expensive chemicals, Chemical and Engineering News, May. 31, 1982.
(8) T. Kleinert, Ethanol-water delignification of wood-rate constants and activation energy, Tappi, Vol. 58, no. 8, August 1975, p. 170.
(9) S. Peter, Chemical Engineering Applications of Supercritical Solvents, Ber. Bunsenges. Phys. Chem. 88, 875 - 882 (1984).

Supercritical Fluid Technology, edited by J.M.L. Penninger, M. Radosz, M.A. McHugh and V.J. Krukonis
1985 Elsevier Science Publishers B.V., Amsterdam — Printed in The Netherlands

BIOMASS HYDROLYSIS WITH SULFUR DIOXIDE AND WATER IN THE REGION OF THE CRITICAL POINT

JOHN R. VICK ROY AND ALVIN O. CONVERSE

Thayer School of Engineering, Dartmouth College, Hanover, NH 03755

ABSTRACT

Solutions of 1% to 33% water in sulfur dioxide, above or near the critical point, were used to hydrolyze the cellulose in wood to glucose. Sulfur dioxide was chosen because its critical temperature, 157.5°C, is within the range where the acid-catalyzed hydrolysis of cellulose proceeds rapidly. Furthermore, dilute sulfur dioxide solutions catalyze cellulose hydrolysis. We know of no other attempts to carry out cellulose hydrolysis with a supercritical fluid.

Batch experiments were conducted in a one-liter, stirred Parr reactor which, due to its mass, has long heat-up (50 min) and cool-down (20 min) times. The water content, for supercritical work, was constrained to less than 10% by the reactor pressure limitations. The maximum glucose yields (glucose obtained/potential glucose x 100) were 42% from delignified wood (Solka-Floc) and 26% from mixed hardwood flour under subcritical conditions. Under supercritical conditions, a maximum glucose yield of 2% was obtained and the residual solids were charred, indicating over-reaction.

To overcome the batch reactor's limitations, a flow-through percolation reactor has been built and experiments with shorter reaction times are being performed.

1. INTRODUCTION

1.1. Background

Two major problems in economically producing glucose from biomass are:

1) Obtaining higher glucose yields.
2) Obtaining a higher concentration of glucose in the product streams in order to reduce the cost of concentrating the products.

A supercritical fluid, with the right solvent properties, might be found to dissolve all or part of the biomass. Hydrolysis might proceed rapidly and give higher yields than other methods under these conditions. The solvent could be flashed off after the reaction and leave behind a concentrated product stream.

Cellulose and hemicellulose can be converted to glucose and other sugars through hydrolysis using acids, alkalis, enzymes derived from bacteria or fungi, or combinations of these methods.

In acid hydrolysis, the cellulose (a polymer of glucose) is split catalytically into oligosaccharides and finally into glucose [refs. 20 & 25]. The hydrolysis of hemicellulose to hexose (galactose, glucose, mannose) and pentose (xylose, arabinose) sugars proceeds in a similar manner. These sugars may further decompose to furfural, hydroxymethyl furfural, levulinic and formic acids, and tars.

1.2 Previous Acid Hydrolysis Work

Hydrolysis with concentrated sulfuric acid [refs. 7 & 17], hydrogen chloride [refs. 2 & 10], and hydrogen fluoride [ref. 21] give 80-95% glucose yields from wood when followed by dilute acid hydrolysis. Mild dilute acid hydrolysis [refs. 1 & 14] gives better than 95% glucose yields when followed by enzymatic hydrolysis. Dilute acid hydrolysis, alone, gives 55-60% glucose yields under more severe conditions [ref. 1].

1.3 Effect of Supercritical Fluids on Biomass

Supercritical carbon dioxide and nitrous oxide extract volatile materials, such as essential oils, spices, and caffeine, from biomass leaving the substrates intact [refs 3, 4, 8, 9, & 13]. Calimli & Olcay [ref. 5] and Koell, et. al. [ref. 15] delignified wood with supercritical esters, ethers, ketones, and alcohols, but a substantial fraction of the cellulose also degraded. Wood completely dissolves and degrades to gases and oils, without charring, after 30-60 minutes in supercritical water or acetone [refs. 16 & 18].

1.4 Effects of Sulfur Dioxide on Wood

The paper and pulp industry uses 2-25% solutions of sulfur dioxide in water to delignify paper and rayon pulp. Typical processes react the wood at 95°C for 6 hours with 22% sulfur dioxide followed by cooking at 115°C for 2-3 hours with 5% sulfur dioxide. The lower initial temperature prevents the lignin from condensing and precipitating from the solution [ref. 20]. Herdle, et. al. [ref. 12] used 30-80% sulfur dioxide in water at 120-145°C to delignify wood in 1.5-2.5 hours. They obtained 32% pulp yields and 5% residual lignin at 135°C with 30% sulfur dioxide for 2 hours at 505 psig. No pulp yields for 80% sulfur dioxide are reported.

Mild hydrolysis of wood with 0.5% sulfur dioxide [ref. 6] and 13-14% sulfur dioxide [ref. 23] is effective as a pretreatment for enzymatic hydrolysis. This treatment opens up the pores of the substrate by removing the hemicellulose and some of the cellulose to facilitate enzymatic attack.

1.5 Scope of Work

Supercritical sulfur dioxide with its critical point between that of CO_2 and water (see the following table), might be expected to exhibit solvent properties in between them, too.

Gas	T_c (°C)	P_c (atm)	P_c (g/cm^3)
CO_2	31	73.8	0.468
SO_2	157.5	77.8	0.524
H_2O	374	218	0.3

data from Perry & Chilton [ref. 19]

Furthermore, the critical temperature of sulfur dioxide (157.5°C) is close to the optimum temperature range for dilute acid hydrolysis (185-263°C). Since the critical properties of sulfur dioxide lie between those of carbon dioxide and water, a supercritical mixture of sulfur dioxide and water might simultaneously react with the cellulose in biomass and extract the products from the unreacted solids. The high diffusivity of a supercritical fluid should allow faster penetration of the substrate and accelerate the reaction, possibly dissolving the substrate and making the reaction homogeneous.

Since the boiling point of sulfur dioxide (-10°C) is close to room temperature, efficient catalyst recovery by vaporization and recompression is expected. By using small amounts of water in the biomass, high product sugar concentrations might be achieved after the sulfur dioxide is flashed off.

2. EXPERIMENTAL APPARATUS AND METHODS

The batch apparatus is shown in Figure 1. A 35 ± 2 g sample is placed in a one-liter Parr stirred reactor (Carpenter Cb3-20 alloy) and sealed. The reactor is evacuated to less than 1 mmHg (1), and then the water and sulfur dioxide are charged (2 & 3). The reactor is heated to the desired conditions, held for the desired reaction time, and cooled with water through an internal coil. The sulfur dioxide is recovered by condensing it in a cylinder cooled by crushed ice (4). After collection, the residual sulfur dioxide is scrubbed (5) and the reactor is opened. The products are recovered with a known amount of distilled water, weighed, and the pH of the liquid is measured. These products are extracted with water to remove the soluble sugars.

The flow-through apparatus is shown in Figure 2. About one gram of sample is placed into a 4.00 mm ID x 250 mm long 316 stainless steel column. The desired amount of water is added and the column is sealed. The column is

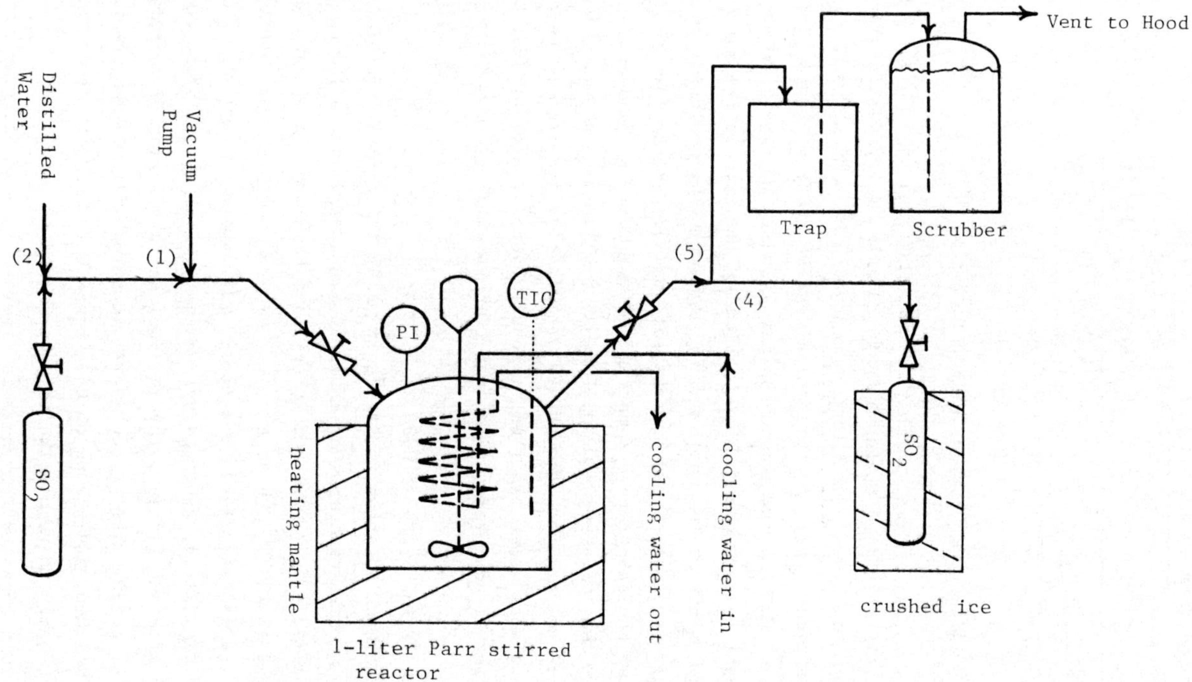

FIGURE 1 - <u>BATCH APPARATUS FOR HYDROLYSIS WITH SUPERCRITICAL SULFUR DIOXIDE AND WATER</u>

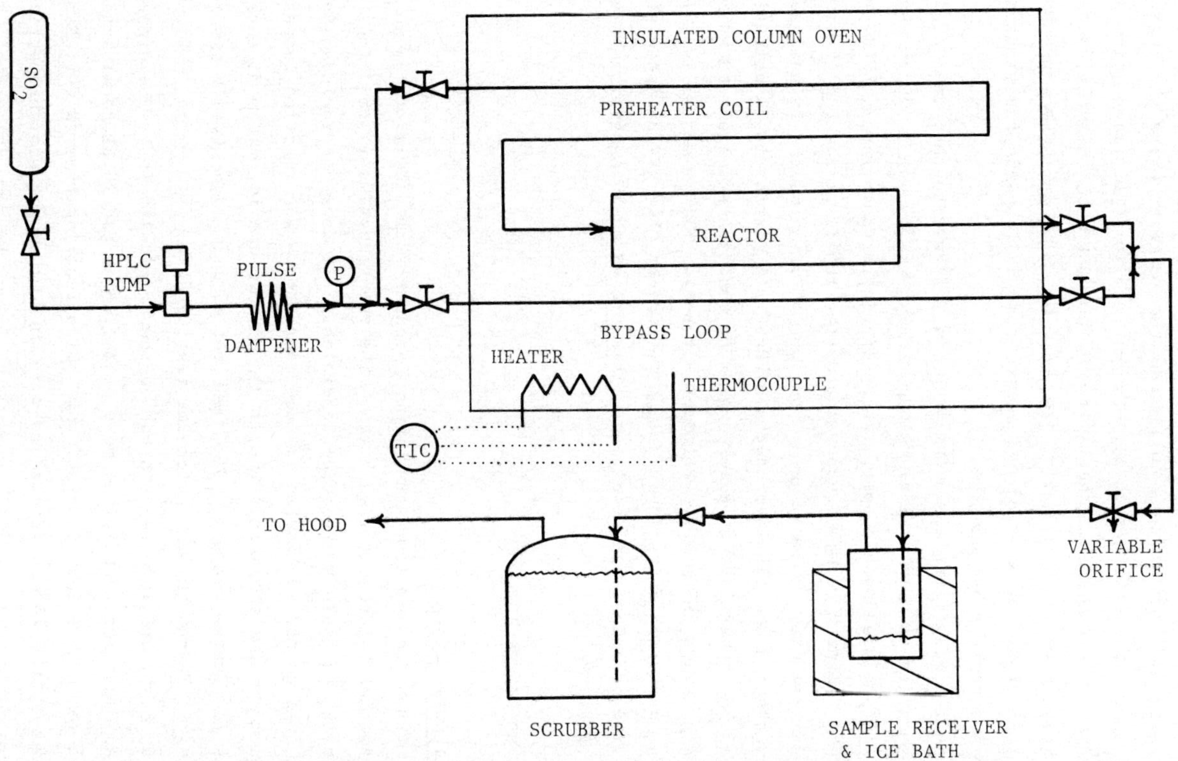

FIGURE 2 - FLOW-THROUGH APPARATUS FOR BIOMASS HYDROLYSIS WITH SULFUR DIOXIDE AND WATER

placed into the oven and the equipment is assembled and tested for leaks. Liquid sulfur dioxide is pumped through a bypass loop until flow conditions are established. The flow is then diverted through the preheater into reactor with the sample. Pressure is maintained by a variable orifice (a fine needle valve) at the column outlet. The products collect in the receiver where the sulfur dioxide is flashed off. Reaction conditions are held for the desired time and then the reactor is isolated and the system is flushed with sulfur dioxide through the bypass loop. The system is purged prior to disassembly, then the column and sample collector are weighed and the residual solids and products are recovered for work up and analysis.

The original sample material is analyzed for moisture content and quantitatively saccharified to determine the potential sugars present. The monomer sugar content is analyzed by high-performance liquid chromatography (HPLC) with a Bio-Rad HPX-85 "Aminex" column, pure water as the mobile phase at 81°C, and a refractive index detector. The soluble fraction of the products is analyzed for monomer sugars and reaction byproducts by HPLC. A portion is quantitatively saccharified to convert the oligomer sugars to monomers and analyzed by HPLC to determine the amount of soluble oligomers present. HPLC analysis of the quantitatively saccharified insoluble solids determines the residual potential sugars. Detailed procedures and equipment lists are given in Appendices A and B of Vick Roy [ref. 24].

3. RESULTS AND DISCUSSION

3.1 Batch Solka-Floc Runs

Table 1 shows the reaction conditions, product classification, sugar yields, and material balances for each run. The reaction time is the time for which the material was held at reaction conditions. It does not include about 50 minutes of heat-up and 20 minutes of cool-down time. The reaction temperatures and pressures are shown relative to the sulfur dioxide/water mixture critical locus in Figures 3 and 4. The feed and product compositions can be found in Table C-1, Appendix C, in Vick Roy [ref. 24].

Runs 2, 5, 6, 9, and 10 were at or above the critical point. The highest yields under supercritical conditions were 1.6% glucose and 10.6% total sugars at 181°C, 0.9% water in sulfur dioxide, and 5 mins reaction time (Run 9, Table 1). In all cases, the water present was at least twice the stoichiometric requirement for hydrolysis of all potential sugars in the Solka-Floc (delignified wood). The reactor pressure limit (130 atm) constrained work to 0-10% water in sulfur dioxide. Longer reaction times (Run 10) and higher temperature (Run 5) charred the material.

Table 1a - Batch Reaction of BW-200 Solka-Floc Cellulose, Water, and Sulfur Dioxide at or Near Supercritical Conditions

RUN NUMBER	2	5	6	9	10	11	12
REACTION CONDITIONS							
Temperature (°C)	161	196	171	181	170	162	161
Pressure (atm.)	82.4	125	91.7	102	81.6	66.1	75.6
Charge (g): Solka-Floc	32.28	32.86	37.58	35.92	34.86	35.41	35.67
H_2O	4.38	7.15	8.17	4.75	4.59	52.87	5.66
SO_2	468	596	539	497	326	440	369
Reaction Time (mins.)	4	5	5	5	30	15	15
Above, at, or below critical point?	above	above	above	above	at	below	below
% OF PRODUCTS AS:							
Solids not extractable with water	97.9	20.2	70.9	69.7	30.1	20.8	77.7
Water-extractables	2.1	79.8	29.1	30.3	69.9	79.3	22.3
Extractable glucose	0	0.2	1.0	1.6	1.5	4.7	1.0
Total extractable sugars	1.6	0.4	13.5	10.6	5.8	10.2	13.0
PRODUCT APPEARANCE	white powder	charred solids, brown sludge	white powder	brown powder	charred solids	brown-yellow sludge	white powder
% YIELD $\left(\frac{\text{extractable sugar in products}}{\text{potential sugar in feed}}\right)$							
Glucose	0	0.3	1.2	1.9	2.0	11.5	1.3
Xylose	0	2.3	7.4	19.7	21.0	59.4	12.1
Total sugars	1.7	0.6	16.5	12.8	7.4	25.1	17.1
POTENTIAL SUGARS CLOSURE (%)	98.3	2.6	97.1	96.9	12.4	75.4	100.0
MASS BALANCE ON SOLKA-FLOC AND WATER (%)	77.1	109.0	92.8	91.6	97.7	84.7	90.8

Table 1b - Batch Reaction of BW-200 Solka-Floc Cellulose, Water, and Sulfur Dioxide at or Near Supercritical Conditions

RUN NUMBER	13	14	15	16	17	18	19
REACTION CONDITIONS							
Temperature (°C)	151	171	181	181	192	182	180
Pressure (atm.)	57.6	72.8	82.3	80.4	94.0	85.0	81.6
Charge (g): Solka-Floc	35.31	36.98	36.37	35.73	36.12	36.70	36.13
H_2O	101.0	114.2	101.1	99.6	114.4	99.4	97.8
SO_2	610	439	326	312	539	383	397
Reaction Time (mins.)	30	5	10	10	13	10	0
Above, at, or below critical point?	below	below	below	below	below	below	below
% OF PRODUCTS AS:							
Solids not extractable with water	14.8	13.0	5.0	6.0	1.6	1.2	17.4
Water-extractables	85.2	87.0	95.0	94.0	98.4	98.8	82.6
Extractable glucose	2.8	2.8	7.4	8.7	0.3	5.4	3.8
Total extractable sugars	4.8	9.0	8.7	9.5	0.7	7.8	7.8
PRODUCT APPEARANCE	brown-yellow slurry	brown-yellow slurry	brown slurry	brown slurry	black sludge	dark brown slurry	tan-yellow slurry
% YIELD $\left(\frac{\text{extractable sugar in products}}{\text{potential sugar in feed}}\right)$							
Glucose	13.0	14.5	36.0	41.6	1.7	27.2	16.6
Xylose	46.3	38.7	16.3	16.0	3.3	5.8	46.7
Total sugars	22.2	45.8	41.9	45.7	3.6	38.5	34.0
POTENTIAL SUGARS BALANCE (% CLOSURE)	88.2	100.0	65.7	74.2	3.6	43.2	100.0
MASS BALANCE ON SOLKA-FLOC AND WATER (%)	95.6	99.7	102.0	101.3	101.7	107.6	93.6

Table 1c - Batch Reaction of BW-200 Solka-Floc Cellulose, Water, and Sulfur Dioxide at or near Supercritical Conditions

RUN NUMBER	20	21	22	23
REACTION CONDITIONS				
Temperature (°C)	180	201	165	203
Pressure (atm.)	80.9	95.6	62.1	99.7
Charge (g): Solka-Floc	35.31	35.28	36.78	35.88
H_2O	204.98	233.97	190.78	29.99
SO_2	568	440	355	298
Reaction Time (mins.)	20	15	5	5
Above, at, or below critical point?	below	below	below	above
% OF PRODUCTS AS:				
Solids not extractable with water	1.8	0.4	11.4	32.1
Water-extractables	98.2	99.6	88.6	67.9
Extractable glucose	5.0	0.1	1.2	0
Total extractable sugars	6.4	0.1	2.3	0
PRODUCT APPEARANCE	brown slurry	black slurry	white sludge	black solids
% YIELD $\left(\frac{\text{extractable sugar in products}}{\text{potential sugar in feed}}\right)$				
Glucose	40.4	0.4	8.9	0
Xylose	44.8	0.4	64.2	0
Total sugars	51.9	0.5	17.1	0
POTENTIAL SUGARS BALANCE (% CLOSURE)	63.9	0.5	89.6	0
MASS BALANCE ON SOLKA-FLOC AND WATER (%)	112	101	110	85.6

FIGURE 3

<u>EXPERIMENTAL RUN TEMPERATURES RELATIVE TO THE SO_2/WATER SYSTEM CRITICAL LOCUS</u>

Solid line is the critical temperature from the data of Spall (1963)

Experimental runs are plotted and numbered

■ Batch Solka-Floc Runs

● Batch Mixed Hardwood Runs

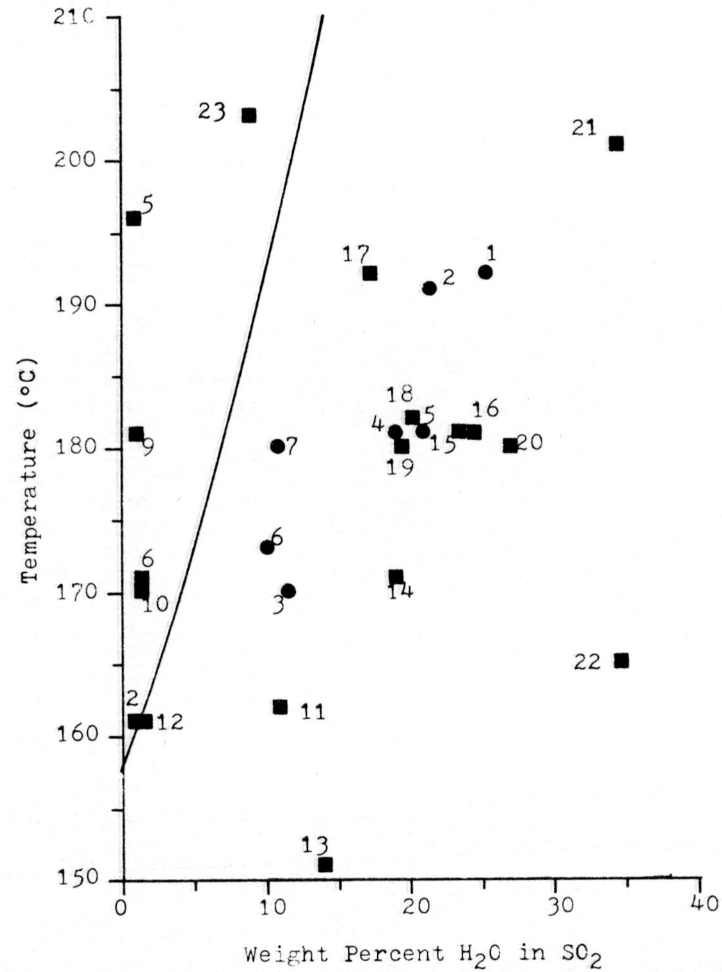

FIGURE 4

<u>EXPERIMENTAL RUN PRESSURES RELATIVE TO THE SO_2/WATER SYSTEM CRITICAL LOCUS</u>

Solid line is the critical pressure from the data of Spall (1963)

Experimental runs are plotted and numbered

■ Batch Solka-Floc Runs

● Batch Mixed Hardwood Runs

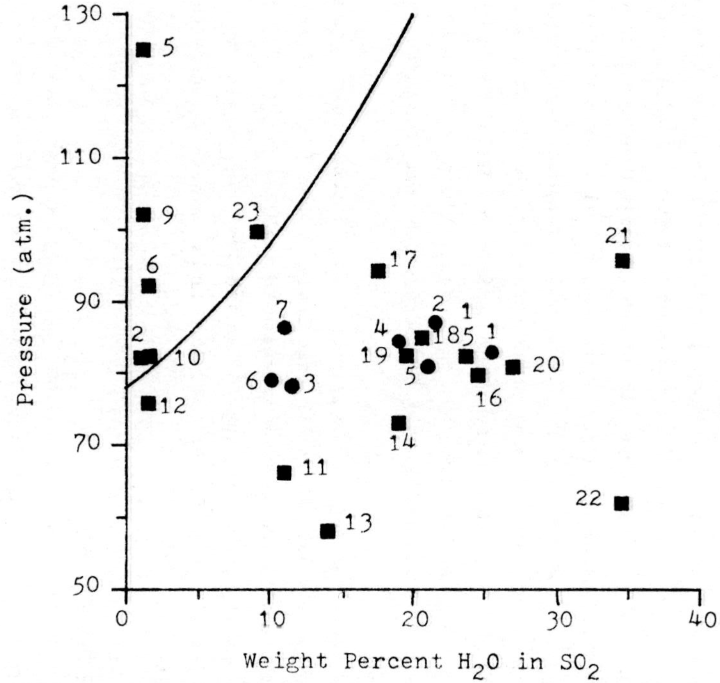

As seen in Figure 5, the best glucose and total sugar yields, 41.6% and 45.7% respectively, are found at 181°C, 80.4 atm, 10 mins reaction time, and 22.3% water (Run 16). 79.2% of the original potential sugars were recovered as unreacted wood, monomer and oligomer sugars, and byproducts. The total weight of the recovered products is 101.3% of the weight of the Solka-Floc and water charged originally. The liquid fraction of the recovered products contains 9.2% glucose and 10.1% soluble sugars. Note that the glucose and total sugar yields in Run 19 (see Table 1) are 16.6% and 34.0%, respectively, even though the reactor was simply heated up and cooled down. The yields at 171°C are 14.5% glucose and 45.8% total sugars. These yields are lower than those at 181°C and 10 minutes reaction time, but there is little or no loss of the potential sugars. Thus the yield might be increased by recycling the unreacted material.

The reproducibilty of the yields is poor. The glucose yield varies from 27.2% to 41.6% under similar conditions (see Table 1, Runs 15, 16, & 18). The sulfur dioxide charged to the reactor is measured within ± 7 g (± 2%), but is not well controlled. The amount charged varied from 300 to 450 g. This variation, plus the long heat-up and cool-down times, probably causes the lack of reproducibility.

3.2 Batch Mixed Hardwood Runs

Hydrolysis runs were made with Wilner 060 Mixed Hardwood Flour (90% birch, 10% maple, 60 mesh) at conditions close to those which gave the optimum sugar yields from Solka-Floc. The results are in Table 2. The best yields, 25.8% glucose and 39.2% total sugars, are from Run 7 at 180°C, 86.3 atm, and 0 mins reaction time. The yields plotted in Figure 6 show a need for shorter reaction times than can be obtained in the batch reactor. The material balance for the runs is 92.3 ± 5.6%. The 60% degradation of potential sugars to byproducts indicates that these conditions are beyond the optimal point for glucose production.

3.3 Flow-Through Mixed Hardwood Runs

Runs were made with Wilner CB4 Mixed Hardwood (4 mesh flat chips) at conditions close to those which gave the maximum batch glucose yields. In each case, liquid products collect in the receiver and solids remain behind in the reactor. Reaction at supercritical conditions (217°C, 122.1 atm, 1.0 mins liquid residence time, 3.1 ml/min liquid flowrate, 15 mins reaction time for the solids, 72% water in charge) gave 4% glucose and 26% total soluble sugars in the final products. Less than 10% potential sugars remain in the residual solids, which are a dark, fibrous, and voluminous material. The yields are unknown due to difficulties in recovering the products.

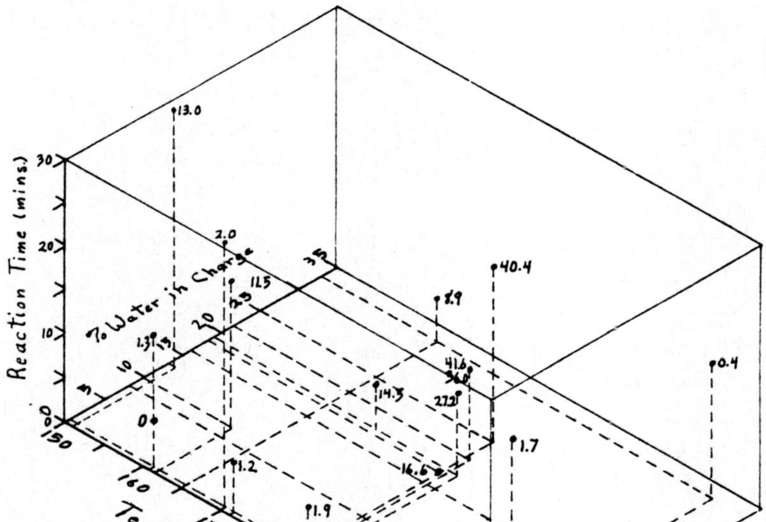

FIGURE 5 - GLUCOSE YIELDS AS FUNCTIONS OF REACTION TIME, TEMPERATURE, AND WATER CONTENT FOR THE BATCH HYDROLYSIS OF SOLKA-FLOC WITH SO_2 AND WATER

Table 2a - Batch Reaction of Wilner 060 Mixed Hardwood Flour, Water, and Sulfur Dioxide at or Near Supercritical Conditions

RUN NUMBER	1	2	3	4	5	6	7
REACTION CONDITIONS							
Temperature (°C)	192	191	170	181	181	173	180
Pressure (atm.)	83.3	86.8	78.2	84.6	81.2	79.3	86.3
Charge (g): Wood	35.5	37.7	35.3	35.4	36.3	35.6	35.3
H_2O	185.3	196.8	58.8	96.7	99.2	59.3	96.3
SO_2	425	539	454	411	369	539	510
Reaction Time (mins.)	5	15	15	10	10	5	0
Above, at, or below critical point?	below	below	below	below	below	below	below
% OF PRODUCTS AS:							
Solids not extractable with water	2.3	3.2	10.8	4.9	5.1	15.3	4.6
Water-extractables	97.7	96.8	89.2	95.1	94.9	84.7	95.4
Extractable glucose	2.4	1.0	5.1	2.9	3.8	4.9	4.6
Total extractable sugars	2.9	1.4	7.8	4.2	5.3	10.1	7.0
PRODUCT APPEARANCE	brown-black slurry	black sludge	black sludge	brown-black sludge	brown-black sludge	dark brown sludge	brown-black sludge
% YIELD $\left(\frac{\text{extractable sugar in products}}{\text{potential sugar in feed}}\right)$							
Glucose	23.9	9.4	19.1	15.0	20.1	17.1	25.8
Xylose	1.9	1.7	2.6	2.9	2.9	16.3	7.3
Total sugars	28.8	12.9	29.2	22.0	28.0	35.7	39.2
POTENTIAL SUGARS BALANCE (% CLOSURE)	34.2	12.9	69.8	26.3	34.3	59.6	43.2
MASS BALANCE ON WOOD AND WATER (%)	101.4	94.5	90.1	88.6	91.1	84.4	96.3

Table 2b - Batch Reaction of Wilner 060 Mixed Hardwood Flour, Water, and Sulfur Dioxide at or Near Supercritical Conditions

RUN NUMBER	8
REACTION CONDITIONS	
Temperature (°C)	184
Pressure (atm.)	12.4
Charge (g): Wood	34.9
H_2O	95.4
SO_2	0
Reaction Time (mins.)	10
Above, at, or below critical point?	below
% OF PRODUCTS AS:	
Solids not extractable with water	19.2
Water-extractables	80.8
Extractable glucose	0
Total extractable sugars	4.6
PRODUCT APPEARANCE	brown paste
% YIELD $(\frac{\text{extractable sugar in products}}{\text{potential sugar in feed}})$	
Glucose	0
Xylose	15.0
Total sugars	24.6
POTENTIAL SUGARS BALANCE (% CLOSURE)	88.2
MASS BALANCE ON WOOD AND WATER (%)	91.0

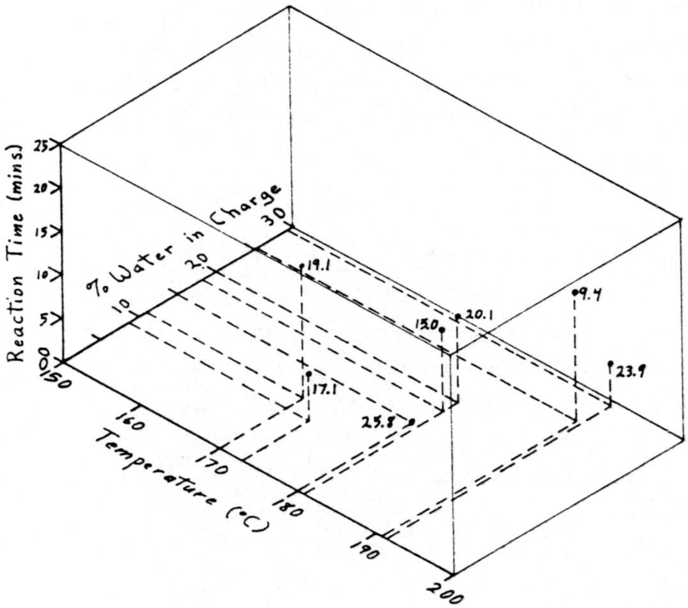

FIGURE 6 - GLUCOSE YIELDS AS FUNCTIONS OF REACTION TIME, TEMPERATURE, AND WATER CONTENT FOR THE BATCH HYDROLYSIS OF MIXED HARDWOODS WITH SO_2 AND WATER

4. CONCLUSIONS

1. 41.6% glucose and 45.7% total sugar yields can be obtained from cellulose by hydrolysis with sulfur dioxide at 180°C, 80.4 atm, and 22.3% water held for 10 minutes in a batch reactor. There are 9.2% glucose and 10.1% total sugars in the products. 25.8% glucose and 39.2% total sugars can be obtained from mixed hardwoods at 180°C, 86.3 atm, and 15.0% water when held for 0 minutes at conditions. The products contain 4.8% glucose and 7.3% total sugars.

2. Flow-through runs indicate that better than 90% of the potential sugars in wood can be reacted and the products removed within 15 minutes with supercritical sulfur dioxide and water. Product concentrations of 4% glucose and 26% total sugars have been obtained. Because of product recovery problems, the yields are not known.

3. So far, yields from sulfur dioxide are less than those from dilute H_2SO_4 or concentrated HCl or HF. Given the current yields and catalyst recovery, the process would not be economical. Glucose yields of 95% or better and sulfur dioxide recoveries of 98% or more are needed.

5. FUTURE WORK

1. Pumping both sulfur dioxide and water through the system will give better control over the water reacting with the wood and help to flush out products that are not soluble in supercritical sulfur dioxide, alone. Corrosion of the precision, high-pressure pumps could be avoided by using two pumps; one for sulfur dioxide and the other for water. Both pumps should be kept at the same pressure to ensure accurate, proportional blending of the two streams.

2. Modify the apparatus to collect fractions and thus obtain kinetic data. Sampling could be done with a multiple-port valve leading to different receivers, such as those in Figure 2. The kinetic constants for a percolation reactor [ref. 11], can be obtained from the initial cellulose concentration, the liquid residence time, the solids reaction time, and the sugar yield. This model assumes very small particles ($\leq$ 1 mm) so that mass transfer within the particle is of little concern. In practice this may be difficult to achieve, since a column filter with pores approximately 100 times smaller than the particles must be used to prevent the filter from plugging during the reaction.

ACKNOWLEDGEMENT

This work was performed under DOE/SERI Contract XX-4-040484 and the authors gratefully acknowledge their support.

REFERENCES

1. Allen, D.C., H.E. Grethlein, & A.O. Converse, "Process Studies for Enzymatic Hydrolysis Using High Solids Slurries of Acid Pretreated Mixed Hardwood," Biotech. & Bioengg., Symp. No. 13, Gatlinburg, TN, 10-13 May 1983, pp. 99-110.
2. Antonoplis, R.A., et. al., "Production of Sugars from Wood Using High-Pressure Hydrogen Chloride," Biotech. & Bioengg., v. 25, n. 11, Nov. 1983, pp. 2757-2773.

3. Brunner, G. & S. Peter, "State of Art of Extraction with Compressed Gases," Ger. Chem. Engg. v. 5, n. 3, 1982, pp. 181-195.
4. Calame, J.P. & R. Steiner, "CO_2 Extraction in the Flavour and Perfumery Industries," Chem. & Ind., n. 12, 19 June 1982, pp. 399-402.
5. Calimli, A. & A. Olcay, "Supercritical-Gas Extraction of Spruce Wood," Holzforschung, v. 32, n. 1, 1978, pp. 7-10.
6. Conner, A.H., "Effects of Aqueous Sulfur Dioxide on Cellulose," Biotech. Letters, v. 2, n. 10, 1980, pp. 439-444.
7. Dunning, J.W. & E.C. Lathrop, "The Saccharification of Agricultural Residues," Ind. & Engg. Chem., v. 37, n. 1, Jan. 1945, pp. 24-29.
8. Eggers, R., "Large-Scale Industrial Plant for Extraction with Supercritical Gases," Angew. Chem. Intl. English Ed., v. 17, n. 10, 1978, pp. 751-754.
9. Friedrich, J.P., G.R. List, & A.J. Heakin, "Petroleum-Free Extraction of Oil from Soybeans with Supercritical CO_2," J. Amer. Oil Chem. Soc., v. 59, n. 7, July 1982, pp. 288-292.
10. Goldstein, I.S., et. al., "The Hydrolysis of Cellulose with Superconcentrated Hydrochloric Acid," Biotech. & Bioengg. Symp. No. 13, Gatlinburg, TN, 10-13 May 1983, pp. 81-97.
11. Greenwald, C.G., J.M. Nystrom, & C.S. Lee, "Yield Predictions for Various Types of Acid Hydrolysis Reactors," Biotech. & Bioengg. Symp. No. 13, Gatlinburg, TN, 10-13 May 1983, pp. 27-39.
12. Herdle, L.E., A. Walker, Jr., & F.L. Wells, "Wood Pulping-Delignification in the Presence of High Sulfur Dioxide Concentrations," U.S. Patent No. 3,523,060, 4 August 1970.
13. Hubert, P. & O.G. Vitzhum, "Fluid Extraction of Hops, Spices, and Tobacco with Supercritical Gases," Angew. Chem. Intl. English Ed., v. 17, n. 10, 1978, pp. 710-715.
14. Knappert, D., H. Grethlein, & A. Converse, "Partial Acid Hydrolysis of Poplar Wood as a Pretreatment for Enzymatic Hydrolysis," Biotech. & Bioengg. Symp. No. 11, Gatlinburg, TN, 12-15 May 1981, pp. 67-77.
15. Koell, P., B. Broenstrup, & J. Metzger, "Thermischer Abbau von Birkenholz mit superkritischen Gasen in einer Hochdruck-, Hochtemperatur-Stroemmungsapparatur," Holzforschung, v. 33, 1979, pp. 112-116.
16. Koell, P. & J. Metzger, "Thermal Degradation of Cellulose and Chitin in Supercritical Acetone," Angew. Chem. Intl. English Ed., v. 17, n. 10, 1978, pp. 754-755.
17. Ladisch, M.R., M.C. Flickinger, & G.T. Tsao, "Fuels and Chemicals from Biomass," Energy, v. 4, n. 2, Apr. 1979, pp. 263-275.
18. Modell, M., "Gasification and Liquefaction of Forest Products in Supercritical Water," AIChE 89th Natl. Meeting, Portland, OR, 20 Aug. 1980.
19. Perry, R.H., & C.H. Chilton, eds., Chemical Engineers' Handbook (5th ed.), McGraw-Hill, NY, 1973.
20. Rydholm, S.A. Pulping Processes, Wiley-Interscience, NY, 1965.
21. Selke, S.M. & M.C. Hawley, "Chemicals from Wood via HF," Ind. Engg. Chem - Prod. Res. Dev., v. 21, n. 1, March 1982, pp. 11-18.
22. Spall, B.C., "Phase Equilibria in the System Sulphur Dioxide-Water from 25-300 C," Can. J. Chem. Engg., v. 41, n. 2, April 1963, pp.79-83.
23. Springer, E.L. & K.A. Libkie, "Prehydrolysis of Birch Wood with Sulphur Dioxide," TAPPI, v. 63, n. 7, July 1980, pp. 119-120.
24. Vick Roy, J.R., "Biomass Hydrolysis with Sulfur Dioxide," M.E. Thesis, Thayer School of Engineering, Dartmouth College, Hanover, NH, June 1984.
25. Wenzl, H.F.J., The Chemical Technology of Wood, Academic Press, New York, 1970.

Supercritical Fluid Technology, edited by J.M.L. Penninger, M. Radosz, M.A. McHugh and V.J. Krukonis
1985 Elsevier Science Publishers B.V., Amsterdam — Printed in The Netherlands

HIGH PRESSURE SOLID POLYMER-SUPERCRITICAL FLUID PHASE BEHAVIOR

I. S. LIAU and M. A. McHUGH
Department of Chemical Engineering, University of Notre Dame
Notre Dame, Indiana 46556

ABSTRACT

An experimental apparatus is constructed for the measurement of polymer swelling in gases and gas sorption in polymers up to 5000 psia. The apparatus is used to determine the swelling of polymethyl methacrylate in the presence of carbon dioxide and the sorption of carbon dioxide in polymethyl methacrylate at 41.8°, 58.1°, and 68.0°C and at pressures up to 4,000 psia. The resultant data are modelled using the Flory-Huggins equation.

INTRODUCTION

One separation technique which has not yot been extensively tested for polymer processing is supercritical fluid (SCF) solvent extraction. An SCF solvent is a solvent at a temperature above its critical temperature and at a pressure above its critical pressure (1). SCF solvents have unique solvent characteristics. For example, the solvent density controls the solubilities of constituents in the supercritical solvent phase. As a consequence, operating near the pure solvent critical point, solvent power can be varied over a wide range with small changes in pressure as well as temperature. Using an SCF solvent with a low to moderate critical temperature, it may be possible to isothermally extract polymer impurities from a solid polymer at a temperature below which any extensive thermal degradation of the polymer would occur (2).

It would be very advantageous to model this proposed SCF solvent extraction process. However, while there are several models for calculating polymer-solvent phase behavior, there are very few data at high pressures to test these models (3). The polymer-solvent data available in the literature are summarized in Table 1 (3). This scarcity of data makes it extremely difficult to develop a reliable model for this process.

TABLE 1. GAS/POLYMER EQUILIBRIUM DATA FOR MOLTEN POLYMERS (3).

POLYMER	GAS	T, °C	$W_1 x 10^2$	Pressure, Atm
Polyethylene	CH_4	140.03	0.36-4.1	51-238
(High Density)		162.78	0.36-4.1	21-317
		188.31	0.36-3.3	24-253
Polyethylene	C_2H_4	124-300	0	>1
(Low Density)		143.2	0	69.4
		157.3	0	69.4
		200.5	0	69.4
		237.0	0	69.4
	CH_4	125.4	0.8-5.0	60-661
		227.0	1.0-6.4	69-661
	N_2	125.0	1.0-1.25	0.69
		155.4	0.6-3.9	62-540
		188.4	0.6-3.9	66-669
		226.1	0.9-4.5	64-547
Polyisobutylene	CH_4	101.86	0.4-2.4	23-341
		126.85	0.3-2.7	23-341
		155.41	0.4-3.0	23-341
		188.46	0.4-3.1	23-341
Polystyrene	CH_4	100.20	0.6-1.6	67-320
		125.42	0.5-2.0	50-668
		155.42	0.4-1.6	48-241
		188.40	0.5-2.2	67-312

W_1 - Weight Fraction of Gas Dissolved in Polymer

The objective of this work is to expand the experimental data base for solid polymer-SCF mixtures. For this study, two sets of experiments are conducted. First, the carbon dioxide induced swelling of polymethyl methacrylate (PMMA) at temperatures ranging from 40°C to 70°C and over a pressure range of 0 to 4000 psia is determined. Next, the sorption of carbon dioxide in PMMA over the same temperature and pressure ranges is determined. Shown in Table 2 and Table 3 are the properties of PMMA and carbon dioxide. Carbon dioxide is chosen as the SCF solvent since it is nontoxic, nonflammable, and inexpensive. The critical temperature of carbon dioxide is also well below the glass transition of PMMA. Therefore, by operating at mild temperatures the swelling and sorption experiments can be performed with supercritical CO_2 at mild temperatures sufficiently removed from the glass transition temperature of PMMA. Polymethyl methacrylate is chosen as the polymer for these studies since it is a well-characterized, representative, amorphous, solid polymer.

TABLE 2. PROPERTIES OF POLYMETHYL METHACRYLATE USED IN THIS STUDY.

Weight average molecular weight, M_w	60600
Number average molecular weight, M_n	33200
Weight/Number average ratio, M_w/M_n	1.82
Glass transition temperature	105°C

TABLE 3. PROPERTIES OF CARBON DIOXIDE.

Molecular weight	44.01
Critical temperature	304.19°K
Critical pressure	1070.6 psia

EXPERIMENTAL

Two types of experiments are performed in this study. In one, the swelling of solid polymer as it sorbs SCF solvent is measured. In the other, the high pressure equilibrium sorption of SCF solvent into solid polymer is determined. Each of these experiments is described separately.

Swelling experiments

Described in the literature are two basic experimental techniques for measuring the swelling of polymers in solvents at high pressures. In one method, the swelling is visually determined using a camera to record the change in length of the polymer sample (4). In the other technique, the swelling of the polymer sample is monitored by attaching the electrical resistance strain gauge to the polymer sample (5).

For this study, the swelling of solid polymer in the presence of supercritical carbon dioxide is visually determined. Amorphous solid PMMA is chosen to minimize any morphological changes which can occur as it sorbs supercritical solvent; carbon dioxide is chosen as a supercritical solvent because of its convenient critical properties.

Experimental apparatus

As shown in Figure 1, the experimental apparatus consists of four basic parts: gas/liquid compressor, pressure gauge, holding tank, and high pressure view cell. Carbon dioxide ("bone dry," Linde Company) is charged at ambient temperature to an air-driven gas/liquid compressor (Superpressure Inc., Model 46-14025-1) where it is compressed and delivered to a holding tank (316 Stainless Steel, 6.4 cm. O.D., 2.5 cm. I.D., volume of 123.33 ± 0.28 CC.) located in a forced convection air bath. The temperature of the air is maintained constant to within ± 0.2°C (American Scientific Products, YSI 63 RC Controller) and is measured with an accuracy of ± 0.26°C using a platinum resistance element (Degussa, Incorporated) calibrated on the 1968 IPTS scale.

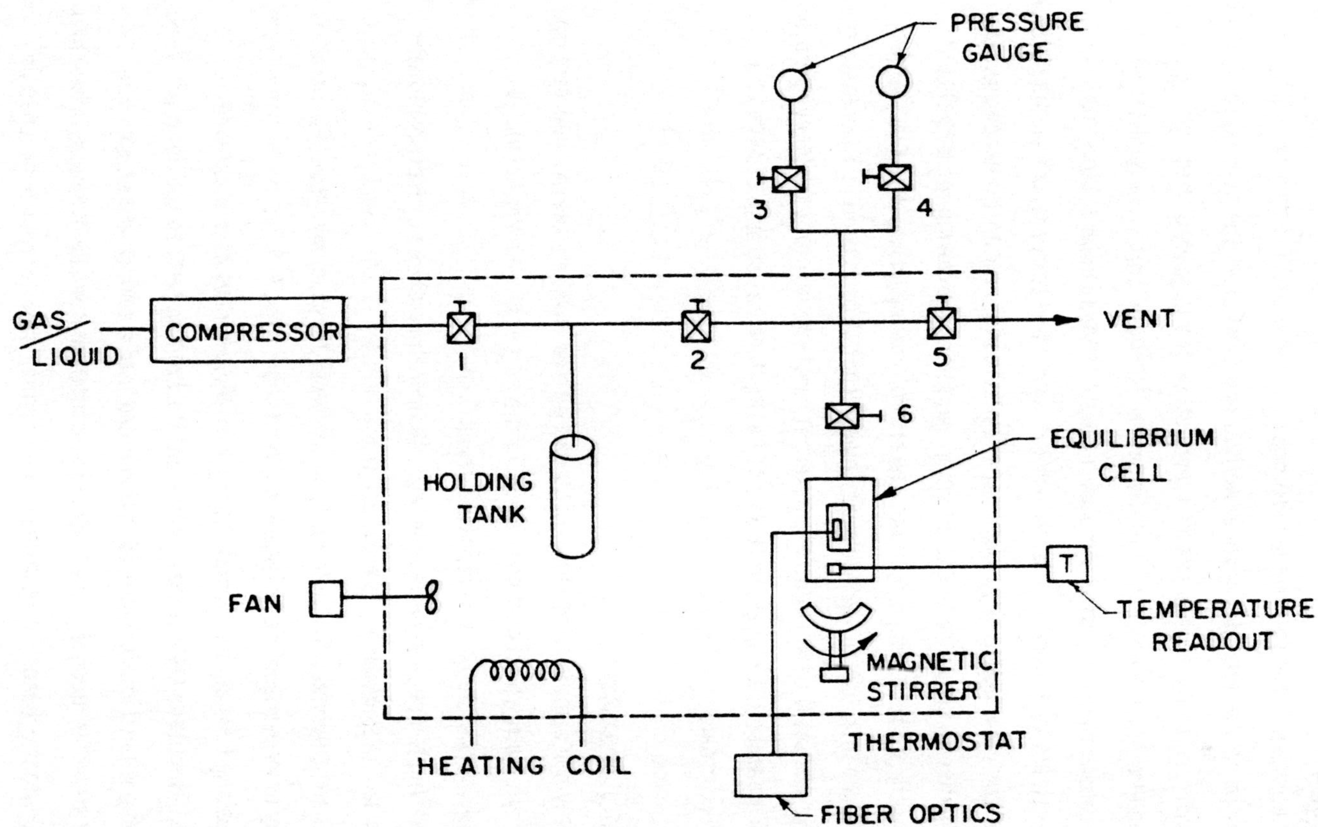

Figure 1. Schematic diagram of the experimental apparatus used in this study.

The pressure of carbon dioxide in the holding tank is measured with a digital Heise gauge (Dresser Industries, Model 715R receiver and Model 715TA 0-1500 psi transmitter and Model 715TA 0-20000 psi transmitter, both accurate to within 0.1% of their respective full-scale readings).

After the carbon dioxide reaches thermal equilibrium in the holding tank, it is then transferred to the high pressure view cell which is loaded with a sample of pretreated PMMA (Aldrich Chemical Company). The view cell, a Jerguson Sight Gauge (Jerguson Gauge & Valve Company, Model 11T51, volume 37.30 ± 0.06 cc.), is charged with PMMA molded into a cylindrical form. The temperature of the outside surface of the view cell is measured using a platinum resistance element similar to that previously mentioned. The polymer sample is weighed using a Mettler balance, accurate to within ± 0.0004 g (Mettler Company, Model AE 160). A 0.64 cm. fiber light pipe (Dolan-Jenner Industries, Model 180) is used to illuminate the contents of the view cell. A cathetometer, accurate to within ± 0.008 centimeter, is used to measure the change in length of the cylindrical sample of PMMA as it sorbs carbon dioxide.

Experimental procedure

The polymer sample is initially pretreated to remove any residual solvent and to relax any strain which may be present in the as-received sample (4). The pretreatment consists of the following steps :

(1) evacuating the polymer at an elevated temperature to remove any residual solvent or low molecular weight impurities;

(2) soaking the polymer in liquid carbon dioxide to release any frozen-in strain which may be present in the solid sample; and

(3) evacuating the polymer sample to remove any residual carbon dioxide.

Carbon dioxide is then transferred from the holding tank to the Jerguson view cell which has been charged with PMMA molded into a cylindrical shape. The length of the PMMA is determined using the cathetometer. Equilibrium is reached when a constant pressure is obtained. The polymer sample usually reaches an equilibrium length within approximately 50 minutes after introducing the carbon dioxide. At equilibrium the system pressure and temperature are recorded along with the overall length of the polymer sample. For these experiments it is assumed that the polymer swells isotropically (4). Hence, the change in length of the PMMA sample is converted directly to a change in volume. Carbon dioxide at a higher pressure is then introduced into the view cell for the next data point. After every few data points the polymer sample is depressurized to atmospheric pressure to check for any permanent strains in the solid polymer. The resultant volume of solid polymer at atmospheric pressure is compared to the volume of the solid polymer after the pretreatment step. If the difference between the volume of depressurized solid and the original pretreated sample is greater than 0.3% then the depressurized volume is used as the original volume.

<u>Sorption experiments</u>

Various techniques for determining the sorption of supercritical solvents are described in the literature. The high pressure experimental techniques can be classified as either static (5-10) or flow techniques (11-13). A simple static system similar to that described by Ozawa et al. (7) is used for these sorption studies. The experimental technique is described in the following section.

<u>Experimental apparatus</u>

The experimental apparatus for the sorption studies is similar to that previously described for the swelling experiments. However, a different view cell is used for these studies. The view cell, a 1.3 cm I.D., 7.6 cm O.D. cell (316 stainless steel, volume of 9.62 ± 0.04 cm^3), is packed with a known amount of PMMA. Pressures and temperatures are measured and monitored as previously described.

<u>Experimental procedure</u>

The PMMA sample in the view cell is pretreated as previously described in the swelling experiments. The total void space in the cell packed with PMMA is then determined. With valves 1, 2, and 4 open (see Figure 1), helium is introduced into the holding tank while maintaining a vacuum inside the sample cell. Valve 1 is closed once the desired system pressure is obtained. The mass of helium in the holding tank is calculated knowing the volume of the holding tank and connecting lines, the helium pressure, and the temperature of the cell. After thermal equilibrium is reached, helium is then transferred into the view cell. At equilibrium, assuming no helium sorbs into PMMA (14), the void volume of the packed sample cell including the polymer void space is determined from the new equilibrium pressure, the volume of the holding tank and lines, the volume of the cell, and the temperature of the cell (7). The density of PMMA can also be calculated knowing the mass and the volume of the PMMA sample. In fact, the experimental measurements of the density of PMMA are usually within ± 0.5% of literature values (15).

The system is then depressurized and evacuated at the elevated temperature to remove any residual helium. Initially the view celll is isolated from the system while the holding tank is pressurized with carbon dioxide by opening valves 1, 2, and 4. Valve 1 is closed once the desired pressure is obtained. Carbon dioxide is then transferred from the holding tank to the view cell. The pressure of the view cell eventually reaches an equilibrium pressure after approximately 50 minutes. The amount of carbon dioxide sorbed into PMMA is obtained from the difference between the initial and final mass of carbon dioxide in the void space in the cell, the holding tank, and the lines. The mass of carbon dioxide is determined from the volume of the vessel and the lines and the local carbon dioxide density. Initial and final temperature and pressure readings in each location are used to obtain the density of carbon dioxide using the equation of state described by Michels and Michels (16). The reduced void volume in the cell due to polymer swelling is also incorporated into the sorption calculation. Isothermal sorption data at successive pressures are determined in a cumulative manner. After every few data points, the polymer is depressurized to atmospheric pressure to minimize any irreversible swelling which may occur.

RESULTS

The experimental data on the carbon dioxide induced swelling of PMMA obtained in this study at 41.8°C, 58.1°C, and 68.0°C are shown in Figure 2. In general, the accumulated error for these experiments is estimated to be less than 3%. The swelling of PMMA in the presence of carbon dioxide exhibits very similar behavior at each isotherm. The swelling increases linearly with pressure for pressures up to about 800 psia. As the pressure increases above 800 psia, the swelling increases more slowly until it reaches a limiting value at very high pressures. As shown in Figure 2, the swelling of PMMA decreases with increasing temperature.

It should be noted, however, that the surface of the PMMA rod appears to melt at 68.0°C for pressures greater than about 1500 psia. This melting takes place only on the surface of the PMMA rod since the rod remains dimensionally stable throughout the experiment. It is quite possible that the glass transition temperature of PMMA has been lowered from 105°C to 68°C by the sorption of high pressure carbon dioxide. This behavior is similar to the melting point depression of crystalline solids such as naphthalene when pressurized by a high pressure gas such as carbon dioxide (17).

The experimental data on the sorption of carbon dioxide in PMMA obtained in this study at the same temperatures as the previously mentioned swelling isotherms are shown in Figure 3. In this instance, the two isotherms at 41.8°C and 58.1°C exhibit similar behavior. The sorption of carbon dioxide in PMMA increases linearly with pressures for pressures up to about 1000 psia. At pressures greater than 1000 psia, the sorption increases slowly until it reaches a limiting value at a pressure of about 3600 psia. The sorption of carbon dioxide is also higher at 41.8°C than at 58.1°C.

The sorption of carbon dioxide in PMMA at 68.0°C exhibits slightly different behavior from that at 41.8°C and 58.1°C. The soprtion isotherm at 68.0°C

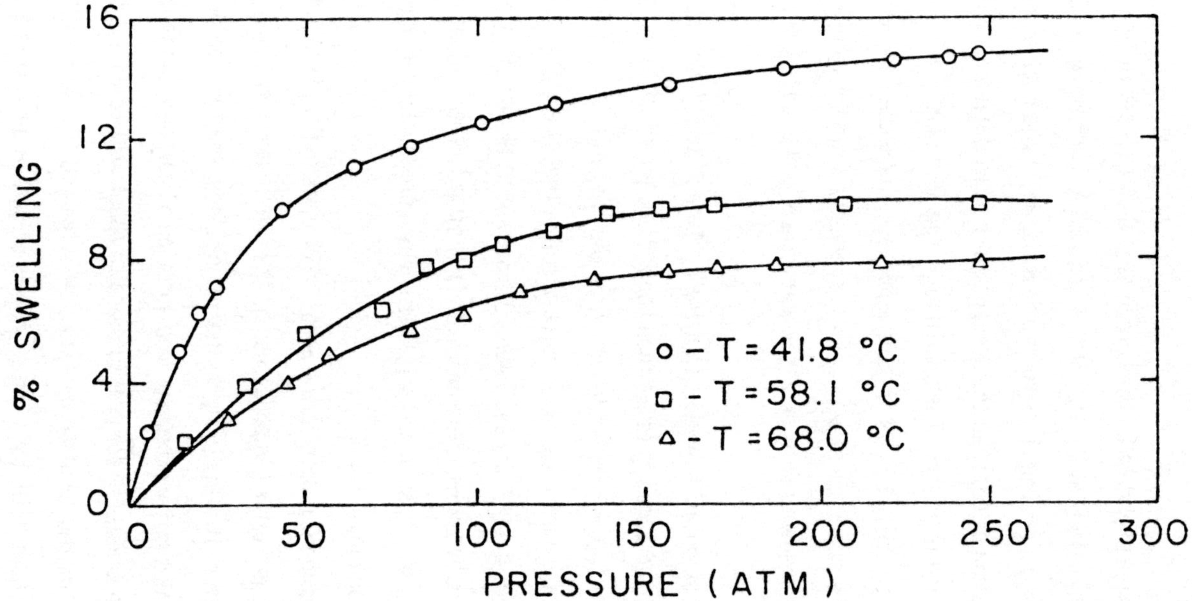

Figure 2. Swelling of PMMA in the presence of carbon dioxide.

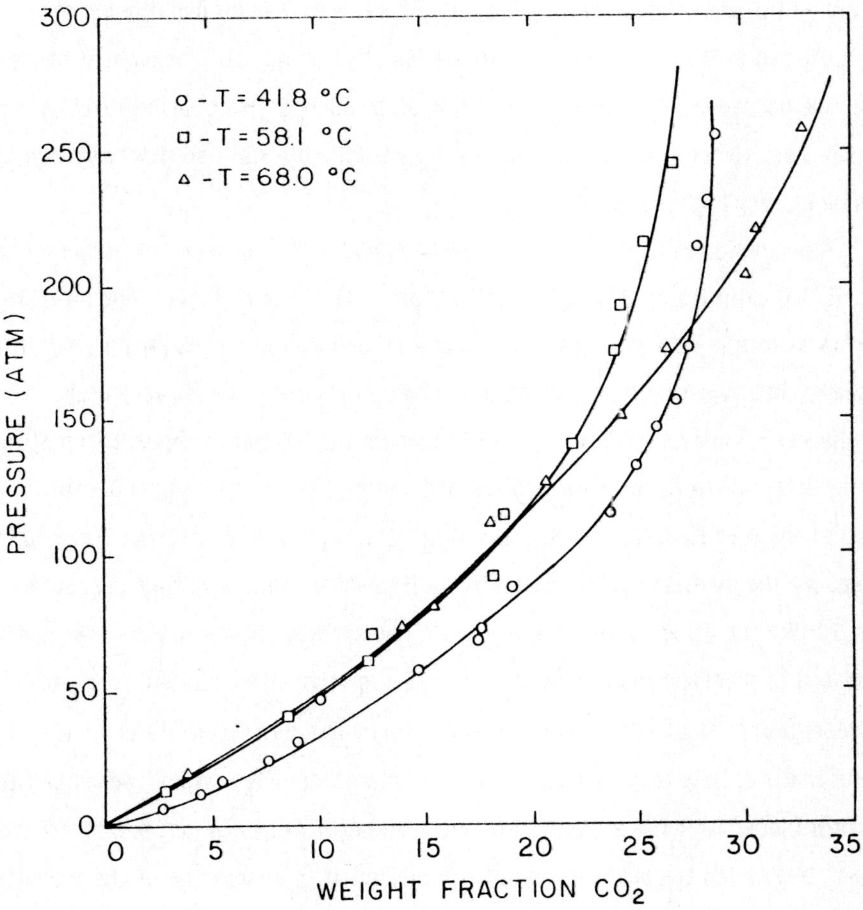

Figure 3. Sorption of carbon dioxide in PMMA.

of PMMA in the presence of carbon dioxide and the sorption of carbon dioxide into PMMA reach limiting values at very high pressures at 41.8°C.

The sorption of carbon dioxide in PMMA and the swelling of PMMA at 58.1°C are less than those at 41.8°C. At 58.1°C carbon dioxide is further from its critical temperature. The density of carbon dioxide becomes more "gas-like" which makes it less soluble in PMMA at 58.1°C than at 41.8°C. Also, PMMA still possesses the rigid, solid, amorphous state at 58.1°C. Therefore, less carbon dioxide is solubilized in PMMA and the swelling of PMMA is also less. The amount of carbon dioxide sorbed in PMMA at this isotherm also reaches a saturation level at high pressures. Hence, the swelling of PMMA at this isotherm also reaches a limiting value at very high pressures.

As shown in Figure 2, the swelling of PMMA in the presence of carbon dioxide at 68.0°C exhibits similar behavior to that at 41.8°C and 58.1°C. The swelling of PMMA at this isotherm is less than at the other two isotherms obtained in this study. This swelling isotherm also reaches a limiting value at very high pressures. However, as previously mentioned, the PMMA rod appears to melt at this temperature for pressures greater than 1500 psia. The experimental technique used to determine the swelling of solid PMMA is not sensitive enough to measure the swelling of the partially melted PMMA sample at high pressures.

Unlike the other isotherms, at 68.0°C the sorption of carbon dioxide in PMMA appears to increase linearly with pressure over the entire range of pressure investigated. At 68.0°C, carbon dioxide is very far away from its critical temperature, thus it possesses "gas-like" characteristics. Also, the temperature is now closer to PMMA's glass transition temperature. For most polymers, the glass transition temperature is not a point, but it is an average of the transition interval of about 10° or 20°C (18). At high carbon dioxide pressures, the glass transition temperature of PMMA may be lowered due to the high solubility of carbon dioxide in the polymer. At high pressures and 68.0°C PMMA loses some of the properties characteristic of the rigid, solid state and acquires some of the properties of a dense liquid. As such the sorption results at this temperature are linearly with pressure over the entire range of pressures investigated. Hence, a limiting value is not observed for this isotherm.

As shown in Figure 3, the sorption of carbon dioxide in PMMA increases linearly at low pressures for all the isotherms investigated in this study. This behavior is in agreement with the work done by Durrill and Griskey (9).

Figure 4 shows the comparison of the swelling of PMMA in helium and in carbon dioxide at 58.1°C. As shown in this figure, PMMA does not swell in the presence of helium at this temperature.

Figure 5 shows the comparison of the sorption of helium and of carbon dioxide in PMMA also at 58.1°C. In contrast to carbon dioxide, very little helium is sorbed in PMMA even at very high pressures.

DISCUSSION

As shown in Figure 2, the swelling of PMMA in the presence of carbon dioxide at 41.8°C is greater than the swelling at 58.1°C and 68.0°C. The sorption of carbon dioxide in PMMA at 41.8°C, which is shown in Figure 3, is also greater than the sorption at 58.1°C and 68.0°C. At 41.8°C PMMA is very far removed from its glass transition temperature (T_g = 105°C). Below the glass transition temperature the polymer segments do not have sufficient energy to move (18). Therefore, PMMA is a very rigid, solid polymer at this condition. However, at 41.8°C carbon dioxide is very close to its critical temperature (T_c = 31.05°C). Hence, at pressures greater than approximately 1100 psia carbon dioxide possesses a "liquid-like" density which makes it more soluble in PMMA than at the higher temperatures of 58.1°C and 68.0°C. The large amount of carbon dioxide being solubilized into PMMA leads to a significant amount of polymer swelling.

As shown in Figure 3 the amount of carbon dioxide sorbed into polymethyl methacrylate at very high pressures reaches a saturation point. This saturation point is reached since increasing the pressure has very little effect on the properties of carbon dioxide at very high pressures. Therefore, both the swelling

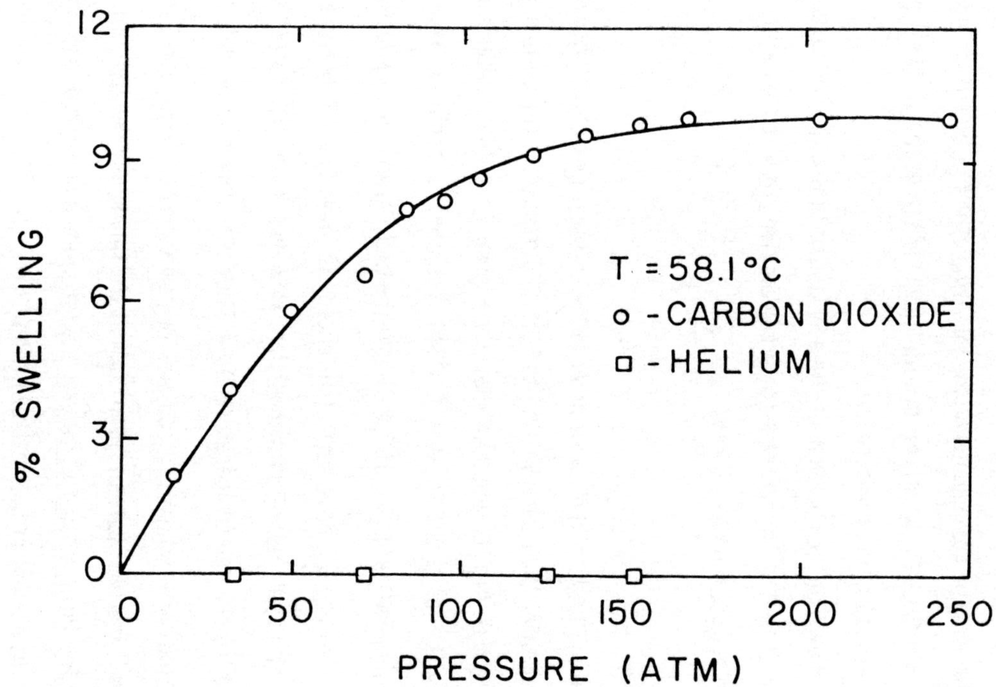

Figure 4. Comparison of the swelling of PMMA in the presence of carbon dioxide and helium.

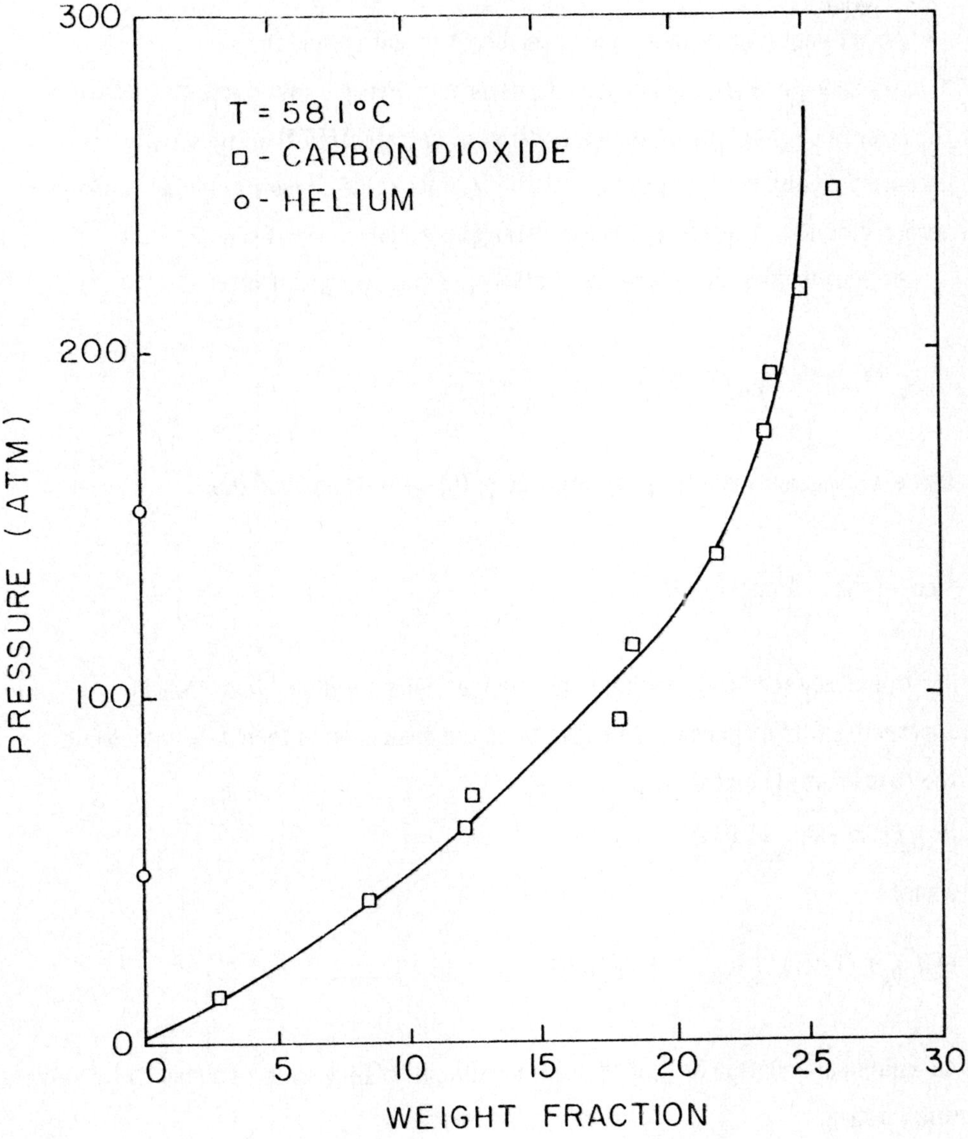

Figure 5. Comparison of the sorption of carbon dioxide and helium in PMMA at 58.1°C.

probably only valid up to 1500 psia.

MODELLING

An attempt was made to model the sorption data using the Kricheveskv-Kasarnovsky equation to determine Henry's Law constants. However, the results of this type of analysis were not satisfactory (i.e., physically unrealistic values were obtained for the infinite dilution partial molar volumes of carbon dioxide). Therefore, the Flory-Huggins equation was used.

At equilibrium the following relationship must be satisfied

$$f^g_{CO_2}\ (T,P,y) = f^s_{CO_2}(T,P,x)$$

Since we are dealing with an amorphous solid we will assume that

$$f^s_{CO_2}(T,P,x) = f^l_{CO_2}(T,P,x)$$

For this study it is assumed that the polymer does not dissolve in the supercritical fluid phase. The fugacity of the gas phase is then determined using the fundamental relationship

$$f^g_{CO_2}(T,P) = \emptyset_{CO_2}(T,P)\ P$$

where

$$\ln \emptyset_{CO_2} = (1/RT) \int_0^P [v_{CO_2} - RT/P]\, dP$$

An empirical equation of state for carbon dioxide (16) is used to numerically solve this integral.

The liquid phase fugacity is determined using

$$f^l_{CO_2}\ (T,P,X) = X_{CO_2} \gamma_{CO_2} f^{ol}{}_{CO_2} \quad \text{(hypothetical)}$$

where

$$\ln[X_{CO_2}\gamma_{CO_2}] = \ln\varphi_{CO2} + \varphi_{PMMA}(1-1/M) + \chi\varphi^2_{PMMA}$$

The pure component liquid phase fugacity of (hypothetical) carbon dioxide is found using the correlation of Prausnitz and Shair (19).

Therefore, the following equation can be solved for the Flory parameter

$$\ln[(\emptyset_{CO_2}P) / f^{ol(hypo)}_{CO_2}] = \ln\varphi_{CO_2} + \varphi_{PMMA}(1-1/M) + \chi\varphi^2_{PMMA}$$

For these calculations v_{CO_2} = 55 cm^3/gmol, v_{PMMA} = 25,560 cm^3/gmol, and

w_{CO_2} = [(cc (STP)/g PMMA)/[(22,261.5 cc/gmol)(gmol/44.01g)]]/[(cc (STP)/ g PMMA)/[(22,261.5 cc/gmol)(gmol/44.01g)] + 1.0]

where

$$W_{PMMA} = 1 - W_{CO_2}$$

$$\varphi_{CO_2} = [(w_{CO_2}/v_{CO_2}) / (w_{CO_2}/v_{CO_2} + w_{PMMA}/v_{PMMA})]$$

$$\varphi_{PMMA} = 1 - \varphi_{CO_2}$$

Shown in Figure 6 is χ as a function of composition. At 58.1°C and 68.1°C χ remains relatively constant to volume fractions of about 0.3. At 58.1°C χ exhibits a maximum at a volume fraction of about 0.26. Also, contrary to theory, χ increases with increasing temperature. Hence, this method of modelling the sorption data can only be considered as a correlational tool for interpolation of existing experimental results.

CONCLUSIONS

Based on the experimental results obtained in this study it may be feasable to

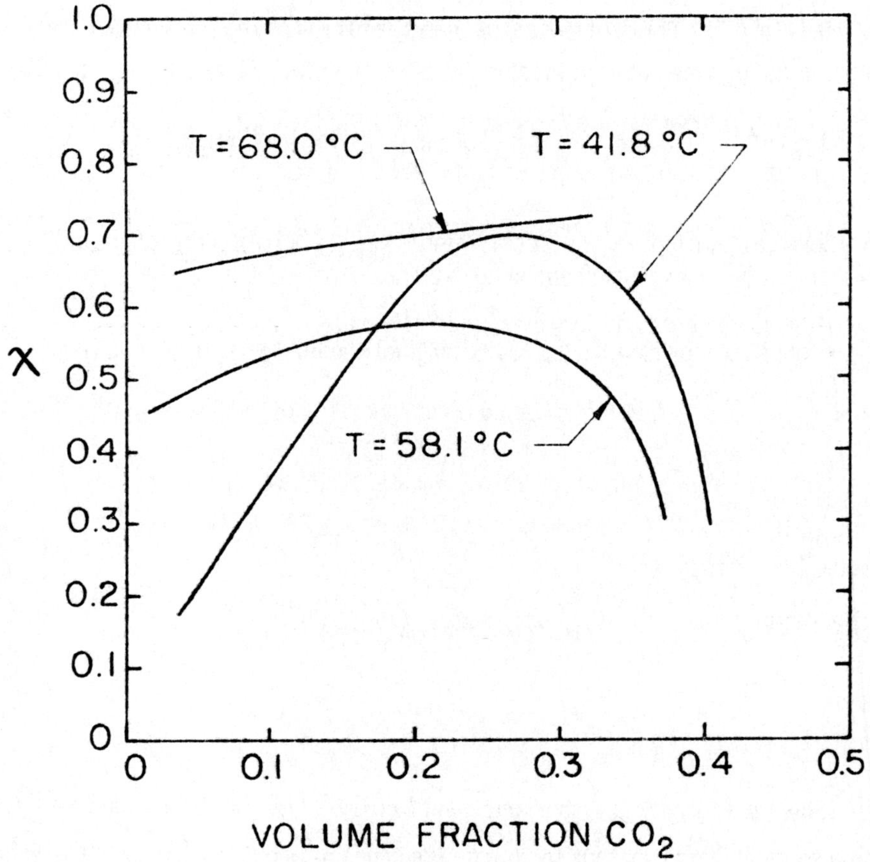

Figure 6. The effect of concentration and temperature on the Flory parameter.

extract lower molecualr weight impurities from solid, amorphous polymers. Subcritical and supercritical CO_2 swells PMMA considerably. Hence, one may envision an extraction process in which spent catalyst, entrapped solvent, or low molecular weight oligomers are extracted from swollen, solid polymers which retain their dimensional integrity. Caution must be exercised, however, to avoid melting the polymer at high loadings of solvent in the polymer at high operating pressures. In the case of PMMA the glass transition temperature is decreased by as much as 35°C at CO_2 pressures in excess of 1500 psia.

Simple polymer solution models such as the Flory-Huggins model fail to model this high pressure sorption and swelling data. More sophisticated models are probably needed such as Flory's equation-of-state.

NOTATION

f	fugacity
M	ratio of the molar volume of CO_2 to PMMA
STP	standard temperature and pressure, 273.2 K and 1 atm
v	molar volume
w	weight fraction

superscripts

g	gas phase
l	liquid phase
o	pure component
s	solid phase

greek symbols

φ	volume fraction
$\emptyset$	fugacity coefficient
γ	activity coefficient
χ	Flory interaction parameter

REFERENCES

1. Paulaitis, M. E., J. M. L. Penninger, R. D. Gray, Jr., and P. Davidson, "Chemical Engineering at Supercritical Fluid Conditions," Ann Arbor Science, Ann Arbor, Michigan, 1983.
2. Hunter, E. and R. B. Richards, U.S. Patent 2,457,238.
3. Bonner, D. C., Polym. Eng. Sci., 17, 65 (1977).

4. Ham, J. S., M. C. Bolen, and J. K. Hughes, J. Polym. Sci., 57, 25 (1962).
5. Moffat, D. H., and K. E. Weale, Fuel (London), 34, 449 (1955).
6. Menon, P. G., in "Advance in High Pressure Research," R. S. Bradley, eds., Vol. 3, pg. 313, Academic Press, New York.
7. Ozawa, S., S. Kusumi, and Y. Ogino, Proceedings of the Fourth Int. Conf. on High Pressure, Kyoto, Japan (1974).
8. Lundberg, J. L., M. B. Wilk, and M. J. Huyett, J. Polym. Sci., 57, 275 (1962).
9. Durill, P. L., and R. G. Griskey, AIChE J., 12, 1147 (1966).
10. Bonner, D. C., and Y. L. Cheng, J. Polym. Sci. Polym. Lett. Ed., 13, 259 (1975).
11. Steil. L. I., and D. F. Harnich, AIChE J., 22, 117 (1976).
12. Covitz, F. H., and J. W. King, J. Polym. Sci., 10, A-1, 689 (1972).
13. Newman, R. D., and J. M. Prausnitz, J. Phys. Chem., 76, 1492 (1972).
14. Franklin, R. E., Trans. Faraday Soc., 45, 275 (1949).
15. Heydemann, P., and H. D. Guicking, Kolloid Z., 193, 16 (1963).
16. Michels, A., and C. Michels, Proc. Roy. Soc. London, 160A, 348 (1937).
17. McHugh, M. A., and T. J. Yogan, J. Chem. Eng. Data, 29, 112 (1984).
18. Tager, A., "Physical Chemistry of Polymers," MIR Publishers, Moscow, 1968.
19. Prausnitz, J. M., and F. H. Shair, AIChE J., 7, 682 (1961).

ACKNOWLEDGEMENT

This material is based upon work supported by the National Science Foundation under Grant No. CPE-8204738.

Supercritical Fluid Technology, edited by J.M.L. Penninger, M. Radosz, M.A. McHugh and V.J. Krukonis
1985 Elsevier Science Publishers B.V., Amsterdam — Printed in The Netherlands

REDUCING CAPILLARY CONDENSATION OF PARAFFINIC DISTILLATE IN ZEOLITE WITH SUPERCRITICAL FLUID

P. BARTON and T. J. KASUN

Department of Chemical Engineering, The Pennsylvania State University, University Park, PA 16802

ABSTRACT

A method is developed to simultaneously dewax lube oils and recover n-paraffins by volatilizing the oil into supercritical gas where it contacts 5A zeolite, followed by ammonia regeneration. Wax distillate vaporized into supercritical 2,2,4-trimethylpentane exhibited an order of magnitude reduction in capillary condensation, resulting in 95% purity n-paraffin and 1-2% n-paraffin oil.

INTRODUCTION

Supercritical fluids are used to increase the volatility of low-volatility materials by orders of magnitude in the vicinity of the critical point of the supercritical fluid. This is a universal effect, irrespective of the nature of the supercritical fluid or the material to be vaporized. Superimposed on this, the extent of the volatility (or solubility) amplification of the material and its individual constituents is influenced by polar contributions to bonding. The material to be vaporized can be in the adsorbed state, such as on synthetic zeolite. In this case, the presence of supercritical fluid reduces capillary condensation in the macropores of the zeolite.

In separating n-paraffins from petroleum fractions with 5A zeolite, the advantages of adsorption from the vapor-phase as contrasted to adsorption from the liquid-phase include faster kinetics and, during desorption, less contamination of the n-paraffins recovered from the micropores with feed material trapped in the macropores. This is dependent on the proximity of the pressure during adsorption to the capillary condensation pressures of the feed constituents at the temperature employed. Supercritical fluid as a carrier for the petroleum fraction can be used to increase the molar concentration of the petroleum constituents in the vapor-phase, or to lower the condensation tendency of these constituents, at a given temperature.

Preparation of 95-98% purity n-paraffin fractions and, in separate batch experiments, 96% yield of denormal (i.e., dewaxed) oil with 266 K pour point from C_{16}-C_{32} wax distillate was reported earlier by Barton and Hajnik (ref. 1). These experiments were performed with Linde 5A zeolite in the absence of supercritical fluid and in the presence of supercritical 2,2,4-trimethylpentane or 2,2-dimethylbutane; desorptions were performed with either nitrogen or ammonia.

The objectives of the present work are to determine, using C_{14}-C_{33} wax distillate and Linde 5A zeolite in the presence of supercritical fluids, the operating conditions needed to attain high yields of high-purity n-paraffins, the effect of higher adsorption temperature using supercritical C_{12} branched paraffin, the performance of cyclical adsorption-desorption in a flow-through zeolite bed, and quantitative data on capillary condensation.

LITERATURE

Literature surveying the application of supercritical fluids and zeolites to processing hydrocarbon fractions has been discussed earlier (ref. 1, 2). General surveys of supercritical fluid topics have been presented by Randall (ref. 3) and Paulaitis et al. (ref. 4, 5). Thermodynamic theories to model phase equilibrium with supercritical fluids (ref. 6, 7, 8, 9, 10) have been demonstrated for mixtures with few components; petroleum fractions have been modeled by Pederson et al. (ref. 11). In the present work, pseudocritical properties were calculated using methods presented in Daubert and Danner (ref. 12), while using gas chromatography to characterize boiling range; fugacity coefficients were based on Hougen and Watson (ref. 13).

Adsorption isotherms for n-paraffins with 5A zeolites have been presented (ref. 14, 15) for low pressure applications as plots of adsorbate capacity versus pressure. The capacity of the zeolite to adsorb n-paraffins at equilibrium increases with decreasing temperature. For example, with n-dodecane at 100 kPa, the capacity changes from 7 to 15 wt % upon decreasing temperature from 636 to 504 K. At 589 K, equilibrium loading increases from 6 wt % to 0.7 kPa to 9 wt % at 500 kPa. With C_{13} to C_{20} n-paraffins at 673 K, there is little pressure effect on equilibrium capacity in the pressure range of 20 to 40 kPa, with capacity ranging from 6.5 to 8.9 g/100 g zeolite for these homologs, respectively. How much these curves, in the form of adsorbate capacity versus fugacity of n-paraffin in the fluid-phase, would change due to the presence of supercritical fluid at high pressure is not known.

Diffusion of hydrocarbon in 5A zeolite has been discussed by Eberly (ref. 16). Adsorption times on the order of several minutes are required to saturate the zeolite. Because of the nonlinearity of the isotherm, desorption diffusivities are considerably lower than adsorption diffusivities, requiring desorption times near 20 minutes. Thermal effects during diffusion are accounted for in a model presented by Ruthven (ref. 17).

Tamon et al. (ref. 18) discussed models for diffusion in the presence of capillary condensation. The onset of capillary condensation of pure vapors at low pressures can be predicted by the Kelvin equation:

$$\ln(P/P_s) = 2\sigma V_L \cos\alpha/(rRT) \qquad (1)$$

The relationship for fluid mixtures at high pressures would be more complex. Condensation occurs progressively from the smaller pores to the larger pores as the gas-phase pressure of the adsorbate is increased. The high pressure side of the porous plug is filled with capillary condensate more readily than the low pressure side. The capillary suction pressure is expressed as:

$$P_r = -2\sigma \cos\alpha/r \qquad (2)$$

Once the condensed phase is formed, the capillary force due to surface tension is higher on the low pressure side than the high pressure side. The capillary force gradient combines with any gas-phase total pressure gradient to act as the driving force in transfer of capillary condensate.

Tamon et al. (ref. 18) described a flow mechanism in capillary condensation that combines surface flow by a hopping mechanisms with viscous flow of condensed fluids. This model doesn't appear applicable to the n-paraffin - zeolite system because there is no selective adsorption of the paraffin on the binder which forms the macropores, and because the hopping model is rather simplified for long chain paraffin molecules.

EXPERIMENTAL

The zeolite used was Linde type 5A 3.2-mm calcium form. The petroleum fraction used was a Kendall wax distillate (ref. 1) that contained C_{14}-C_{33} (C_{23} average) hydrocarbons with normal boiling points of 525 to 755 K and 16 wt % n-paraffins predominantly in the range of C_{14} to C_{29} with a C_{21} average. This oil had a density of 0.822 g/cc at 305 K and a pour point of 296 K. A C_{14} to C_{27} (C_{20} average) distilled fraction of this oil was used in the cycled pilot plant experiments. It contained 17.5 wt % n-paraffins predominantly in the range C_{14} to C_{25} with a C_{19} average, and had a density of 0.831 g/cc at 298 K. The first batch adsorption experiment was performed in the absence of supercritical fluid at 577 K and 0.25 MPa. The supercritical fluids employed were: 2,2,4-trimethylpentane (T_c=544 K, P_c=2.57 MPa) at 524 to 577 K and 4.4 to 7.4 MPa; branched C_{12}

paraffin (T_c=631 K, P_c=1.87 MPa) at 631 to 688 K and 1.4 to 5.3 MPa. The branched 12-carbon atom paraffin was obtained from Exxon Chemical Company as a 12-carbon atom branched diolefin, Tetramer M. This diolefin was hydrogenated in a rocking autoclave with 64% nickel pelletized catalyst at 439 K and 12.4 MPa hydrogen pressure. The density of the paraffin product was 0.754 g/cc at 298 K. Ammonia (T_c=406 K, P_c=11.3 MPa) was used for zeolite desorption at 534 to 723 K and 0.1 to 27 MPa; nitrogen was used for purging before and after desorption.

Two apparatuses were used in this work: (1) a batch unit charged with fresh zeolite in each experiment, and (2) a pilot plant operated cyclically with a single zeolite charge. The batch experimental apparatus and operating procedures were described earlier (ref. 1). In this batch apparatus, the zeolite was suspended in a basket in the vapor space over the oil and supercritical fluid was added to vaporize the oil, or alternatively oil-solvent blend was pumped into the hot vessel. Details of modifications in experimental procedures and description of pilot plant equipment and operation can be found in Kasun (ref. 19). In all of the tests with supercritical fluids, the operating conditions were such that feed oil was completely vaporized.

The zeolite handling procedures were improved over the previous work in order to determine more accurately the material balances during adsorption and desorption. Vacuum baking at 0.007 Pa and 561 K served as zeolite pretreatment. Zeolite handling and weighing were performed in nitrogen glove boxes. All transfers between units were made under nitrogen. In the batch experiments, supercritical solvent/oil feed weight ratio ranged from 19 to 24, and zeolite/oil feed weight ratio ranged from 5.8 to 7.1. Adsorption run times ranged from 1.3 to 5.0 hr with 0.5 to 2.1 hr above the critical temperature. Desorption times ranged from 1.9 to 12 hr. Ammonia treats were 0.6 to 1.3 g/g zeolite.

Desorptions were performed at atmospheric and elevated pressures, with throttling of the exit gas into a partial condenser. Because adsorbate mist recovery from the exit gas can be problematical, several partial condenser designs were tested. The best designs employed downflow through an air-cooled 60 cm length of 1.8 cm OD glass tubing or 1.9 cm OD stainless steel tubing, into an ice-cooled cup followed by upflow through a demisting chamber. This chamber contained a 10-cm diameter by 13-cm high bed of Metex demisting wire mesh or a 7-cm diameter by 15-cm high bed of fine glass wool. When using ammonia, the gas exited into an ice-cooled stirred titrator containing sulfuric acid.

The pilot plant for testing a flow-through zeolite bed in cyclical adsorption-desorption is sketched in Figure 1. The zeolite bed was 7.7 cm in

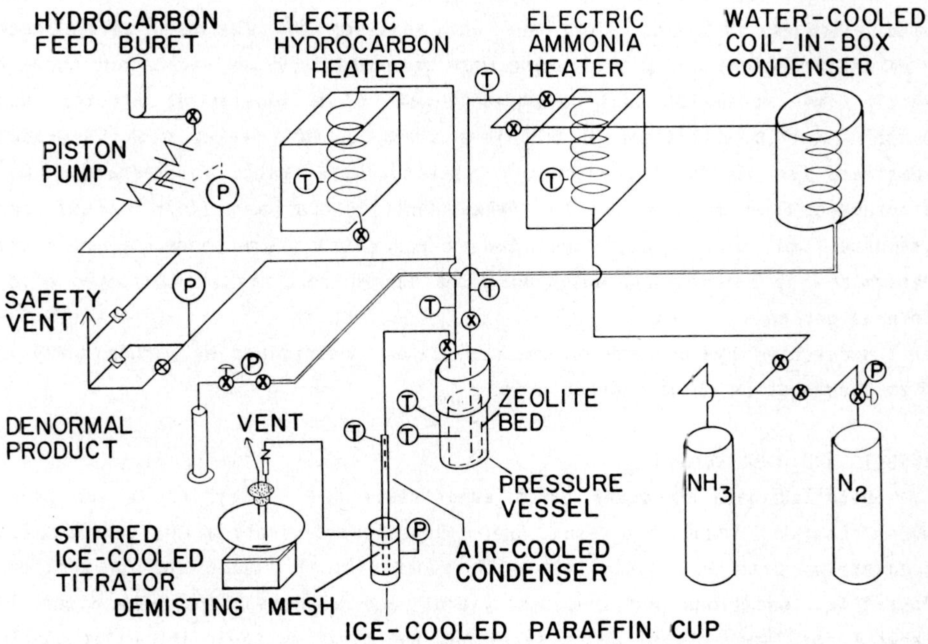

Fig. 1. Pilot plant for cyclical adsorption-desorption of zeolite under supercritical conditions.

diameter, 22.7 cm in height, 1053 cc in volume, and contained 736 g Linde 5A zeolite. The bed was operated in downflow during adsorption and upflow during desorption. For the adsorption step, the feed oil was blended with C_{12} solvent at 24/1 to 27/1 solvent/oil weight ratio. The oil-solvent blend was pumped into the system continuously at 20 cc/min with a reciprocating pump, through a heating coil cast in an electrically-heated aluminum block, through the zeolite bed, through a cooling coil immersed in water, and discharged through a back-pressure reducing valve. Zeolite/oil feed weight ratio was 6.2. Adsorption run time ranged from 1.6 to 6.3 hours. At the end of the adsorption step, the system was depressurized. A 20-minute nitrogen purge initiated the desorption step followed by ammonia flow at 4-20 g/min for 0.9 to 2.6 hours, and a final nitrogen purge. Ammonia treats ranged from 0.9 to 1.1 g per g zeolite. Ten adsorption-desorption cycles were performed.

Solvents and oils were separated by distillations finished under vacuum with nitrogen stripping. Oil and wax analyses for residual solvent and n-paraffin content and distribution were determined by gas chromatography. A Perkin-Elmer Model 900 dual column unit with flame ionization detectors was used. Two 1.8-m, 1.5 mm ID stainless steel columns packed with 5% Dexsil substrate on 60/80 Chromasorb P packing were used. Temperature was programmed from 343 to 668 K. Peak identification was aided by internal standards of C_{14}, C_{16}, and C_{20} n-paraffin. The peak areas were determined by cutting and weighing. The weight percents were assumed equal to area percents.

Undesorbed hydrocarbon on the zeolite was determined by carbon analysis with a combustion-chromatograph unit.

RESULTS AND DISCUSSION

Detailed data on seven batch experiments and ten cycles in the pilot plant can be found in Kasun (ref. 19) for separation of C_{14} to C_{33} n-paraffins from wax distillate with supercritical fluids and 5A zeolite. Operating conditions and product yields and analyses are summarized in Table 1 for three of the batch tests and the first cycle in the pilot plant. The yields may not add up to 100% because they are based on different experimentally determined numbers.

Run 2 is a blank run made in the absence of supercritical fluid. The zeolite was suspended in the vapor space over feed oil at 577 K and 250 kPa for a couple of hours. The zeolite was weighed and then regenerated with ammonia at 658 K and 100 kPa. The content of n-paraffin was reduced from 16 wt % in the feed oil to 5 wt % in the denormal oil. This is insufficient removal to be considered as satisfactory "dewaxing." The n-paraffin content of the desorbed product was only 6 wt %, which is lower than even the feed oil. The extent of capillary condensation in the macropores exceeded the amount of n-paraffin adsorbed in the micropores by more than an order of magnitude. The amount of feed carbonized or not recoverable from the zeolite represented 6 wt % of the feed, and probably consisted of predominantly n-paraffin.

Run 5 was made using supercritical 2,2,4-trimethylpentane to volatilize the oil for contact with the zeolite. The average temperature and pressure during adsorption were 566 K and 4.2 MPa. The system was hotter than the critical temperature of the solvent-oil blend, as calculated by Kay's rule, for two hours. The presence of the supercritical fluid reduces the capillary condensation by an order of magnitude, to one-fifth the adsorbed n-paraffin level. Desorption was done with ammonia under the same conditions as run 2. The decrease in fractional desorption from 88% in run 2 to 59% in run 5 confirms that capillary condensate from the macropores is much easier to recover than n-paraffins from the macropores. The denormal oil was "dewaxed" to a n-paraffin content of 2 wt %, with a yield that represents 88% of the non-n-paraffin in the feed. The desorbed product contained 85% of the n-paraffin in the feed, at a concentration level of 77 wt %. This n-paraffin purity is not as high as desired because the capillary condensation had not been entirely eliminated. The amount of feed carbonized or not recoverable from the zeolite remained at 6 wt %.

TABLE 1
Separation of C_{14}-C_{33} wax distillate with supercritical fluid and 5A zeolite.

Solvent	None	2,2,4-trimethyl-pentane	C12 branched Paraffin	
Run	2	5	6	Cycle 1 [a]
Adsorption				
g zeolite/g oil	5.8	6.4	6.0	6.2
g solvent/g oil	-	23.0	24.0	24.0
Temperature, K	577	566	650	640
Pressure, MPa	0.25	4.2	5.2	4.4
Density, g/cc	0.002	0.42	0.26	0.40
Time, h				
Total	5.0	5.0	5.0	0.7 [b]
Above T_c	-	2.1	1.7	3.7
g/100 g zeolite				
Adsorbed n-paraffin	0.9	2.4	6.0	5.2
Capillary condensate	5.8	0.5	1.2	0.6
Desorption				
g NH_3/g zeolite	0.6	0.75	0.56	0.92
Temperature, K	658	658	658	628-703
Time, h	1.9	2.0	1.7	1.8
% desorbed	88	59	67	50
Denormal product				
Yield, wt % of feed	44	77	73	67
n-paraffin content, wt %	4.9	2.3	1.2	12.0
Desorbed product				
Yield, wt %	38	12	15	18
n-paraffin content, wt %	6	77	61	81
Feed carbonized, wt %	6	6	29	18

[a] Feed oil is C_{14}-C_{27} fraction.
[b] Space time.

Run 6 was made using C_{12}-branched paraffin as supercritical solvent at 650 K and 5.2 MPa. This temperature is about the same as used in ammonia desorption. The amount of capillary condensate is one-fifth the amount of n-paraffin adsorbed in the micropores. The denormal product yield and composition is near that of run 5. The desorbed product yield and n-paraffin content are significantly reduced because the amount of feed carbonized or unrecoverable from the sieves increased to 29%. Since the desorption step in runs 2, 5 and 6 were similar, the additional degradation in run 6 occurred during the high temperature adsorption step.

In the pilot plant apparatus with the flow-through bed, the zeolite is exposed to a different environment, as it becomes loaded, than in the batch experiments. The zeolite at the adsorption front is unsaturated and has the capability of reducing the n-paraffin content of the denormal oil to a lower value than in the batch experiments. The zeolite at the back of the bed is exposed to higher n-paraffin partial pressure, providing potential for increasing the n-paraffin content of the material within the zeolite. In cycle 1 shown in Table 1, the space time was limited to 0.7 hour in order to reduce the fraction of feed carbonized. As a result, the fraction of n-paraffins that diffused from the vapor phase was not as high as in the batch experiments, which had total adsorption times up to 5 hours.

The feed carbonized or unrecoverable by ammonia desorption in cycle 1 in the pilot plant operation represented 18% of the wax distillate pumped through the bed. Based on a comparison with run 5 which had the same desorption conditions but a lower adsorption temperature, possibly two-thirds of this loss occurred during the 640 K adsorption step. The capacity of the zeolite to adsorb n-paraffin decreased exponentially from 5.2 g/100 g zeolite in the first cycle to 0.3 g/100 g zeolite after the fifth cycle. This is due to carbonization at the high temperatures (628-703 K) employed in the adsorption and desorption steps and to the inability of ammonia to completely desorb the n-paraffins. Operation of both steps at 570 K would have been preferable.

The use of ammonia at supercritical pressures during desorption did not improve the desorption efficiency.

<u>Controlling product properties</u>

The n-paraffin content of desorbed product is seen in Figure 2 to be directly proportional to the density of the supercritical fluid-oil phase. Thus, increasing the density is one way to lower capillary condensation.

The magnitude of n-paraffin adsorption in the zeolite micropores was found to be a strong function of the proximity of the operating temperature to the critical temperature of the vapor blend. In Figure 3, a spike in

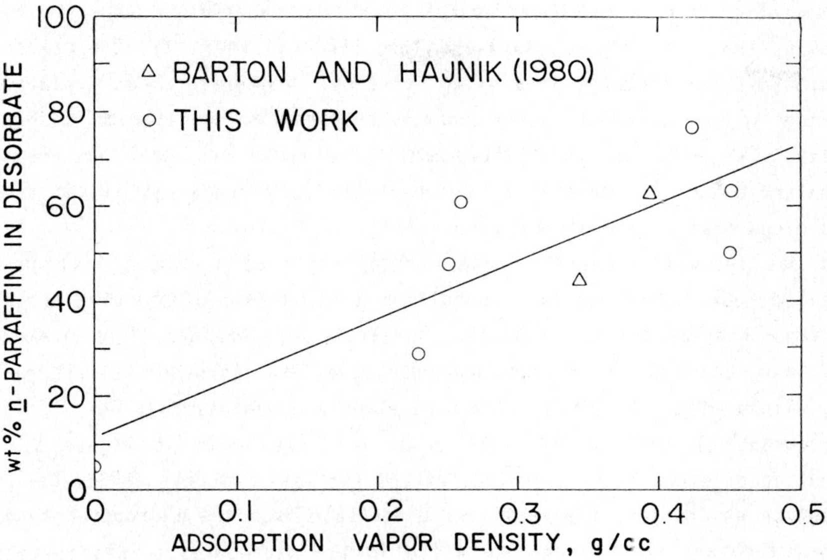

Fig. 2. Effect of adsorption vapor density on n-paraffin content of desorbate.

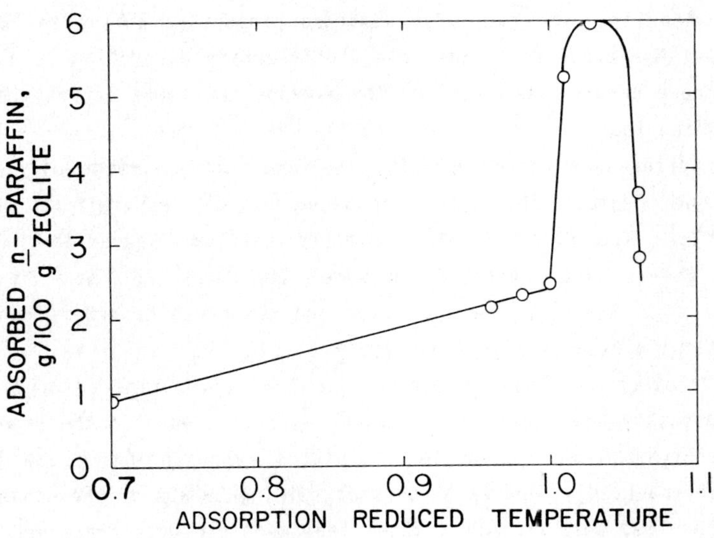

Fig. 3. Effect of reduced temperature on amount of C_{14}-C_{33} n-paraffins adsorbed on 5A zeolite.

adsorption is noted at reduced temperatures between 1.0 and 1.05. Since all the adsorption runs with supercritical fluids were performed under operating conditions that completely volatilized the feed oil (ref. 1), the reason for the improved performance near the critical temperature is related to diffusion in the zeolite pores and/or changes in equilibrium adsorption capacity. A spike in volatility amplification occurs near the critical temperature (ref. 1); this spike reduces capillary condensation and removes liquid plugs that retard diffusion.

A gas chromatograph of the wax distillate feed is shown in Figure 4. The spiked peaks represent the n-paraffins and the rest of the area above the base line represents the branched paraffins, naphthenes, and aromatics. Significant amounts of hydrocarbons in the C_{14} through C_{33} range are noted, with a midpoint at C_{23}. The n-paraffin midpoint is C_{21}. A chromatograph of the denormal oil from run 5 is shown in Figure 5. The elimination of n-paraffins is noted, except for small amounts of C_{17}-C_{19}.

During desorption, the capillary condensate from the macropores tends to be removed before the adsorbed paraffin in the micropores. This results in fractionation of the total n-paraffin fraction recovered. Chromatographs from early and late fractions from run 5 are compared in Figures 6 and 7. The early cut contained 56 wt % n-paraffin predominantly C_{14} to C_{25} with a midpoint of C_{19}-, contaminated by non-n-paraffins with a midpoint near C_{24}. The late cut contained 93 wt % n-paraffin predominantly C_{14} to C_{27} with a C_{19}+ midpoint. The composite desorbed product, reported in Table 1, contained 77 wt % n-paraffin. Some selectivity during desorption is noted in that the desorbed paraffin contains on the average two fewer carbons than the n-paraffin in the feed.

Similar performance was noted with the lower carbon range oil used in the pilot plant tests. This feed contained C_{14}-C_{25} n-paraffins with an average of C_{19}. The denormal oil contained C_{15}-C_{25} n-paraffins with an average of C_{17}+. The desorbed product contained C_{14}-C_{23} n-paraffin with an average of C_{18}. The early, late, and composite desorbate contained 77, 87, and 81 wt % n-paraffin, respectively.

In this project, the use of supercritical fluids during adsorption and selective condensation and demisting during desorption enabled the production of 93-95% purity C_{14}-C_{27} n-paraffins in yields representing 29-33% of the n-paraffins in the feed. Denormaled oils containing 88-90% of the non-n-paraffin in the feed were produced; these "dewaxed" products contained 0.9 to 2.3 wt % n-paraffin and had pour points of 266 to 280 K, respectively.

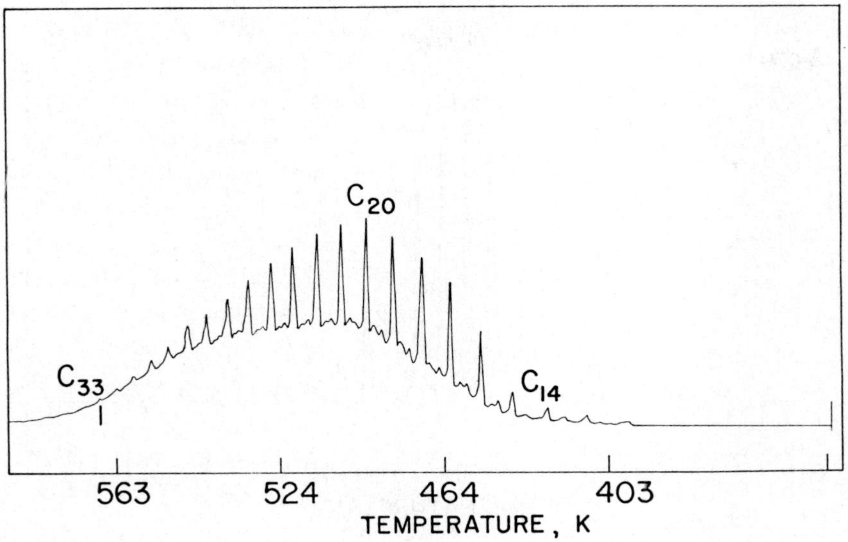

Fig. 4. Temperature programmed gas chromatogram of Kendall wax distillate.

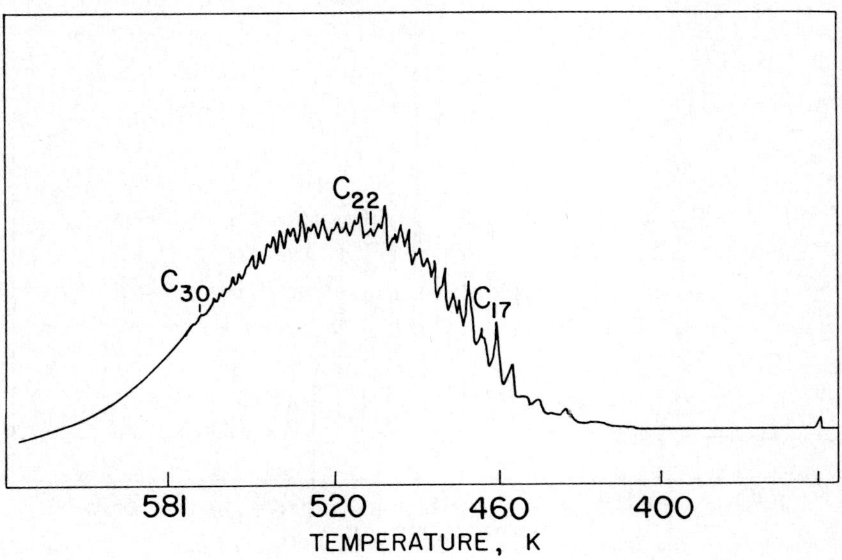

Fig. 5. Chromatogram of Run 5 denormal oil.

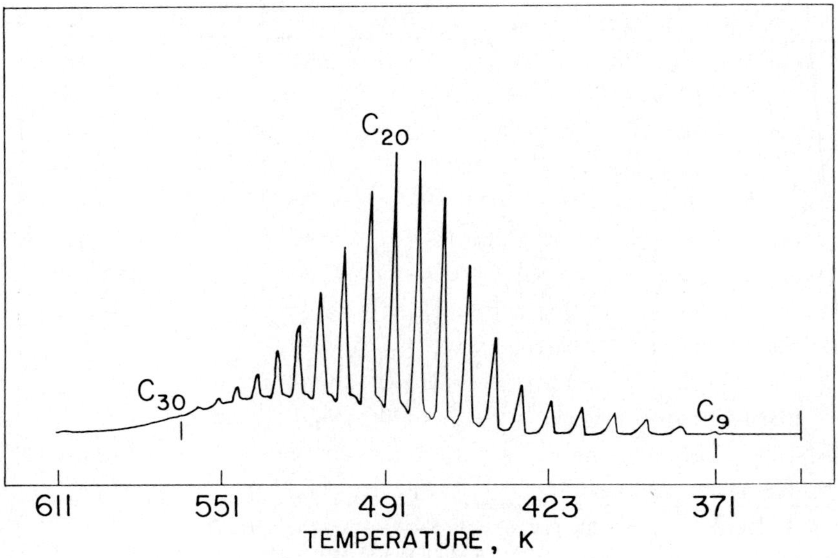

Fig. 6. Chromatogram of early n-paraffin desorbate from Run 5.

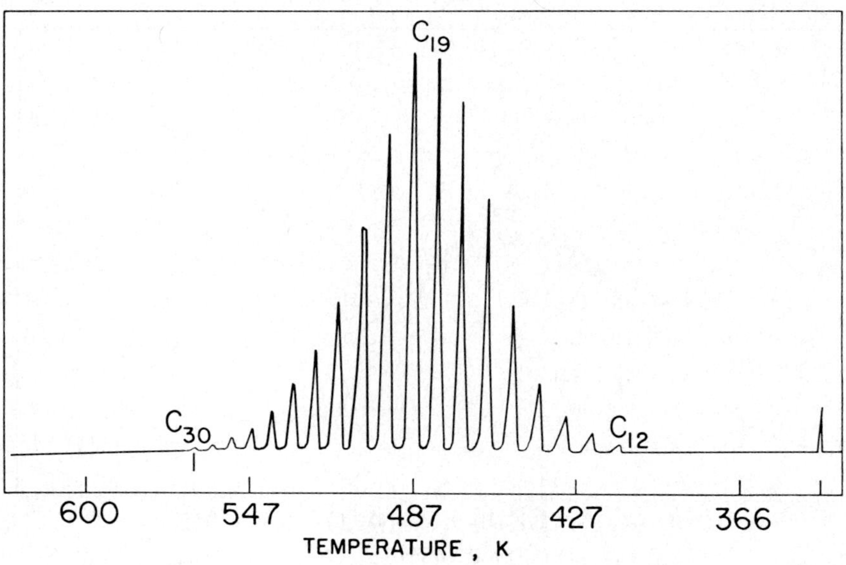

Fig. 7. Chromatogram of late n-paraffin desorbate from Run 5.

Capillary condensation

The amount of capillary condensation in the macropores of the zeolite was calculated from the total hydrocarbon, n-paraffin, and carbon measured in the desorption step. The n-paraffin content of the capillary condensed oil was assumed to approach the n-paraffin content in the recovered denormal oil; the amount of capillary condensation calculated is not very sensitive to this assumption. The amount of capillary condensation correlated well with the fugacity of the oil in the vapor-phase (see Figure 8). The fugacity coefficient of the oil in the vapor was estimated from its pseudocritical properties and corresponding states, while assuming ideal solution between the oil and solvent. Values of fugacity coefficient near 0.1 were obtained for wax distillate in supercritical fluids.

Capillary condensation is seen in Figure 8 to decrease with increased temperature. Reducing the end point of the wax distillate from C_{33} to C_{27} reduced capillary condensation by one-half. The point in the upper right of Figure 8 is in the absence of supercritical fluids; the points in the lower left are in the presence of supercritical fluid. Supercritical fluids are seen to reduce capillary condensation by an order of magnitude; this effect is accounted for by the change in fugacity coefficient of the oil upon adding supercritical solvent.

CONCLUSIONS

Separation of n-paraffins from petroleum distillate by vapor-phase contact with 5A zeolite is greatly improved by performing the adsorption step, but not the desorption step, with supercritical fluid. The distillation end point of the petroleum fraction should be limited to less than C_{25} to allow the use of lower adsorption and desorption temperature to reduce carbonization loss. A maximum operating temperature of 570 K is recommended for the adsorption and ammonia desorption cycles. Carbonization loss is less than 6% per cycle at 570 K but 18% at 640 K.

For maximum adsorption capacity, the adsorption should be performed at a reduced temperature between 1.0 and 1.05. Vapor-phase density should be maintained at 0.4 g/cc or higher; this reduces capillary condensation and increases purity of recovered n-paraffin. Production of 93-95% purity of C_{14}-C_{27} n-paraffins is attained in yields representing 29-33% of the n-paraffins in C_{24}-C_{33} feedstock. "Dewaxed" oils with n-paraffin content of 0.9 to 2.3 wt % and pour points of 266 to 280 K are produced in yields representing 88-90% of the non-n-paraffin in the feed.

The amount of capillary condensation correlates with the fugacity of the oil in the supercritical vapor mixture. Addition of supercritical solvent reduces capillary condensation by an order of magnitude, a change that can be

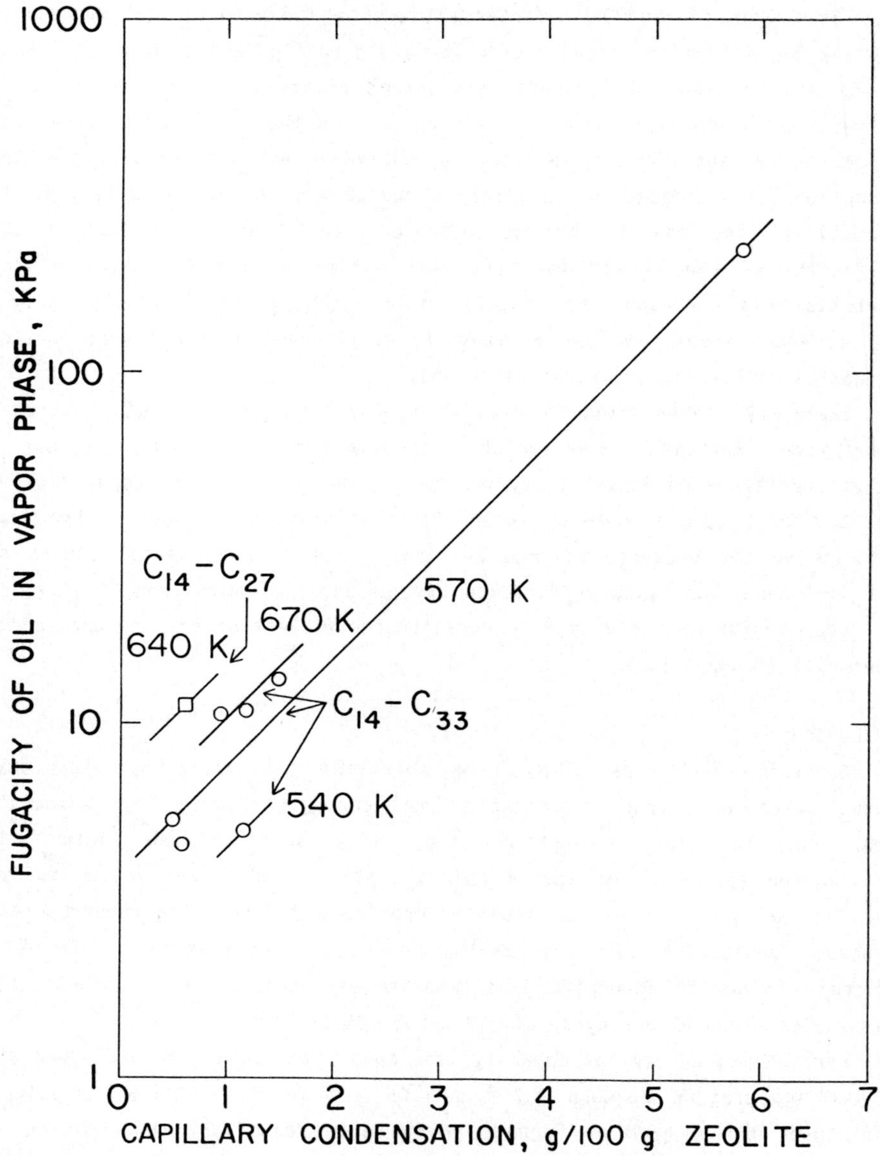

Fig. 8. Capillary condensation of wax distillate hydrocarbons in 5A zeolite.

attributed to the magnitude of the fugacity coefficient (~0.1) of the oil in the vapor-phase.

ACKNOWLEDGEMENT

The financial support of the Pennsylvania Research Corporation is appreciated.

NOTATION

P = pressure

P_c = critical pressure

P_r = capillary suction pressure

P_s = saturated vapor pressure

r = pore radius

R = gas constant

T = temperature

T_c = critical temperature

α = contact angle

σ = surface tension

REFERENCES

1. P. Barton and D. F. Hajnik, in W. H. Flank (Ed.), Adsorption and Ion Exchange with Synthetic Zeolites, ACS Symposium Series 135, American Chemical Society, Washington, DC, 1980.
2. P. Barton, Ind. Eng. Chem. Process Des. Dev., 22 (1983) 589.
3. L. G. Randall, Separation Sci. Technol., 17 (1982) 1.
4. M. E. Paulaitis, J. M. L. Penninger, R. D. Gray, Jr., and P. Davidson, Chemical Engineering at Supercritical Fluid Conditions, Ann Arbor Science, Ann Arbor, MI, 1983.
5. M. E. Paulaitis, V. J. Kurkonis, R. T. Kurnik, and R. C. Reid, Rev. Chem. Eng., 1 (1983) 179.
6. J. Chrastil, J. Phys. Chem., 86 (1982) 3016.
7. G. T. Hong, M. Modell, and J. F. Tester, in M. E. Paulaitis, J. M. L. Penninger, R. D. Gray, Jr., and P. Davidson (Eds.), Chemical Engineering at Supercritical Fluid Conditions, Ann Arbor Science, Ann Arbor, MI, 1983.
8. M. E. Mackay and M. E. Paulaitis, Ind. Eng. Chem. Fundam., 18 (1979) 149.
9. K. W. Won, Fluid Phase Equilibria, 10 (1983) 191.
10. D. H. Ziger and C. A. Eckert, Ind. Eng. Chem. Process Des. Dev., 22 (1983) 582.
11. K. S. Pedersen, P. Thomassen, and A. Fredenslund, Thermodynamics of Petroleum Mixtures Containing Heavy Hydrocarbons; I. Phase Envelope Calculations Using the SRK Equation of State, Report No. SEP 8204, Instituttet for Kemiteknik, Danmarks Tekniske Hojskole, 1982.
12. T. E. Daubert and R. P. Danner (Eds.), Technical Data Book - Petroleum Refining, 4th ed., American Petroleum Institute, Washington, DC, 1983.
13. O. A. Hougen and K. M. Watson, Chemical Process Principle Charts, Wiley, New York, 1946.
14. C. G. V. Burgess, R. H. E. Duffett, G. J. Minkoff, and R. G. Taylor, J. Appl. Chem., 14 (1964) 350.
15. D. E. Cooper, H. E. Griswold, R. M. Lewis, and R. W. Stokeld, Chem. Eng. Prog., 62(4), (1966) 69.
16. P. E. Eberly, Jr. in J. A. Rabo (Ed.), Zeolite Chemistry and Catalysis, ACS Monograph 171, American Chemical Society, Washington, DC, 1976.

17. D. M. Ruthven in R. P. Townsend (Ed.), The Properties and Applications of Zeolites, Spec. Publ. 33, Royal Society of Chemistry, London, 1980.
18. H. Tamon, M. Okasaki, and R. Toei, AIChE Journal, 27 (1981) 271.
19. T. J. Kasun, Influence of Critical Fluids in Separating n-Paraffins from Wax Distillate Using Molecular Sieves, M.S. Thesis, The Pennsylvania State University, University Park, PA, 1983.

Supercritical Fluid Technology, edited by J.M.L. Penninger, M. Radosz, M.A. McHugh and V.J. Krukonis
1985 Elsevier Science Publishers B.V., Amsterdam — Printed in The Netherlands

ENTRAINER SELECTION IN SUPERCRITICAL EXTRACTION

A.K. Sunol, Hagh B. and Chen S.
U. of S. Florida, Tampa, Florida

Abstract

Key interactions in entrainer selection are discussed. An experimental system for screening entrainers is described and preliminary results in application of entrainers to supercritical extraction of coal near ambient temperatures are shown.

Introduction

Entrainer-aided supercritical extraction is becoming an attractive alternative for carrying out difficult separations. Entrainers could be used effectively in supercritical extraction (SCE) of natural products, lignin from wood [Peter], or in supercritical extraction of coal [Sunol].

As shown by Stahl, we can extract neither sugars nor proteins through supercritical extraction at reasonably low pressures and below their decomposition temperatures. In such cases entrainers provide an added dimension to supercritical extraction. They are especially effective if solubility of heavy components in supercritical solvents is very low, and/or high selectivity is desired. Furthermore, they could be used as reactants effectively. As shown in the works of Brunner and Peter or Kurnik, solubilities of low volatile matter increase in the presence of an entrainer. Furthermore, selectivity could be introduced through an entrainer which has strong affinity for the desired extractant.

The temperature and pressure sensitivity of solubility and/or selectivity increases in the presence of an entrainer. This would provide a sharper fractionation as well as solvent regeneration by mere temperature increase. In some applications, entrainers could be used to slurry the solute to the extractor and/or carry the extract from regeneration column.

The entrainers could also be used as reactive components as in extractive distillation with reaction. One of the many potential applications of reactive entrainers is processing of heavy hydrocarbons such as coal. Coal can be extracted in the presence of donor solvents such as tetralin for improved yield. Then, probably a new logical constraint would be an upper ceiling on donor solvent concentrations at which pore structure collapses. Alternatively, entrainers could be used to desulfurize coal near ambient temperatures. We would provide some preliminary results towards such goals.

Although entrainers bring some added advantages, it also brings much more complicated flowsheets. Figure 1 which is a conceptual flowsheet for entrainer-aided supercritical extraction of coal aims to show the added complexity in the solvent recovery stages. The separations of extract from entrainer and entrainer from supercritical solvent are not as easy and as sharp as in supercritical extractions without entrainer.

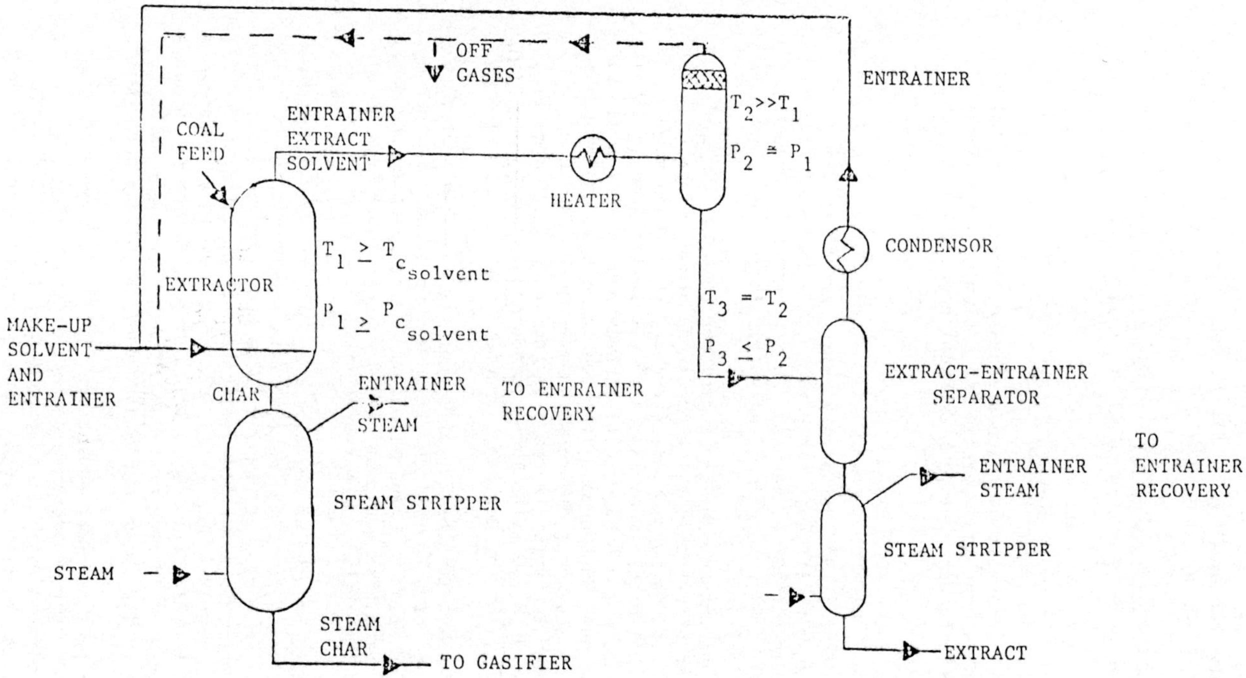

FIGURE 1: A DESIGN FOR SCE OF COAL WITH THE AID OF ENTRAINERS

Entrainer Selection

The entrainer selection should be based on thermodynamic considerations. The description of the phase behavior of the system throughout the entire temperature and pressure range of extraction and regeneration stages is desirable for both operating parameter and solvent selection. Unfortunately, we do not have a way of predicting the phase behavior a priori. Although binary parameters regressed from data and an equation of state could represent multi-component behavior, these parameters need to be determined from experiment with multi-components. There is not much data for multi-component mixtures and the available data is restricted to supercritical solvents such as carbon dioxide and ethylene. The need for more data prompted, in part, the construction of the experimental system to be described later in the paper.

One way of simplifying the task of solvent or entrainer selection is to concentrate on the key steps and interactions. We will not dwell much on supercritical solvent selection. Different solvents with same critical properties was shown to have different solvency power [Brunner and Peter]. The supercritical solvent selection depends on extraction temperature which is preferably ambient and is very often bounded by solute decomposition temperature. Then, cost and safety considerations are often overriding since almost any solvent is somewhat effective near its critical temperature. This statement is quite correct unless fractionation (selectivity) is the primary objective. Thus, we assume that if there is a need for entrainer, then the solute-entrainer interaction is the dominant interaction in the extraction stage. One needs to superimpose the solvent-entrainer interaction for the regeneration stage. The solvent selection guidelines for liquid extraction would provide an initial selection for entrainer. The supercritical solvent-entrainer interaction could than be checked for an effective regeneration strategy.

Information on entrainer-solute interactions may be obtained by extrapolating our accumulated knowledge of liquid systems. Table 1 [Schweitzer] show these interactions in terms of activity coefficient. Hydrogen bond forming tendencies of different classes of solvents and resulting molecular interactions [Henley and Seader] that are shown as Table 2 and Table 3 respectively, complement the broader picture shown in Table 2. Ideally, one would prefer to regenerate the solvent through a temperature raise at constant pressure. The solvent-entrainer binary should have a wide two phase region at that temperature and pressure. The extraction and regeneration stages are shown by Figure 2 from Peter. The phase behavior shown in Figure 2 show transition from Type 1 behavior to Type 2 and there is a miscibility gap between entrainer and solvent. The non-volatile matter loadings are rather high. This type of behavior may be seen more with very close boiling system fractionations. The phase behavior more common in high pressure extraction of natural products or hydrocarbons is shown in Figure 3. The solute is usually made up of soluble and insoluble fractions. The soluble fraction would be extracted if the entrainer amount is above a minimum and the combined solvent temperature and pressures are just above the mixture critical point. The solvent regeneration step is shown in a second ternary diagram. More energy efficient regenerations could be obtained by raising the temperature at constant pressures as in route A. Thus, two phase envelope widens resulting in solute rich (III) and lean (II)

TABLE 1

Solute-Solvent Group Interaction

Group	Solute	Solvent 1	2	3	4	5	6	7	8	9
1	Acid, aromatic OH (phenol	0	-	-	-	-	0	+	+	+
2	Paraffin OH (alcohol), water, imide or amide with active-H	-	0	+	+	+	+	+	+	+
3	Ketone, aromatic nitrate, tertiary amine, pyridine, sulfone, trialkyl phosphate, or phosphine oxide	-	+	0	+	+	-	0	+	+
4	Ester, aldehyde, carbonate, phosphate, nitrite, or nitrate, amide without active-H, intramolecular bonding, e.g., o-nitro-phenol	-	+	+	0	+	-	+	+	+
5	Ether, oxide, sulfide, sulfoxide, primary and secondary amine or imine	-	+	+	+	0	-	0	+	+
6	Multihalo paraffin with active H	0	+	-	-	-	0	0	+	0
7	Aromatic, halogen aromatic, olefin	+	+	0	+	0	0	0	0	0
8	Paraffin	+	+	+	+	+	+	0	0	0
9	Monohalo paraffin or olefin	+	+	+	+	+	0	0	+	0

Effect of interaction groups on solute activity coefficient:
+ increases activity coefficient, minimum-boiling-point azeotropic
0 mild interaction, seldom azeotropic
- decreases activity coefficient, maximum-boiling-point azeotropic

TABLE 2

Classification of molecules based on potential for forming hydrogen bonds

Class	Description	Examples
I	Molecules capable of forming dimensional networks of strong H-bonds	Water, glycols, glycerol, amino, alcohols, hydroxylamines, hydroxyacids, polyphenols, and amides
II	Other molecules containing both active hydrogen atoms and donor atoms (O, N, & F)	Alcohols, acids, phenols, primary and secondary amines oximes, nitro and nitrile compounds with -hydrogen atoms, ammonia, hydrazine, hydrogen fluoride, and hydrogen cyanide
III	Molecules containing donor atoms but no active hydrogen	Ethers, ketones, aldehydes, esters, tertiary amines (including pyridine type), and nitro and nitrile compounds without -hydrogen atoms
IV	Molecules containing active hydrogen atoms but no donor atoms that have two or three chlorine atoms on the same carbon atoms as a hydrogen atoms, or one chlorine on the same carbon atom and one or more chlorine atoms on adjacent carbon atoms	$CHCl_3$ CH_2Cl_2, CH_3CHCl_2, CH_2ClCH_2Cl, $CH_2ClCHClCH_2Cl$, and $CH_2ClChCl_2$
V	All other molecules having neither active hydrogen atoms nor donor atoms	Hydrocarbons, carbon disulfide, sulfides, mercaptans, and halohydrocarbons not in Class IV

TABLE 3

Molecule interactions causing deviations from Raoult's law

Type of Deviation	Classes	Effect on Hydrogen Boding
Always negative	III + IV	H-bonds formed only
Quasi-ideal; always positive or ideal	III + III III + V IV + IV IV + V V + V	No H-bonds involved
Usually positive, but some negative	I + I I + II I + III II + II II + III	H-bonds broken and formed
Always positive	I + IV (frequently limited solubility) II + IV	H-bonds broken and formed, but dissociation of Class I or II is more important effect
Always positive	I + V II + V	H-bonds broken only

phases. Solvent regeneration through pressure reduction is the other alternative (Route B). Although this route is much less energy efficient, the separations are generally much sharper (the two phase envelope is wider and/or the tie-line end points are further apart).

The solvent and entrainer regeneration for recirculation and possible product purification could be achieved through pressure reduction, or temperature increase, or compositional changes, or any combination. The temperature increase route is generally more appealing if not the sharpest. As shown in Figure 4 for a binary separation, the sensitivity of solubility to temperature is greatest just above the critical pressure of the supercritical solvent. Unfortunately, for ternary plus systems one would encounter liquid phase formation and sometimes very complex phase behavior if he tries to operate near the critical temperature and pressure of the solvent. Furthermore, as can be seen from Figures 5 and 6 [Paulatis] increased pressures would improve the yield and at times the selectivity. Thus one ends up with an extraction pressure 2-4 times more than the critical pressure of solvent. The temperature should ideally be close to critical temperature of the solvent where rate of entrainer-solvent concentration change with temperature would be greater. In other words entrainer-solvent mixture should be just supercritical.

We have tried to simplify the task of entrainer selection by limiting our analysis to binary interactions and coupling them later through pressure and

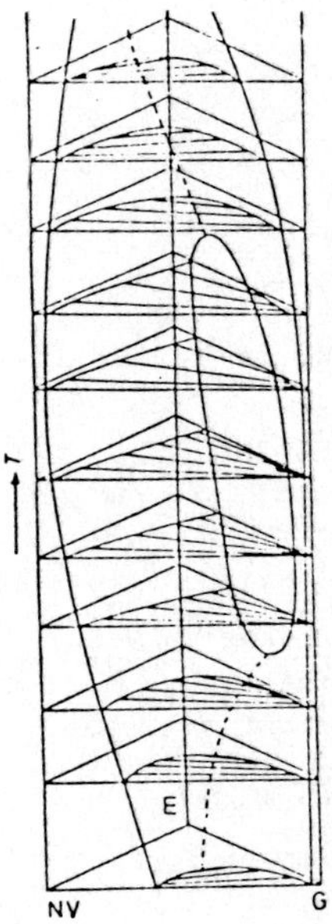

Temperature dependence of phase equilibria of a quasi-ternary system consisting of substances of low volatility (NV), an entrainer (E), and a supercritical component (G) (Peter)

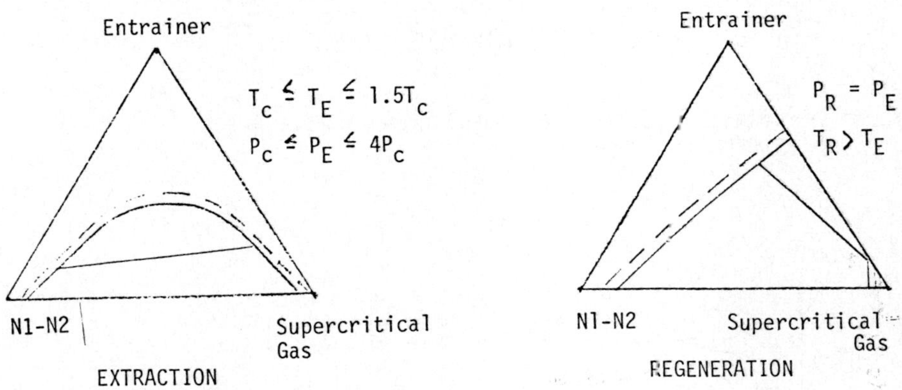

FIGURE 2: EXTRACTION AND SOLVENT REGENERATION STEPS SHOWN IN TERNARY DIAGRAM. N1, N2 ARE NONVOLATILE SUBSTANCES 1 AND 2

temperature selection. There may be several specialized cases our observations would not be accurate nor correct. A review by Paulatis discusses some of these ternary behavior. We hope to publish a more complete treatise based on experimental results and parametric studies in very near future. A flexible extraction system was constructed and is currently being operated to study effectiveness of an ambitious variety of solid or liquid entrainers.

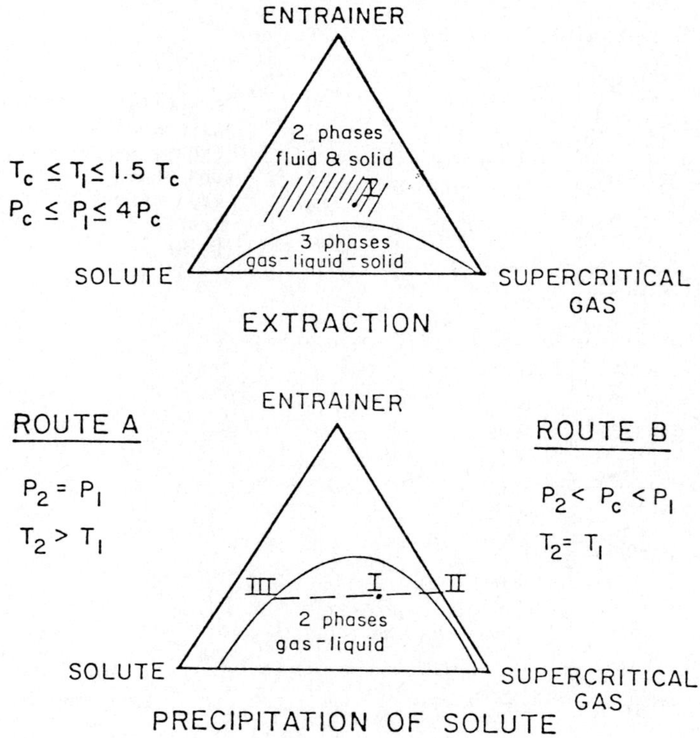

FIGURE 3: TERNARY DIAGRAMS FOR REPRESENTATION OF SCE IN THE PRESENCE OF AN ENTRAINER

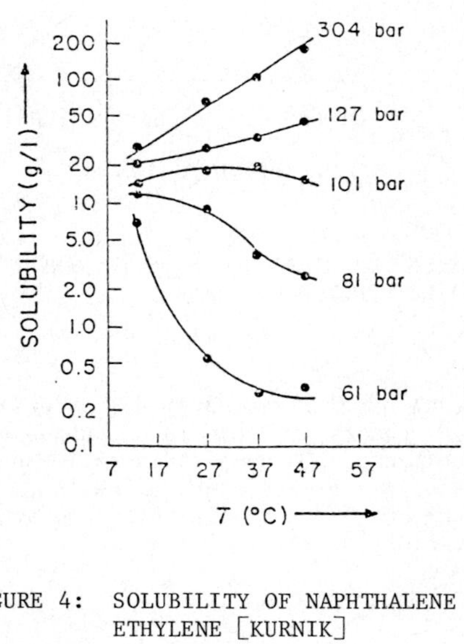

FIGURE 4: SOLUBILITY OF NAPHTHALENE IN COMPRESSED ETHYLENE [KURNIK]

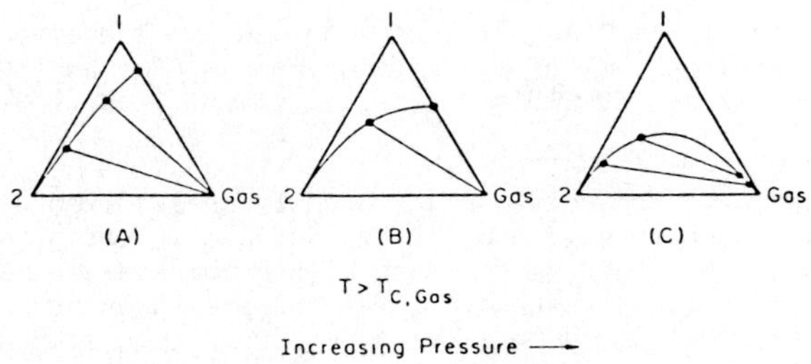

FIGURE 5: ROLE OF PRESSURE FOR A SIMPLE TERNARY SYSTEM [PAULATIS]

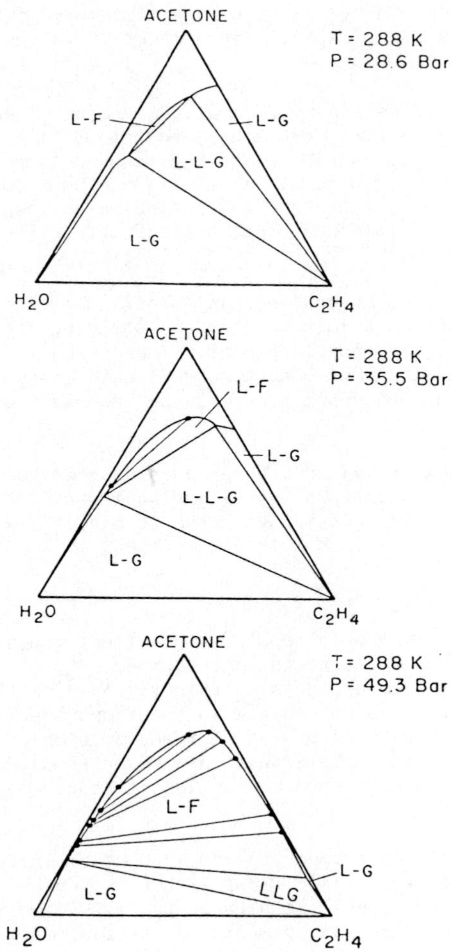

FIGURE 6: PRESSURE EFFECT ON H_2O/ACETONE/C_2H_4 SYSTEM [PAULATIS]

Description of the Supercritical Extraction (SCE) System

The dynamic system shown in Figure 7 is being used to study solubility in supercritical solvents or solvent mixtures. Temperatures as high as 450 C and pressures up to 3500 psia are possible. Multiple solvent feeds could be used.

Solvents that are gaseous at feed conditions are usually pressurized with an air operated compressor. Alternatively, a metering pump and inverted feed gas cylinder could be used. This feed system make higher compression ratios possible and is necessary for gases that have low vapor pressures at room temperature (e.g. Ammonia). The same pump is used to pump liquid solvents.

For most extractions, a 1" OD 10" long SS 304 extractor is used. The stationary solute is entrapped with wire mesh screens and/or glass wool at both ends. Several alternating layers of solute and glass wool are used to prevent entrainment. A coil is used to bring the solvent within 0.5 C of a sand bath heater. A heavy duty crane is used to raise the sand bath up or down for extractor change and allow minimum heat-up time of solute. For example, coal can be heated to 400 C in 2 minutes and extractor can be changed for a new experiment in 5 minutes.

When more than 50 grams of solute is necessary for reliable results, we use the autoclave as the extractor. A wire mesh basket mounted on the rotating shaft contains the solid solute and the supercritical solvent passes upwards. Otherwise, autoclave is used either as a surge tank for supercritical gas or as the entrainer saturator. A modified basket cage is designed for liquid supercritical fluid contacting, but its effectiveness has not been tested yet.

The entrainer and solvent can be mixed at desired ratios prior to extraction section. However, unless care is taken a multi-phase feed would be created. A multi-phase feed would not be very desirable except in cases where reaction kinetics favor use of an entrainer concentrated liquid phase. Alternatively, autoclave prefilled with entrainer could be used as the supercritical fluid saturator.

The extract and entrainer are precipitated in side-armed glass flasks. The escaped liquid and solid phases are entrapped in a glass wool packed flask immersed in an ice-bath. The gas solvent is vented after it passes a water saturator and wet-test meter.

Experimental

The objective of the experimental program is to study the effectiveness of entrainers in supercritical extraction. Although entrainer-solvent interactions are important, the emphasis at this point is in the study of the extraction step where entrainer-solute interactions are more important. Future modifications for alternate ways of fractionation of the extracted material is anticipated. The systems which are being studied, the validation of the experimental procedures, and the implementations are discussed below.

In the solubility studies, the extractor used was the 1" OD tube. Between experiments, the extraction section was purged first with nitrogen and then with carbon dioxide. Although the transfer line between the extractor and back pressure regulator was kept as short as possible, at initial stages some precipitation occurred at this line. Through temperature control this problem was eliminated. Nevertheless, the system is run for 5 to 10 minutes prior to data gathering. The amount of solid solute collected per sample is approximately 3 grams.

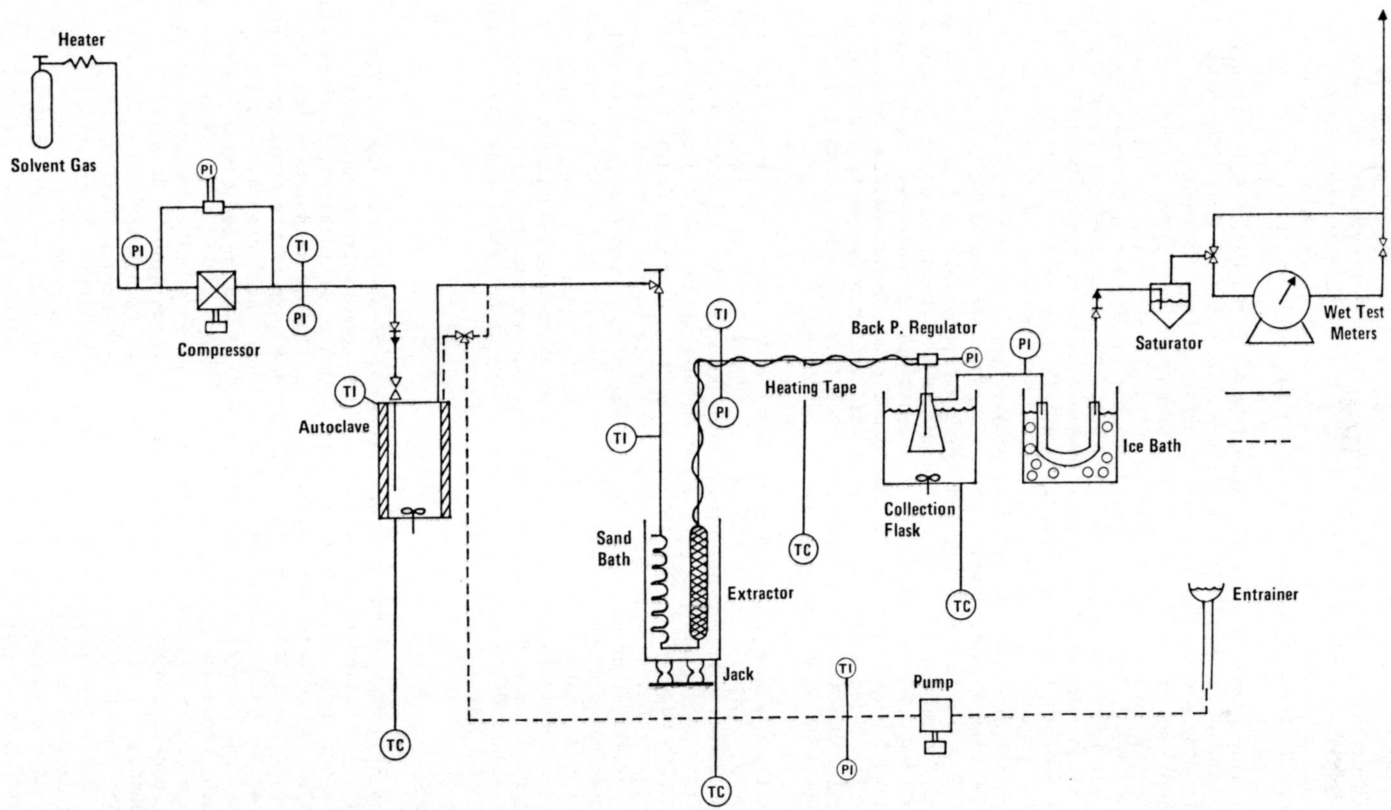

FIGURE 7: SUPERCRITICAL EXTRACTION SYSTEM

The validation for experimental system and procedures were achieved through comparisons of results with published data. Solubility of benzoic acid in supercritical carbon dioxide at 1500 psia and 35, 50, 65 degrees centigrade was compared with regressed data from Kurnik and the results do agree (within 5% of estimated values). The solubility data on napthalene-benzoic acid mixture in supercritical carbon dioxide compares well with Kurnik's data both in terms of solubility as well as selectivity.

Solubility of liquids in supercritical gases are studied using the autoclave as the saturator. Initially samples were collected in a sampling bomb (the sampling bomb and arrangement is not shown in the flowsheet). These experiments by nature does not give information on solubility of supercritical solvent in the liquid until the stationary liquid phase gets saturated with supercritical fluid. We had mixed success in the sense that we can get approximate answers. More work involving recirculation of both phases is in progress. Iso-octane/Carbon Dioxide system is used in these studies and results are checked with King's data.

The high temperature coal extractions are carried using the tubular extractor with mixed solvents. The solvents contained small amount of donor components. The feed for high temperature experiments is liquid and is mixed prior to feed pump. The feed mixture state is determined using an equation such as Peng Robinson. The extract amount could be increased by increasing the extraction severity as well as composition of the reactive components. However, the pore structure collapses as extraction severity and donor component composition increase. The pore structure collapse does seem to depend on coal rank also.

Most of the entrainer-aided supercritical extractions are carried in the tubular extractor while others are carried batch in a basket. For the tubular extractor runs, the autoclave is used as the entrainer saturator to ensure flow and phase uniformity. Percent extraction (on dry basis) at low temperatures were above 10% with highly potent solvents such as diamineformamide. A typical result is shown in Table 4. Prewetting of coal with entrainer, temperature or other forms of pretreatment of coal, and extraction temperature increase enhance extraction by either breaking the physical bonds between coal and its volatile fraction or by providing enough activation energy. Higher extraction temperatures change selection of supercritical solvent and polar solvents such as ammonia could be used. If ammonia is used, one has the option of using amines which would have reacted with carbon dioxide.

Summary

Entrainers provide an added dimension to supercritical extraction. They are especially effective if solubility of heavy components in supercritical solvents are very low, and/or high selectivities are desired. Furthermore, they could be used as reactants effectively. The description of the phase behavior of the system throughout the entire temperature and pressure range of extraction and regeneration stages is desirable for both operating parameter and solvent selection. Unfortunately, we do not have a way of describing the phase behavior a priori. Although binary parameters regressed from data and an equation of state could represent multi-component behavior, these parameters need to be determined from experiments with multi-components.

TABLE 4

AN EXAMPLE FOR ENTRAINER-AIDED SCE

	PSOC 425
Entrainer	DMF
Temperature [C]	70
Pressure [psia]	1500
Entrainer Amount [cc]	100
SCE solvent	Carbon Dioxide
SCE solvent amount [SCF]	8.7
Coal Amount [gr.]	42.85
Extraction Time [min]	60

PERCENT EXTRACTION (Dry basis) = 21.85 %

* PSOC-425 is a low sulfur (0.95) Texas lignite with 28.6 % moisture, 10.03 % ash, 35.66 % volatile matter, and 25.71 % Fixed Carbon.

The solute-entrainer interaction is the key interaction in the extraction stage while the solvent-entrainer interaction is important in the regeneration stages. The solvent selection guidelines for liquid extraction would provide an initial selection. The supercritical solvent-entrainer interaction could than be checked for an effective regeneration strategy. Ideally, one would prefer to regenerate the solvent through a temperature raise at constant pressure. The solvent entrainer binary should have a wide two phase region at that temperature and pressure.

A flexible extraction system was constructed and is currently being operated to study effectiveness of an ambitious variety of solid or liquid entrainers.

The preliminary experiments show that entrainers are a very effective way of extracting low volatile matter from coal near ambient temperatures with rather inexpensive supercritical solvents. With careful selection of entrainers one could simultaneously treat coal chemically. Presence of supercritical fluids provide a fast processing time and virtually complete recovery of rather specialized entrainers. Furthermore, reactive entrainers may be a vehicle for improving the extraction yields in high temperature supercritical gas extraction processes without altering the pore structure significantly.

BIBLIOGRAPHY

Brunner, G., and H. Hederer; High Pressure Science Technology AIRAPT Conf. 6th., 1977/pub 1979 Vol. 1, 527 Timmerhaus (ed.)

Brunner, G., and S. Peter; Gas Extraction in the Presence of Entrainer, presented in the 2nd World Congress of CHEs Montreal, October, 1981

Henley, E.J., and J.D. Seader; Equilibrium-Stage Separation Operations in Chemical Engineering, Wiley, (1981)

King M.B., and T.R. Bott, Separation Science and Technology, 17 (1), 119-150, 1982

Kurnik, R.T., and R.C. Reid; Solubility of Solid mixtures in Supercritical Fluids Paper presented in 2nd International CHE world Congress Montreal, Canada Oct 1981

Peter S.; High Pressure Extraction of Lignin from Wood, presented at AICHE annual meeting in San Francisco, 1984 Nov.

Paulatis, M.E., V.J. Krukonis, R.T. Kurnik, and R.C. Reed; Supercritical Fluid Extraction in Reviews in Chemical Engineering, vol 1, No.2.

Schwaitzer; Handbook of Separation Techniques for Chemical Engineers, McGraw Hill 1979

Sunol, A. K.; Supercritical Extraction of Coal (Phd Dissertation), Virginia Polytechnic Institute & State University, 1982 Dec.

Stahl et.al., Paper presented at the symposium "Extraction with Supercritical Gases" Essen Germany, (June 6 1976)